PRATYLENCHUS (NEMATODA: PRATYLENCHIDAE): DIAGNOSIS, BIOLOGY, PATHOGENICITY AND MANAGEMENT

PRATYLENCHUS (NEMATODA: PRATYLENCHIDAE): DIAGNOSIS, BIOLOGY, PATHOGENICITY AND MANAGEMENT

Pablo Castillo and Nicola Vovlas

David J. Hunt and Roland N. Perry (Series Editors)

NEMATOLOGY MONOGRAPHS AND PERSPECTIVES

VOLUME 6

BRILL
LEIDEN-BOSTON
2007

This book is printed on acid-free paper.

Library of Congress Cataloging-in-Publication Data

The Library of Congress Cataloging-in-Publication Data is available from the Publisher.

Cover illustration: The cereal and legume root-lesion nematode *Pratylenchus thornei* migrating into the cortical parenchyma of chickpea roots.

ISBN-10: 90 04 15564 3
ISBN-13: 978 90 04 15564 0

Dedication

The authors dedicate this book to our parents, *José Mª* and *Gregoria* and *Athanasio* and *Polixena*, as a way to express our innermost gratitude for their efforts and sacrifices during their lives and the permanent support they demonstrated for us.

Nematology Monographs & Perspectives, 2007, Vol. 6, vii-xii

Contents

About the authors

Pablo Castillo was born in Castillo de Locubín, province of Jaén, southern Spain, in June 1960. He received his B.Sc. in Biology at the University of Granada, Spain, in June 1984 and his Ph.D. in Nematology at the University of Granada in December 1988. He joined the Department of Crop Protection at the Centre for Agricultural Research and Formation in Granada from the regional government of Andalusia in 1989. In 1990 he was recipient of the Phytopathological Research Award 'Antonio Ciccarone', from the Mediterranean Phytopathological Union, and in 1992 moved to Córdoba (Spain) to the Department of Crop Protection at the Institute of Sustainable Agriculture belonging to the Spanish Council for Scientific Research, where he is now a research scientist.

Castillo's research has focused primarily on plant-nematode and nematode-soil-borne fungi interactions, as well as diagnosis of nematode diseases, host-parasite relationships and nematode control by means of strategies compatible with sustainable agriculture. He has produced more than 65 peer-reviewed research articles and recently co-authored a comprehensive monograph of the plant-parasitic nematode genus *Rotylenchus*.

Nicola Vovlas was born in Distraton-Arta, Greece, in October, 1944. He received his university degree in Agricultural Sciences in Italy from the University of Naples in 1968. From 1977 to 1991 he held the position of researcher at the Istituto di Nematologia Agraria di Bari where he currently serves as a research leader. His research interests include nematode taxonomy, biology and host-parasite relationships. Currently his responsibilities include teaching nematology, taxonomy, biology and nematode interactions. He is an active member of the Society of Nematologists, the Organisation of Tropical American Nematologists and the European Society of Nematologists (he also served for 4 years as active member of the governing board of this society) and currently is on the editorial board of the International Journal of Nematology and Helminthologia. His research activities have resulted in nine contributions to national and 30 to international meetings, and 95 national journal and 116 international journal articles. He contributed to the CABI Publishing Crop Protection Compendium and recently co-authored a comprehensive monograph of the plant-parasitic nematode genus *Rotylenchus*.

Preface

It is a pleasure for us, as Series Editors, to introduce the second monograph on plant-parasitic nematodes by Pablo Castillo and Nicola Vovlas, the first being on the genus *Rotylenchus*. This volume, the sixth in the *Nematology Monographs and Perspectives* series, follows the same, highly successful pattern as its predecessor and, in dealing with *Pratylenchus*, a genus of such worldwide economic importance, will be of great assistance to nematologists, agronomists and plant pathologists. This volume covers the morphology and systematics of the genus and provides full descriptions of all 68 species considered to be valid by the authors. All species are illustrated by line drawings and, in some cases, light micrographs. In addition, distribution data and associated hosts are also provided together with a comprehensive bibliography. Dichotomous and tabular keys to the valid species are provided. Management options for these nematodes are discussed in detail and include biological methods as well as rotation strategies, resistant/tolerant varieties and chemical control.

In the Preface to the authors' previous monograph on *Rotylenchus*, Michel Luc commented: "It is to be hoped that the present work will serve as a model for future monographs on other genera of plant-parasitic nematodes". That hope has now been fulfilled.

David J. Hunt and Roland N. Perry
Harpenden, UK
March 2007

Nematology Monographs & Perspectives, 2007, Vol. 6, xvii-xviii

Foreword

Root-lesion nematodes of the genus *Pratylenchus* are recognised worldwide as one of the major constraints of crops of primary economic importance, including banana, cereals, coffee, corn, legumes, peanut, potato and many fruits. *Pratylenchus* species rank second only to root-knot and cyst nematodes as having greatest economic impact on crops worldwide. This is not only due to their wide host range, but also to their distribution in almost every cool, temperate and tropical environment. They are migratory endoparasites that cause severe root damage on a wide range of crops whilst feeding primarily in the cortical parenchyma.

The genus *Pratylenchus* comprises 68 nominal species of worldwide distribution that parasitise a wide variety of plants. Nevertheless, the majority of economic damage to herbaceous, vegetables and fruit crops throughout the world is attributable to a dozen of the most common species, including *P. brachyurus* in corn, cotton, peanut, pineapple, potato and tobacco; *P. coffeae* in coffee, citrus, sugarcane and tea; *P. goodeyi* in banana; *P. neglectus* in cereals and legumes; *P. penetrans* in grasses, forages, fruit trees and strawberries; *P. pratensis* in cereals, grasses, ornamentals and strawberries; *P. scribneri* in potato, soybean and strawberries; *P. thornei* in cereals and legumes; *P. vulnus* in pome and stone fruit trees, ornamentals and roses; and *P. zeae* which parasitises corn, rice, sugarcane and wheat.

The book comprises ten chapters and presents summarised and specialised information on various aspects of the root-lesion nematodes belonging to the genus *Pratylenchus*. Chapter 1 describes the importance of *Pratylenchus* species in agricultural crops and their world distribution. Chapter 2 presents general morphology and diagnostic traits of *Pratylenchus* spp. and their usefulness in taxonomy. This chapter also includes morphometric and cluster analyses to separate groups of species in order to facilitate identification. Chapter 3 analyses the taxonomy and systematic position of *Pratylenchus* and related genera, including a list of nominal species. Chapter 4 provides complete descriptions and morphometrics of all populations characterised for each species, as well as their world distribution. Chapter 5 includes comprehensive tabular and dichotomous keys for species identification. Whilst recognising the complexity of creating a system to identify such a large number of species, we are nevertheless confident that, by using both systems, the reader has the necessary information to identify the species of *Pratylenchus* with which they are dealing. Chapter 6 summarises new diagnostic technologies based on biochemical and molecular analyses, which are becoming increasingly important for practical identifications. Chapter 7 includes numerous aspects of the biology of *Pratylenchus* spp., life cycle, epidemiology and ecology. Chapter 8 comprises a complete revision of the host-parasite relationships between *Pratylenchus* spp. and herbaceous and woody plants. Chapter 9 presents the pathogenicity and damage of *Pratylenchus* spp. to crops, as well as their interactions with beneficial and pathogenic fungi and other nematodes. Finally, Chapter 10 illustrates different management strategies for *Pratylenchus* species, including crop rotation, host-plant resistance, chemical control, soil solarisation and biological control by means of nematophagous fungi, entomopathogenic nematodes, the hyperparasitic bacterium *Pasteuria penetrans* and nematicidal plants.

Pablo Castillo and Nicola Vovlas
Córdoba, Spain
November, 2006

Acknowledgements

Since the major portion of this book was completed at home, we would like to express our deep gratitude and indebtedness to our wives and the rest of our families for the neglect and loss they endured during the preparation of this second book in the last 3 years and for all the support they provided. We know that words are not enough to make up for this loss, but it is the only thing that we can do here and now. Likewise, we thank our colleagues and friends from the Institute of Sustainable Agriculture, Consejo Superior de Investigaciones Científicas (CSIC), Córdoba (Rafael M. Jiménez Díaz, Juan A. Navas Cortés and Blanca B. Landa del Castillo) and the Institute for Plant Protection, Nematology Section, Consiglio Nazionale delle Ricerche (CNR), Bari (Alberto Troccoli and Franco Elia), for their help, suggestions and interest in this book. We also thank Dr R.N. Inserra for his help and Dr Deliang Peng for his contribution to the translation of papers in the Chinese language.

We particularly thank the Series Editors for their great assistance with the English and for editorial suggestions and comments.

Finally, we are grateful to the journals for allowing reproduction of illustrations and to the many nematologists who, through their work on this genus, have contributed to a better knowledge of diagnosis, biology, parasitism, ecology and management of the many species included in the genus *Pratylenchus*.

Pablo Castillo[1] and Nicola Vovlas[2]

[1]*Institute of Sustainable Agriculture, Consejo Superior de Investigaciones Científicas (CSIC), Córdoba, Spain*
E-mail: ag1cascp@uco.es or *pcastillo@ias.csic.es*

[2] *Institute for Plant Protection, Section of Bari Consiglio Nazionale delle Ricerche (CNR), Bari, Italy*
E-mail: n.vovlas@ba.ipp.cnr.it

Nematology Monographs & Perspectives, 2007, Vol. 6, 1-7

Chapter 1

Introduction

Root-lesion nematodes of the genus *Pratylenchus* Filipjev, 1936 rank second only to root-knot and cyst nematodes in terms of their worldwide economic impact on crops (Sasser & Freckman, 1987). This is due not only to their wide host range, but their distribution in almost every cool, temperate and tropical environment. *Pratylenchus* species are obligate biotrophic, soil-inhabiting parasites and are found in all agricultural regions of the world. In natural environments data from some northern temperate grassland ecosystems confirm that species of *Pratylenchus* (84% incidence) are actually more frequently encountered than those of cyst (21%) or root-knot (19%) nematodes in plant-species rich, not recently cultivated, grassland soils (De Goede & Bongers, 1998). Although *Pratylenchus* species are polyphagous, there are clear differences in host preferences amongst the species of the genus. They are recognised worldwide as one of the major constraints of crops of primary economic importance, including banana, cereals, coffee, corn, legumes, peanut, potato, vegetables and fruit trees. Being migratory endoparasites, *Pratylenchus* spp. cause severe root damage on a wide range of crops while feeding primarily in the cortical parenchyma. Both plant penetration and migration within the plant are probably facilitated by a combination of stylet thrusting and enzymatic softening of the cell walls of the host. Although *Pratylenchus* nematodes may also be found feeding ectoparasitically, damage to host plants is more directly related to endoparasitic activity. In any case, *Pratylenchus* spp. assume a hit-and-run parasitic strategy, remaining migratory throughout their life cycle. The wide host range of *Pratylenchus* spp. suggests that their parasitism is a less specialised (*i.e.*, more primitive) form of plant parasitism. They may have evolved from ancient fungivorous ancestors (*e.g.*, *Aphelenchus*) as part of the evolutionary sequence leading to ectoparasitism (*e.g.*, *Xiphinema*) and migratory endoparasitism (*e.g.*, *Pratylenchus*) and finally to obligate sedentary plant parasitism (*e.g.*, *Meloidogyne*) (Siddiqi, 2000). Nevertheless, endoparasitism of *Pratylenchus* spp. suggests a coevolution with higher plant hosts and also involves an advantage with regard to ectoparasitism as the nematode is enclosed and protected in a relatively stable environment within the root tissue.

The common name of these nematodes, *root-lesion nematodes*, is derived from the often conspicuous necrotic lesions they cause on host roots, although in the past, the term *meadow nematodes* was used occasionally because of their abundance in this habitat and it is from where the first species was described. The genus *Pratylenchus* comprises about 68 valid species of worldwide distribution that parasitise a wide variety of plants. Nevertheless, the majority of economic damage induced to herbaceous and fruit crops throughout the world is attributable to but a dozen of the most common species, including *Pratylenchus brachyurus* (Godfrey, 1929) Filipjev & Schuurmans Stekhoven, 1941, which parasitises corn, cotton, peanut, pineapple, potato and tobacco; *Pratylenchus coffeae* (Zimmermann, 1898) Filipjev & Schuurmans Stekhoven, 1941, which parasitises coffee, citrus, sugarcane and tea; *Pratylenchus goodeyi* Sher & Allen, 1953, which parasitises banana; *Pratylenchus loosi* Loof, 1960 infecting tea; *Pratylenchus neglectus* (Rensch, 1924) Filipjev & Shuurmans Stekhoven, 1941, which parasitises cereals and legumes; *Pratylenchus penetrans* (Cobb, 1917) Filipjev & Schuurmans Stekhoven, 1941, which parasitises forages, fruit trees, grasses and strawberries; *Pratylenchus pratensis* (de Man, 1880) Filipjev, 1936, which parasitises cereals, grasses and ornamentals; *Pratylenchus scribneri* Steiner *in* Sherbakoff & Stanley, 1943, which parasitises potato, soybean and strawberries; *Pratylenchus thornei* Sher & Allen, 1953, which parasitises cereals and legumes; *Pratylenchus vulnus* Allen & Jensen, 1951, which parasitises pome and stone fruit trees, ornamentals and rose; and *Pratylenchus zeae* Graham, 1951, which parasitises corn, rice, sugarcane and wheat (Barker, 1998; Manzanilla-López *et al.*, 2004).

The taxonomic separation of the various species in the genus is very difficult, *Pratylenchus* being stenomorphic because of the small number of diagnostic features available at the species level and the intraspecific variability of some of these characters. As a result, taxonomic difficulties often arise from underestimation of intraspecific variability of certain morphological characters that are presently used for species diagnosis. Taxonomy of plant-parasitic nematodes, including that of the genus *Pratylenchus*, received increasing attention after the discovery of the first efficient nematicides (*e.g.*, 1,2 dichloropropane (1,2-D)-1,3 dichloro-

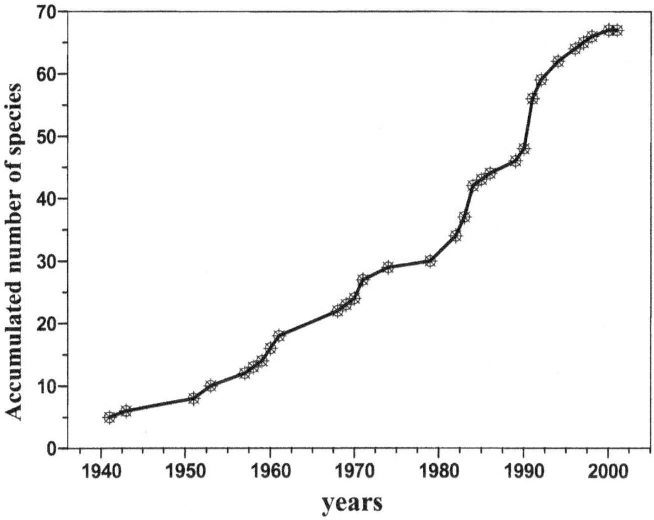

Fig. 1. *Cumulative numbers of recognised species described in* Pratylenchus *from 1940 to 2006.*

propene (1,3-D), ethylene dibromide (EDB), dibromochloro-propane). Nevertheless, due to the difficulties in separating species, the number of new species proposed in *Pratylenchus* has increased in an almost linear pattern with a slope of 1.1 species per year between 1940 and 2006 (Fig. 1); in fact, a valid species was added to the list of nominal species during the course of preparing this monograph. Although morphology continues to be the basis for identification of *Pratylenchus* spp., new technologies based on biochemical and molecular analyses are becoming increasingly important for nematode systematics and practical diagnostics (Al-Banna *et al.*, 1997; Duncan *et al.*, 1999; Andrés *et al.*, 2000; De Luca *et al.*, 2004).

As migratory endoparasites, these nematodes destroy tissues of the root system, causing surface cracking and internal rotting of tubers and predisposing the parasitised tissues to secondary infections by fungi and bacteria. However, somewhat unexpectedly, *P. coffeae* has also been reported in the bark and underlying wood of the mahogany tree, *Swietenia mahagoni* (L.) Jacq., by Goodey (1937). Symptoms of damage to crops by *Pratylenchus* spp. are non-specific and can be easily overlooked or mistaken for damage caused by other soil pathogens, or attributed to other causes such as nutrient deficiency or lack of water. Reduction in the quantity of feeder roots usually results in a general loss of vigour and yield in young and mature fruit trees (Nyczepir & Becker, 1998). The effects of *Pratylenchus* spp. on plant growth, and hence yield, are largely the result of the disruption these organisms cause to the normal process of root growth and exploration of the soil for water and nutrients (Loof, 1991). In fruit-tree orchards it is common to observe loss of vigour and reduction of yield in *Pratylenchus*-infected trees, which commonly require replanting. However, in replanted soils severely infested by *Pratylenchus* spp., newly planted trees may fail to establish, and exhibit suppressed growth and shortening of productive life (Mai & Abawi, 1981). Pathogenicity studies on *Pratylenchus* species indicated that they are very well adapted to parasitism as extremely high populations in soils do not kill their host plants. Nevertheless, damage thresholds are quite different among *Pratylenchus*-host plant combinations and range from 0.05 to 30 nematodes/cm^3 of soil.

Although *Pratylenchus* nematodes are mobile organisms and have the capacity to move short distances from the root zone that they infect, many agricultural operations favour a more rapid and widespread dispersion. Nevertheless, migration does occur in soil and around plant roots, but only when sufficient moisture is available, the soil texture is suitable and soil temperatures lie between the lower and the upper thresholds for biological activities (Wallace, 1973). The life cycle of *Pratylenchus* is simple and all species develop several generations in a growing season. Although large numbers of specimens can be detected in infected roots early in the growing season, nematode soil populations may be rather low, especially in the absence of a host crop (Loof, 1991).

Although no races have been described in *Pratylenchus*, evaluation of host suitability and pathogenicity in vegetable and fruit tree crops conducted in different geographical areas suggests that differences exist in both host range and pathogenicity of isolates within a species (Pinochet *et al.*, 1994; France & Brodie, 1995).

To survive in soil *Pratylenchus* must be able to circumvent biotic and abiotic obstacles (*e.g.*, voracious predators, changes in soil temperature, moisture and pH and the death of its host plant). *Pratylenchus* escape these constraints by employing a combination of behavioural and physiological survival strategies (*e.g.*, anhydrobiosis). Similarly, several weeds have been reported as good hosts of *Pratylenchus* spp. and help to maintain and disperse *Pratylenchus* spp. within and among fields.

Apart from direct damage to roots, in some situations *Pratylenchus* spp. may also predispose plants to other pathogens. The best known of such interactions is that between *Pratylenchus* spp. and *Verticillium* wilt (Rowe & Powelson, 2002; Rotenberg *et al.*, 2004). Interactions between *Pratylenchus* spp. and several *formae speciales* of *Fusarium oxysporum* Schltdl. have also been described on a number of crops (Mauza & Webster, 1982; Sumner & Minton, 1987). These studies demonstrated that infection by *Pratylenchus* spp. increased the incidence or severity of wilt diseases on susceptible cultivars. However, such published results were not confirmed in other compatible plant-nematode fungus combinations (Castillo *et al.*, 1998a) and it appears that modification of wilt disease incidence or severity may be related to the specific nematode-fungus combination. Furthermore, such apparently contradictory results indicate that interactions between soil-borne fungi and *Pratylenchus* spp. are biological and physiological, rather than physical, in nature.

Geographical distribution of *Pratylenchus*

The geographic distribution of *Pratylenchus* species is mostly dependent on the prevalence of host plants supporting reproduction, abiotic factors (mainly temperature) and their introduction to new areas by means of infected plant material. Nevertheless, the distribution of *Pratylenchus* species seems to be worldwide as they occur on every continent, including Antarctica, where *Pratylenchus andinus* Lordello *et al.*, 1961 was recently recorded by Ryss *et al.* (2005) on the Nunatak Base, East Antarctica (Table 1; Fig. 2). Nevertheless, some species present a distribution restricted by climate; *e.g.*, some species occur only in the tropics and others in temperate zones. The highest biodiversity of the genus occurs in Asia where 40 species have been reported, followed by Europe with 32, North America with 27, Central and South America 22, Africa with 16, Oceania with 12 and Antarctica with but a single species. The most widely distributed and commonest species are *P. neglectus*, *P. penetrans*, *P. thornei* and *P. vulnus* which have been reported on every continent with the exception of Antarctica (Table 1). Other species with a wide geographical distribution (Table 1) include: *P. brachyurus*, *Pratylenchus crenatus* Loof, 1960, *P. pratensis*, *P. scribneri* and *P. zeae*, whilst other species are restricted to one continent, *e.g.*, *Pratylenchus acuticaudatus* Braasch & Decker, 1989, *Pratylenchus angulatus* Siddiqi, 1994,

Table 1. *Occurrence in each continent of* Pratylenchus *species arranged in alphabetical order.*

Continent (no. of species)	*Pratylenchus* species
Europe (32)*	*acuticaudatus, alleni, bolivianus, brachyurus, brzeskii, coffeae, convallariae, crenatus, dunensis, estoniensis, fallax, flakkensis, goodeyi, hexincisus, kasari, kralli, macrostylus, mediterraneus, neglectus, penetrans, pinguicaudatus, pratensis, pratensisobrinus, pseudopratensis, silvaticus, scribneri, sudanensis, teres, thornei, vulnus, wescolagricus, zeae*
Africa (16)	*coffeae, delattrei, elamini, flakkensis, goodeyi, neglectus, penetrans, pinguicaudatus, pratensis, pseudopratensis, scribneri, sudanensis, thornei, vulnus, yassini, zeae*
Asia (40)	*alleni, artemisiae, bhattii, brachyurus, coffeae, convallariae, crassi, crenatus, cruciferus, delattrei, ekrami, fallax, gibbicaudatus, japonicus, kumaoensis, loosi, manaliensis, mediterraneus, microstylus, mulchandi, neglectus, okinawensis, penetrans, pratensis, pratensisobrinus, pseudocoffeae, pseudopratensis, roseus, scribneri, septincisus, subpenetrans, subranjani, sudanensis, teres, thornei, typicus, unzenensis, vulnus, yamagutii, zeae*
North America (27)	*alleni, arlingtoni, bolivianus, brachyurus, coffeae, crenatus, fallax, flakkensis, hexincisus, hippeastri, loosi, macrostylus, morettoi, neglectus, penetrans, pratensis, pratensisobrinus, pseudocoffeae, pseudopratensis, scribneri, sensillatus, subpenetrans, tenuis, thornei, ventroprojectus, vulnus, zeae*
South and Central America (22)	*alleni, andinus, bolivianus, brachyurus, coffeae, crenatus, flakkensis, hexincisus, jaehni, loosi, neglectus, neobrachyurus, panamaensis, penetrans, pratensis, pseudofallax, pseudopratensis, scribneri, teres, thornei, vulnus, zeae*
Oceania (12)	*coffeae, crenatus, curvicauda, goodeyi, neglectus, penetrans, pinguicaudatus, scribneri, teres, thornei, vulnus, zeae*
Antarctica (1)	*andinus*

* Number of species recorded in each continent.

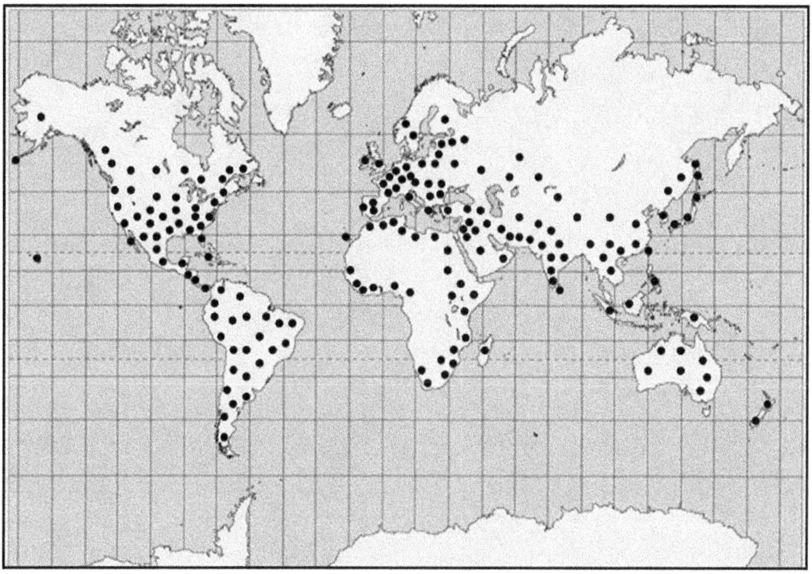

Fig. 2. *World distribution of the genus* Pratylenchus.

Pratylenchus curvicauda Siddiqi, Dabur & Bajaj, 1991, *Pratylenchus gibbicaudatus* Minagawa, 1982 and *Pratylenchus japonicus* Ryss, 1988.

Thirty-seven species (54% of the 68 nominal species) in the genus have only been reported from a single continent whilst the remaining 31 species (46%) have been reported from two or more continents (see Table 1). Nevertheless, despite the global distribution of the genus, some 32 of the described species have so far only been recorded from their type locality. These are:

a) Europe: *acuticaudatus*, *dunensis*, *estoniensis*, *kasari*, *kralli* and *silvaticus*.
b) Africa: *angulatus*, *elamini* and *yassini*.
c) Asia: *artemisiae*, *bhattii*, *crassi*, *cruciferus*, *ekrami*, *japonicus*, *kumaoensis*, *manaliensis*, *microstylus*, *okinawensis*, *roseus*, *septincisus*, *typicus*, *unzenensis* and *yamagutii*.
d) North America: *arlingtoni*, *hippeastri*, *morettoi* and *ventroprojectus*.
e) South and Central America: *jaehni*, *panamaensis* and *pseudofallax*.
f) Oceania: *curvicauda*.

Nematology Monographs & Perspectives, 2007, Vol. 6, 9-36

Chapter 2

Morphology of *Pratylenchus* species

The identification of *Pratylenchus* species is usually based on female morphology as they possess more diagnostic characters than the male, which in any case is rare or unknown for a substantial number of species (Loof, 1991). In fact, males are common in only 23 of the 68 species that we regard as valid. We concur with Baldwin and Perry (2004) that nematode morphology in the 21st century must be integrated with new technologies such as 4D and confocal microscopy and diagnostic biochemical and molecular approaches (see Chapter 5). Consequently, the integration of these new technologies with established morphological and host range data for assessment of nematode species and parasitic variants remains an important concern. In this chapter, we describe the morphology and structure of *Pratylenchus* spp. with special attention to diagnostic traits that are useful for identification.

Body wall

The body wall is composed of a cuticle, a hypodermis or epidermis and the somatic musculature. The cuticle presents a superficial ornamentation consisting of transverse annulation, the annuli being *ca* 1 μm apart, and lateral longitudinal lines forming the lateral fields. Kisiel *et al.* (1972) showed that the ultrastructure of the cuticle in *P. penetrans* was similar to the general pattern amongst nematodes and comprised three major zones: the cortical, matrix and basal layers. The cuticle has two main functions: *i*) it protects the nematode from, and interacts with, the external environment; and *ii*) it is crucial to movement as it prevents distortion of the body during muscle contraction (Bird & Bird, 1998). It is not known whether the hypodermis or epidermis is cellular or syncytial (Loof, 1978).

The body annuli are separated from each other by transverse striae, the depth of which appears to vary widely. The distance between these striae corresponds to the width of the annuli. Corbett and Clark (1983) studied body annulation with scanning electron microscopy (SEM) and found a continuous series from the widest annulus width in *P. andinus* (1.6 μm) and *P. fallax* (1.4 μm) to the narrowest in *P. zeae* (0.9 μm). However, the low depth of the body annulation may give the appearance of finer annuli, *e.g.*, *P. thornei* probably has the finest annulation; indeed it sometimes appears to have a smooth cuticle, the annuli being 1.3-1.6 μm apart yet only 0.22-0.33 μm deep. Corbett and Clark (1983) showed with SEM studies that annulus width in 14 species was unexpectedly variable. This was found to be due to the occurrence in several species of double annuli (annuli of normal width with an extra line in the middle). This, together with the almost complete overlap of the range of annuli widths in all species (except for *Pratylenchus pseudopratensis* Seinhorst, 1968 which has very narrow annuli), makes this character unacceptable for distinguishing species (Corbett & Clark, 1983).

Body length and shape

Body length in *Pratylenchus* varies from a mean value of *ca* 0.51 mm in *P. neglectus* to as short as 0.36 mm in *P. angulatus* and as large as 0.74 mm in *Pratylenchus morettoi* Luc, Baldwin & Bell, 1986. Used alone, this character is not very useful as an identification aid as a great number of species have a rather similar body length (Fig. 3) and, as can be seen in the figure, it is difficult to separate groups of species based upon this character. Furthermore, within a species body length may be modified by biotic and abiotic factors, *e.g.*, Duncan *et al.* (1998) found female body length of *P. coffeae* on Florida citrus to be seasonal and correlated with concentration of starch in the fibrous roots of the host. Body habitus upon relaxation is quite homogeneous within the genus and can be described as slightly curved ventrally to almost straight. Body length, as well as other morphometric values in fixed specimens, were usually found to be smaller than those of live specimens of *Pratylenchus hippeastri* Inserra *et al.*, 2007 (Inserra *et al.*, 2007).

Labial region

The labial region in *Pratylenchus* is not well defined, but is sometimes offset by a narrowing of the body contour. Its shape is characteristic for certain species, *e.g.*, *P. brachyurus* and *P. thornei*. This feature may be

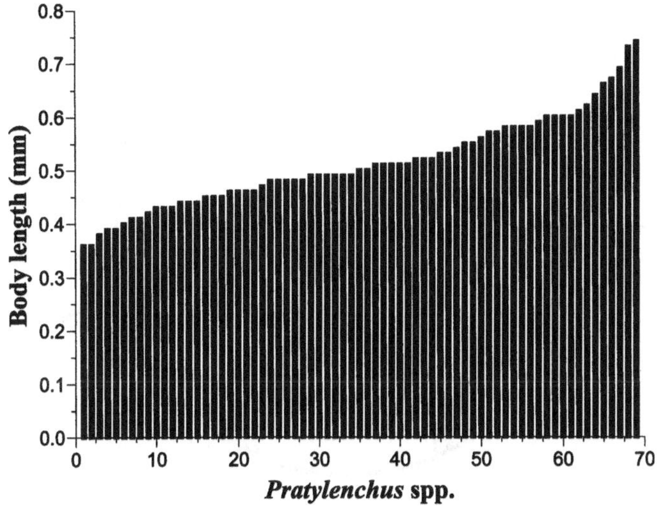

Fig. 3. *Means of body length (L in mm) in the 68 nominal species of the genus* Pratylenchus.

best appreciated under the light microscope when the labial region is seen in profile by transmitted light. The form of the labial region in SEM studies is also characteristic for certain species and represents a useful aid to identification (Corbett & Clark, 1983). The basic *en face* view pattern has an oval oral aperture, which is centrally located in the labial disc and surrounded by six inner labial sensillae arranged as three on each side of the oral aperture (Fig. 4). The oral disc may be separated from the first lip annulus; subdorsal and subventral segments are not easily differentiated and are variable in shape. The amphidial apertures are oval-shaped. They vary in size according to species and lie between the labial disc and the lateral sectors (Corbett & Clark, 1983). *En face* views have been studied by SEM in approximately a third of the species in the genus. Corbett and Clark (1983) distinguished three different face patterns:

Group 1 is distinguished by having a plain, undivided face with no division between the submedian and lateral segments, presumably due to fusion of the first lip annulus to the oral disc. Some species belonging to this group are: *P. brachyurus*, *P. coffeae*, *P. crenatus*, *P. goodeyi*, *P. loosi* Loof, 1960, *P. pseudopratensis* and *P. zeae*.

Group 2 is characterised by having the submedian segments fused to the oral disc and narrower at their inner extremity but widening towards

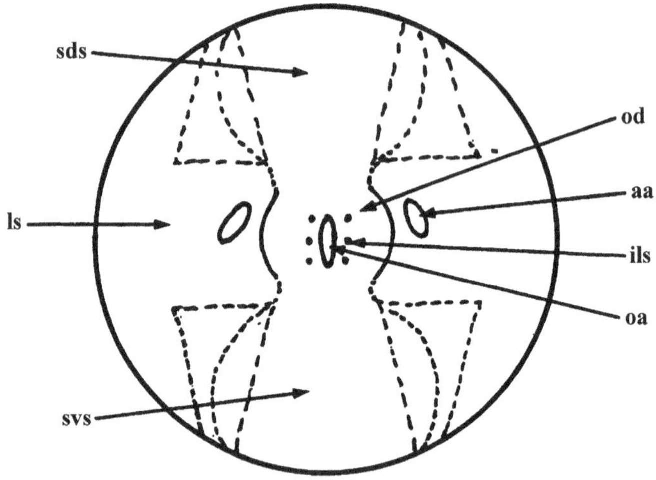

Fig. 4. *Diagram of* en face *view showing oral disc and variations of first lip annulus. Abbreviations: aa = amphidial aperture; ils = inner labial sensilla; ls = lateral segment; oa = oral aperture; od = oral disc; sds = subdorsal segment; svs = subventral segment. After Corbett and Clark (1983).*

the outer edge of the face: they are separated from the lateral segments that complete the circular face, the amphidial apertures being on their inner edges. Two species belonging to this group are *P. neglectus* and *P. thornei*.

Group 3 is characterised by the distinctive dumbbell-shaped pattern of the submedian segments with slightly smaller lateral segments to complete the circle. The amphidial apertures are once again on the inner edges of the lateral segments. Species belonging to this group include *P. andinus*, *P. fallax* Seinhorst, 1968, *P. penetrans*, *P. pinguicaudatus* Corbett, 1969, *P. pratensis*, *P. scribneri* and *P. vulnus*.

The number of lip annuli as a diagnostic character was first introduced by Allen and Jensen (1951) and has become an important diagnostic aid in several subsequent revisions of the genus (Sher & Allen, 1953; Loof, 1960, 1978, 1991; Corbett & Clark, 1983; Frederick & Tarjan, 1989; Handoo & Golden, 1989). The number of annuli varies between species but also occasionally within a species where there may be an extra annulus on one side of the labial region, *e.g.*, *P. crenatus*, *P. neglectus*, *P. pinguicaudatus* and *P. thornei* (Corbett & Clark, 1983). Nevertheless, lip annulus number is constant enough to form a good specific taxonomic character and its importance in diagnosis has been confirmed and illustrated in all of the most recent revisions.

The labial region of *Pratylenchus* shows a typical and characteristically strong labial framework, representing a marked morphological adaptation to the endoparasitic behaviour of nematodes that need to pierce cell walls as they migrate within a root.

Lateral fields

Laterally the cuticle is provided with longitudinal grooves, very often incorrectly called incisures, which delimit the lateral fields. The lateral fields extend from near the median pharyngeal bulb to the tail and usually have four equidistant straight lines forming three bands or ridges of the same width (Fig. 5A). This structure functions as a skeletal axis which makes the typical undulating movement in the dorsoventral plane possible. The outer lines may occasionally be slightly crenate in the tail region where areolation may also occur. As pointed out by Román and Hirschmann (1969a), deviations from the basic pattern are observed within a given species. These deviations usually extend along most of the length of the lateral fields but in some cases are restricted to the vulval region. Thus, the lines may not be equidistantly spaced and the inner band may be narrower or wider than the outer bands which are usually of the same width (Fig. 5B, C). Occasionally, the inner band is marked by short, equally spaced, diagonal lines (Fig. 5D). Some specimens of *P. penetrans* have straight rows of punctations in the centre of the outer bands, these being in addition to short, diagonal, lines in the inner band (Fig. 5E). In *P. penetrans* some small, additional, ridge(s) may arise, thereby giving supernumerary lines that can be hidden by a larger ridge and consequently not observed in the usual lateral view (Geraert, 2006). Specimens of *P. coffeae* may have punctations scattered in the outer bands (Fig. 5F). The lateral fields may exhibit five lines when an additional line, which may be straight or slightly wavy, continuous or broken at intervals, is present in the centre of the middle band (Fig. 5G, H). Six lines may occur when the outer or inner lines are doubled. When the outer lines are doubled they are usually continuous (Fig. 5I). When the inner lines are doubled, they may be continuous (Fig. 5J, L), or broken at intervals with their ends directed outward or inward (Fig. 5K). Occasionally, short striae may cross diagonally through the centre band

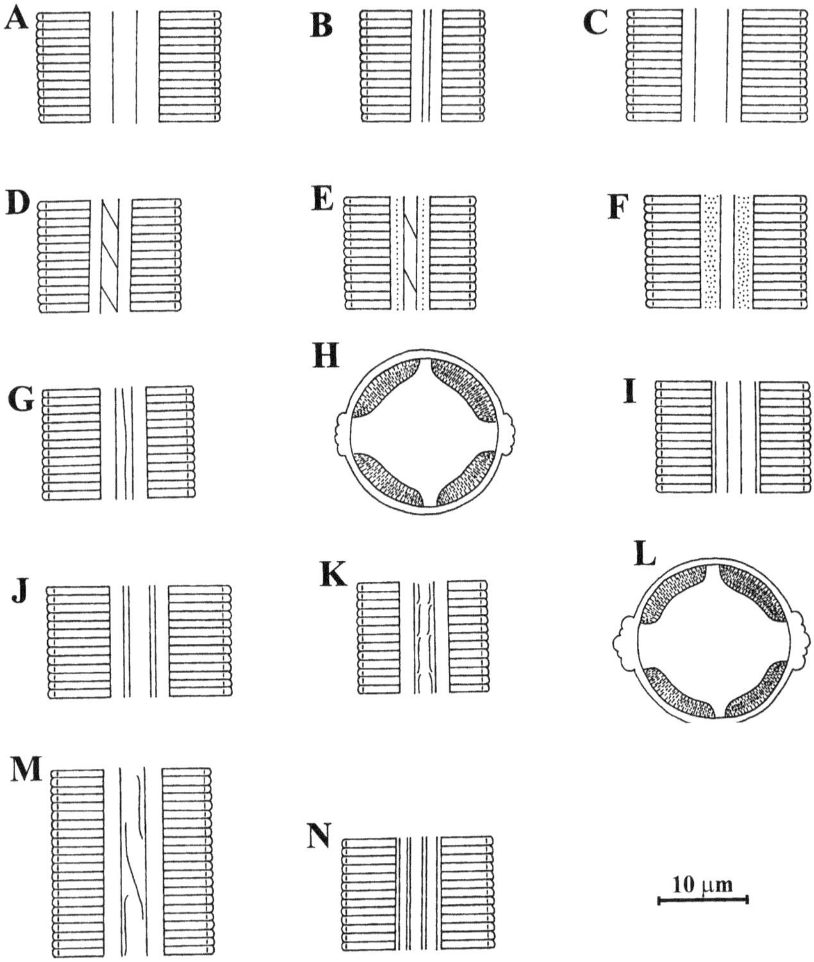

Fig. 5. *Variation in lateral field of* Pratylenchus *spp. A: Four lines forming three equally spaced bands; B: Four lines with inner band narrower; C: Four lines with inner band wider; D: Four lines with diagonal lines in inner band; E: Four lines with diagonal lines in inner band and straight lines of punctations in outer bands; F: Four lines with scattered punctations in outer bands; G: Five lines; H: Cross-section of specimen with five lines; I: Six lines, outer lines doubled; J: Six lines, inner lines doubled; K: Six lines, innermost lines broken and bent at their ends; L: Cross-section of specimen with six lines; M: Six lines with diagonal line in middle band; N: Eight lines. After Román and Hirschmann (1969a).*

(Fig. 5M). The lateral field may also exhibit four doubled lines, thereby making a total of eight lines (Fig. 5N). These lines are generally straight and continuous although the innermost may occasionally be broken or bent.

The structure of the lateral fields has been used to distinguish species of *Pratylenchus* and its validity as a taxonomic character was discussed by Corbett and Clark (1983). SEM studies by Corbett and Clark (1983) showed that the lateral fields in *Pratylenchus* basically consist of four lines and that a variety of ornamentation of the lateral field is found in so many species as to make its form a poor distinguishing character. For example, oblique striae in the lateral fields have been used as supplementary differentiating characters and Loof (1960) found *P. neglectus*, *P. penetrans*, *P. thornei*, *P. loosi* and *P. coffeae* to have them. Other studies have reported this character in *P. neglectus* and *P. hexincisus* Taylor & Jenkins, 1957 (Sher & Allen, 1953; Taylor & Jenkins, 1957; Fortuner, 1973; Townshend & Anderson, 1976; Corbett, 1983). Corbett and Clark (1983) found other ornamentation, especially areolation of the outer bands of the lateral fields, to be common. Thus, lateral field structure is not a particularly good taxonomic character for differentiating *Pratylenchus* species. No other longitudinal striations, as exist in other tylenchids such as *Rotylenchus* spp., occur along the body (Castillo & Vovlas, 2005).

Stylet

The stylet of *Pratylenchus* is short and stout with well developed basal knobs. The conical part (conus or metenchium) is strongly sclerotised as illustrated by Chen and Wen (1972). The average size of the stylet in *Pratylenchus* is *ca* 16 μm (Fig. 6) but, depending on species, may be as short as 11.5 μm (*Pratylenchus microstylus* Bajaj & Bhatti, 1984) or as large as 23 μm (*Pratylenchus macrostylus* Wu, 1971). The range of stylet length within a species usually varies by 1-3 μm, depending on species and population (Loof, 1991). Román and Hirschmann (1969a) found ranges of 1.2 μm in *P. vulnus*, 1.8 μm in *P. brachyurus* and *P. penetrans*, 2.4 μm in *P. coffeae* and *P. scribneri* and 3.0 μm in *P. zeae*. Several studies have reported even higher variability in stylet length in some populations, but such a wide range may be due to difficulties in

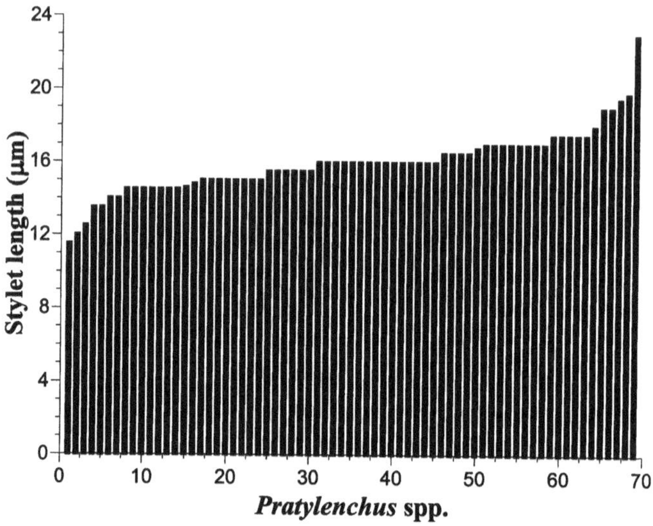

Fig. 6. *Means of stylet length (in μm) in the 68 nominal species of the genus* Pratylenchus.

measuring the stylet accurately, the tip being particularly difficult to discern through the heavy head skeleton (Loof, 1991).

The shape of the stylet knobs within a *Pratylenchus* species may be stable or may show great variation as pointed out by Román and Hirschmann (1969a). *Pratylenchus brachyurus* (Fig. 7A), *P. scribneri* (Fig. 7B) and *P. vulnus* (Fig. 7C) show little variation in shape whereas *P. zeae*, *P. penetrans* and *P. coffeae* exhibit pronounced variation in knob shape. *Pratylenchus zeae* (Fig. 7D-G) and *P. penetrans* (Fig. 7H-O) have knobs that vary from more or less rounded to markedly cup-shaped anteriorly, whereas *P. coffeae* shows a variation from a rounded to a narrow knob type (Fig. 7P-T). This reduction in knob width is most pronounced in the males (Fig. 7T). Mizukubo *et al.* (1990) proposed a transition series in the stylet knob shape of *Pratylenchus* species ranging from high, narrow, narrowly rounded and fused to the shaft (Fig. 8A) to acutely indented and forwardly pointing (Fig. 8J).

Ultrastructural observations by Endo *et al.* (1997) showed that the protractor and anterior somatic muscles structurally interact with the labial framework, stylet knobs and mitochondria-rich sarcoplasm of the protractor muscles (Fig. 9).

We studied the relationship between body and stylet length by linear regression analysis. The regression analysis was performed using mean

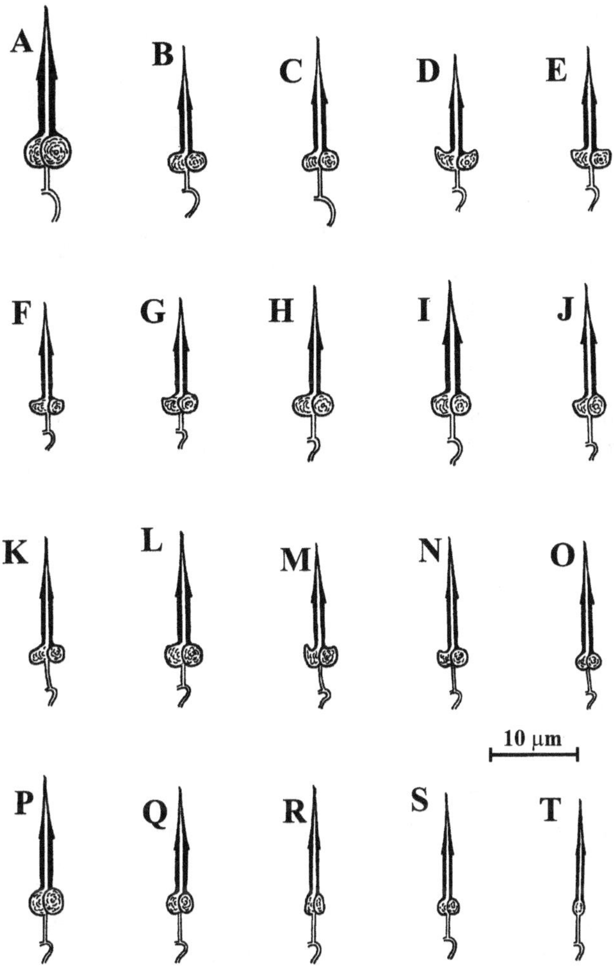

Fig. 7. *Variation in shape of stylet knobs in* Pratylenchus *spp. A:* P. brachyurus*; B:* P. scribneri*; C:* P. vulnus*; D-G:* P. zeae*; H-O:* P. penetrans*. H-M: Female; N, O. Male. P-T:* P. coffeae*. P-R: Female; S, T: Male. After Román and Hirschmann (1969a).*

values of both characters for each valid species. Results showed that stylet length did not increase according to increase in body length (Fig. 10), as expressed by the general equation: $Y = 8.57X + 11.56$; $R^2 = 0.165$. The coefficient of determination (R^2), R^2 adjusted for one degree of freedom (R_a^2) and the patterns of residuals plotted against expected values confirmed that both parameters are not correlated

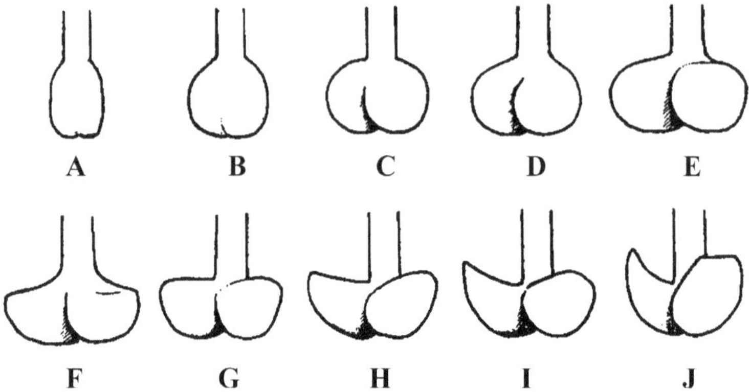

Fig. 8. *Assumed transition of the stylet knobs indexed as: A: High, narrow; narrowly rounded, fused to shaft; B: High, rounded, spherical; C: Rounded; D: Rounded sloping; E: Broadly rounded, well separated, concave; F: Laterally directed; G: Flattened anteriorly, broadly flattened, cup-shaped, rounded posteriorly; H: Sub-indented, forwardly directed; I: Indented anteriorly, angular, tulip-shaped; J: Acutely indented, forwardly pointing. After Mizukubo* et al. *(1990).*

(Campbell & Madden, 1990). A good example of this is *P. macrostylus* which possesses the highest stylet length but a medium body length. These data agree with those reported by Geraert (2006) who also

---

Fig. 9. *Ultrastructure of anterior region of* Pratylenchus penetrans. *A: General diagrammatic reconstruction showing stylet, protractor muscles and amphidial sensilla; B: Longitudinal tangential section through stylet region showing retracted stylet (St) and protractors muscles (Pm) extending from base and anterior surfaces of stylet knobs (Sk) to attachment sites at stomatal wall (SW), labial framework (CF); C: Cross-section slightly posterior to stylet knobs showing electron-dense accumulations (EDA) associated with electron-lucent filaments that are continuous with main protractor muscle elements. Portions of amphidial gland sensilla (Se), a pair of microvillus nerve processes (NP), and pharyngeal lumen (EL). Abbreviations: AC = amphidial canal; ADP = amphidial dendritic process; ANT = accessory neuron terminal; AP = amphidial pouch; ASH = amphidial sheath cell process; ASO = amphidial socket cell process; BP = basal plate; CU = cuticle; ID = dendrite giving rise to processes that remain within the amphidial sheath cell process; ILS = inner labial sensilla; PM = protractor muscle; S = stoma/prestoma; SM = somatic muscle; ST = stylet; STO = stomatal socket cell process. (Scale bars = 1 μm.) After Trett and Perry (1985b); Endo* et al. *(1997).*

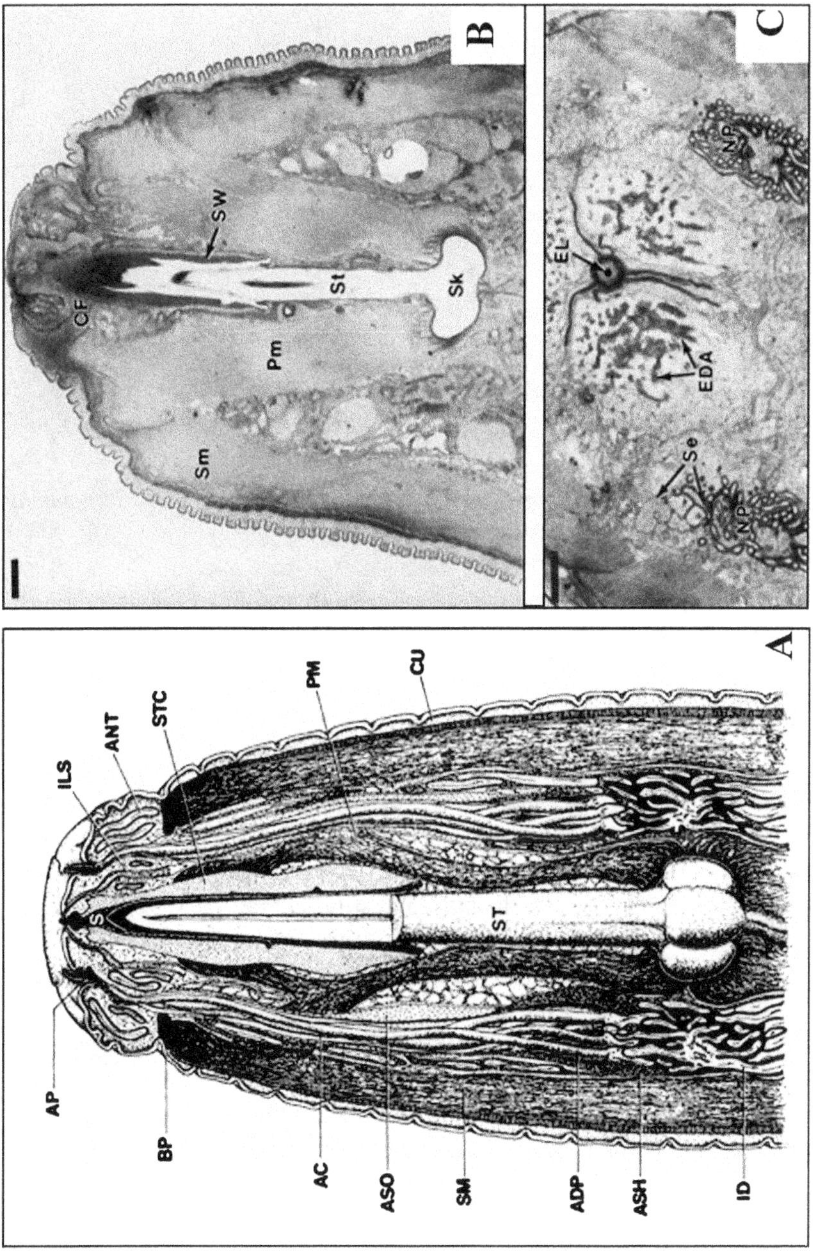
A
AP
BP
AC
ASO
SM
ADP
ASH
ID
ILS
ANT
STC
PM
CU
S
ST
B
CF
SW
St
Sk
Pm
Sm
C
EL
EDA
Se
NP

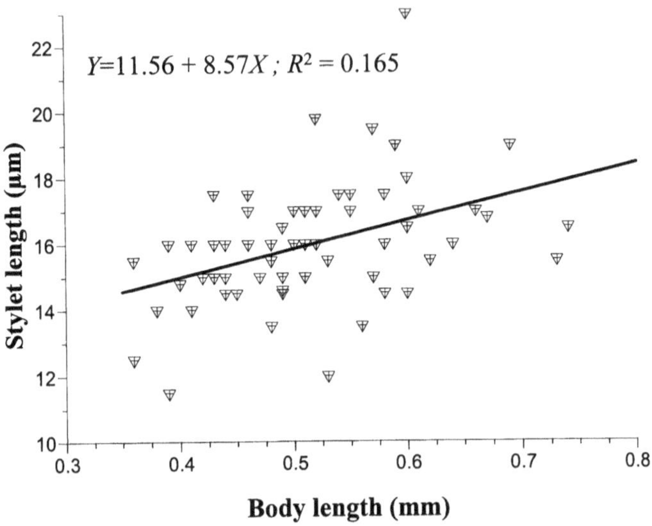

Fig. 10. *Linear regression between body and stylet length in the 68 nominal species of the genus* Pratylenchus.

found a poor correlation between body length and stylet length amongst *Pratylenchus* spp.

Pharynx

The pharynx (= oesophagus) in *Pratylenchus* is typical for tylenchids and is composed of three parts: procorpus, metacorpus (median bulb) and postcorpus (containing the three pharyngeal glands). The long pharyngeal gland lobe overlaps the intestine laterally and ventrally and contains three gland nuclei (Fig. 11). Ultrastructural studies by Chen and Wen (1972) and Endo *et al.* (1997) on *P. penetrans* showed that the procorpus consists of six cells and that the lumen of the pharynx is circular through the procorpus until the valves in the median bulb and then continues posteriorly as a triradiate form. The posterior region of the pharynx contains three unicellular glands, one dorsal and two subventral. Each of these glands possesses its own duct. The dorsal duct runs through the pharyngeal lumen and empties into the pharyngeal lumen 2-3 μm posterior to the stylet knobs, whereas the subventral ducts empty into the lumen a short distance posterior to the valves of the median bulb (Seinhorst, 1971). The gland lobe is syncytial, each

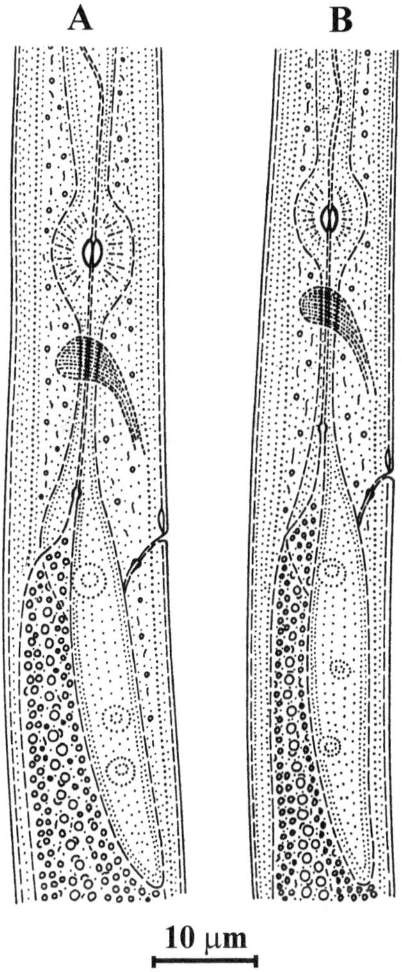

Fig. 11. *Pharyngeal region of* Pratylenchus brachyurus. *A: Female; B: Male. After Román and Hirschmann (1969a).*

gland containing two types of nuclei, one being large and the others smaller. Some of the smaller nuclei are associated only with glandular tissue, whereas others are part of nerve cells within the pharynx (Kisiel *et al.*, 1976). The length of the pharyngeal lobe (*i.e.*, from the pharyngo-intestinal junction to its posterior end) was used for the first time by Corbett (1969) to characterise *P. pinguicaudatus*. The average length of the pharyngeal lobe in *Pratylenchus* is *ca* 39 μm (Fig. 12) but, depending on species, may be as short as 14 μm (*P. hexincisus*) or as long as

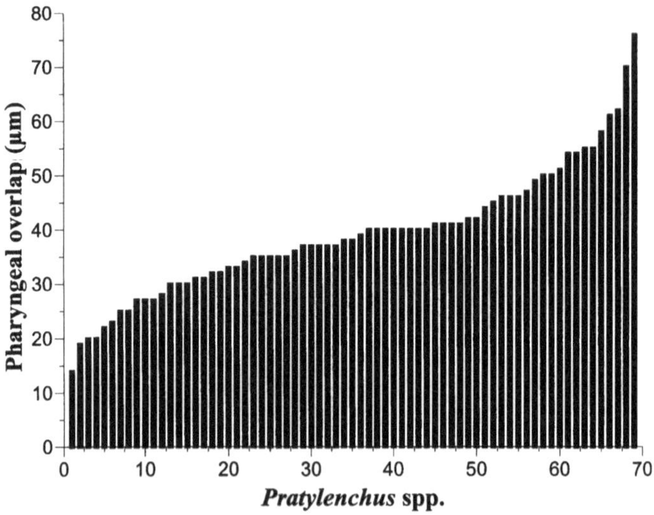

Fig. 12. *Means of pharyngeal gland overlap (in µm) in the 68 nominal species of the genus* Pratylenchus.

76 µm (*P. morettoi*). Nevertheless, we agree with Loof (1978) that this character shows high intraspecific variation and thus the character can only be reliably used in specimens that have been carefully processed and mounted.

Excretory pore

The excretory pore (sometimes referred to as the secretory-excretory pore) is located ventrally near the pharyngo-intestinal junction. The excretory duct has a cuticularised thickening extending *ca* 5 µm from the pore. In some specimens the duct is visible up to the middle of the body where it may join the excretory canal. In *P. crenatus*, Karssen and Bolk (2000) observed a swelling of the secretory-excretory duct in fresh females and juveniles, irrespective of geographical origin of the population.

Nervous system

The centre of the nervous system consists of a circum-pharyngeal nerve ring with associated ganglia from which longitudinal nerves run anteriorly and posteriorly. The ganglia at the anterior side of the nerve ring are connected to the labial sense organs, those at the posterior side mostly being connected to the dorsal, lateral and ventral nerves that run through the respective hypodermal chords. The only parts of the nervous system readily recognisable under the light microscope are the nerve ring itself and the somatic receptor organs (phasmids). Cephalids have never been reported in *Pratylenchus* although Loof (1978) indicated their presence in some species (*e.g.*, *P. goodeyi*, *P. neglectus* and *P. thornei*). From the nerve ring, nerves extend anteriorly to the amphids and labial sensilla and posteriorly to the tail; one nerve connects to the hemizonid, the major lateroventral commissure near the excretory pore and located just anterior to the excretory pore, whilst another connects to the hemizonion, which is located slightly posterior to the hemizonid. Each amphid comprises a glandular sheath cell, a supporting socket cell and a number of dendritic processes that are bathed in secretions, apparently produced by the sheath cell (Perry, 1996). The cells of the amphids produce secretions that appear to have multiple roles. Trett and Perry (1985a) pointed out that secretions in the amphidial duct may serve to maintain electrical continuity between the bases and tips of the dendritic process. Moreover, ultrastructural observations of the amphids of *P. penetrans* after exposure to concentrations of the nematicide aldicarb equivalent to field rate (5 and 10 ppm), showed hypertrophy of the internal dendrite terminals within the sheath cell, indicating that sheath cell metabolism was adversely affected. Phasmids are similar in general structure to the amphids, each consisting of an external pore, a cuticle lined duct, a socket cell, a sheath cell and a dendritic receptor. They are usually located about midway along the tail and open to the surface within the lateral fields. In the male they are located slightly posterior to the middle of the tail and may extend into the bursa in some species.

Deirids

Deirids are small pore-like structures usually located within the lateral field in the pharyngeal region. Although these structures have been reported in many descriptions of *Pratylenchus* species, there is uncertainty about their existence. Sher and Allen (1953) described the genus as having deirids, but Román and Hirschmann (1969a) could not find them on the six species they studied and Loof (1978) and Corbett and Clark (1983) concluded that it is therefore probable that they do not actually occur in *Pratylenchus*.

Intestine

The intestine is a simple tube with a distinct lumen. It is lined by a cellular, one-layered epithelium containing many lipid globules (Kisiel *et al.*, 1974).

Reproductive system

The reproductive system of the adult female consists of a single, anteriorly directed, gonad and a post-vulval uterine sac (Fig. 13). The relative position of the vulva (*i.e.*, V%) in *Pratylenchus* spp. varies, being about 78% in *P. pratensis*, but as anterior as 70% in *Pratylenchus teres* Khan & Singh, 1974 or as posterior as 87% in *P. macrostylus*. As with other characters, when used alone, relative vulval position is not very useful for identifying species as a large number share a rather similar V value (Fig. 14). The ovary is typically outstretched and only rarely reflexed (and then because of excessive growth). It is lined with epithelium that terminates in a single apical cap cell. The ovary is shorter in monosexual than in bisexual species. It is followed by the oviduct, which appears as a narrow, folded tube in young females, but is usually obscured by large oocytes in older, actively reproducing females. The oviduct is represented by a relatively short constriction between the ovary and spermatheca and lacks a visible lumen. It consists of two rows of four cells (Bert *et al.*, 2003). Román and Hirschmann (1969a) showed that in orcein-stained specimens the nuclei of these cells appear elongated and deeply stained (Fig. 13). Spermatozoa are distinctly visible in formalin-preserved specimens, particularly in those with few spermatozoa in the spermatheca (Fig. 13C). The spermatheca of *P. scribneri*, *P. zeae* and *P. brachyurus* is small and empty and is often obscured by a large oocyte. In orcein-stained specimens of *P. scribneri* the spermatheca is composed of about ten epithelial cells. Bert *et al.* (2003) showed that the spermatheca was formed from a total of 12 cells, the arrangement of which is species-specific. The spermatheca of *P. crenatus* comprises ten, more or less rounded, cells in variable positions

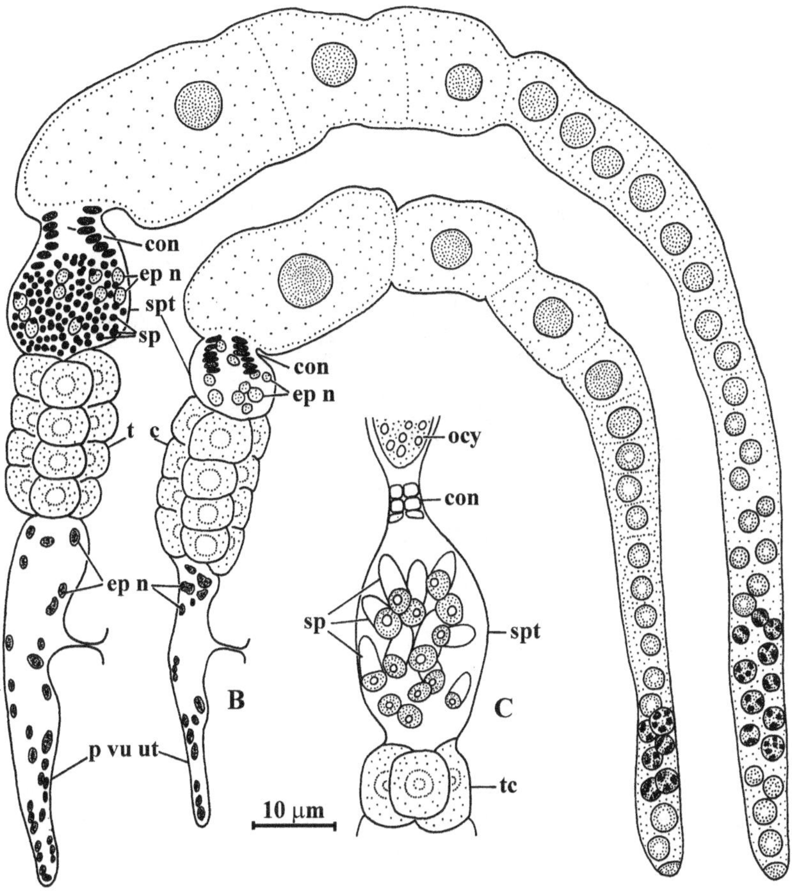

Fig. 13. *Female reproductive system of* Pratylenchus. *A: Reproductive system of* P. coffeae *stained with orcein; B: Reproductive system of* P. scribneri *stained with orcein; C: Spermatheca of* P. vulnus *in formalin preserved material (gonads are typically straight, but were curved for convenience in illustration). Abbreviations: con = constriction; sp = spermatozoa; ep n = epithelial nuclei; spt = spermatheca; ocy = oocyte; sty kn = stylet knobs; p vu ut = post-vulvar uterine branch; tc = tricolumella. After Román and Hirschmann (1969a).*

with two, more elongate, cells making the connection to the uterus. The spermatheca of *P. thornei* is partly offset, the offset portion comprising four cells on average, the spermatheca cells being more or less rounded and equally sized. The spermatheca of *P. penetrans* is asymmetrical and the oviduct is not connected entirely axially to the spermatheca.

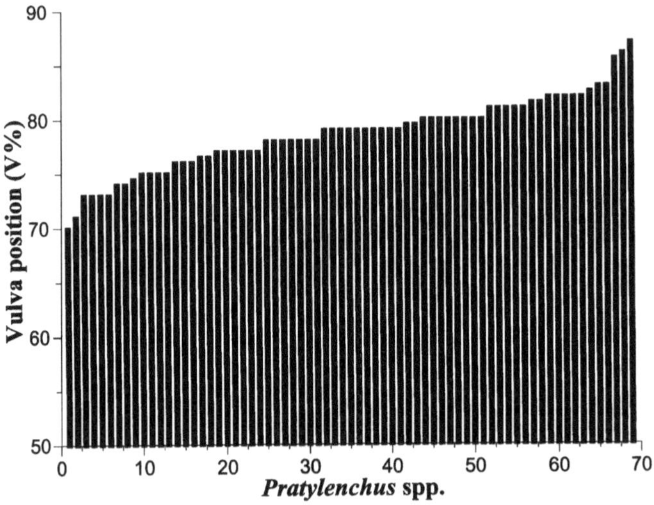

Fig. 14. *Mean vulva position (%) in the 68 nominal species of the genus* Pratylenchus.

The spermatheca is followed by the uterus which consists of a two-part tube lined with flat cells and is muscular and extensile. The distal part is a specialised region for secretion which is composed of 12 cells arranged in three rows of four cells each (*i.e.*, a tricolumella). The proximal part is a short tube lined with flat epithelium. The posterior genital branch comprises a post-vulval uterine sac which may carry a tricolumellate crustaformeria and a vestigial ovary (as in *P. coffeae* and *P. zeae*) (Román & Hirschmann, 1969a). The post-vulval uterine branch has a mean length of 23 μm in *P. acuticaudatus* or *Pratylenchus wescolagricus* Corbett, 1983, but may be as short as 9 μm in *P. angulatus* or as long as 59 μm in *P. morettoi* (Fig. 15). Nevertheless, it is often difficult to determine the exact length of this structure and also whether differentiated cellular tissue is present at the distal end or not.

Román and Triantaphyllou (1969) studied the gametogenesis and chromosomes in three bisexual species of *Pratylenchus* (*P. penetrans*, *P. vulnus* and *P. coffeae*) and four monosexual species (*P. brachyurus*, *P. neglectus*, *P. scribneri* and *P. zeae*). Bisexual *Pratylenchus* spp. multiplying by amphimixis have low chromosome numbers, *e.g.*, *P. penetrans* 2n = 10, *P. vulnus* 2n = 12, *P. coffeae* 2n = 14. The mitotic parthenogenetically reproducing species appear to be polyploids, *e.g.*, *P. neglectus* 2n = 20; *P. zeae* 2n = 21-26; *P. brachyurus* 2n = 30-

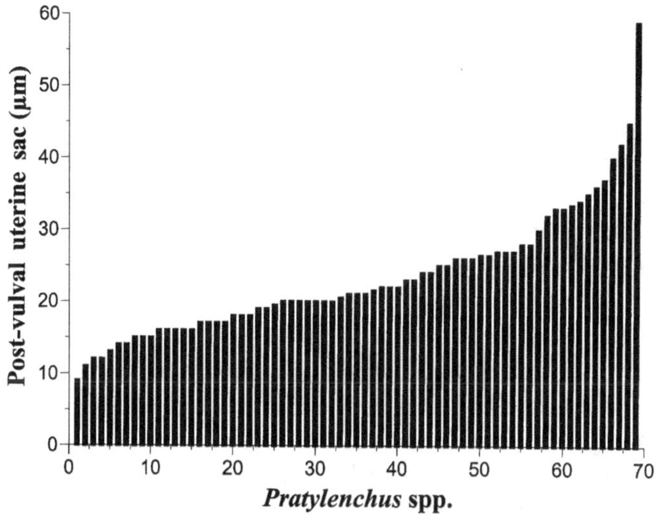

Fig. 15. *Mean post-vulval uterine sac length (μm) in the 68 nominal species of the genus* Pratylenchus.

32, and seem to have evolved from the amphimictic bisexual species. *Pratylenchus scribneri*, with 2n = 12 or 25-26 depending on meiotic or mitotic parthenogenesis, appears to be in a state of rapid evolution with regard to gametogenesis and mode of reproduction (Román & Triantaphyllou, 1969). The variation in chromosome number in bisexual species indicates that the genus is heterogeneous and that the basic chromosome number in this group is difficult to determine (Román & Triantaphyllou, 1969). These studies demonstrated that the cytogenetic situation in the genus is extremely complex.

Spermatogenesis apparently follows the normal pattern for amphimictic nematodes with two maturation divisions resulting in spermatids with the haploid chromosome complement. Román and Triantaphyllou (1969) pointed out that the observed variation in chromosome number and mode of reproduction among members of *Pratylenchus* may explain the difficulties taxonomists have encountered in characterising individual species of this genus.

The male reproductive system is monorchic and, not offering such a variety of shape and structure as the female system, is of lesser taxonomic importance (Loof, 1978). The testis consists of a multiple row of spermatogonia. The *vas deferens* is generally filled with spermatozoa and joins ventrally to the cloaca. The slightly arcuate spicules rest on a simple trough-shaped gubernaculum, which is immovable and not supplied with muscles. The caudal alae have crenate margins and enclose the tail tip. The spicules are simple, equal and curved, whilst the gubernaculum is also of simple shape. The ultrastructure of the spicules in *P. penetrans* was studied by Wen and Chen (1976) and Endo *et al.* (1997) who showed that each spicule is composed of a blade, shaft and base (Fig. 16). The spicules originate from the thickening and invagination of the posterior walls of the spicular pouch. At the tip of each spicule are two small pores, a dendritic process being associated with each. In the shaft portion, the spicule has a thick body with a central core filled with nerve tissue. Dorsal and ventral wings project from the body. When protruded, the two wings of one spicule curve toward the corresponding wings of the other, thus forming a complete tube for sperm conduction. The base of each spicule is composed of an electron-dense, sclerotised ring with nerve tissue in the central core. Each spicule is mobilised by two protractor and two retractor muscles. The posterior lips (= hypoptygma) of the cloaca are prominent and innervated. The *vas deferens* and rectum join near the cloacal opening (Fig. 16B). A pair of sensilla is located at the posterior lip of the cloaca. The gubernaculum is similar to the spicules in electron density and does not communicate with the spicular pouch but is embedded and fixed in its cuticular wall.

Geraert and De Grisse (1981) pointed out that spicule structure can be used in tylenchid taxonomy, although most species descriptions only mention its length. Studies using transmission electron microscopy (TEM) showed that the spicule tip in *Pratylenchus* has a longer ventral side with the pore(s) apparently opening to the dorsal side (Geraert & De Grisse, 1981). In addition, comparison of transverse sections of several tylenchids indicated that the spicules in *Pratylenchus* are much more incurved than in other genera such as *Rotylenchus* or *Scutellonema* and that their length is a constant ratio to body length (2.5-6.0%) (Geraert & De Grisse, 1981).

Tail

The shape and degree of annulation or crenation of the female tail is included in most descriptions of *Pratylenchus* species. Frederick and Tarjan (1989) differentiated seven types of tail shape and tail tip, including: bluntly pointed, digitate, finely pointed, hemispherical,

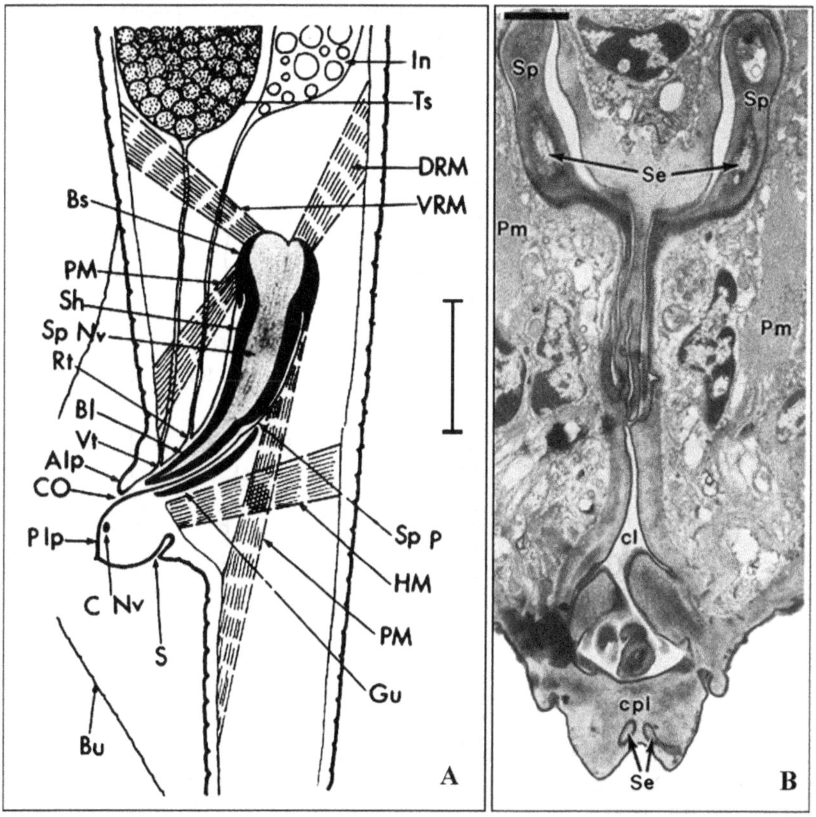

Fig. 16. *A: Schematic drawing of posterior end of* Pratylenchus penetrans *male; B: Tangential section showing retracted spicules with sensilla, protractor muscles, and part of pathway for protrusible spicules. Abbreviations: Alp = Anterior lip of cloaca; BI = blade; Bs = base; Bu = bursa; C or cl = cloaca; CC = central core; CNv = cloacal nerve; CO = cloacal opening; cpl = cloacal posterior lips; DRM = dorsal retractor muscle; DW = dorsal wing; GB = guiding bar; Gu = gubernaculum; HM = H-shaped muscle; In = intestine; Plp = posterior lip of cloaca; PM = protractor muscle; Rt = rectal terminal; S = space posterior to the posterior lips of cloaca; Se = sensilla; Sh = shaft; Sp = spicules; SpNv = spicular nerve; SpP = spicular pouch; Ts = testis; VD =* vas deferens*; VRM = ventral retractor muscle; Vt =* vas deferens *terminal; VW = ventral wing. (Scale bars: A = 10 μm; B = 1 μm.) After Wen and Chen (1976); Endo* et al. *(1997).*

subdigitate, subhemispherical, and truncate; the tip being annulated, cleft or smooth (Fig. 17). However, Loof (1991) distinguished only two basic

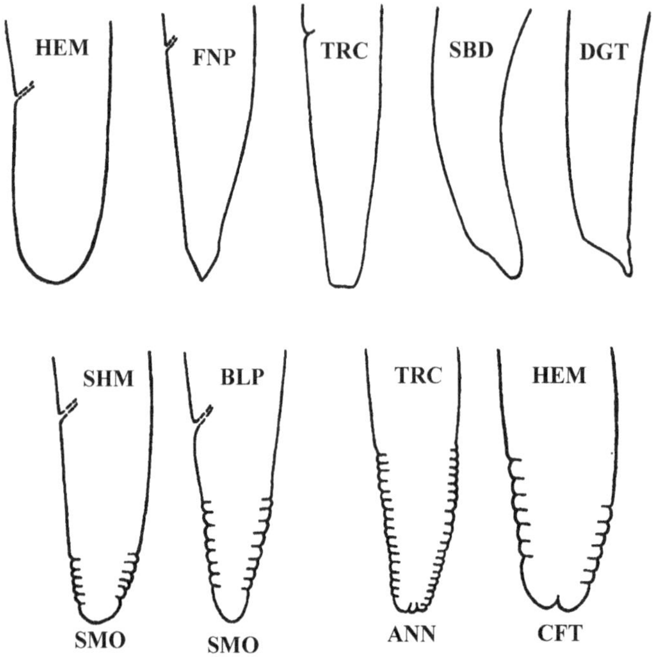

Fig. 17. *Female tail shape and tail tip shape in* Pratylenchus*: BLP = bluntly pointed; DGT = digitate; FNP = finely pointed; HEM = hemispherical; SBD = subdigitate; SHM = subhemispherical; TRC = truncate. Tail tip annulation: ANN = annulated; CFT = cleft; SMO = smooth. After Frederick and Tarjan (1989).*

different tail shapes in *Pratylenchus*: conoid (tails tapering strongly, with a narrow terminus) and cylindrical (tails slightly tapering with a broad terminus). SEM studies by Corbett and Clark (1983) showed that there is some variability but that tail shape and presence or lack of annulation and its form around the tail tip are sufficiently characteristic for each species to be a good taxonomic character. The least variation is present in *P. brachyurus* (Fig. 18A), *P. scribneri* (Fig. 18B) and *P. coffeae* (Fig. 18C), all of which have broadly rounded, smooth tails. *Pratylenchus brachyurus* (Fig. 18A) has a thicker cuticle and the hyaline portion of the tail tip is longer than in the two other species. *Pratylenchus zeae*, *P. penetrans* and *P. vulnus* exhibit a wide variety of tail shapes. Different types of pointed, smooth, tail tips have been observed in *P. zeae* (Fig. 18D-F). One specimen had a pointed, annulated, dorsally curved terminus (Fig. 18G) whilst another had a rounded, smooth terminus

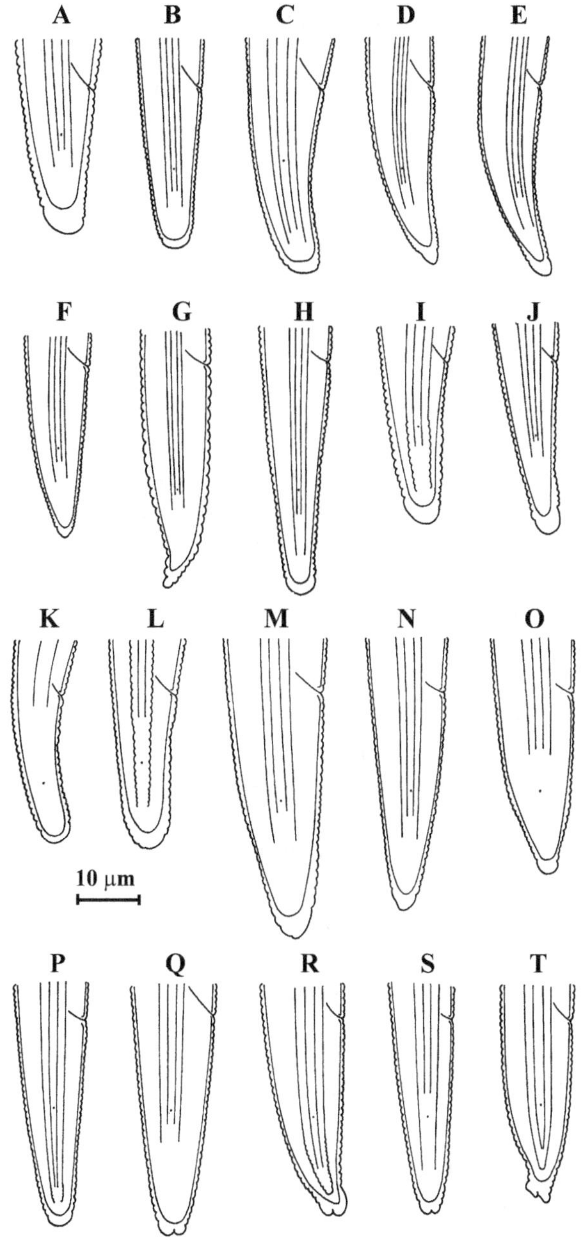

Fig. 18. *Variability in female tail shape of* Pratylenchus *spp. A:* P. brachyurus*; B:* P. scribneri*; C:* P. coffeae*; D-H:* P. zeae*; I, J-M:* P. penetrans*; N-T:* P. vulnus. *After Román and Hirschmann (1969a).*

(Fig. 18H). In *P. penetrans* the tail terminus is more or less rounded and smooth (Fig. 18I, J-M) and the hyaline portion of the terminus is long, although specimens with a short hyaline terminus were also found (Fig. 18K). The annulation may continue further posteriorly on the dorsal surface than on the ventral (Fig. 18L, M), although the annuli of this species never extend completely around the terminus. *Pratylenchus vulnus* tails were the most variable (Fig. 18N-T) with pointed tails (Fig. 18N) being most common but with variations including rounded, smooth (Fig. 18O, P), annulated (Fig. 18Q, S) and truncate (Fig. 18T). However, Tarté and Mai (1976) showed that the form of the female tail terminus may vary in *P. penetrans* depending on environmental factors, particularly the host plant, so the character needs to be interpreted with care.

The number of annuli along the ventral surface of the tail between the anus and the tail tip was used by Seinhorst (1968) and endorsed by Loof (1978) to differentiate species of *Pratylenchus*. However, Corbett and Clark (1983) and the present monograph indicate that the number of tail annuli varies widely within species and that there was also variation between species. The number of tail annuli was reported in the original description of 39 out of the 68 valid species with mean number varying from 12-15 for *P. crassi* to 32-44 annuli for *P. kasari*. A range between 18-25 annuli is the most frequent among *Pratylenchus* spp., *i.e.*, *P. acuticaudatus*, *P. angulatus*, *P. artemisiae*, *P. bolivianus*, *P. crenatus*, *P. flakkensis*, *P. goodeyi*, *P. hexincisus*, *P. mediterraneus*, *P. penetrans*, *P. pseudofallax*, *P. pseudopratensis*, *P. yamagutii*, *etc*. However, in some species, the number of tail annuli reportedly differed depending on the geographic origin of the population, *e.g.*, *P. zeae* (see description). In conclusion, the number of tail annuli is not a particularly good diagnostic character for differentiating *Pratylenchus* species.

Morphometrics and cluster analysis

To characterise and identify *Pratylenchus* spp. it is necessary to study the main diagnostic characters, namely the demanian indexes (de Man, 1884), body length, number of lip annuli, shape of labial region, stylet length, shape of stylet knobs, lateral fields, shape of spermatheca, structure and length of the post-vulval uterine sac, shape of tail terminus and presence or absence of males (Loof, 1991). Although some reliable diagnostic biochemical and molecular approaches have recently been used to identify certain root-lesion nematode species (see Chapter 5), accurate measurements are also essential in the description and correct identification of nematode populations (Shurtleff & Averre, 2000). However, differences have been detected due to the effect of the fixative used and the processing method (Olowe & Corbett, 1983), environmental factors (Olowe & Corbett, 1984a), geographical location (Olowe & Corbett, 1984b), host (Loof, 1960) and, last but not least, the accuracy of the operator (Frederick & Tarjan, 1978; Brown, 1981). Nevertheless, several of these studies have demonstrated that some measurements, such as stylet length and V ratio, were less modified by these biotic or abiotic factors and proved to be reliable diagnostic characters in the genus.

Knowledge of the intraspecific variability of many of the morphological and morphometrical characters used to identify *Pratylenchus* spp. is often lacking although some studies have been done on important pathogenic species such as *P. penetrans*, *P. coffeae* and *P. vulnus* (Tarté & Mai, 1976; Doucet *et al.*, 2001; Lax *et al.*, 2004). Tarté and Mai (1976) demonstrated that in *P. penetrans* body length, body diameter, pharynx length, stylet length, V value, a and b ratios, as well as qualitative characters, such as tail terminus, growth of ovary and shape of the median bulb, were influenced by environmental factors, particularly the host plant. This study also showed that nematodes reared on pea and cabbage had a higher percentage of females with a crenate tail terminus than those from tomato, rye, beet and alfalfa callus culture. In addition, populations from different geographical locations also exhibited variability in morphological characters. Rashid and Khan (1976) also found significant differences in body length, c ratio and shape of tail in *P. coffeae* specimens extracted from within chrysanthemum root tissues in comparison to those collected from soil around the roots. Moreover, the length of the pharyngeal gland lobe and post-vulval uterine sac were influenced significantly by environmental factors such as temperature, nutrient medium and overcrowding (Olowe & Corbett, 1984a), implying that these characters are probably unacceptable for consistently accurate species diagnosis. Analysis of the degree of inter-population variability of morphometrical characters in *P. vulnus* has demonstrated that not all characters have the same taxonomic significance (Lax *et al.*, 2004). The least intrapopulation variability in the isolates and populations studied were female stylet length and the ratio V, these values therefore having high taxonomic significance. However, characters with a high inter-population variability included body length, body diameter and ratios b and b′ in females and males. Similarly, Doucet *et al.* (2001) studied the influence of temperature on several morphometrical characters (body length, ratio a, body diam. and body diam. at anus) in isolates of *P. vulnus* and, except for stylet length and ratio V, found significant differences between the temperatures.

Using the main characters previously described for the 68 valid species of *Pratylenchus*, we did a hierarchical cluster analysis in order to study the association of species in the genus. The cluster analysis was achieved by the Joining (Tree Clustering) procedure in the software package Statistica (Version 5.0, StatSoft®, 1997) using Ward's method (Sneath & Sokal, 1973) with the percent disagreement distances between species defined by the average of the multiple characters. The analysis was based upon the following main characters: number of lip annuli, body length, stylet length, vulva position (V%), pharyngeal lobe, shape of spermatheca, post-vulval uterine sac length, female tail shape and tail tip, presence or absence of males and the ratios a, b and c.

The 68 species of *Pratylenchus* were separated by the cluster analysis into two main groups (Fig. 19A). Cluster I included a total of 43 species and Cluster II consisted of 25 species.

Cluster I is separated into two subclusters: **a** and **b**. Subcluster **a** comprises 30 species. Subcluster **a** is itself separated into two subgroups: $\mathbf{a}_1$ and $\mathbf{a}_2$. Subgroup $\mathbf{a}_1$ comprises 15 species: *pratensis*, *kasari*, *pratensisobrinus*, *manaliensis*, *pseudofallax*, *okinawensis*, *yamagutii*, *estoniensis*, *pseudopratensis*, *ventroprojectus*, *gibbicaudatus*, *loosi*, *morettoi*, *vulnus* and *zeae*, and is characterised by bisexual species with two or three lip annuli and moderate to long stylet. Subgroup $\mathbf{a}_2$ comprises 15 species: *convallariae*, *flakkensis*, *roseus*, *elamini*, *fallax*, *macrostylus*, *kralli*, *brachyurus*, *goodeyi*, *hippeastri*, *neglectus*, *neobrachyurus*, *silvaticus*, *brzeskii* and *scribneri*, and is characterised mostly by a striated tail terminus. Subcluster **b** comprises 13 species: *alleni*, *bhattii*, *penetrans*, *mediterraneus*, *thornei*, *sudanensis*, *unzenensis*, *artemisiae*, *coffeae*, *pseudocoffeae*, *hexincisus*, *dunensis* and *panamaensis*, and is mostly characterised by the cylindrical to subcylindrical female tail.

Cluster II is separated into two subclusters: **a** and **b**. Subcluster **a** comprises 18 species: *acuticaudatus*, *japonicus*, *angulatus*, *crassi*, *tenuis*, *ekrami*, *mulchandi*, *subpenetrans*, *typicus*, *andinus*, *wescolagricus*, *bolivianus*, *pinguicaudatus*, *cruciferus*, *sensillatus*, *delattrei*, *microstylus*

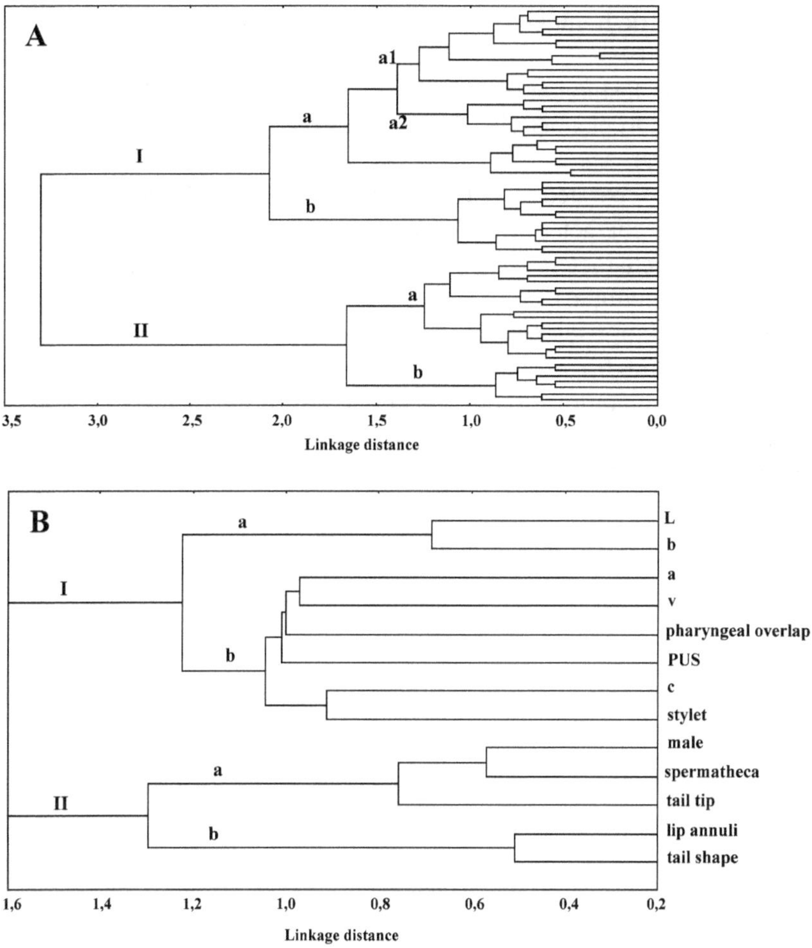

Fig. 19. *A: Horizontal hierarchical tree plot with 68 nominal species of* Pratylenchus. *Euclidean distances are plotted against linkage distance expressed as percent of disagreement. Major species clusters, and their constituent subclusters, are indicated by numbers and letters, respectively; B: Horizontal hierarchical tree plot of 13 main characters used to separate the 68 nominal species of* Pratylenchus. *Euclidean distances are plotted against linkage distance expressed as percent of disagreement.*

and *jaehni*, and is mostly characterised by a long stylet. Finally, subcluster **b** comprises seven species: *arlingtoni*, *yassini*, *curvicauda*, *subranjani*, *teres*, *crenatus* and *kumaoensis*, and is mostly characterised by a conoid female tail. These results demonstrate the high morphological and morphometric homogeneity occurring within the nominal species of the genus (Fig. 19).

In a similar fashion, hierarchical cluster analysis of the 13 main characters used to characterise the 68 valid species produced two main clusters (Fig. 19).

Cluster I is separated into two subclusters: **a** and **b**. Subcluster **a** comprises two characters: body length and b ratio; and subcluster **b** comprises six characters: pharyngeal overlap, post-vulval uterine sac length, stylet length, a, c and V ratios.

Cluster II comprises five characters: presence of males, shape of spermatheca, tail tip and tail shape and lip annulation. Thus, these data confirm the complexity previously noted for the identification and separation of the 68 valid species of *Pratylenchus* into homogeneous groups (see tabular and dichotomous keys).

Nematology Monographs & Perspectives, 2007, Vol. 6, 37-50

Chapter 3

Taxonomy and systematics

Species of *Pratylenchus* are very similar in gross morphology and most specific differences can only be detected using high magnifications. This means that, although the genus is easily recognisable, it is extremely difficult to construct satisfactory keys for species determination. Moreover, the uncertain state of the taxonomy of the genus is amply illustrated by the widely diverging synonymies that have been proposed in previous revisions of the genus (Loof, 1978, 1991; Ryss, 1988; Frederick & Tarjan, 1989).

Taxonomy

The genus *Pratylenchus* was established by Filipjev (1934), but the true generic diagnosis was published by Filipjev (1936) as a group of nematodes possessing a tylenchoid pharynx overlapping the anterior portion of the intestine and an uniovarial gonad in adult females. The first species of the genus, *Tylenchus pratensis* de Man, 1880, was described by de Man (1880) and was designated by Filipjev (1934) as the type of his new genus *Pratylenchus*. The etymology of the genus is derived from first three letters of the type species (***pra****tensis*) and *Tylenchus*. The first comprehensive treatment of the genus was by Filipjev and Schuurmans Stekhoven (1941), although T. Goodey (1932) had already discussed most of the species. Subsequently, the impact of the work by Thorne (1949), Allen and Jensen (1951) and Sher and Allen (1953) was considerable and marked the beginnings of the modern era in studying this genus. In that period, the taxonomic position of the genus was modified by several authors. Thorne (1949) proposed the subfamily Pratylenchinae under Tylenchidae with five genera: *Pratylenchus* (type genus), *Radopholus*, *Chitinotylenchus*, *Nacobbus* and *Rotylenchulus*. Chitwood (1950), apparently unaware of Thorne's action, placed *Pratylenchus* and *Radopholus* in the subfamily Hoplolaiminae, together with other genera having vermiform females (*Hoplolaimus*, *Helicoty-*

lenchus, *Rotylenchus*, *Tylenchorhynchus*). He recognised other subfamilies as having saccate or globose females: Heteroderinae (*Heterodera*, *Meloidogyne*) and a new subfamily Nacobbinae (*Nacobbus*, *Rotylenchulus*). J.B. Goodey (1963) placed under Hoplolaimidae the subfamilies Pratylenchinae (*Pratylenchus*, *Pratylenchoides*, *Hoplotylus*, *Radopholus*, *Hirschmanniella*) and Nacobbinae (*Nacobbus* together with *Rotylenchulus*). Siddiqi (1963) was the first to provide a clear concept of the Pratylenchidae when he elevated Pratylenchinae to family rank, recognised two subfamilies and placed five genera in the Pratylenchinae (*Pratylenchus*, *Radopholus*, *Pratylenchoides*, *Hirschmanniella*, *Zygotylenchus*) and one in Nacobbinae (*Nacobbus*). He also excluded *Chitinotylenchus* from Pratylenchidae, regarding it as a doubtful genus, and *Hoplotylus* as showing more affinities with Hoplolaimidae. Following Allen (1960), he also suggested that *Rotylenchulus* could be removed from Pratylenchidae as having more affinities with the Hoplolaimidae. Since the actions of Siddiqi (1963), *Pratylenchus* has maintained consensus as a well recognised taxon, although the division of pratylenchids into subfamilies and/or families has been more controversial (see Siddiqi, 1971, 1986; Andrássy, 1976), each author employing various characters and hierarchies to define the suprageneric taxa.

Morphologically, *Pratylenchus* species are differentiated from the other genera in the family Pratylenchidae by the number of gonads forming the female genital system (*i.e.*, one *vs* two), whether or not the caudal alae reach to the tip of the tail, sexual dimorphism, presence or absence of deirids and by the orientation of the pharyngeal gland lobe overlap (Luc, 1987). These diagnostic features, applicable at the species or genus level, are not based on the presence of synapomorphies as determined by careful studies of ancestor-descendant relationships (*i.e.*, phylogeny), but on phenetic or morphological similarities. Luc (1987) referred to the genus *Pratylenchus* as a "... stenomorphic genus in which the species are difficult to separate because of the small number of characters diagnostic at species level and because of the intraspecific variability of some of these characters", accepting 60 valid species in the genus. Other taxonomists have accepted quite different numbers of valid species, *viz.*, 45 spp. (Ryss, 1988), 46 spp. (Loof, 1991), 49 spp. (Frederick & Tarjan, 1989), 63 spp. (Handoo & Golden, 1989), 69 spp. (Siddiqi, 1986) and 89 spp. (Siddiqi, 2000). The difference in the numbers reflects, at least partially, variations in the number of synonyms, thereby indicating that the various authors do not agree on the diagnostic characters (and may even assign different weights to the same diagnostic characters), suggesting an uncertainty as to the specific diagnostic characters for the genus. Studies on the variability of diagnostic characters of *Pratylenchus* species have demonstrated the vast intraspecific variability of many of the characters utilised to decide whether a species is valid or not. This situation calls for careful examination of many characters, including those with current diagnostic values or those hitherto not considered when discriminating among closely related species.

Systematics

The fossil record for nematodes is poor (Poinar, 1983) and not until recently have molecular methods permitted reliable linking of groups of nematode species into phylogenetic clades (Blaxter *et al.*, 1998). Parasitism is an acquired trait and one unlikely to have evolved before evolution of the host, which for vascular plants is 400 million years ago. Presumably most of the genes in existing parasites share a common origin with the majority of genes in present-day free-living forms. The discovery of horizontal gene transfer of parasitic functionality from microbes to nematodes is probably the most significant finding in plant nematology in the past quarter century and it is likely that the small suite of nematode genes with apparent bacterial or fungal origins identified thus far will prove not to be unusual or rare anomalies. Evolution of parasitism by means of a horizontal gene transfer mechanism helps to explain how plant parasitism has apparently arisen on multiple, independent occasions, and it will be especially interesting to examine in detail the candidate gene(s) identified in other plant-nematode species (Bird & Koltai, 2000; Jones *et al.*, 2005).

The pharynx is one of the most important organs in nematode taxonomy. Its morphology has apparently evolved together with the nature of parasitism. In Pratylenchidae the asymmetric arrangement of the subventral glands appears related, whereby some species of *Pratylenchoides* and the genera *Pratylenchus* and *Zygotylenchus* form a range of increasing deviation from the primitive pharynx by lengthening of the subventral glands (Seinhorst, 1971). Analysis of the D3 expansion region of 26S rDNA showed *P. neglectus* (= *P. minyus*) to be the sister taxon to the rest of the Pratylenchidae that were analysed (Al-Banna *et al.*, 1997). This species has the smallest degree of overlap in the pharynx compared with all other species (*P. coffeae*, *P. scribneri*, *P. brachyurus*, *P. hexincisus*, *P. thornei* and *P. penetrans*) examined in the study (Al-Banna *et al.*, 1997). Therefore, the short overlap of the pharyngeal glands over the intestine is probably a primitive character that may be used to infer phylogeny based on morphological characters (Al-Banna *et al.*, 1997).

Baujard *et al.* (1990) stated that ornamentations of the lateral fields were very variable and therefore could not be used in the systematics of the genus. The number of lip annuli is also variable yet remains a good character for species identification.

The number of gonads forming the female reproductive system may give indications to phyletic relationships (Triantaphyllou & Hirschmann, 1980). In Tylenchida two opposed gonads in the female represents the primitive condition (Husain, 1976). The derived condition is represented by one gonad, which is the end product of various steps of evolution. The loss of the posterior gonad is a step-wise process including the reduction in size of the posterior ovary and then the complete degeneration of the ovary leaving the uterus as a sac-like structure (post-vulval uterine sac) (Triantaphyllou & Hirschmann, 1980), as occurs in *Pratylenchus*. The oogonial cells undergo an additional division during early embryogenesis, so that second-stage juveniles have a genital primordium with two oogonial cells, which give rise to either didelphic or monodelphic forms. In *Pratylenchus* the two oogonial cells remain together in the anterior part of the gonad; thus, adult females develop with only one gonad (Román & Hirschmann, 1969b). Nevertheless, in some females of *P. zeae* the gonads follow a didelphic pattern of development up to the fourth moult, the posterior vestigial ovary being formed with a few non-functional oogonial cells (Román & Hirschmann, 1969b). In *P. penetrans* development up to the fourth moult is similarly didelphic, but the posterior oogonial cell does not divide and eventually degenerates (Triantaphyllou & Hirschmann, 1980). Therefore, the presence of two ovaries must be considered ancestral in all Nematoda and one ovary may be reduced or lost since it may be easier to lose a complex biological function than to develop it several times. It has been shown that two gonads form in the fourth-stage juveniles of all species of *Pratylenchus* after which the posterior gonad is reduced (with or without remnants of ovary cells) in the adult females (Román & Hirschmann, 1969b; Luc, 1987).

Similarly, the cytogenetic situation in *Pratylenchus* is extremely complex because at present it is difficult to identify the basic chromosome number in the genus and to explain how the existing variation among species has been derived. The fact remains that changes in the haploid chromosomal complement are associated with, and have probably played an important role in, speciation of *Pratylenchus* (Triantaphyllou & Hirschmann, 1980). However, considering that *Radopholus* is morphologically more primitive than *Pratylenchus*, it may be assumed that the five chromosomes of the citrus race of *Radopholus similis* (Cobb, 1893) Thorne, 1949 may be the basic chromosome number of the ancestral form from which both *Radopholus* and *Pratylenchus* have evolved (Triantaphyllou & Hirschmann, 1980). In addition, there is no doubt that species that reproduce by meiotic parthenogenesis have evolved from amphimictic ancestors, whereas mitotic parthenogenetic species have evolved either from amphimictic or from facultative meiotic parthenogenetic ancestors (Triantaphyllou & Hirschmann, 1980).

Al-Banna *et al.* (1997) performed a phylogenetic analysis on species of *Pratylenchus* based on ribosomal DNA (rDNA) sequences. Evolution of rDNA is relatively independent of changes in morphology and analysis of these genetic data has been shown to provide good phylogenetic resolution. A gene tree of lesion nematodes and relatives, based on the analysis of the large subunit (LSU) 26S rRNA gene, suggested that *Pratylenchus* does not represent a monophyletic group (Al-Banna *et al.*, 1997). Parsimony analyses of nucleotides of the D3 expansion region of the 26S rDNA of *Pratylenchus* spp. showed that the taxon represents a paraphyletic assemblage (Al-Banna *et al.*, 1997). It was apparent that the out-group taxon *Hirschmanniella belli* Sher, 1968 shared a common ancestor with the clade that included *P. vulnus* and *P. crenatus* and that *Nacobbus aberrans* (Thorne, 1935) Thorne & Allen, 1944 and *R. similis* shared a common ancestor with *P. coffeae*, *P. scribneri*, *P. brachyurus* and *P. hexincisus*, whilst *P. thornei*, *P. penetrans* and *P. neglectus* (= *P. minyus*) branched from the main *Pratylenchus* clade, each sharing a common ancestor with the rest of the species in the tree. Although this proposal for polyphyly is intriguing, the bootstrap values for the most parsimonious tree were not definitive. Frequently, the addition of more taxa (Forey *et al.*, 1992) or different outgroups (Milinkovitch & Lyons-Weiler, 1998) improves the resolution of phylogenetic trees. Understanding the phylogenetic relationships among species of nematodes can clarify the biogeographic history of these nematodes. In the

Al-Banna *et al.* (1997) study, four species of *Pratylenchus* (*P. scribneri*, *P. coffeae*, *P. brachyurus* and *P. hexincisus*) are included in one clade and are serious pathogens of many crops (banana, citrus, coffee, corn, pineapple, potato) in tropical and subtropical regions. This group shares a common ancestor with *N. aberrans* and *R. similis*, perhaps indicating an origin in Gondwanaland (Al-Banna *et al.*, 1997).

Phylogenetic position based on molecular characters indicates that *Pratylenchus* is paraphyletic, suggesting that lineage sorting is incomplete (Al-Banna *et al.*, 1997; Duncan *et al.*, 1999). The existence of cryptic nematode species that are morphologically indistinguishable yet genetically divergent, thereby suggesting the existence of species complexes, has recently been reported in other nematode genera (*e.g.*, Courtright *et al.*, 2000; Reid *et al.*, 2003). Duncan *et al.* (1999) studied the genome variation between 32 nematode isolates belonging to *P. coffeae*, *P. loosi*, *P. panamaensis* and *P. pseudocoffeae* (using the D2/D3 expansion segment of the large subunit of nuclear rDNA) to estimate phylogenetic relationships among them. This study established the phylogenetic relationships of the species studied and indicated that *P. coffeae* may represent a species complex.

Carta *et al.* (2001) provided gene sequences for the D3 segment of the large subunit rDNA gene in *P. scribneri*, *P. hexincisus*, *P. teres* and *P. zeae*. The sequences were aligned with the closest comparable previously published molecular sequences and evaluated by parsimony, distance and maximum-likelihood methods. Their results indicated a gradual progression from four to two lip annuli. The best supported group was essentially tropical, as noted before (Al-Banna *et al.*, 1997) and all its members had two lip annuli. Another well supported intermediate temperate group included the sequences for *P. arlingtoni*, *P. convallariae* and *P. fallax* (Handoo *et al.*, 2001), *P. penetrans*, *P. hexincisus* and *P. zeae*, which was equivalent to the '*pratensis*-group' (Frederick & Tarjan, 1989), except for *P. hexincisus* with two lip annuli and *P. penetrans*. Similarly, Carta *et al.* (2002) studied the phylogenetic relationships among selected *Pratylenchus* spp. (*e.g.*, *P. bolivianus*, *P. crenatus*, *P. goodeyi*, *P. morettoi*, *P. neglectus*, *P. teres*, *P. wescolagricus*, *P. yassini* and *P. zeae*) based on morphological and morphometrical characters (*en face* view, V, tail shape, stylet length, L, a, b, c, post-vulval uterine sac, lip annuli, *etc.*). The phylogenetic tree showed three main branches, with the ancestral character states present in the outgroup *Hirschmanniella*, which included four lip annuli, lateral fields with four lines, acute tail tip, undivided flat *en face* view, presence of males, two female gonads and stylet >16 μm. Consequently, these data support the paraphyletic origin of *Pratylenchus* spp.

Ryss (2002b) carried out a phylogenetic analysis including 49 species of the genus *Pratylenchus* based on 26 morphological and morphometrical characters. *Pratylenchus kasari* was shown to be the closest to the conventional ancestor (outgroup). Similarly, the group of *P. vulnus*, *P. pseudocoffeae* and *P. ekrami* was also close to the ancestral type, *P. vulnus* being the most primitive species within this group. The next group, which originated from the ancestor species of *Pratylenchus*, is a phylogenetic stock of all other species of the genus, with *P. goodeyi* being the most primitive species. According to the phylogenetic analysis carried out by Ryss (2002b), the following species groups could be outlined within the genus *Pratylenchus*: *i*) alleni: *P. alleni*, *P. neglectus*; *ii*) crenatus: *P. crenatus*, *P. estoniensis*, *P. teres*; *iii*) subranjani: *P. zeae*, *P. subranjani*; *iv*) flakkensis: *P. flakkensis*, *P. convallariae*; *v*) penetrans: *P. crassi*, *P. penetrans*, *P. pseudopratensis*; *vi*) pseudopratensis: *P. pseudopratensis*, *P. andinus*, *P. sensillatus*; *vii*) bolivianus: *P. bolivianus*, *P. wescolagricus*; *viii*) scribneri: *P. scribneri*, *P. hexincisus*, *P. neglectus*; *ix*) *delattrei*: *P. delattrei*, *P. microstylus*, *P. mulchandi*, *P. thornei*; *x*) brachyurus: *P. brachyurus*, *P. macrostylus*, *P. japonicus*, *P. zeae*; *xi*) pinguicaudatus: *P. pinguicaudatus*; *xii*) pratensis-coffeae: *P. pratensis*, *P. coffeae*, *P. fallax*, *P. loosi*, *P. sudanensis*, *P. gibbicaudatus*; and *xiii*) vulnus: *P. vulnus*, *P. ekrami*, *P. pseudocoffeae*. As indicated by Ryss (2002b), *Pratylenchus* displays a great biological progress because of its large number of species, distribution over a large area of geographic area and the general development of its morpho-biological features. Based on the phylogenetic analysis and the morphological changes in evolution in the genus *Pratylenchus*, Ryss (2002b) concluded that these adaptations have been developed independently in different phylogenetic lines of pratylenchs.

However, Subbotin *et al.* (2006) did a phylogenetic analysis of the order Tylenchida Thorne, 1949 as inferred from D2 and D3 expansion fragments of the 28S rRNA gene sequences. In this study, the authors found data supporting a relationship of *Meloidogyne* with Pratylenchidae (*Pratylenchus* and *Hirschmanniella*). These results were congruent with phylogenetic studies based on the small-subunit (SSU) rDNA (Holterman *et al.*, 2006) and with the hypothesis suggested by Ryss (1988) that, among Hoplolaimoidea, pratylenchids are most closely related to *Meloidogyne* in details of the labial region and pharyngeal structure. Ryss believed these morphological similarities to be indicative of common ancestry between Meloidogynidae and Pratylenchidae. Consequently, this study, even with the limited sampling of *Pratylenchus* spp., indicated that the classical interpretation of Pratylenchidae as a monophyletic group is not supported (Subbotin *et al.*, 2006).

Finally, phylogenetic analyses carried out by Inserra *et al.* (2007) on some *Pratylenchus* species (*Pratylenchus coffeae*-group, *P. gutierrezi*-group, *P. jaehni*- and *P. loosi*-group, *P. hexincisus*, *P. hippeastri*, *P. scribneri* and *P. zeae*) revealed paraphyletic relationships between amphimictic (*Pratylenchus coffeae*-group, *P. gutierrezi*-group, *P. jaehni*, *P. loosi*-group) and non-amphimictic species (*P. hexincisus*, *P. hippeastri*, *P. scribneri* and *P. zeae*) as well as between root-lesion nematodes with two lip annuli and *P. zeae* which has three lip annuli (Inserra *et al.*, 2007).

Classification (systematic scheme according to Siddiqi, 2000)

Phylum Nematoda Rudolphi, 1808
  Class Secernentea von Linstow, 1905
    Subclass Tylenchia Inglis, 1983
      Order Tylenchida Thorne, 1949
        Suborder Tylenchina Thorne, 1949
          Superfamily Hoplolaimoidea Filipjev, 1934
            Family Pratylenchidae Thorne, 1949
              Subfamily Pratylenchinae Thorne, 1949
                Genus *Pratylenchus* Filipjev, 1936

Genus *Pratylenchus* Filipjev, 1936

= *Pratylenchus* Filipjev, 1934 (= *nomen nudum*)

DIAGNOSIS (MODIFIED AFTER SIDDIQI, 2000)

Female

Pratylenchinae. No marked sexual dimorphism in anterior region. Lateral fields with four to six lines, occasionally with oblique median markings. Deirids absent. Phasmids near mid-tail. Labial region low, flattened anteriorly or rarely rounded, continuous with body contour; sclerotisation massive; labial disc inconspicuous, in SEM dumbbell-shaped, with six labial pits around a minute oral aperture; amphidial apertures pore-like, near labial disc, indistinct. Stylet 20 μm or less long, with round, anteriorly flat or indented basal knobs. Median bulb oval to round, very muscular. Pharyngeal glands usually less than two body diam. long, extending over intestine mostly ventrally. Vulva in posterior region (usually at 70-80%). Pseudo-mono-prodelphic, with only anterior ovary functional. Post-vulval uterine sac present, with or without rudiments of posterior ovary. Spermatheca large, rounded, usually axial. Female tail subcylindrical to conoid, usually *ca* 2-3 anal body diam. long; terminus smooth or annulated, devoid of a process or mucron. Bursa enclosing tail terminus. Spicules with subterminal pore on dorsal side. Gubernaculum simple, trough-like, fixed.

TYPE SPECIES

1. *Pratylenchus pratensis* (de Man, 1880) Filipjev, 1936
   = *Tylenchus pratensis* de Man, 1880
   = *Anguillulina pratensis* (de Man, 1880) Goffart, 1929
   = *Pratylenchus helophilus* Seinhorst, 1959
   = *Pratylenchus irregularis* Loof, 1960

OTHER VALID SPECIES

2. *Pratylenchus acuticaudatus* Braasch & Decker, 1989
3. *Pratylenchus alleni* Ferris, 1961
4. *Pratylenchus andinus* Lordello, Zamith & Boock, 1961
5. *Pratylenchus angulatus* Siddiqi, 1994
6. *Pratylenchus arlingtoni* Handoo, Carta & Skantar, 2001
7. *Pratylenchus artemisiae* Zheng & Chen, 1994
8. *Pratylenchus bhattii* Siddiqi, Dabur & Bajaj, 1991
9. *Pratylenchus bolivianus* Corbett, 1983
   = *Pratylenchus australis* Valenzuela & Raski, 1985
10. *Pratylenchus brachyurus* (Godfrey, 1929) Filipjev & Schuurmans Stekhoven, 1941
    = *Tylenchus brachyurus* Godfrey, 1929
    = *Anguillulina brachyurus* (Godfrey, 1929) Goodey, 1932
    = *Pratylenchus leiocephalus* Steiner, 1949
    = *Pratylenchus pratensis* apud Thorne, 1940
    = *Pratylenchus steineri* Lordello, Zamith & Boock, 1954
11. *Pratylenchus brzeskii* Karssen, Waeyenberge & Moens, 2000

12. *Pratylenchus coffeae* (Zimmermann, 1898) Filipjev & Schuurmans Stekhoven, 1941
    = *Tylenchus coffeae* Zimmermann, 1898
    = *Anguillulina coffeae* (Zimmermann, 1898) Goodey, 1932
    = *Tylenchus musicolai* Cobb, 1919
    = *Anguillulina musicolai* (Cobb, 1919) Goodey, 1932
    = *Pratylenchus musicolai* (Cobb, 1919) Filipjev, 1936
    = *Tylenchus mahogani* Cobb, 1920
    = *Anguillulina mahogani* (Cobb, 1920) Goodey, 1932
    = *Pratylenchus mahogani* (Cobb, 1920) Filipjev, 1936
    = *Pratylenchus brassicae* (Shahina & Maqbool, 1996) Siddiqi, 2000 n. syn.
    = *Radopholus brassicae* Shahina & Maqbool, 1996
13. *Pratylenchus convallariae* Seinhorst, 1959
14. *Pratylenchus crassi* Das & Sultana, 1979
15. *Pratylenchus crenatus* Loof, 1960
    = *Pratylenchus clavicaudatus* Baranovskaya & Haque, 1968
    = *Pratylenchus pratensis* apud Thorne, 1949
16. *Pratylenchus cruciferus* Bajaj & Bhatti, 1984
17. *Pratylenchus curvicauda* Siddiqi, Dabur & Bajaj, 1991
18. *Pratylenchus delattrei* Luc, 1958
    = *Pratylenchus singhi* Das & Sultana, 1979
    = *Pratylenchus portulacus* Zarina & Maqbool, 1998
    = *Pratylenchus graminis* Subramaniyan & Sivakumar, 1991 n. syn.
19. *Pratylenchus dunensis* de la Peña, Moens, van Aelst & Karssen, 2006
20. *Pratylenchus ekrami* Bajaj & Bhatti, 1984
21. *Pratylenchus elamini* Zeidan & Geraert, 1991
22. *Pratylenchus estoniensis* Ryss, 1982
23. *Pratylenchus fallax* Seinhorst, 1968
24. *Pratylenchus flakkensis* Seinhorst, 1968
25. *Pratylenchus gibbicaudatus* Minagawa, 1982
26. *Pratylenchus goodeyi* Sher & Allen, 1953
    = *Tylenchus musicolai* apud Goodey, 1928
27. *Pratylenchus hexincisus* Taylor & Jenkins, 1957
28. *Pratylenchus hippeastri* Inserra, Troccoli, Gozel, Bernard, Dunn & Duncan, 2007
29. *Pratylenchus jaehni* Inserra, Duncan, Troccoli, Dunn, dos Santos, Kaplan & Vovlas, 2001

30. *Pratylenchus japonicus* Ryss, 1988
    = *Pratylenchus macrostylus japonicus* Ryss, 1988
31. *Pratylenchus kasari* Ryss, 1982
32. *Pratylenchus kralli* Ryss, 1982
33. *Pratylenchus kumaoensis* Lal & Khan, 1990
34. *Pratylenchus loosi* Loof, 1960
    = *Pratylenchus coffeae* apud Loos, 1953
35. *Pratylenchus macrostylus* Wu, 1971
36. *Pratylenchus manaliensis* Khan & Sharma, 1992
37. *Pratylenchus mediterraneus* Corbett, 1983
38. *Pratylenchus microstylus* Bajaj & Bhatti, 1984
39. *Pratylenchus morettoi* Luc, Baldwin & Bell, 1986
40. *Pratylenchus mulchandi* Nandakumar & Khera, 1970
    = *Pratylenchus muli* Nandakumar & Khera, 1969 (= *nomen nudum*)
41. *Pratylenchus neglectus* (Rensch, 1924) Filipjev & Schuurmans Stekhoven, 1941
    = *Aphelenchus neglectus* Rensch, 1924
    = *Tylenchus neglectus* (Rensch, 1924) Steiner, 1928
    = *Anguillulina* (*Pratylenchus*) *neglecta* (Rensch, 1924) W. Schneider, 1939
    = *Pratylenchus minyus* Sher & Allen, 1953
    = *Pratylenchus capitatus* Ivanova, 1968
    = *Pratylenchus neocapitatus* Khan & Singh, 1975
    = *Pratylenchus similis* Khan & Singh, 1975
    = *Pratylenchus gotohi* (Mizukubo & Minagawa, 1991) Pourjam, Kheiri, Geraert & Alizadeh, 1999
42. *Pratylenchus neobrachyurus* Siddiqi, 1994
43. *Pratylenchus okinawaensis* Minagawa, 1991
44. *Pratylenchus panamaensis* Siddiqi, Dabur & Bajaj, 1991
    = *Pratylenchus gutierrezi* Golden, López & Vilchez, 1992
45. *Pratylenchus penetrans* (Cobb, 1917) Filipjev & Schuurmans Stekhoven, 1941
    = *Tylenchus penetrans* Cobb, 1917, male, *nec* female
    = *Anguillulina* (*Pratylenchus*) *penetrans* (Cobb) Goodey, 1932
    = *Tylenchus gulosus* Kühn, 1890 (= *nomen oblitum*)
    = *Pratylenchus gulosus* (Kühn, 1890) Filipjev & Schuurmans Stekhoven, 1941
    = *Pratylenchus globulicola* Romániko, 1960
46. *Pratylenchus pinguicaudatus* Corbett, 1969

47. *Pratylenchus pratensisobrinus* Bernard, 1984
48. *Pratylenchus pseudocoffeae* Mizukubo, 1992
49. *Pratylenchus pseudofallax* Café-Filho & Huang, 1989
50. *Pratylenchus pseudopratensis* Seinhorst, 1968
    = *Pratylenchus sefaensis* Fortuner, 1973
51. *Pratylenchus roseus* Zarina & Maqbool, 1998
52. *Pratylenchus scribneri* Steiner *in* Sherbakoff & Stanley, 1943
    = *Tylenchus penetrans* Cobb, 1917
    = *Pratylenchus agilis* (Thorne & Malek, 1968) Hernández, Jordana, Goldaracena & Pinochet, 2000
    = *Pratylenchus crossandrae* Subramaniyan & Sivakumar, 1991
53. *Pratylenchus sensillatus* Anderson & Townshend, 1985
54. *Pratylenchus silvaticus* Brzeski, 1998
55. *Pratylenchus subpenetrans* Taylor & Jenkins, 1957
    = *Pratylenchus gongjuensis* Choi, Lee, Park, Han & Choi, 2006 n. syn.
56. *Pratylenchus subranjani* Mizukubo, Toida, Keereewan & Yoshida, 1990
57. *Pratylenchus sudanensis* Loof & Yassin, 1971
58. *Pratylenchus tenuis* Thorne & Malek, 1968
59. *Pratylenchus teres* Khan & Singh, 1974
60. *Pratylenchus thornei* Sher & Allen, 1953
    = *Pratylenchus ranjani* (Khan & Singh, 1975) Pourjam, Kheiri, Geraert & Alizadeh, 1999
    = *Pratylenchus peelari* in Chawla & Prasad, 1973 (= *nomen nudum*)
    = *Pratylenchus allius* (Shahina & Maqbool, 1996) Siddiqi, 2000 n. syn.
    = *Radopholus allius* Shahina & Maqbool, 1996
61. *Pratylenchus typicus* Rashid, 1974
62. *Pratylenchus unzenensis* Mizukubo, 1992
63. *Pratylenchus ventroprojectus* Bernard, 1984
64. *Pratylenchus vulnus* Allen & Jensen, 1951
65. *Pratylenchus wescolagricus* Corbett, 1983
66. *Pratylenchus yamagutii* Minagawa, 1991
67. *Pratylenchus yassini* Zeidan & Geraert, 1991
68. *Pratylenchus zeae* Graham, 1951
    = *Pratylenchus cubensis* Razjivin & O'Relly, 1976

= *Pratylenchus zeae* Steiner *in* Clayton & McMurtrey, 1950 (= *nomen nudum*)
= *Pratylenchus impar* Khan & Singh, 1974
= *Pratylenchus jordanensis* Hashim, 1984

*SPECIES INQUIRENDAE**

Pratylenchus barkati Das & Sultana, 1979
Pratylenchus bicaudatus Meyl, 1954
= *Pratylenchus pratensis bicaudatus* Meyl, 1954
Pratylenchus brevicercus Das, 1960
Pratylenchus cerealis Haque, 1966
Pratylenchus chrysanthus Edward, Misra, Rai & Peter, 1969
Pratylenchus codiaei Singh & Jain, 1984
Pratylenchus coffeae brasiliensis Lordello, 1956 (? syn. of *P. zeae*)
Pratylenchus dasi Fortuner, 1985 (*nom. nov.* for *P. capitatus* Das & Sultana, 1979 nec *P. capitatus* Ivanova, 1968)
= *Pratylenchus hyderabadensis* Singh & Gill, 1986 (junior *nom. nov.* for *P. capitatus* Das & Sultana, 1979)
Pratylenchus dioscoreae Yang & Zhao, 1992
Pratylenchus emarginatus Eroshenko, 1978
Pratylenchus exilis Das & Sultana, 1979
Pratylenchus heterocercus (Kreis, 1930) Andrássy, 1960
= *Dolichodorus heterocercus* Kreis, 1930 (syn. of *P. penetrans* for Andrássy, 1960)
Pratylenchus indicus Das, 1960
Pratylenchus kolourus Fortuner, 1985 (was a *nom. nov.*)
= *Tylenchus* (*Chitinotylenchus*) *coffeae brevicauda* Rahm, 1928
Pratylenchus loofi Singh & Jain, 1984
Pratylenchus manohari Quraishi, 1982
Pratylenchus montanus Zyubin, 1966
Pratylenchus nizamabadensis Maharaju & Das, 1981
Pratylenchus obtusicaudatus Romániko, 1977
Pratylenchus obtusus (Bastian, 1865) Goodey, 1951
= *Tylenchus obtusus* Bastian, 1865
= *Anguillulina obtusa* (Bastian, 1865) Goodey, 1932

*Due to poor descriptions or the lack of information adequate comparisons could not be made and, therefore, the listed species are considered as *species inquirendae*.

= *Rotylenchus obtusus* (Bastian, 1865) Filipjev, 1936
= *Tylenchorhynchus obtusus* (Bastian, 1865) Filipjev & Schuurmans Stekhoven, 1941
Pratylenchus pratensis tenuistriatus Meyl, 1953
Pratylenchus sacchari (Soltwedel, 1888) Filipjev, 1936
= *Tylenchus sacchari* Soltwedel, 1888
= *Anguillulina sacchari* (Soltwedel, 1888) Goodey, 1932
Pratylenchus septincisus Chang, 1991
Pratylenchus stupidus Romániko, 1977
Pratylenchus tulaganovi Samibaeva, 1966
Pratylenchus tumidicepsi Merzheevskaya, 1951
Pratylenchus uralensis Romániko, 1966
Pratylenchus variacaudatus Romániko, 1977

NOMINA NUDA

Pratylenchus angelicae Kapoor, 1983
Pratylenchus himalayaensis Kapoor, 1983
Pratylenchus menthae Kapoor, 1983
Pratylenchus rhizasinus Sher, 1948

NOTE

None of the *Pratylenchus* species proposed in the thesis of Kapoor (1983) and by Sher (1948) is available as they are not considered to have been validly published (Fortuner, 1985).

SPECIES DESCRIBED IN *PRATYLENCHUS* AND LATER TRANSFERRED TO OTHER GENERA

Original name	Present combination
Pratylenchus aberrans Thorne, 1935	*Nacobbus aberrans* (Thorne, 1935) Thorne & Allen, 1944
Pratylenchus dendrophilus Marcinowski, 1909	*Neoditylenchus dendrophilus* (Marcinowski, 1909) Meyl, 1961
Pratylenchus graminophilus Goodeyi, 1933	*Subanguina graminophila* (Goodey, 1933) Brzeski, 1981
Pratylenchus tumifaciens Cobb, 1932	*Anguina tumifaciens* (Cobb, 1932) Filipjev & Schuurmans Stekhoven, 1941

Nematology Monographs & Perspectives, 2007, Vol. 6, 51-280

Chapter 4

Diagnosis and descriptions of *Pratylenchus* species

In this chapter are included the measurements of most of the studied populations of the 68 valid species belonging to the genus *Pratylenchus*, a complete morphological description of each species, a diagnostic characterisation, including the matrix code for the polytomous key, the geographic distribution and host plants of each valid species arranged in alphabetical order. The number cited before each species is the same as that in the foregoing species list where any synonyms are listed in full.

1. *Pratylenchus pratensis* (de Man, 1880) Filipjev, 1936
(Fig. 20)

MEASUREMENTS

- Female lectotype (after Loof, 1960): L = 0.46 mm; a = 28; b = 3.6; c = 20; V = 77; stylet = 15 μm.
- 27 females (after Loof, 1960): L = 0.49 (0.39-0.61) mm; a = 22.3-33.0; b = 5.0-6.4; c = 12.8-22.0; V = 76-80; stylet = 12-15 μm.
- 10 females (after Seinhorst, 1959): L = 0.48-0.62 mm; a = 27-38; b = 5.0-7.1; c = 17-22; V = 77-80; stylet = 16-17 μm.
- 10 males (after Seinhorst, 1959): L = 0.46-0.60 mm; a = 31-43; b = 5.4-8.1; c = 16-25; T = 36-61; stylet = 14-16 μm.
- 15 females (after Loof, 1974): L = 0.47-0.56 mm; a = 24-32; b = 4.7-6.3; c = 16-22; c′ = 2.1-2.7; V = 75-79; stylet = 13-15 μm.
- 10 females (after Loof, 1974): L = 0.59-0.69 mm; a = 26-34; b = 6.3-7.8; c = 15-21; c′ = 2.4-3.1; V = 75-78; stylet = 14-15 μm.
- 5 females (after Loof, 1974): L = 0.54-0.62 mm; a = 25-33; b = 5.5-6.9; c = 17-24; c′ = 2.5-2.6; V = 71-80; stylet = 14-15 μm.
- 3 males (after Loof, 1974): L = 0.44-0.57 mm; a = 28-32; b = 5.2-6.3; c = 17-19; c′ = 2.4-2.6; T = 34-53; stylet = 14 μm.

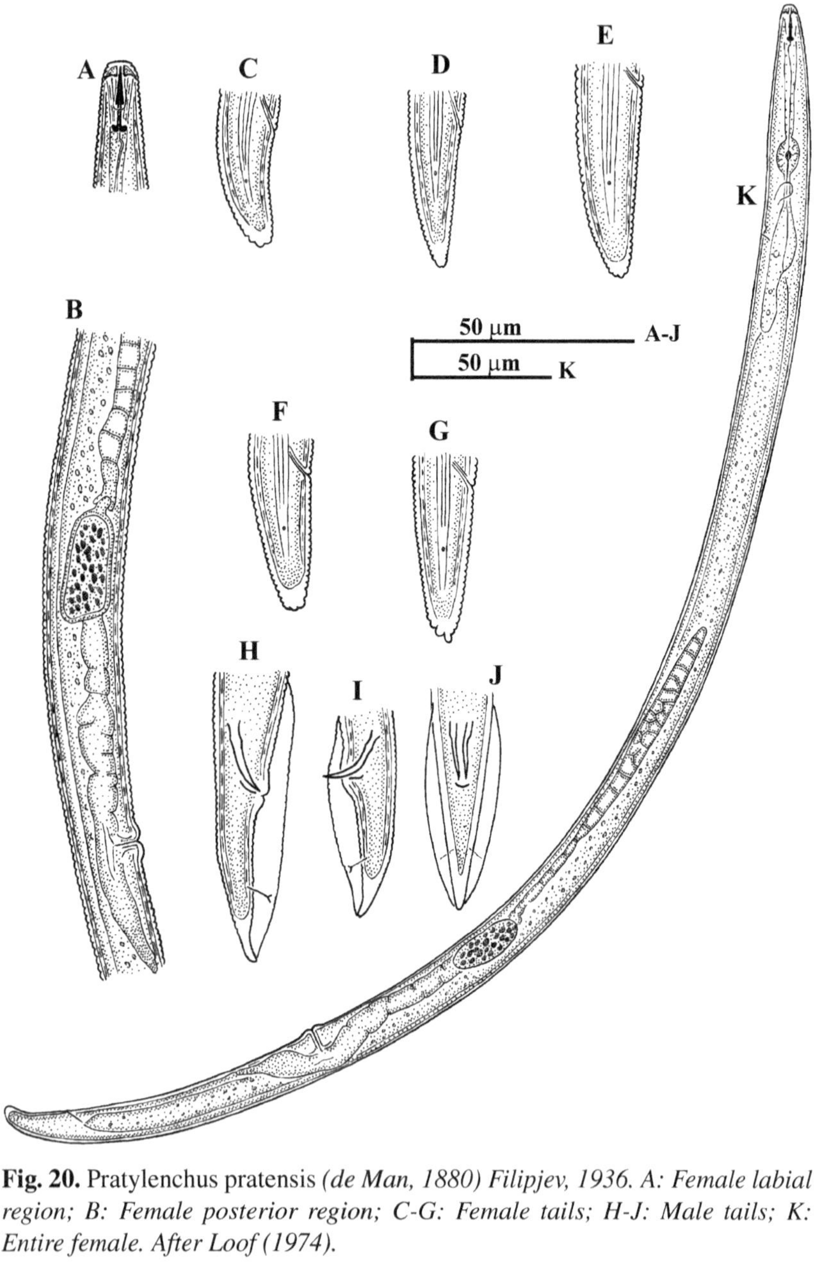

Fig. 20. Pratylenchus pratensis *(de Man, 1880) Filipjev, 1936. A: Female labial region; B: Female posterior region; C-G: Female tails; H-J: Male tails; K: Entire female. After Loof (1974).*

- 5 males (after Loof, 1974): L = 0.48-0.63 mm; a = 28-36; b = 5.3-6.6; c = 17-21; T = 38-60; stylet = 13-14 μm.
- 10 females (after Ryss, 1988): L = 0.50 (0.43-0.60) mm; a = 23 (25-30); b = 5.0 (4.7-6.8); b′ = 4.0 (3.4-4.6); c = 19 (15-23); c′ = 2.4 (2.0-2.6); V = 77 (75-79); stylet = 14 (13-15) μm.
- 4 males (after Ryss, 1988): L = 0.58 (0.50-0.60) mm; a = 30 (26-33); b = 5.8 (5.2-6.4); b′ = 3.9 (3.4-4.3); c = 19 (18-19); c′ = 2.5 (2.2-2.7); stylet = 14 μm; spicules = 18 (17-19) μm; gubernaculum = 6-7 μm.

Description

Female

Body slender, almost straight when relaxed. Cuticular annulation fine (0.9 μm at mid-body), inconspicuous, sometimes very difficult to observe. Lateral fields with four longitudinal lines. Labial region almost continuous with body contour, composed of three annuli; its edges rounded. Basal labial sclerotisation extending posteriorly *ca* one annulus. Stylet stout, with well separated basal knobs. Orifice of dorsal pharyngeal gland *ca* 2.5 μm posterior to base of stylet. Median pharyngeal bulb broadly oval. Pharyngeal glands overlapping intestine ventrally to ventrolaterally. Excretory pore generally just anterior to pharyngo-intestinal junction. Hemizonid immediately anterior to excretory pore. Vulva transverse. Uterus with large oval to rectangular spermatheca filled with sperm. Ovary with oocytes in single row except in a short multiplication zone. Post-vulval uterine sac slightly longer than body diam., its length 14-28 μm or 22-32% of vulva-anus distance; generally undifferentiated, occasionally with one separate cell. Tail with 20-28 annuli excluding those around tip; annulation continuing around terminus which is variable in shape, usually oblique, sometimes more symmetrically conoid or appearing slightly mucronate. Phasmids *ca* mid-tail.

Male

Similar to female. Testis single, outstretched. Spicules curved, 17-19 μm long. Gubernaculum 6-7 μm long. Bursa enveloping tail, margin faintly crenate. Phasmids extending into bursa, located posterior to mid-tail.

DIAGNOSIS AND RELATIONSHIPS

Pratylenchus pratensis is characterised by: cuticle finely annulated, labial region with three annuli, oval to rectangular spermatheca, post-vulval uterine sac length similar to body diam., tail with 20-28 annuli, annulated until terminus.

The matrix code is: A2, B2, C2, D4, E2, F3, G3, H2, I1, J1, K1.

It is close to *P. convallariae*, *P. fallax*, *P. kasari*, *P. pratensisobrinus* and *P. pseudopratensis* from which it differs by stylet length, the position of the vulva, shape of spermatheca, shape of tail, tail annuli, tail tip and presence of males (see corresponding descriptions).

DISTRIBUTION

Apart from the type locality (Leiden, The Netherlands), it has been recorded in several host and localities in The Netherlands, occurring mainly in clay and peat soils (Loof, 1974). Seinhorst (1959) reported it from clay soil under grass at Wageningen, The Netherlands, and it also occurs in brackish habitats in the Wadden district (Loof, 1974). It has been recorded in several European countries: Belgium from moist loamy meadow soil (Coomans, 1962); Bulgaria (Stoyanov, 1979); Finland (Kurppa, 1988); Germany from saline meadow soil (Paetzold, 1955, 1958; Decker & Dowe, 1974); Italy on cereals (Inserra *et al.*, 1979); Moldavia on solanaceous crops (Dement'eva, 1971) and corn (Nesterov & Lizogubova, 1972); Poland on fruit trees (Kozlowska & Wasilewska, 1972); Russia on several hosts (Ryss, 1988), fodder grass (Chizhov, 1978), *Valeriana officinalis* L. (Pavlyuk, 1972), conifer seedlings (Gubina, 1973); Slovakia on cereals (Liskova *et al.*, 1988); Slovenia on corn (Urek *et al.*, 2003); and from several hosts and localities in Spain (Gómez Barcina *et al.*, 1989; Peña Santiago *et al.*, 2004). It has been recorded in some countries of Africa: in Algeria (Troccoli *et al.*, 1992); in Libya on ornamental plants (Saadabi, 1993); South Africa (Van den Berg, 1971). It has been recorded from several hosts and localities in Asia: Azerbaidzhan on wheat (Kasimova & Atakishieva, 1981); China on longan, *Dimocarpus longan* Lour (Liu & Zhang, 1999; Yin, 1991); India on sugarcane (Mehta & Sundararaj, 1990; Rashid, 1997), strawberry (Khan, 2003), kiwi fruit, *Actinidia deliciosa* Chev. (Khan, 2000), carnation (Khanna & Jeevan-Jyot, 1999), apple (Khan & Sharma, 1990) and chrysanthemum (Khanna & Khan, 1990); Pakistan on vegetable crops (Anwar *et al.*, 1992a), nurseries (Khan *et al.*, 1989), tobacco (Saeed *et al.*, 1986); Uzbekistan on vegetable crops (Ailarova, 1986; Karimova, 1986), potato (Usmanova & Adylova, 1976), grapevine (Azizova, 1972) and flowers (Khakimova, 1978). It has been recorded from North America in Arkansas, USA (Wehunt *et al.*, 1989), and in Canada (Townshend *et al.*, 1978a); and also from Cuba (Fernandez Díaz-Silveira & Ortega Herrera, 1998) and Mexico on corn (Revelo-Moran *et al.*, 1993).

2. *Pratylenchus acuticaudatus* Braasch & Decker, 1989
(Fig. 21)

MEASUREMENTS

- Female holotype (after Braasch & Decker, 1989): L = 0.47 mm; a = 29; b = 4.3; c = 14; V = 79; stylet = 15 μm.
- 20 females (after Braasch & Decker, 1989): L = 0.51 (0.43-0.60) mm; a = 30 (22-35); b = 4.8 (4.0-5.9); c = 16.9 (13-26); V = 78 (76-79.5); stylet = 16 (15-16.5) μm.

DESCRIPTION

Female

Habitus slightly curved or almost straight after heat relaxation and fixation. Body gradually tapering posterior to vulva to tail tip. Body annuli 1.0 μm apart at mid-body; annuli around anal region wider than those at mid-body. Labial region low with two annuli and flat at front, not offset from body; 2 μm height $\times$ 7-8 μm wide. Labial framework well developed. Stylet short, robust, with anteriorly flattened rounded knobs. Dorsal pharyngeal gland outlet 2-3 μm posterior to stylet base (o = 13-18). Lateral fields with four lines, 4-5 μm wide at mid-body, decreasing to 2.5-3 μm posterior to vulva. Inner lateral lines converging to a single line from posterior to vulva to near tail tip. Phasmids inconspicuous. Median bulb oval. Pharyngeal gland 42 (32-50) μm long, overlapping intestine ventrally. Isthmus encircled by nerve ring. Excretory pore *ca* one body diam. posterior to median bulb, located opposite pharyngo-intestinal junction, 85 (75-95) μm from anterior end. Hemizonid just anterior to excretory pore. Anterior branch of gonad outstretched, ovary usually consisting of single row of oocytes, spermatheca inconspicuous, not functional. Post-vulval uterine sac 23

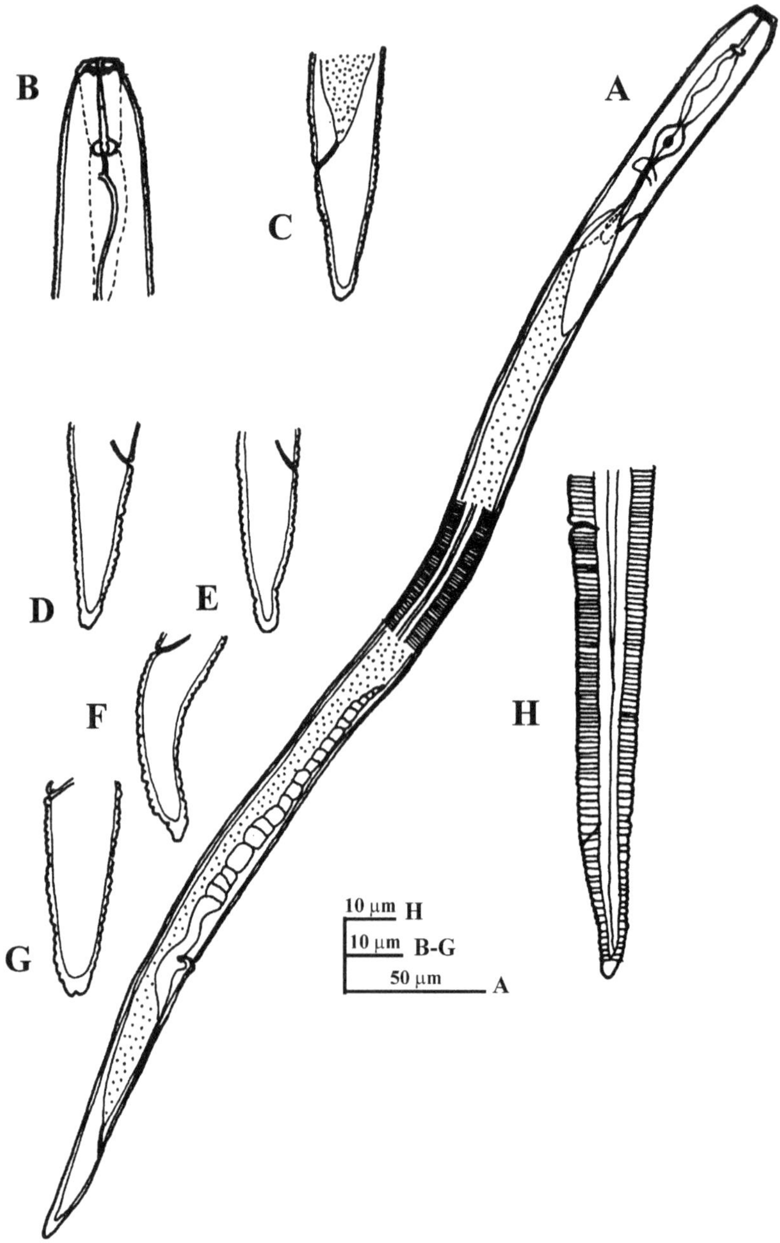

Fig. 21. Pratylenchus acuticaudatus *Braasch & Decker, 1989. A: Entire female; B: Female labial region; C-G: Female tails; H: Female posterior region. After Braasch and Decker (1989).*

(17-30) μm long, undifferentiated, occupying *ca* 30 (23-38)% of vulva-anus distance. Tail subacute, bearing 20 (17-26) annuli, hyaline area 2.5 μm long. Tail tip conoid-rounded, smooth, tail tip often indented, curved in some specimens.

Male

Not found.

DIAGNOSIS AND RELATIONSHIPS

Pratylenchus acuticaudatus is characterised by: a flat labial region with two annuli, not offset from body, subacute tail with smooth rounded tail tip and lateral fields with four lines with inner lateral lines converging to a single one posterior to vulva to near tail tip.

The matrix code is: A1, B1, C3, D1, E2, F3, G3, H1, I3, J1, K1.

It is close to *P. bolivianus*, *P. brachyurus*, *P. neglectus* and *P. vulnus* from which it differs in the number of lip annuli, stylet length, position of the vulva, tail shape (see the corresponding descriptions).

DISTRIBUTION

It has been recorded only from the type locality at Mühlhausen, Germany, parasitising roots of *Ctenanthe oppenhaimiana* (E. Morr.) K. Schum.

3. *Pratylenchus alleni* Ferris, 1961
(Fig. 22)

MEASUREMENTS

- Female holotype (after Ferris, 1961): L = 0.41 mm; a = 22; b = 5.7; c = 22; V = 79; stylet = 14 μm.
- 10 females (after Ferris, 1961): L = 0.38 (0.33-0.44) mm; a = 23 (19-27); b = 5.4 (4.7-6.1); c = 20 (15-25); V = 80 (78-83); stylet = 14 (13.5-15) μm.
- 10 males (after Ferris, 1961): L = 0.37 (0.35-0.40) mm; a = 26 (22-34); b = 5.3 (5.1-5.7); c = 20 (18-22); T = 38-52; stylet = 13.6 (13.5-14.5) μm.
- 6 females (after Ryss, 1988): L = 0.36 (0.34-0.43) mm; a = 24 (19-28); b = 5.8 (4.9-6.3); c = 21 (15-26); c′ = 1.8-2.3; V = 80 (77-83); stylet = 14 (13.5-14.5) μm.

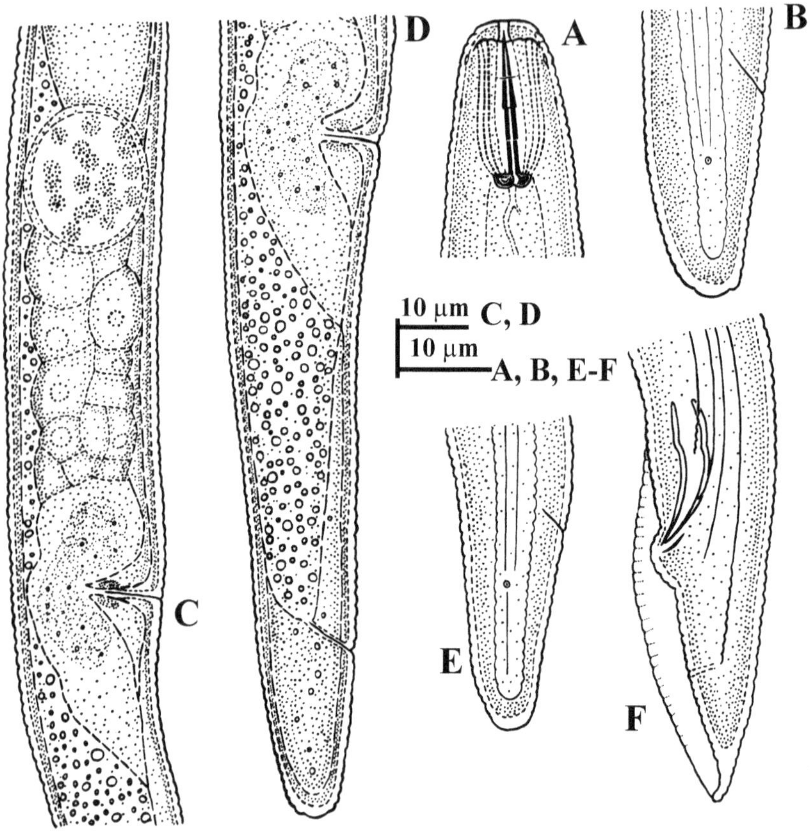

Fig. 22. Pratylenchus alleni *Ferris, 1961. A: Female labial region; B, E: Female tails; C, D: Female posterior regions; F: Male tail. After Ferris (1961).*

– 6 males (after Ryss, 1988): L = 0.36 (0.33-0.42) mm; a = 28 (24-35); b = 5.5 (4.7-6.4); c = 21 (18-24); c′ = 2.0-2.5; stylet = 13.5 (13-14) μm; spicules = 13-16 μm; gubernaculum = 5 μm.

DESCRIPTION

Female

Labial region bluntly rounded, with two annuli. Outer margins of sclerotised labial framework extending into body *ca* one body annulus. Lateral fields 25% of body diam., marked by four incisures, middle band wider than outer ones. Stylet-guiding apparatus extending posteriorly from basal plate of labial framework for *ca* three body annuli. Both anterior and posterior cephalids present of which anterior cephalid larger and located at second annulus posterior to labial region. Posterior cephalid located at *ca* 6th body annulus. Stylet *ca* 14 μm long. Basal knobs well developed, flattened anteriorly. Hemizonid just anterior to excretory pore, *ca* two annuli long. Ovary usually consisting of double row of oocytes except for a short single row at either end. Spermatheca round, oviduct cellular from 1.5-3 times as long as spermatheca, uterus usually *ca* as long as spermatheca. Post-vulval uterine sac slightly longer than vulval body diam. Vulva-anus distance equal to *ca* 3 (2.3-4.2) times tail length. Phasmids anterior to mid-tail. Tail rounded with smooth terminus, but sometimes with one or two coarse annuli.

Male

Similar to female. Spermatocytes arranged in double or triple rows. Phasmids, slightly anterior to mid-tail, may extend slightly into bursa. Spicules arcuate, hafted, resting on simple, trough-shaped, gubernaculum.

DIAGNOSIS AND RELATIONSHIPS

Pratylenchus alleni is characterised by: labial region bluntly rounded, with two annuli, lateral fields with four lines, spermatheca round, tail subcylindrical with rounded smooth terminus and presence of males.

The matrix code is: A1, B2, C2, D2, E3, F2, G2, H1, I3, J1, K1.

It can be distinguished from closely related species (*P. flakkensis*, *P. neobrachyurus*) by lateral fields, stylet length, position of excretory pore, position of the vulva, length of the post-vulval uterine sac and number of annuli on tail (see the corresponding descriptions).

DISTRIBUTION

Apart from the type locality (Eldorado, Illinois, USA, from soil around the roots of soybean), it has been recorded from *Chrysanthemum*, India (Singh & Jain, 1984); from vetch and chickpea in Turkey (Di Vito *et al.*, 1994); from Martinique (Van den Berg & Cadet, 1992); on soybean, cotton and wheat, Arkansas (USA) (Robbins *et al.*, 1987, 1989a, b); on corn, Iowa (USA) (Williams, 1982); potato, Ohio (USA) (Brown *et al.*, 1980); Canada (Townshend *et al.*, 1978a); from Russia on *Zygophyllum* sp. (Ryss, 1988); and from Argentina on raspberry (Doucet *et al.*, 2005).

4. *Pratylenchus andinus* Lordello, Zamith & Boock, 1961
(Fig. 23)

MEASUREMENTS

- 15 females (after Lordello *et al.*, 1961): L = 0.54-0.56 mm; a = 21.5-22.0; b = 7.0-7.4; c = 27.0-28.3; V = 81-85; stylet = 17-18 μm.
- 5 females (after Loof, 1978): L = 0.60-0.66 mm; a = 24-33; b = 5.3-6.0; b′ = 3.9-4.4; c = 22-33; V = 82-85; stylet = 17.5-18.5 μm.

DESCRIPTION

Female

Body straight or slightly curved ventrally on death, finely annulated, narrowing slightly posterior to vulva, then nearly cylindrical to broadly smooth tail tip, occasionally indented or notched. Lateral fields with four lines, outer lines irregularly crenate, sparsely areolated on tail. Inner two continuing posterior to phasmid, which is in lower part of middle band, approximately mid-tail opposite 8th to 12th annulus from tail tip. Labial region low with three very narrow annuli, stepped at sides and flat at front, offset by a small constriction. Labial framework moderately sclerotised, extending into body *ca* two annuli. *En face* view, by SEM, first lip annulus with fused subdorsal and subventral sectors forming angular dumbbell shape separated by a groove from lateral segments bearing amphidial apertures on their inner edge; oral aperture oval, surrounded by six small pores in two rows of three. Stylet robust with rounded basal dorsal gland opening *ca* 2 μm posterior to stylet base. Median bulb ovate, well developed; nerve ring encircling isthmus just before basal lobe enlarges; pharyngeal lobe overlapping intestine ventrally and laterally, overlap (28-53 μm), *ca* 1.5 body diam. long. Pharyngo-intestinal junction opposite or slightly anterior to excretory pore which is 65-93 μm from anterior end and immediately posterior to hemizonid. Single anterior reproductive tract with small, rounded, functionless, spermatheca. Post-vulval uterine sac 19-34 μm long, undifferentiated. Tail subcylindrical, with rounded and smooth terminus, ventral surface with 16-19 annuli.

Male

Not found.

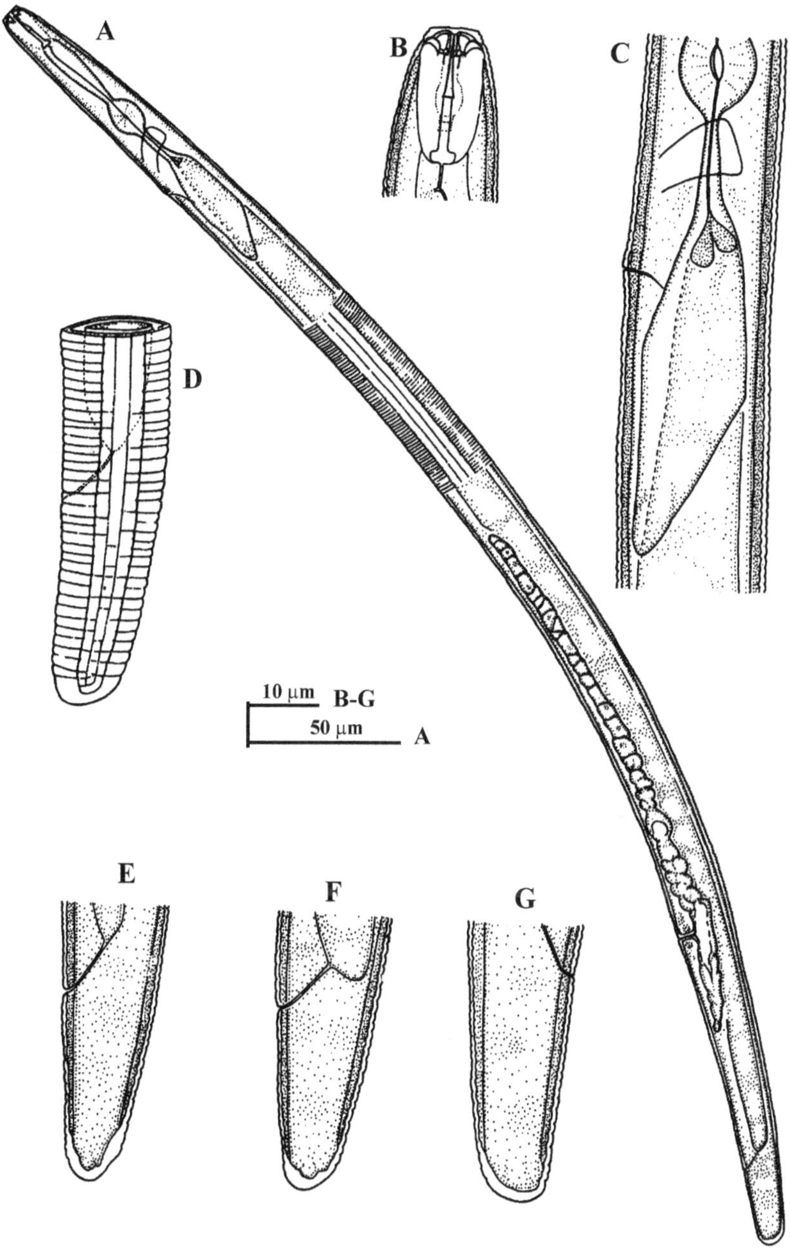

Fig. 23. Pratylenchus andinus *Lordello, Zamith & Boock, 1961. A: Entire female; B: Female labial region; C: Female pharyngeal region; D-G: Female tails. After Lordello* et al. *(1961).*

DIAGNOSIS AND RELATIONSHIPS

Pratylenchus andinus is characterised by: labial region low with three very narrow annuli, offset by a small constriction, lateral fields with four lines and tail subcylindrical, with rounded and smooth terminus.

The matrix code is: A2, B1, C3, D2, E3, F4, G1, H1, I3, J1, K1.

Pratylenchus andinus most closely resembles *P. pinguicaudatus* from which it differs by having a low labial region with three annuli, stylet length, pharyngeal overlap, a broadly rounded tail and SEM *en face* view (see the corresponding description).

DISTRIBUTION

Recorded from the type locality at Koari Fam, Cochabamba Valley, Bolivia, around roots of *Solanum andigenum* Juz. & Buk; in Argentina on grassland (Torres & Chaves, 1999); and on the Nunatak Base, East Antarctica (Ryss *et al.*, 2005).

5. *Pratylenchus angulatus* Siddiqi, 1994
(Fig. 24)

MEASUREMENTS

- Female holotype (after Siddiqi, 1994): L = 0.35 mm; a = 30; b = 5.3; b′ = 3.2; c = 16.6; c′ = 2.9; V = 82; stylet = 13 μm.
- 6 females (after Siddiqi, 1994): L = 0.36 (0.31-0.41) mm; a = 30 (26-32); b = 5.5 (5.0-5.9); b′ = 2.9 (2.7-3.1); c = 18 (15-21); c′ = 2.8 (2.7-3.0); V = 82 (80-83); stylet = 12.5 (12-13) μm.

DESCRIPTION

Female

Body straight to slightly arcuate ventrally upon relaxation; max. diam. 12-13 μm. Cuticle finely annulated; annuli 1.2 (1.1-1.3) μm wide near mid-body. Lateral fields *ca* one-third of body diam., with four lines forming three bands of which middle band slightly narrower than outer ones; not areolated. Labial region low, truncate, almost continuous with body, with two annuli; anterior annulus sharply angular (hence species name); labial framework heavily sclerotised; outer margins of labial framework extending one annulus into body. Cephalids not seen. Stylet

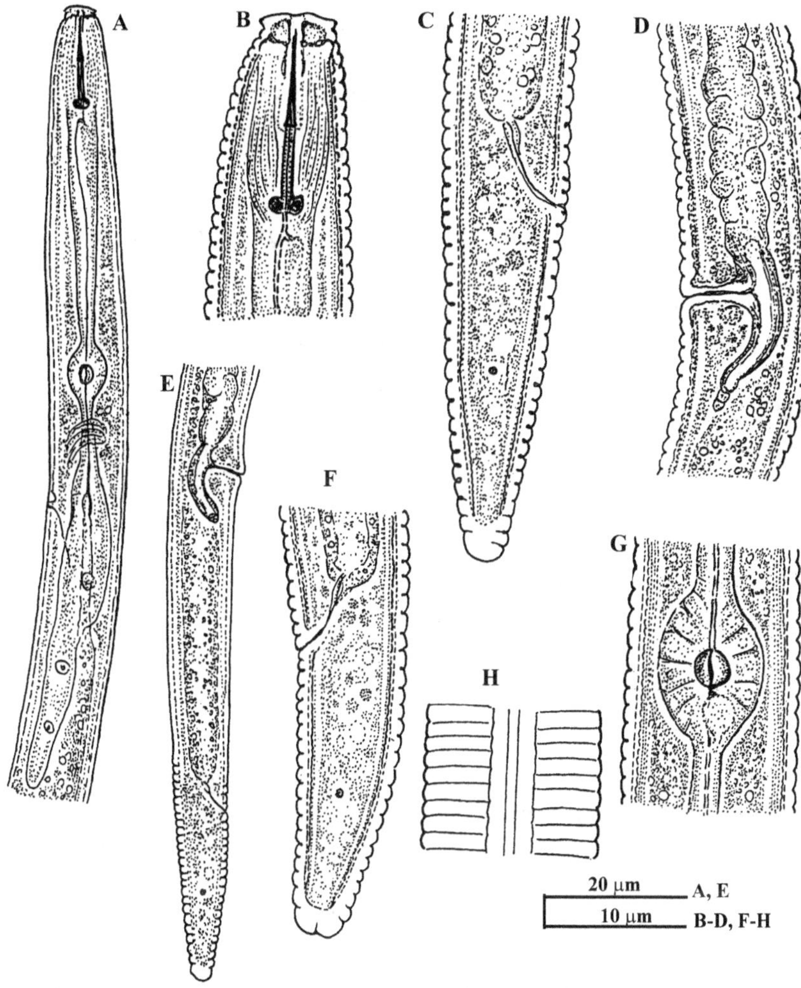

Fig. 24. Pratylenchus angulatus *Siddiqi, 1994. A: Female pharyngeal region; B: Female labial region; C, F: Female tails; D: Vulval region; E: Female posterior region; G: Pharyngeal median bulb; H: Lateral field at mid-body. After Siddiqi (1994).*

of medium strength, 12.5 (12-13) μm or *ca* 2.1-2.2 times labial region diam. long, in two, almost equal, parts; conus thin, tubular; basal knobs rounded, small, *ca* 2.7 μm across and 1.6 μm high. Orifice of dorsal pharyngeal gland 2.0-2.5 μm posterior to stylet base. Median pharyngeal bulb round to oval, *ca* 10-11 $\times$ 7-8 μm; central valvular apparatus conspicuous, 2.5 μm wide. Excretory pore opposite pharyngo-intestinal junction or slightly posterior, 66 (60-70) μm from anterior end of body. Hemizonid distinct, two annuli wide, just anterior to excretory pore. Pharyngeal glands extending mostly on ventral side of intestine; distance from anterior end of body to end of glands = 120 (108-126) μm. Vulva a depressed transverse slit, at 292 (253-310) μm from anterior end; lips rounded but not raised. Vagina two-fifth body diam. long. Ovary short with up to seven oocytes, barely reaching mid-body region. Spermatheca with sperm seen in only one female. Post-vulval uterine sac *ca* one vulval body diam. long, with rudiments of posterior gonad. Rectum *ca* three-quarter anal body diam. long. Tail subcylindroid, with a rounded to conoid-rounded terminus which may be smooth or marked by one striation, 20 (15-25) μm or 2.8 (2.7-3.0) anal body diam. long, with 20 (17-23) annuli. Phasmids pore-like, near middle of the tail.

Male

Not found, but presumably present as one female had a spermatheca with sperm.

DIAGNOSIS AND RELATIONSHIPS

Pratylenchus angulatus is characterised by: labial region low, truncate with two annuli, anterior annulus sharply angular, lateral fields with four lines with smooth margins, tail subcylindroid with terminus smooth or marked by one striation.

The matrix code is: A1, B1, C1, D1, E3, F1, G1, H1, I1, J1, K1.

It can be distinguished from closely related species (*P. brachyurus* and *P. neobrachyurus*) by stylet length and tail shape (see the corresponding descriptions).

DISTRIBUTION

It has been reported only from the type locality in Africa at Korup Forest, South-West Province, Cameroon, from the rhizosphere of the forest tree, *Caloncoba* sp.

6. *Pratylenchus arlingtoni* Handoo, Carta & Skantar, 2001
(Figs 25, 26)

MEASUREMENTS

- Female holotype (after Handoo *et al.*, 2001): L = 0.42 mm; a = 24.8; b = 4.0; c = 20.6; c′ = 2.0; V = 81.5; stylet = 17 μm.
- 20 females (after Handoo *et al.*, 2001): L = 0.46 (0.41-0.54) mm; a = 27 (21-33); b = 4.4 (4.1-5.3); c = 22 (18-28); V = 82 (81-86); stylet = 17 (16-17.5) μm.

DESCRIPTION

Female

Body vermiform with some tapering at extremities. Labial region slightly offset, three annuli (2.5-3.0 μm height, 7.5-9.0 μm diam.). Labial framework extending inward for two or three annuli. Anterior body annuli measuring 1.2 μm, tail annuli measuring 0.9 μm. Stylet knobs broadly rounded, anterior outer edges directed slightly anteriad (2.5-3.0 μm height, 4.5-5.0 μm width). Dorsal pharyngeal gland opening at 2-3 μm posterior to stylet knobs. Lateral fields beginning posterior to stylet as four narrow crenate lines, widening to five by median bulb level and between six to eight by level of anterior intestine through to vulva. Four lines extending from just posterior to vulva to phasmid, after which three lines extending to a few annuli short of tail tip. Lateral fields sometimes areolated at extremities. Pharyngeal glands pyriform in *ca* 25% of specimens or slightly overlapping in others (93-107 μm long). Excretory pore and canal located within an area slightly anterior or posterior to pharyngo-intestinal junction at 70 (48-85) μm from anterior end. Vulva elevated, vagina extending inward *ca* 70% of body diam. Anterior gonad with single row of oocytes, extending anteriad for nearly three times vulva-anus distance. An egg (36-cell) within body measuring 75 × 21 μm; another (4-cell) 64 × 14 μm. Spermatheca oval, 17 μm long with apparent sperm cells of 1.4-2.2 μm diam, located 74 μm anterior to vulva (present in one specimen out of at least 200). Distance from vulva to spermatheca = 120% of vulva-anus distance. Post-vulval uterine sac generally undifferentiated, with discernible cap cell or four columnar cells sometimes present dorsally. Phasmid located 10-15 annuli anterior to tail tip. Tail terminus

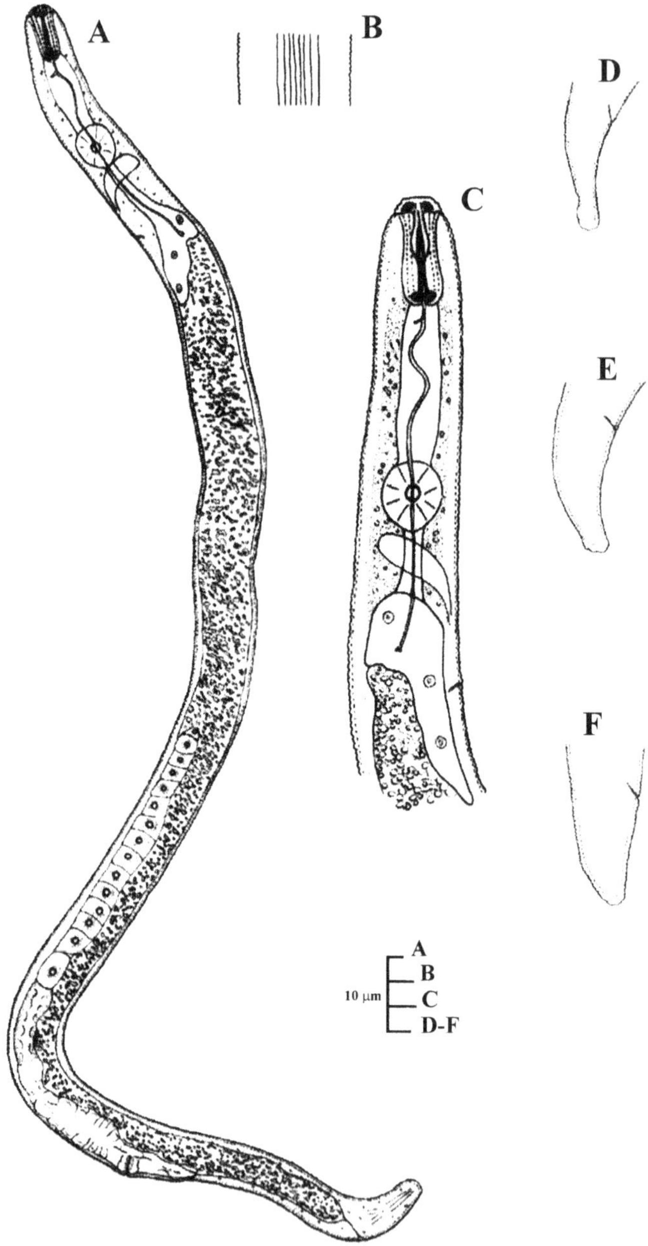

Fig. 25. Pratylenchus arlingtoni *Handoo, Carta & Skantar, 2001. A: Entire female; B: Lateral field at mid-body; C: Female pharyngeal region; D-F: Female tails. After Handoo* et al. *(2001).*

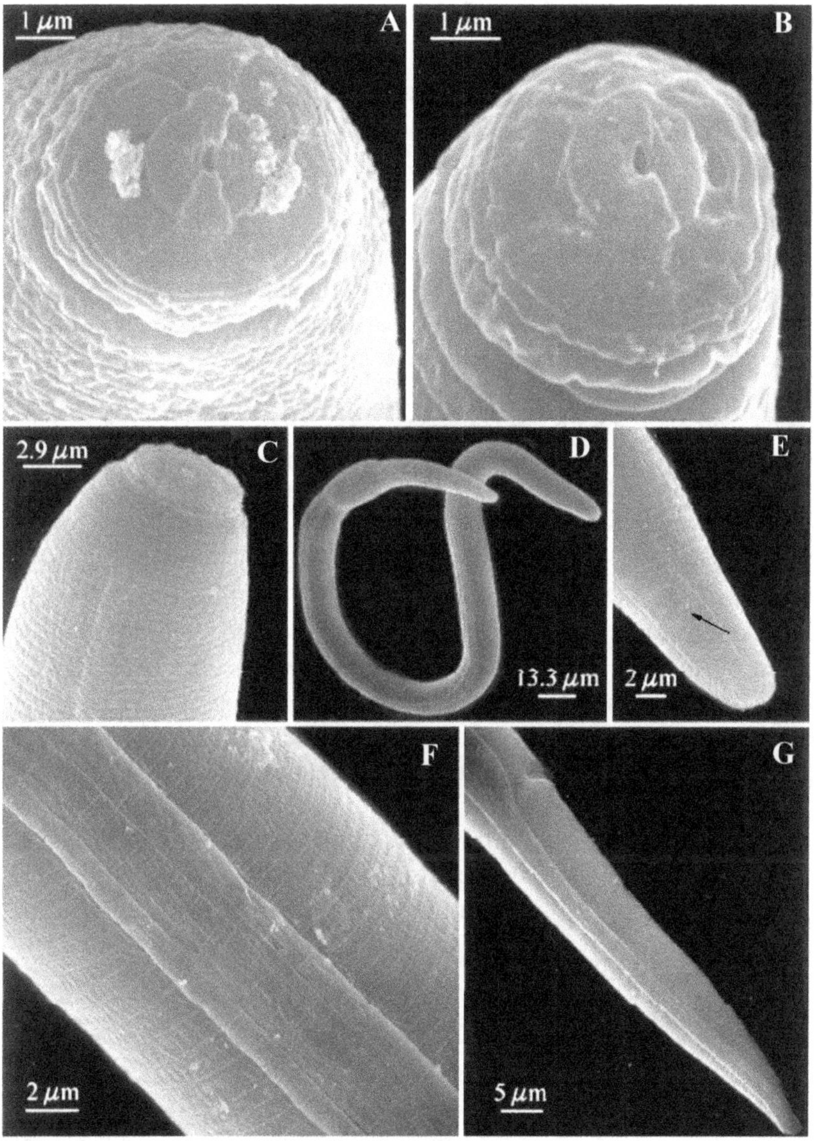

Fig. 26. *SEM micrographs of* P. arlingtoni *Handoo, Carta & Skantar, 2001. A, B:* En face *views; C: Female labial region; D: Entire female; E: Female tail (arrow = phasmid); F: Lateral field at mid-body; G: Female posterior region. After Handoo* et al. *(2001).*

coarsely annulated, with variable shape from conoid, clavate to truncate, sometimes bifid.

Male

Not found.

DIAGNOSIS AND RELATIONSHIPS

Pratylenchus arlingtoni is unique among *Pratylenchus* species in having 6-8 lines in the lateral fields from the pharyngeal to the vulval region, pharyngeal glands with a pyriform basal bulb to a slight overlap, a crenate female tail tip, an elevated vulva and, very rarely, an oval spermatheca.

The matrix code is: A2, B1, C3, D3, E3, F2, G3, H2, I1, J3, K1.

It is close to *P. crenatus*, but differs by longer pharynx (93-107 *vs* 57-78 μm), pyriform to slightly overlapping basal pharyngeal glands (6-31 *vs* 17-38 μm), somewhat lower b value (4.1-5.3 *vs* 4.9-7.0 in *P. crenatus*), smaller post-vulval uterine sac length relative to vulval-anal distance × 100, 28-42 *vs* 40-50, lower c′ value (1.1-1.5 *vs* 1.6-2.9), more elevated vulva and 6-8 lateral lines compared to 4-6. It differs from *P. roseus* by the lack of a vulval membrane, a longer stylet (16-17.5 *vs* 15-16 μm) and a shorter pharyngeal overlap (6-31 *vs* 118-130 μm). It differs from *P. teres* by more posterior vulva (V = 81-86 *vs* 69-78), fewer tail annuli (19-25 *vs* 24-30), shorter pharyngeal overlap (6-31 *vs* 58 μm), longer PUS/VA × 100 (28-42 *vs* 18) and 6-8 lateral lines rather than only six. It differs from *P. convallariae* and *P. fallax* by a lower b value (4.1-5.3 *vs* 6.0-9.0 and 5.2-6.7, respectively) and more posterior vulva (V = 81-86 *vs* 78-81 and 77-81, respectively), lacks males and has 2-4 more lateral lines. In addition, it differs from *P. fallax* and *P. pseudofallax* by stylet length (16.0-17.5 *vs* 15.0-15.5 μm), more offset labial region and coarser tail annulation.

DISTRIBUTION

It has been recorded only from the type locality in the rhizosphere of turf (*Poa pratensis* L., blue grass and *Festuca arundinacea* Schribn., tall fescue) under *Quercus* spp. (oaks) at Arlington Cemetery, Arlington, VA, USA.

7. *Pratylenchus artemisiae* Zheng & Chen, 1994
(Fig. 27)

MEASUREMENTS

- Female holotype (after Zheng & Chen, 1994): L = 0.59 mm; a = 29.3; b = 6.1; b′ = 7.2; c = 21.0; V = 79; stylet = 13.5 μm.
- 32 females (after Zheng & Chen, 1994): L = 0.56 (0.45-0.69) mm; a = 24.3 (21.4-29.3); b = 5.9 (5.3-6.4); b′ = 4.3 (3.5-5.7); c = 22.1 (17.4-25.8); V = 78 (76-81); stylet = 13.5 (11.5-14.5) μm.
- 30 males (after Zheng & Chen, 1994): L = 0.52 (0.43-0.60) mm; a = 26.2 (21.1-31.0); b = 5.6 (5.3-5.9); b′ = 4.2 (3.3-5.7); c = 22.1 (18.3-24.8); T = 45 (34-58); stylet = 13.5 (11.5-14.5) μm; spicules = 15 (13.5-17.0) μm; gubernaculum = 4 (3-5) μm.

DESCRIPTION

Female

Body vermiform with some tapering at tail region, slightly ventrally curved when killed by gentle heat. Labial region flat, slightly offset, with two, rarely three, annuli. Somatic annuli measuring *ca* 1 μm wide at mid-body. Labial framework extending posteriad for 1-2 annuli. Stylet stout, knobs rounded with little variation in shape. Dorsal pharyngeal gland orifice 2-3 μm posterior to stylet base. Median bulb oval. Excretory pore located 86 μm from anterior end. Hemizonid *ca* two annuli long, located 1-3 annuli anterior to excretory pore. Pharyngeal glands overlapping intestine by *ca* 2.5 body diam. Lateral fields with four longitudinal lines incompletely areolated at mid-body, occupying one-third of body diam., starting at 7-9 body annuli from anterior end. Anterior genital tract outstretched. Post-vulval uterine sac 27 μm long, usually as long as mid-body diam. Spermatheca oval to oblong, usually containing sperm. Tail subcylindrical, bearing 16-22 annuli ventrally, terminus broadly rounded, smooth.

Male

Common, with similar appearance to female, except for sexual dimorphism. Stylet knobs rounded. Hemizonid *ca* 2-3 annuli long, located 1-3 annuli anterior to excretory pore. Testis outstretched. Spicules and gubernaculum ventrally curved. Phasmids located posterior to mid-tail.

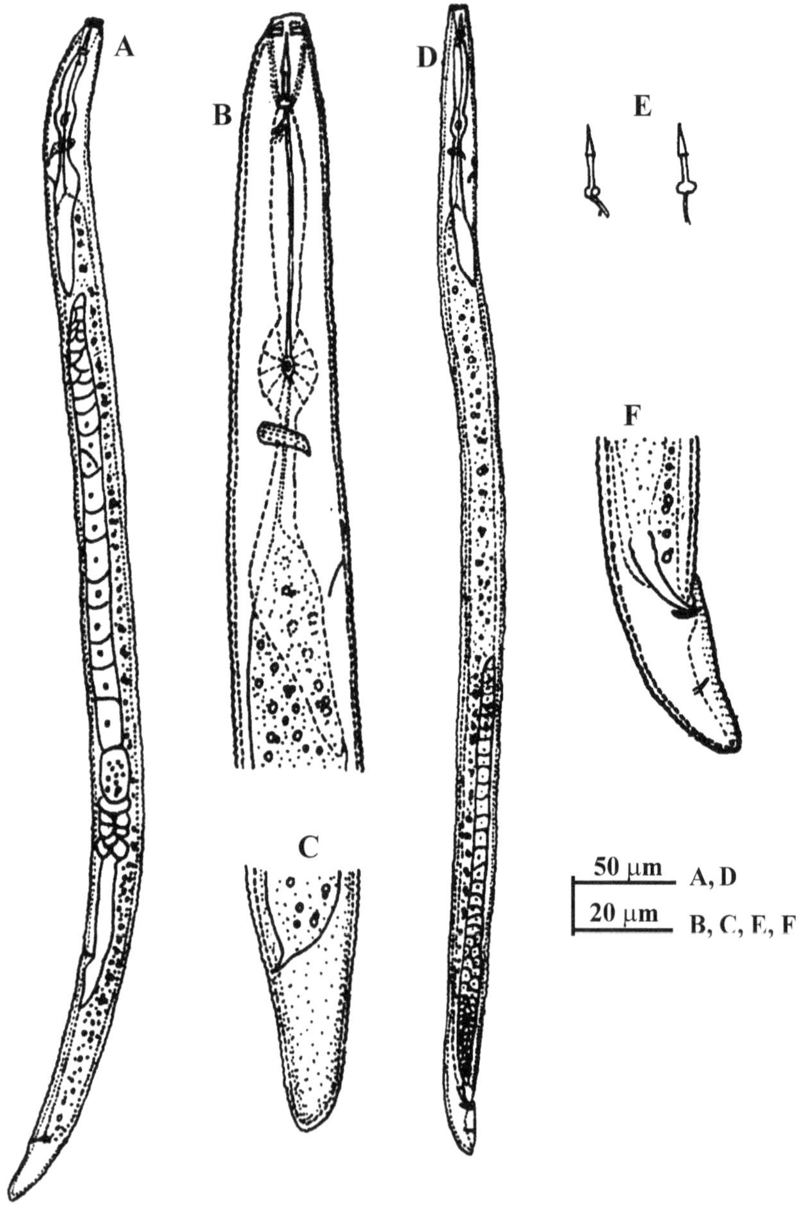

Fig. 27. Pratylenchus artemisiae *Zheng & Chen, 1994. A: Entire female; B: Female pharyngeal region; C: Female tail; D: Entire male; E: Stylets; F: Male tail. After Zheng and Chen (1994).*

DIAGNOSIS AND RELATIONSHIPS

Pratylenchus artemisiae is characterised by: low, offset labial region with two annuli, stylet knobs rounded, spermatheca oval to oblong and a subcylindrical tail with a broadly rounded smooth terminus and abundant males.

The matrix code is: A1, B2, C2, D3, E2, F3, G2, H1, I2, J1, K1.

It is close to *P. alleni*, *P. coffeae* and *P. loosi* from which it can be distinguished by body and stylet length, labial annuli and tail shape.

DISTRIBUTION

It has been recorded only from the type locality at Fengtai suburb, Beijing, China, in the rhizosphere of *Artemisia* sp.

8. *Pratylenchus bhattii* Siddiqi, Dabur & Bajaj, 1991

(Fig. 28)

MEASUREMENTS

- Female holotype (after Siddiqi *et al.*, 1991): L = 0.49 mm; a = 29; b = 5.3; b′ = 4.2; c = 19.3; V = 73; stylet = 14 μm.
- 10 females (after Siddiqi *et al.*, 1991): L = 0.48 (0.42-0.60) mm; a = 28 (25-31); b = 5.7 (4.7-7.4); b′ = 4.4 (3.4-4.9); c = 17 (15-20); c′ = 2.5 (2.3-3.2); V = 73 (69-76); stylet = 13.5 (13-14) μm.
- 11 males (after Siddiqi *et al.*, 1991): L = 0.42 (0.35-0.48) mm; a = 28 (25-29); b = 5.0 (4.7-5.3); b′ = 3.9 (3.4-4.6); c = 18 (16-21); stylet = 13 (12-13.5) μm; spicules = 16 (15-17) μm.

DESCRIPTION

Female

Body posture almost straight or slightly arcuate after fixation. Cuticular annuli *ca* 1 μm wide at mid-body, 1.1 μm wide near pharyngo-intestinal junction. Lateral fields occupying one-quarter to one-third body diam., with four lines, not areolated posterior to pharyngeal region. Labial region continuous with body contour or slightly offset, with three distinct annuli. Labial framework heavily sclerotised, its margins extending posteriad for one body annulus. Stylet rather slender, with 6.7-7 μm long conus and round basal knobs measuring *ca* 3 μm across and with

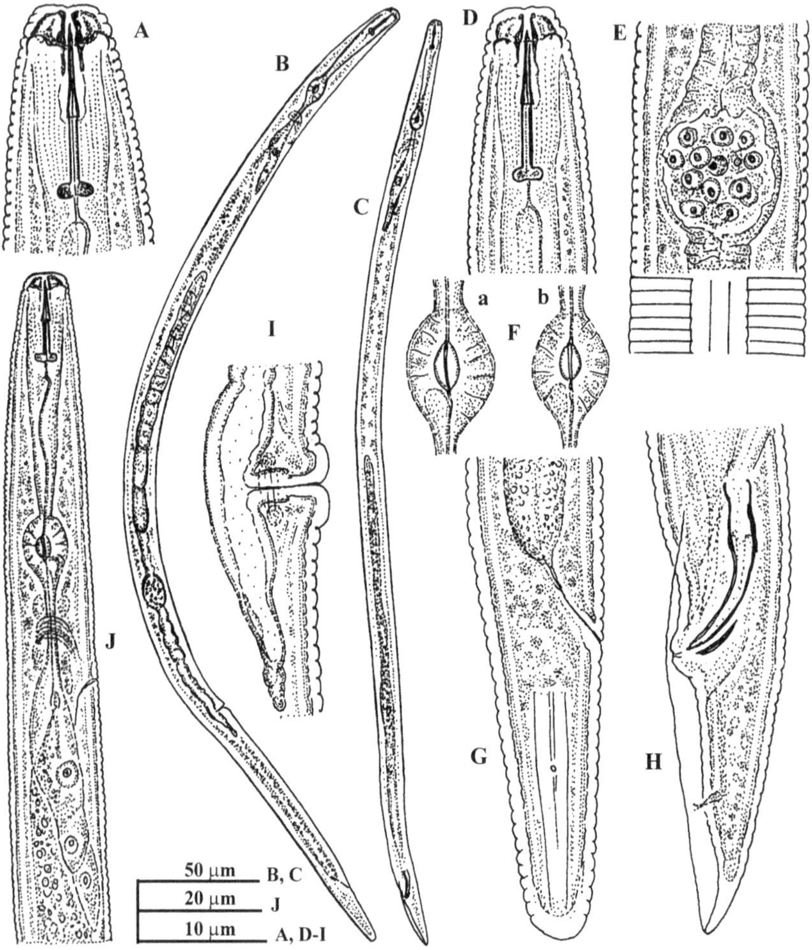

Fig. 28. Pratylenchus bhattii *Siddiqi, Dabur & Bajaj, 1991. A: Female labial region; B: Entire female; C: Entire male; D: Male labial region; E: Spermatheca; F: Pharyngeal median bulb of female (a) and male (b); G: Female tail; H: Male tail; I: Vulval region; J: Female pharyngeal region. After Siddiqi* et al. *(1991).*

flattened to concave anterior surfaces; 17 (16-18) annuli between anterior end of body and stylet base. Opening of dorsal pharyngeal gland 2.5-3 μm posterior to stylet base. Median pharyngeal bulb oval, rarely round, 9-10 μm long, 7.0-7.5 μm diam.; with prominent oval valvular apparatus, 3.3-3.7 μm long, 2.3-3 μm wide. Pharyngeal gland lobe overlapping intestine for 1.5-2.5 body diam. Nerve ring at middle or posterior region of isthmus. Excretory pore 68 (65-73) μm from anterior end. Hemizonid just anterior to excretory pore, 2-4 annuli wide. Ovary outstretched; oocytes arranged in two rows in zone of multiplication, in single row in growth zone. Spermatheca generally rounded (12.5 × 11.5 μm), sometimes longitudinally oval (16 × 11.5 μm) located 37-53 μm anterior to vulva and filled with sperm; 5-6 sperm filling its diam. Vulva located on a prominent protuberance. Post-vulval uterine sac 11-22 μm long, with rudimentary ovarian cells. Tail subcylindrical to conoid with 19 (16-23) annuli on ventral surface. Tail tip rounded and smooth; lateral field lines not reaching tail terminus. Phasmids at mid-tail.

Male

Stylet and labial sclerotisation weak in comparison to female. Stylet conus 6.2-6.7 μm long. Spicules 15.5 (13-16.5) μm long. Gubernaculum trough-shaped, 3-5 μm long. Phasmids near mid-tail, extending into bursa.

DIAGNOSIS AND RELATIONSHIPS

Pratylenchus bhattii is characterised by: three labial annuli, stylet 13-14 μm long, V = 69-76, subcylindrical to conoid tail with smooth terminus and presence of males.

The matrix code is: A2, B2, C2, D2, E1, F2, G2, H1, I2, J1, K1.

It can be distinguished from closely related species (*P. sudanensis*, *P. penetrans*) in having a smaller and less robust stylet, a vulva with raised lips, lateral fields lines not reaching the female tail terminus, which is not crenate, position of vulva and smaller spicules (see the corresponding descriptions).

DISTRIBUTION

It has been recorded only from the type locality in the rhizosphere of wild sugarcane, *Saccharum spontaneum* L., near Gohana, Sonepat district, Haryana State, India.

9. *Pratylenchus bolivianus* Corbett, 1983
(Fig. 29)

MEASUREMENTS

- Female holotype (after Corbett, 1983): L = 0.61 mm; a = 25; b = 5.4; b′ = 4.0; c = 22; V = 82; stylet = 19 μm.
- 15 females (after Corbett, 1983): L = 0.59 (0.53-0.63) mm; a = 27 (26-29); b = 5.2 (3.9-5.9); b′ = 4.1 (3.4-4.9); c = 19 (16-21); V = 81 (80-82); stylet = 19 (17-20) μm.
- 12 females (after Valenzuela & Raski, 1985): L = 0.63 (0.57-0.72) mm; a = 31 (25-33); b = 7.8 (6.4-8.2); c = 18 (16-22); V = 82 (77-83); stylet = 19 (18-20) μm.

DESCRIPTION

Female

Body straight or slightly curved ventrally on death, often twisted through *ca* 180°, tapering posterior to vulva to tail which bends ventrally in death. Lateral fields with four lines, outer two bands narrower than middle one, areolated, occasionally with oblique striae in middle band in mid-body opposite reproductive tract; four lines extending posterior to phasmid which is inconspicuous, on anterior half of tail 10-15 annuli from tip. Labial region usually with three, rarely four, annuli with massive skeleton extending into body for at least two annuli. In *en face* view by SEM subdorsal and subventral segments of first annulus appearing fused to form an almost rectangular shape with convex dorsal and ventral margins and longer, concave lateral sides separated from lateral sectors, each bearing an amphidial opening on inner edge. Oral aperture oval, surrounded by six small pores. Stylet robust, basal knobs rounded, dorsal gland opening 2.7-4.0 μm posterior to stylet base. Median bulb ovate, well developed; nerve ring encircling isthmus just in front of posterior glandular portion of pharynx. Pharyngeal lobe overlapping intestine ventrally and laterally; overlap 18-49 μm (*ca* 1-2 body diam.) long. Pharyngo-intestinal junction usually more than one body diam. posterior to excretory pore which is 89-105 μm from anterior end. Single anterior reproductive tract with functionless, inconspicuous, spermatheca. Post-vulval uterine sac 22-31 μm, *ca* 1.25 body diam. long. Tail convex dorsally, straight to concave ventrally, conoid. Ventral

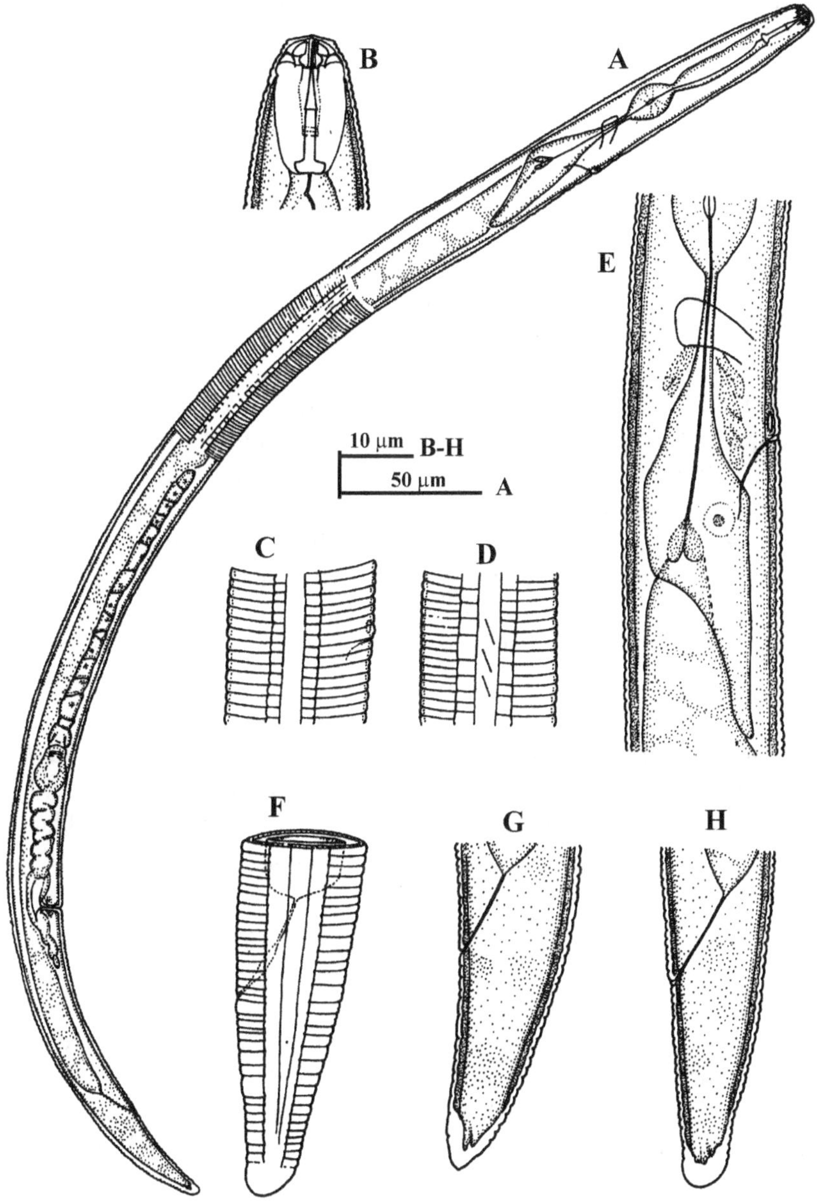

Fig. 29. Pratylenchus bolivianus *Corbett, 1983. A: Entire female; B: Female labial region; C, D: Lateral field at mid-body; E: Pharyngeal region; F-H: Female tails. After Siddiqi* et al. *(1991).*

surface with 15-19, occasionally up to 24, irregular annuli; tip smooth, rounded.

Male

Not found.

DIAGNOSIS AND RELATIONSHIPS

Pratylenchus bolivianus is characterised by: body twisted through *ca* 180° after relaxation, tapering posterior to vulva, labial region with three, rarely four, annuli, stylet 19 μm long and tail tip rounded and smooth.

The matrix code is: A2, B1, C4, D1, E3, F4, G3, H1, I2, J1, K2.

It is close to *P. delattrei*, *P. mulchandi*, *P. pseudopratensis* and *P. zeae* from which it differs by the body and stylet length, lip annulation, position of vulva and female tail shape (see the corresponding descriptions). SEM studies (Valenzuela & Raski, 1985) indicate this species conforms generally with *en face* morphology and belongs in Group 2 of Corbett and Clark (1983).

Pratylenchus australis Valenzuela & Raski, 1985 was described from moist soil beneath deep tundra from two sites at Orange Bay, Hardy Peninsula, Hoste Island, Chile. Frederick and Tarjan (1989) pointed out that the only outstanding difference between *P. australis* and *P. bolivianus* is in the b ratio, which usually shows high intraspecific variability and is considered inadequate to differentiate species. Thus, they proposed *P. australis* as a junior synonym of *P. bolivianus* and this action is maintained here.

DISTRIBUTION

It has been recorded from the type locality, Toralapa Experimental Station, Cochabamba Valley, Bolivia, around roots of oats, *Avena sativa* L., and in potato roots and soil, Koari Farm, in the same valley. Reported from England and Wales on *Alstroemeria* (Cotten *et al.*, 1990; Amsing, 1996); and from Florida, USA, around roots of *Erica* sp. (Lehman, 1996). It has been also recorded in the deep tundra from two sites at Orange Bay, Hardy Peninsula, Hoste Island, Chile.

10. *Pratylenchus brachyurus* (Godfrey, 1929) Filipjev & Schuurmans Stekhoven, 1941

(Fig. 30)

MEASUREMENTS

- Females (after Loof, 1960): L = 0.39-0.75 mm; a = 15-29; b = 5-10; c = 13-28; V = 82-89; stylet = 17-22 μm.
- Males (after Sher & Allen, 1953): L = 0.46-0.56 mm; a = 27-29; b = 6.0; c = 21; T = 51-53; stylet = 19 μm.
- 12 females (after Van den Berg, 1971): L = 0.56 (0.52-0.60) mm; a = 20 (18-23); b = 3.9 (3.4-5.2); c = 19 (16-23); V = 86 (84-87); stylet = 19 (18-21) μm.
- 7 females (after Ryss, 1988): L = 0.55 (0.40-0.70) mm; a = 21 (17-25); b = 7 (5-8); c = 21 (15-26); c′ = 2.0 (1.7-2.2); V = 85 (82-88); stylet = 20 (19-22) μm.

DESCRIPTION

Female

Labial region offset from body, with angular anterior margin and two distinct annuli, apical region undivided, bearing six papillae around oval oral aperture and two large amphidial openings. Under SEM, most specimens with two lip annuli, some with two or three annuli and rare specimens with three annuli on one side and four annuli on other side (Baujard *et al.*, 1990). Labial framework strong. *En face* view characterised by fusion of median and lateral lips with oral disc; labial disc separated from first annulus by an incisure (Baujard *et al.*, 1990; Hernández *et al.*, 2000). Lateral fields usually bearing four lines in mid-body, occasionally dividing into five or six in vulval region, terminating as two lines near tail. Stylet with stout, rounded, basal knobs. Excretory pore just posterior to hemizonid, 57-108 μm from anterior end, usually in region of pharyngo-intestinal junction which is often indistinct. Pharyngeal glands overlapping intestine ventrally and laterally. Gonad outstretched, occasionally reflexed, rarely extending past pharyngeal lobe. Oocytes in single file, occasionally forming a double row. Vulva at 82-89% of body length; post-vulval uterine sac less than one body diam. (10-30 μm) long; spermatheca inconspicuous, not functional. Tail broadly conoid, smooth, with broadly rounded, truncate or spatulate tip, 13-24 ventral annuli; phasmids *ca* mid-tail.

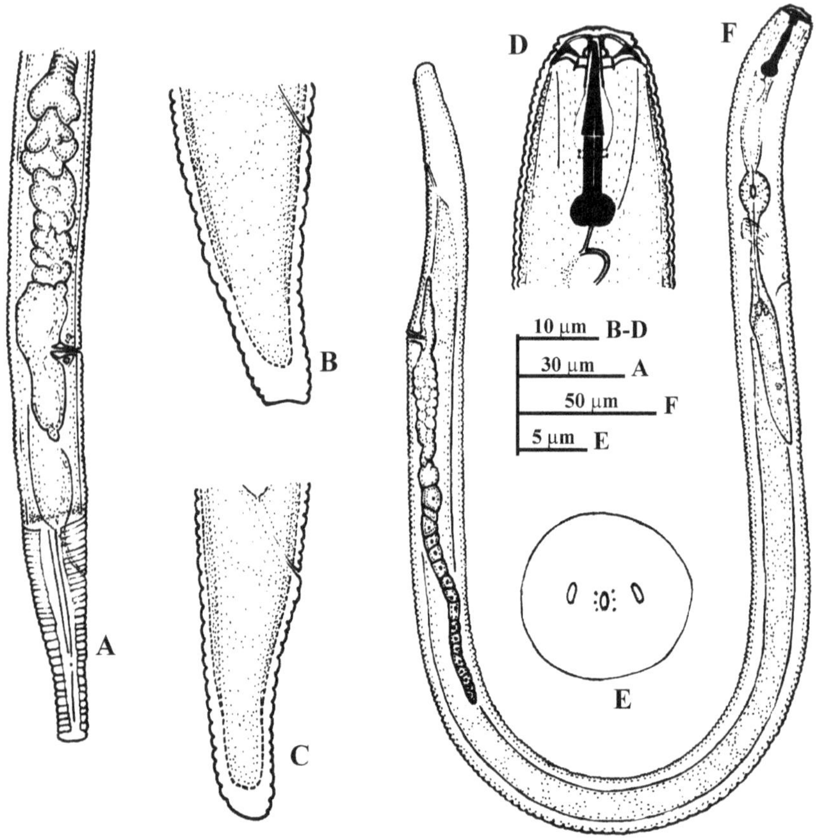

Fig. 30. Pratylenchus brachyurus *(Godfrey, 1929) Filipjev & Schuurmans Stekhoven, 1941. A: Female posterior region; B, C: Female tails; D: Female labial region; E:* En face *view; F: Entire female; G: SEM micrographs of* en face *view; H: SEM micrographs of female tail. Rectangular box indicates phasmid position. (Scale bars: G = 2 μm; H = 5 μm.) After Corbett (1976); Hernández* et al. *(2000).*

Male

Very rare. Similar in form to female. Single outstretched testis. Phasmids slightly posterior to mid-tail, not extending into delicate bursa. Spicules arcuate, gubernaculum simple.

DIAGNOSIS AND RELATIONSHIPS

Pratylenchus brachyurus is characterised by: labial region with two annuli, the anterior one showing an angular contour, stylet with

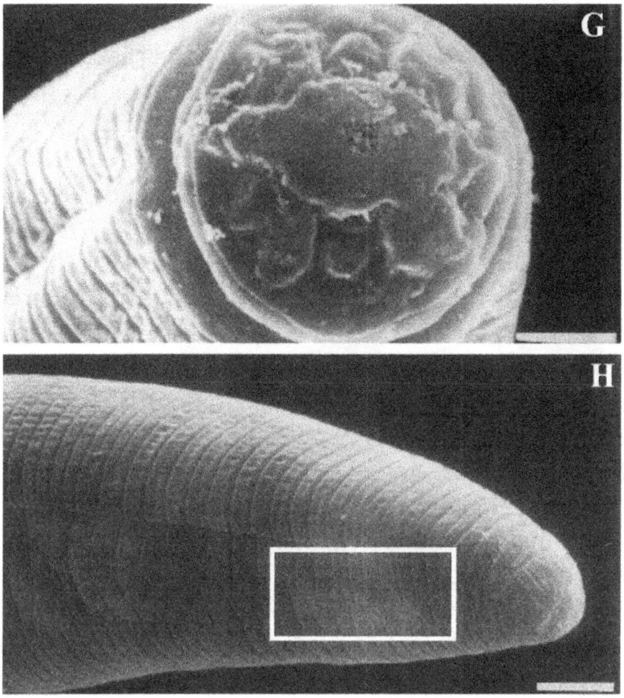

Fig. 30. *(Continued).*

stout, rounded, basal knobs rounded, vulva at 82-89% of body length; post-vulval uterine sac less than one body diam. long, spermatheca inconspicuous, not functional, tail broadly conoid, smooth and broadly rounded, truncate or spatulate at tip. Males are rare.

The matrix code is: A1, B2, C4, D1, E4, F3, G3, H1, I4, J2-3, K1.

It is close to *P. japonicus* and *P. neobrachyurus* from which it differs by body and stylet length, shape of labial region, shape of stylet knobs, position of the vulva and female tail shape (see the corresponding descriptions).

DISTRIBUTION

Pratylenchus brachyurus is widely distributed throughout the tropics. Originally it was described parasitising pineapple in Hawaii (Godfrey, 1929). It is also reported damaging various crops in Africa: Cameroon on several crops (Sakwe & Geraert, 1994); Ethiopia on sweet potato (Agu, 2004); Ivory Coast on rice (Coyne *et al.*, 1999) and plantain

(Adiko, 1988); in Guinea on rice (Coyne *et al.*, 1996); Kenya on corn and soybean (Van den Berg *et al.*, 2001); Madagascar on tobacco (Baudin & Huu-Hai, 1973); Malawi on corn (Khonga & Hillocks, 1996); Mozambique on various crops (Oever & Mangane, 1992); Nigeria on plantains (Speijer *et al.*, 2001), soybean (Ademola *et al.*, 1996), peanut (Khan & Misari, 1992), several crops (Atu, 1991) and corn (Egunjobi, 1974, 1977); Senegal on cowpea, *Vigna unguiculata* (L.) Walp (Baujard *et al.*, 1990); South Africa on soybean (Fourie *et al.*, 2001), peanut (Venter *et al.*, 1992), wheat (Jordaan *et al.*, 1992), rice (Van den Berg & De Waele, 1989), corn (De Waele & Jordaan, 1988a) and potato (Koen, 1967); Uganda on pineapple (Bafokuzara, 1982); West Africa on pineapple (Guerout, 1975); Zambia on potato (Martin, 1972). In Africa it also damages *Dioscorea* sp. (Unny & Jerath, 1965) and cassava (de Guiran, 1965) and is the commonest species of *Pratylenchus* in the Nile delta in Egypt (Oteifa, 1962). It has also been recorded in North America: Alabama on peanuts (Ingram & Rodríguez-Kábana, 1980); Arkansas on soybean (Robbins *et al.*, 1987) and cotton (Robbins *et al.*, 1989a); California on seedling *Pinus ponderosa* Douglas *ex* Lawson & C. Lawson (Viglierchio, 1979); Florida on corn (McSorley & Dickson, 1989), citrus (Kaplan, 1985) and avocado (McSorley & Campbell, 1980); Georgia on triticale (Johnson *et al.*, 1998), crotalaria and corn (Brodie & Murphy, 1975); Hawaii on pineapple (Ko *et al.*, 1995) and sugarcane (Jensen *et al.*, 1959); North Carolina on peach (Barker & Clayton, 1973); Oklahoma on peanuts (Filonow & Russell, 1991); South Carolina on soybean (Lewis *et al.*, 1977); Texas on peanut (Boswell & Grichar, 1983). It damages several crops in North America: citrus (Brooks & Perry, 1967), cotton (Martin *et al.*, 1951), tobacco (Graham, 1951) peaches (Malo, 1963), soybean and corn (Endo, 1967), Sudan grass and millet (Brodie *et al.*, 1970) and several forest trees (Ruehle, 1969, 1971), damaging *Pinus palustris* P. Mill. severely (Ruehle, 1973) and also avocados (Young & Ruehle, 1955).

It is widely distributed in several countries in Central and South America: Belize on several crops (Bridge *et al.*, 1996); Brazil on corn (Lordello *et al.*, 1992), cotton (Tihohod *et al.*, 1991), pineapples (Monteiro & Lordello, 1972); Colombia on pastures (Stanton *et al.*, 1989); Costa Rica on pineapple (Lopez & Salazar, 1990), on rubber (Martinez *et al.*, 1972), coffee (Lordello *et al.*, 1968), sugarcane (Lordello & de Mendoça, 1970), several grasses (Lordello & Mello Filho, 1969a, 1970), *Eucalyptus* spp. (Lordello, 1967), cotton and soybean (Lordello *et al.*, 1958) and is found in the roots of declining bean (Mello Filho & Lordello, 1970); Cuba on coffee (Fernandez Díaz-Silveira & Ortega Herrera, 1998), kenaf, *Hibiscus cannabinus* L. (Alvarez & Fernandez, 1982); Brazil on soybean (Sharma *et al.*, 2002), potato tubers (Kubo *et al.*, 2001), corn (Sharma *et al.*, 2000), common bean (Rossi *et al.*, 2000b), fruit crops (De Souza *et al.*, 1999), coffee (Oliveira *et al.*, 1999a) and fig (Campos, 1997); French Guyana on vegetables (Cadet & Van den Berg, 1995); Honduras on citrus (Pinochet *et al.*, 1978); Mexico on strawberries (Sandoval-Hernandez & Téliz-Ortíz, 1990); Trinidad on several crops (Singh, 1973); Venezuela (Crozzoli, 2002), pineapple (Jimenez *et al.*, 2001), cucurbits (Naveda *et al.*, 1999), sunflower and sesame (Sharma & Amabile, 1998a, b), citrus (Crozzoli *et al.*, 1998), guava (Crozzoli *et al.*, 1991). It has also been recorded from Asia: China on lychee (Yin *et al.*, 1994) and several crops (Yin, 1991); India on sugarcane (Mehta & Sundararaj, 1990), rubber, *Hevea brasiliensis* (Willd.) Muell.-Arg., nurseries (Mukherjee *et al.*, 2000), ginger (Rama & Dasgupta, 2000a), floricultural crops (Khan *et al.*, 1997), peanut (Prasad & Rangappa, 1994), su-babool, *Leucaena leucocephala* (Lam.) de Wit (Azmi, 1993), vegetable, fruit and ornamental crops (Anwar *et al.*, 1992a); Japan on vegetables and sugarcane (Nakasono *et al.*, 1984), on several crops (Gotoh, 1974); Oman on several crops (Mani *et al.*, 1997, 2005), alfalfa (Waller & Bridge, 1978); Pakistan on peanut (Anwar *et al.*, 1993), rose (Saeed *et al.*, 1988); Thailand on soybean (Toida *et al.*, 1996); Turkey on chickpea (Di Vito *et al.*, 1994), peanut (Kepenekci & Ozturk, 2002); Vietnam on peanut (Sharma *et al.*, 1994). It has been recorded from several countries in Europe, Bulgaria on tobacco (Baicheva *et al.*, 1984), strawberry (Katalan-Gateva & Nedelchev, 1983); and Russia on several hosts (Ryss, 1988). It has been reported in Australia on cereals (Riley & Kelly, 2002), peanuts (Broadley, 1981) and on other crops (Colbran, 1968).

11. *Pratylenchus brzeskii* Karssen, Waeyenberge & Moens, 2000
(Fig. 31)

MEASUREMENTS

- 20 females (after Karssen *et al.*, 2000): L = 0.69 (0.63-0.74) mm; a = 28 (26-31); b = 7.7 (7.2-8.4); c = 14.3 (13.2-16.5); V = 77 (75-78); stylet = 19 (18-19) μm.

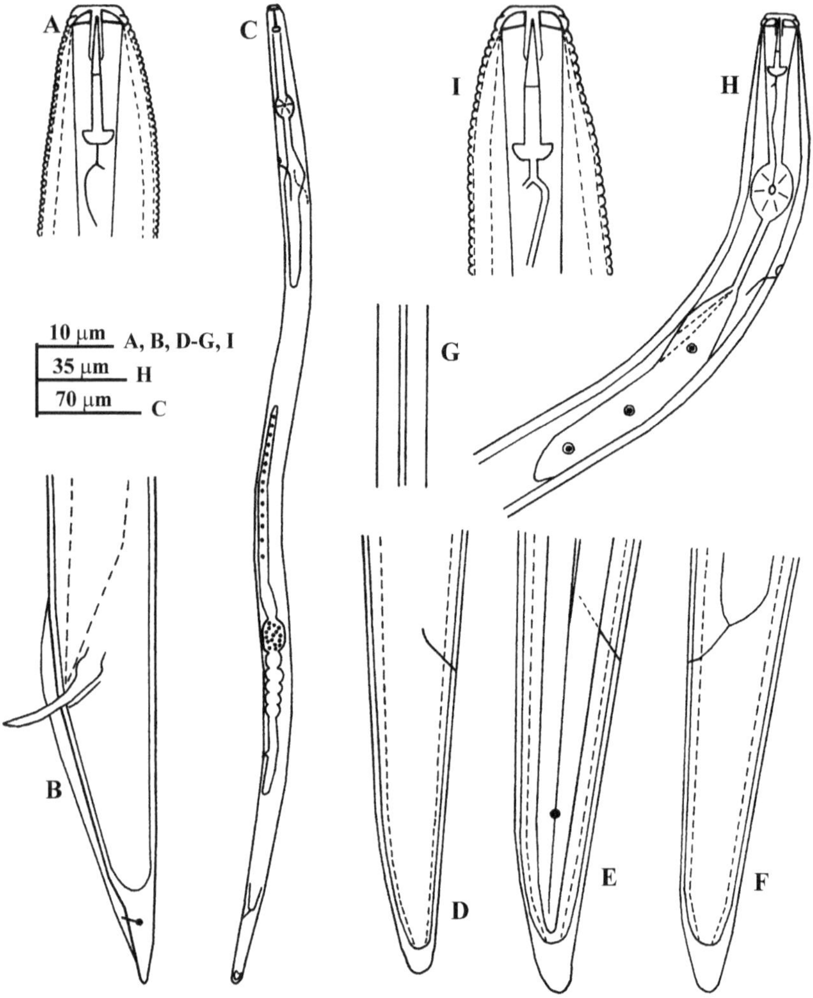

Fig. 31. Pratylenchus brzeskii *Karssen, Waeyenberge & Moens, 2000. A: Male labial region; B: Male tail; C: Entire female; D-F: Female tails; G: Lateral field; H: Female pharyngeal region; I: Female labial region. After Karssen* et al. *(2000).*

– 10 males (after Karssen *et al.*, 2000): L = 0.66 (0.62-0.68) mm; a = 30 (28-32); b = 7.8 (7.2-8.3); c = 14.1 (12.8-14.8); stylet = 18 (18-18.5) μm; spicules = 22 (21.5-23) μm; gubernaculum = 7.0 (6.5-8.0) μm.

DESCRIPTION

Female

Female body vermiform, straight, relatively long (625-735 μm) and slender, slightly tapering at both ends. Cuticular annulation distinct. Lateral fields with four lines, not areolated, two inner lines close together; in anterior and posterior body, two inner lines fusing to one line. Labial region slightly offset and low; diam. more than three times its height, heavily sclerotised labial framework present and vestibule extension distinct. Labial region slightly curved ventrally, composed of two annuli, apical lip flattened with rounded edges, slightly narrower than second annulus. Stylet long, robust, cone straight, shaft cylindrical with three large, anteriorly slightly concave, knobs, offset. Pharyngeal gland orifice close to stylet base. Metacorpus well developed, rounded with small valve. In non-fixed specimens, metacorpal lumen lining often with several small vesicles. Hemizonid positioned between metacorpus and pharyngo-intestinal junction level, 4-5 μm in length, anterior or adjacent to excretory pore. Long pharyngeal gland lobe ventrally overlapping intestine, three gland nuclei present. Vulva positioned posteriorly (75-78%), well developed, lips slightly protruding. Reproductive tract mono-prodelphic, germinative zone outstretched, anterior part always posterior to pharyngeal glands. Growth zone most distinct, with one row of oocytes. Spermatheca always faint in outline, ranging in shape from relatively small oval to large rectangular when filled with small, rounded sperm. Crustaformeria cells in young females arranged into a tricolumella. Posterior uterine branch short (19-35 μm), undifferentiated. Tail conoid, annuli distinct (29-34), terminus smooth and rounded to narrowly rounded. Hyaline tail part distinct, relatively long (4.5-6.5 μm). Three lateral lines present in tail region. Phasmids rounded, small, located mid-tail on inner line of lateral fields.

Male

Males occurring frequently (one or two males per five females). Except for sexual dimorphism, males morphologically comparable with females but morphometrically always smaller. Testis single and usually anteriorly outstretched. Spicules paired, delicate, ventrally curved. Gubernaculum simple, slightly curved. Bursa, faint crenate-edged, enclosing straight tail. Phasmid orifice located on bursa near mid-tail.

Diagnosis and relationships

Pratylenchus brzeskii is characterised by: relatively long body, labial region with two annuli, stylet robust with prominent, anteriorly slightly concave, knobs, lateral fields with four lines with inner lines close together, oval to rectangular spermatheca and tail conoid with smooth terminus and relatively long hyaline region.

The matrix code is: A1, B2, C4, D4, E2, F4, G3, H1, I4, J1, K1.

It can be distinguished from *Pratylenchus* species with two lip annuli and a smooth tail terminus (*P. acuticaudatus*, *P. alleni*, *P. angulatus*, *P. brachyurus*, *P. coffeae*, *P. hexincisus*, *P. loosi*, *P. macrostylus*, *P. neglectus*, *P. neobrachyurus*, *P. pseudocoffeae* and *P. scribneri*) by the following characteristics: an empty spermatheca and males very rare or absent in *P. acuticaudatus, P. alleni*, *P. angulatus*, *P. brachyurus*, *P. hexincisus*, *P. macrostylus*, *P. neglectus* and *P. scribneri*; *P. alleni* has also a short stylet (< 15 μm) and a bluntly rounded, short tail; *P. coffeae* and *P. pseudocoffeae* have a differentiated posterior uterine branch, a broadly rounded to truncate tail terminus and smaller stylet length (< 18 μm); *P. loosi* has a smaller stylet and body length and the inner lateral field lines are not close together; *P. neobrachyurus* is a small species with an angular shaped anterior labial region (see the corresponding descriptions).

Molecular analysis in the original description was based on ITS sequence and confirmed that this species is very different from *P. coffeae*, *P. loosi* and *P. penetrans*.

Distribution

It has been recorded from the type locality in the coastal fore dunes of Oostvoorne, The Netherlands, from the roots of marram grass (*Ammophila arenaria* (L.) Link). It has also been recorded in *Elymus farctus* (Viv.) Runemark and/or *A. arenaria* in the coastal dunes of Koksijde (Belgium), Bray-Dune Plage and Cabourg (France) and Haringvlietdam, Texel and Terschelling (The Netherlands). It was also recorded as an unknown *Pratylenchus* species on *Elymus arenarius* L. near Krynica Morska, Poland, by Brzeski (1998).

12. *Pratylenchus coffeae* (Zimmermann, 1898) Filipjev & Schuurmans Stekhoven, 1941

(Figs 32, 33)

MEASUREMENTS

- Female neotype (Sher & Allen, 1953): L = 0.59 mm; a = 34; b = 6.3; c = 21; V = 82; stylet = 18 μm.
- Females (Sher & Allen, 1953): L = 0.45-0.70 mm; a = 25-35; b = 5-7; c = 17-22; V = 76-83; stylet = 15-18 μm.
- Males (Sher & Allen, 1953): L = 0.45-0.70 mm; a = 26-40; b = 6-7; c = 17-24; T = 45-52; stylet = 15-17 μm.
- 69 females (after Loof, 1960): L = 0.53 (0.37-0.70) mm; a = 23.7 (17.7-30.5); b = 6.8 (5.0-7.8); c = 19.0 (13.7-23.9); V = 80 (76-84); stylet = 14-17 μm.
- 10 males (after Loof, 1960): L = 0.48 (0.41-0.56) mm; a = 27.4 (23.8-31.4); b = 6.5 (5.9-7.7); c = 19.1 (17.6-23.3); T = 48 (37-58); stylet = 14-15 μm.
- 30 females (after Bajaj & Bhatti, 1984): L = 0.62 (0.55-0.70) mm; a = 28 (23-38); b = 7.3 (6.1-7.8); b′ = 4.8 (4.0-5.5); c = 20 (17-26); V = 81 (79-84); stylet = 15 (14-17) μm.
- 10 females (after Ryss, 1988): L = 0.62 (0.49-0.65) mm; a = 28 (25-31); b = 5.9 (5.1-6.6); c = 21 (19-22); c′ = 2.9 (2.5-3.2); V = 79 (76-82); stylet = 17 (15-18) μm.
- 10 males (after Ryss, 1988): L = 0.52 (0.44-0.60) mm; a = 28 (25-32); b = 6.2 (5.6-6.9); c = 20 (18-23); c′ = 2.6 (2.3-2.8); stylet = 16 (15-17) μm; spicules = 17 (16-18) μm; gubernaculum = 6 (5-7) μm.
- 141 females (after Mizukubo, 1992): L = 0.51 (0.36-0.66) mm; a = 26.5 (19.3-30.6); b = 6.2 (4.1-8.0); b′ = 3.8 (2.3-5.5); c = 19.0 (15.7-23.6); c′ = 2.2 (1.6-2.8); V = 80 (74-83); stylet = 16 (13-18) μm.
- 20 females (after Inserra *et al.*, 2001): L = 0.60 (0.52-0.72) mm; a = 28.7 (23.4-34.0); b = 6.7 (5.6-7.2); c = 20.9 (17.0-31.0); V = 81 (76-83); stylet = 16.9 (16.5-17.0) μm.
- 20 males (after Inserra *et al.*, 2001): L = 0.59 (0.56-0.65) mm; a = 30.8 (28.7-33.9); b = 6.7 (6.0-7.3); c = 21.8 (19.4-25.4); T = 43 (32-48); stylet = 15.0 (14.5-15.5) μm; spicules = 17.5 (16-18) μm; gubernaculum = 5.3 (5.0-5.5) μm.

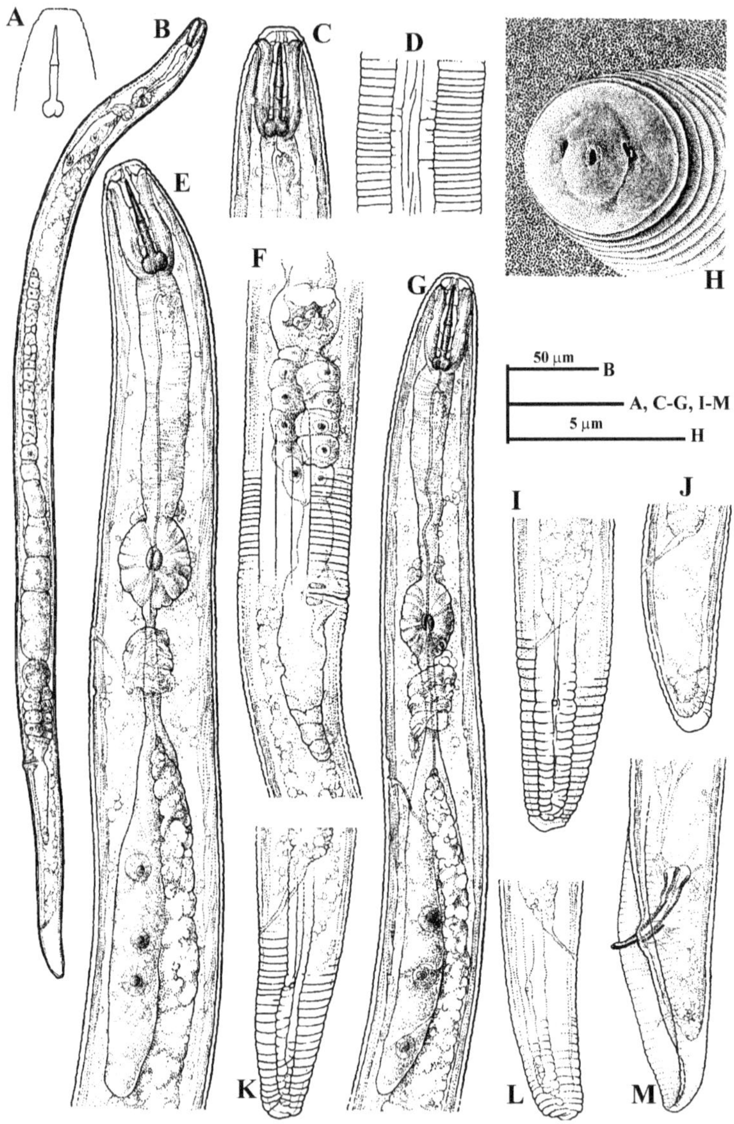

Fig. 32. Pratylenchus coffeae *(Zimmermann, 1898) Filipjev & Schuurmans Stekhoven, 1941. A: Female stylet; B: Female entire body; C: Anterior portion of female body; D: Female lateral field. Note striae and crenate incisures; E: Female pharyngeal region; F: Vulval region; G: Male pharyngeal region; H: Female head pattern (SEM); I-L: Female tail variations; M: Male tail. Note the faintly crenate bursa margins. After Inserra* et al., *2001.*

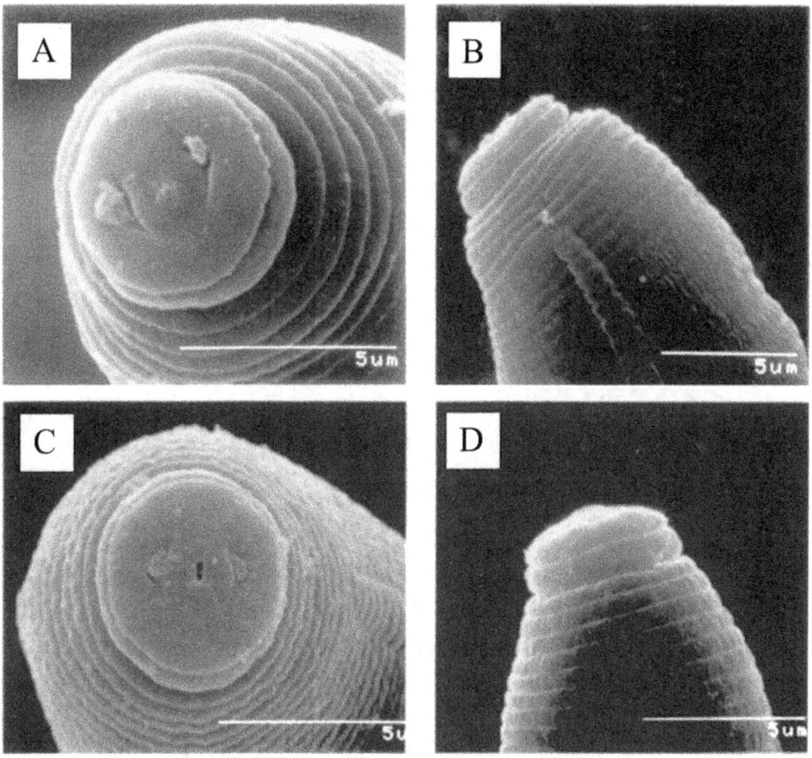

Fig. 33. *SEM micrographs of* P. coffeae *(Zimmermann, 1898) Filipjev & Schuurmans Stekhoven, 1941. A: Female* en face *view; B: Female labial region; C: Male* en face *view; D: Male labial region. After Inserra* et al. *(2001).*

DESCRIPTION

Female

Body rather slender in young and fatter in older specimens, cuticle distinctly annulated. Lateral fields normally with four, sometimes with five or six lines. Labial region slightly offset from body, with two distinct annuli; occasionally three annuli on one side of labial region. *En face* view characterised by fusion of subdorsal, subventral and lateral lips with oral disc (Corbett & Clark, 1983; Hernández *et al.*, 2000).

Fig. 32. *(Continued). E: Female pharyngeal region; F: Vulval region; G: Male pharyngeal region; H: Female head pattern (SEM); I-L: Female tail variations; M: Male tail. Note the faintly crenate bursa margins. After Inserra* et al. *(2001).*

Basal knobs of stylet round to oblong. Post-vulval uterine sac 1.0-1.5 times body diam. long but may be up to 90 μm long, with a terminal rudimentary ovary which sometimes has distinct oocytes. Spermatheca large, broadly oval to nearly rounded, usually with sperm. Intra-uterine eggs may contain embryos. Tail 2.0-2.5 times anal body diam. in young females, 1.5-2.0 times anal body diam. in older females; terminus indented, sometimes appearing smoothly rounded, truncate or irregularly crenate.

Male

Abundant. Spicules slender, with well marked manubria and ventrally arcuate shaft, 16-20 μm long; gubernaculum 4-7 μm in length; hypoptygma prominent; bursal margins faintly crenate.

DIAGNOSIS AND RELATIONSHIPS

Pratylenchus coffeae is characterised by: labial region slightly offset with two distinct annuli, stylet > 15 μm, vulva position > 79%, post-vulval uterine sac 1.0-1.5 times body diam. long with a terminal rudimentary ovary, female tail tip truncate or hemispherical and abundant males.

The matrix code is: A1, B2, C2, D3, E3, F6, G2, H1, I2, J1-2-3, K1.

It can be distinguished from the closely related species (*P. loosi*) by body annulation, position of the vulva, the shape of the female tail and tail tip which is truncate or hemispherical in *P. coffeae vs* bluntly or finely pointed in *P. loosi* (see the corresponding description).

SEM observations and statistical analysis using ten selected characters (number of lip annuli, stylet length, stylet knobs width, labial region diam., distance from stylet to dorsal pharyngeal gland opening, position of excretory pore, ratio V, ratio c, post-vulval uterine sac length divided by vaginal body diam. and number of tail annuli) by Mizukubo (1992) suggest that *P. coffeae* in Japan is a complex composed of at least three independent species. SEM studies by Mizukubo (1992) demonstrated that the *en face* pattern of *P. coffeae* of different populations from Japan was mostly similar to that described by Corbett and Clark (1983) and it was not affected by different host plants (*e.g.*, rice or sweet potato), but was affected by the geographical origin of the population (Mizukubo, 1992). In fact, this study showed clear differences in a population with shorter stylet and *en face* view identical to Group 2 of Corbett and

Clark (1983), which was later described as *Pratylenchus pseudocoffeae* Mizukubo, 1992. Conversely, as pointed out by Mizukubo (1992), differences between other populations in the study were dependent on minor characters that are not used in the identification of *Pratylenchus* species and therefore are maintained as intraspecific variability in *P. coffeae*.

Shahina and Maqbool (1996) described two species within the genus *Radopholus* in which the female genital posterior branch was reduced to a uterine sac (*R. allius* Shahina & Maqbool, 1996 and *R. brassicae* Shahina & Maqbool, 1996). Siddiqi (2000) re-examined the paratypes of *R. allius* and *R. brassicae* and found them to belong to the genus *Pratylenchus*, since the specimens were mounted with the anterior region slightly twisted so that the glands appeared to lie dorsal to the intestine and erroneously placed in *Radopholus* at the original description. After the study of paratypes Siddiqi (2000) pointed out that *P. brassicae* (Shahina & Maqbool, 1996) Siddiqi, 2000 could be a synonym of *P. coffeae*. After a precise comparison of both descriptions, we concur with that opinion and consequently *P. brassicae* is considered here as a new junior synonym of *P. coffeae*.

DISTRIBUTION

It has been recorded from the type locality at Bogor (western Java), Indonesia, within roots of coffee (*Coffea* sp.). *Pratylenchus coffeae* parasitises coffee in many countries of the world, and citrus in India, Japan and USA. Other hosts are *Cassia lora* L., *Oxalis acetosella* L. and several weeds in India; bamboo, *Albizia falcata* (L.) Back., *Cinchona succirubra* Pavon., *Hevea brasiliensis* Muell.-Arg., *Leucaena glauca* Benth., *Theobroma cacao* L., *Potamogeton* sp. and *Nitella* sp. in Indonesia; yam tubers in Puerto Rico; tomato, caladium, *Rumex acetosella* L., *Juglans regia* L. and *Prunus salicina* Lindl. in Australia; *Chrysanthemum frutescens* L., *Ligularia kaempferi* Sieb. & Zucc. and *Musa acuminata* M. in Hawaii; bark of mahogany trees in Barbados (Cobb, 1920); apple, banana, camellia, Chinese evergreen, chrysanthemum, coffee, cotton, dahlia, grapevine, alfalfa, marigold, red clover and strawberry (Siddiqi, 1972).

It is widely distributed in Asia: China from root and rhizosphere soil of Chinese olive, *Canarium album* (Lour.) Raeusch. (Zhang *et al.*, 2002), sweet potato (Gao *et al.*, 2000), mango (Yin, 1994); several crops (Yin,

1991); India on coffee (Giribabu & Saha, 2003), soil around the roots of citrus orchards (Bajaj & Bhatti, 1984), banana (Sundararaju *et al.*, 2002), on oil palm, *Elaeis guineensis* Jacq. (Sundararaju & Ratnakaran, 2002), coconut and arecanut (Rama & Dasgupta, 2000b), ginger (Rama & Dasgupta, 1998), medicinal and aromatic plants (Sivakumar & Vadivelu, 1997), pigeon pea and peanut (Sharma *et al.*, 1992b), chrysanthemum (Rashid & Khan, 1975); Iran on coffee and other hosts (Pourjam *et al.*, 1997, 1999b) and chrysanthemum (Lee *et al.*, 2006); in Japan severely damaging sweet potato (Gotoh, 1964; Iwahori & Sano, 2003), on various crops (Gotoh, 1974; Orui & Mizukubo, 1999b) and coniferous seedlings (Zinno & Igarashi, 1972); Korea (Kornobis & Dobosz, 1997); Oman on fruit crops (Mani *et al.*, 1997) and date palm (*Phoenix dactylifera* L.) (Mani *et al.*, 2005); Pakistan on rose (Saeed *et al.*, 1988); Papua New Guinea on various crops (Bridge & Page, 1984); Taiwan on tea (Wu *et al.*, 2002), strawberries (Wu *et al.*, 2002), root and tuber crops (Tsay *et al.*, 1994), citrus (Huang & Chiang, 1976); Thailand on pummelo (Toida *et al.*, 1996); Turkey on poppy crop, *Papaver somniferum* L. (Akgul & Okten, 2001); Vietnam on banana (Chau *et al.*, 1997; Van den Berg *et al.*, 2006), coffee and banana (Ryss & Fam-Tkhan, 1989).

It is also widespread in Central and South America, being recorded from several countries (Brazil, Florida, Honduras, Java, Martinique and Puerto Rico) and hosts (cocoyam, yam, *Aglaonema*, citrus, banana and coffee) by Inserra *et al.* (2001). It has been reported in Belize on yams and plantain (Bridge *et al.*, 1996); Brazil on yam, *Dioscorea cayennensis* Lam. (Moura *et al.*, 2001), potato tubers (Kubo *et al.*, 2001), soursop, *Annona muricata* L. (Moura *et al.*, 1998), sweet potato (Charchar, 1997); Colombia on banana (Grislaes-Lopez & Lescot, 1999); Costa Rica on ornamentals (Marbán-Mendoza & Flores, 1993); Cuba on rice (Fernandez Díaz-Silveira & Ortega Herrera, 1998); Ecuador on abaca, *Musa textilis* L. Née (Bridge, 1976); Honduras on coffee (Pourjam *et al.*, 1999b), citrus (Pinochet *et al.*, 1978); Martinique on cultivated ornamentals (Quénéhervé *et al.*, 1997); Mexico (Knobloch & Laughlin, 1973); Puerto Rico on banana (Oramas & Román, 1982); Trinidad on anthurium, *Anthurium andraeanum* Linden (Bala & Hosein, 1996); Venezuela on banana (Montiel *et al.*, 1997).

It has been reported from Africa: Cameroon on plantain (Bridge *et al.*, 1995); Ivory Coast on banana (Adiko, 1988; Adiko & N'Guessan, 2001); Ghana on coffee (Pourjam *et al.*, 1999b); Mozambique on various crops (Oever & Mangane, 1992); Nigeria on plantain (Speijer *et al.*, 2001);

South Africa on banana (Daneel *et al.*, 2003); Tanzania on banana (Rajab *et al.*, 1999). It also occurs in some parts of North America: Florida on citrus (Duncan *et al.*, 1998; Inserra *et al.*, 1998), aster (Inserra *et al.*, 1998); North Carolina on peach (Barker & Clayton, 1973); and Canada (Townshend *et al.*, 1978a). Finally, it has been recorded from several countries in Europe: Bulgaria on fruit tree nurseries (Choleva *et al.*, 1984), *Rosa damascena* Miller (Katalan-Gateva & Budurova, 1979); Italy on carnation (Ambrogioni & Rapetti, 1992); Russia on several crops (Ryss, 1988); Slovenia in vineyard soils (Urek *et al.*, 2003); and several localities in Spain (Jiménez-Millán *et al.*, 1965; Arias & Romero, 1979; Gómez-Barcina *et al.*, 1989) parasitising banana.

13. *Pratylenchus convallariae* Seinhorst, 1959

(Fig. 34)

MEASUREMENTS

- Female holotype (after Seinhorst, 1959): L = 0.62 mm; a = 31; b = 7; c = 24; V = 81; stylet = 17 μm.
- 10 females (after Seinhorst, 1959): L = 0.58-0.61 mm; a = 23-27; b = 6-9; c = 17-28; V = 78-81; stylet = 16-17 μm.
- 10 males (after Seinhorst, 1959): L = 0.49-0.58 mm; a = 27-36; b = 5.2-6; c = 20-23; T = 25-51; stylet = 15-17 μm.
- 51 females (after Loof, 1960): L = 0.40-0.60 mm; a = 21-30; b = 5.6-7.6; c = 17-23; V = 76-81; stylet = 14-17 μm.
- 14 males (after Loof, 1960): L = 0.38-0.52 mm; a = 24-34; b = 5.1-6.4; c = 17-22; T = 36-55; stylet = 14-16 μm.
- 20 females (after Brzeski & Sczcygiel, 1977): L = 0.54 (0.49-0.59) mm; a = 29 (24-34); b = 6.8 (6.2-7.8); c = 18 (17-22); c′ = 2.2 (1.9-2.5); V = 79 (77-81); stylet = 16-18 μm.
- 10 males (after Brzeski & Sczcygiel, 1977): L = 0.50 (0.44-0.58) mm; a = 34 (31-40); b = 6.3 (6.1-6.5); c = 21 (19-24); stylet = 16-17 μm; spicules = 15-16 μm; gubernaculum = 4-6 μm.
- 4 females (after Ryss, 1988): L = 0.53 (0.52-0.59) mm; a = 27 (24-30); b = 6.6 (6.3-6.8); c = 18 (17-25); c′ = 2.1 (1.7-2.4); V = 79 (77-81); stylet = 16-17 μm.

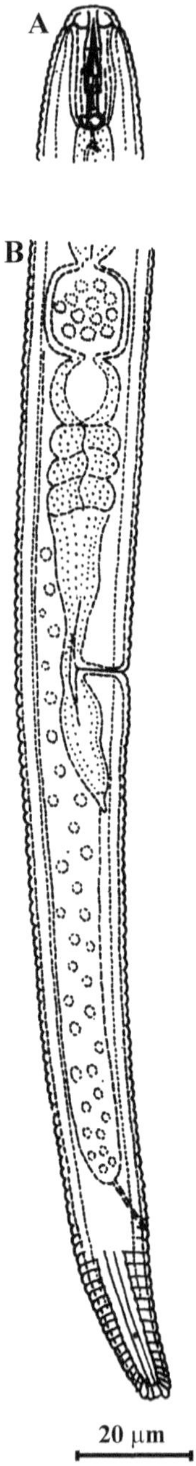

Fig. 34. Pratylenchus convallariae *Seinhorst, 1959. A: Female labial region; B: Female posterior region. After Seinhorst (1959).*

Description

Female

Cuticle marked by fine striae, *ca* 1.5 μm apart. Lateral fields with four lines. Labial region slightly offset from body, bearing three annuli. Lateral margins of labial framework extending posteriorly for *ca* half a body annulus. Stylet 17 (16-18) μm long. Dorsal pharyngeal gland opening into lumen of pharynx *ca* 3 μm from stylet base. Excretory pore at level of junction between pharynx and intestine. Hemizonid situated just anterior to excretory pore. Ovary single, outstretched, with single row of oocytes. Spermatheca round. Distance between vulva and spermatheca = 44-66% of vulva-anus distance. Uterus short. Post-vulval uterine sac extending posteriorly 1.4-2 vulval body diam. and *ca* as long as tail. Tail truncate with 16-19 (rarely more than 20) annuli. Tail tip truncate, coarsely and often irregularly annulated, often divided into two irregular lobes, connection between lateral fields around tip being shorter than that between the dorsal and ventral surface of nematode. Phasmids at mid-tail. Inner two lines of lateral fields extending beyond phasmid.

Male

Not differing anatomically from male of *P. penetrans*. In some specimens tail tip appearing bifid; in others, bursal margin slightly irregular immediately anterior to tail tip. Bursal edge coarsely crenate. Testis shorter than *vas deferens*; in specimens studied, demarcation between these two parts was more distinct than in other species of *Pratylenchus*.

Diagnosis and relationships

Pratylenchus convallariae is characterised by: labial region with three annuli, males common, spermatheca round and female tail tip truncate and irregularly annulated.

The matrix code is: A2, B2, C3, D2, E2, F6, G2, H2, I3, J1, K1.

It is close to *P. pratensis*, *P. crenatus* and *P. penetrans*. From *P. pratensis* it may be distinguished by the truncate tail and rounded spermatheca, from *P. crenatus* by the presence of a spermatheca and numerous males, the weaker body annulation, presence of four lines in lateral fields, more anterior position of vulva and tail shape; and from *P. penetrans* by the annulated tail tip and vulva position. This last difference is small, but on examination of 50 females of both species was found to be highly significant (difference = six times its standard error).

DISTRIBUTION

It has been recorded from the type locality at Wassenaar, The Netherlands, parasitising roots of German lily of the valley (*Convallaria majalis* L.). Additionally, it has been recorded in Germany (Decker & Dove, 1974); Poland (Brzeski & Szczygiel, 1977); Portugal (Abrantes *et al.*, 1987); Russia (Ryss, 1988); Slovenia on corn (Urek *et al.*, 2003); several localities and hosts in Spain (Espárrago & Navas, 1995; Talavera & Tobar-Jiménez, 1997; Talavera & Navas, 2002); Japan on German lily of the valley (Saigusa & Ando, 1973); and on sweet potato (Huan & Xu, 1985) and peach orchards (Wu, 1993) in Zhejiang Province, China.

14. *Pratylenchus crassi* Das & Sultana, 1979
(Fig. 35)

MEASUREMENTS

- Female holotype (after Das & Sultana, 1979): L = 0.43 mm; a = 20.3; b = 8.8; c = 18.5; V = 73; stylet = 17 μm.
- 9 females (after Das & Sultana, 1979): L = 0.41-0.45 mm; a = 20-26; b = 7.7-8.8; c = 18-24; V = 72-77; stylet = 17-18 μm.

DESCRIPTION

Female

Cuticle finely annulated, coarsest striae near labial region being only 1.1 μm apart, while near mid-body they are less than 1 μm apart. Lateral fields with four crenate lines occupying less than one-third body diam. Labial region low, flattened, strongly sclerotised, *ca* three times as wide as high, bearing two annuli. Stylet 17-18 μm long with anteriorly directed cup-shaped knobs. Stylet shaft 8.6 μm long. Orifice of dorsal pharyngeal gland situated 2.1 μm posterior to stylet base. Pharynx with cylindrical procorpus, small spheroid metacorpus and pharyngeal glands in form of ventral lobe, extending over intestine for *ca* 2 body diam. Excretory pore 75.1 μm (67-78.1 μm in paratypes) from anterior end, posterior to nerve ring, excretory duct strongly cuticularised. Hemizonid

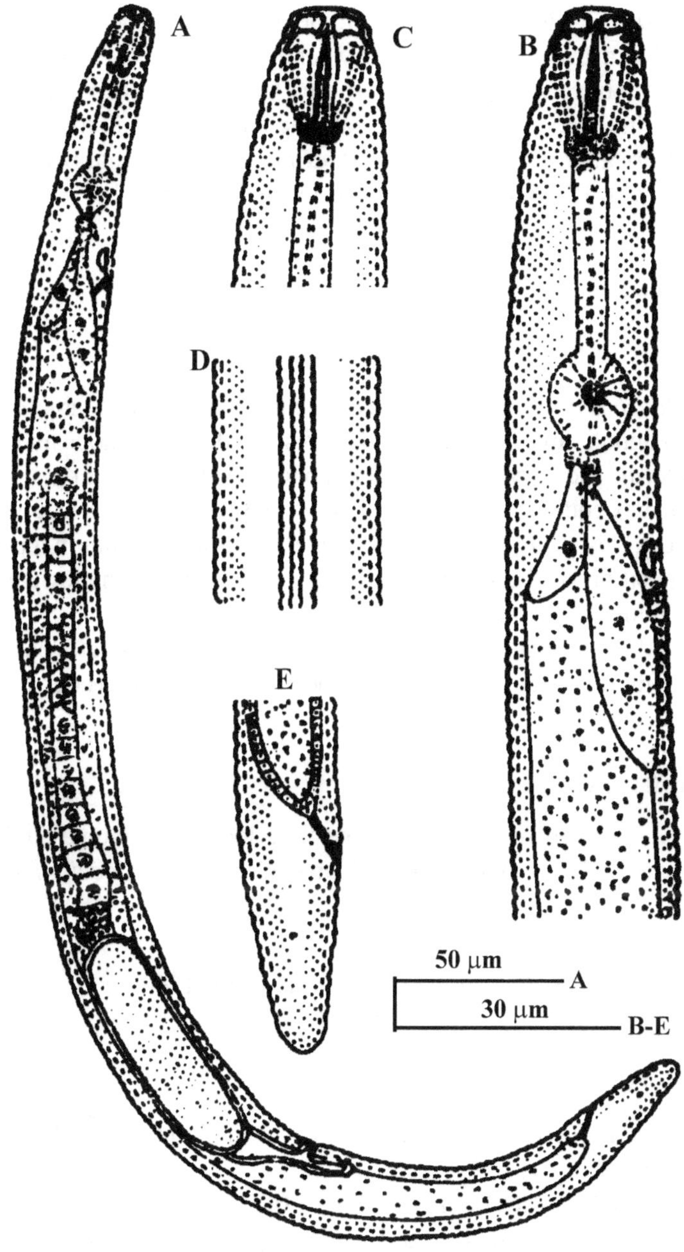

Fig. 35. Pratylenchus crassi *Das & Sultana, 1979. A: Entire female; B: Female pharyngeal region; C: Female labial region; D: Lateral field at mid-body region; E: Female tail. After Das and Sultana (1979).*

large, 2 μm in length, located immediately anterior to excretory pore. Vulva a transverse slit. Ovary single, prodelphic, outstretched. Spermatheca large, oval, filled with sperm. Crustaformeria present. Post-vulval uterine sac very small, less than one vulval body diam. in length. Tail cylindrical with smooth rounded terminus, *ca* two anal body diam. in length, with 12-15 annuli. Phasmids located in mid-tail.

Male

Not found.

DIAGNOSIS AND RELATIONSHIPS

Pratylenchus crassi is characterised by: low and flattened labial region with two annuli, anteriorly directed, cup-shaped, stylet knobs and post-vulval uterine sac very small (less than one vulval body diam.).

The matrix code is: A1, B1, C3, D3, E1, F3, G1, H1, I3, J1, K1.

It is very close to *P. loosi* and *P. coffeae*, but differs mainly in the body length, position of the vulva, tail shape and absence of males (see the corresponding descriptions).

Loof (1991) considered *P. crassi* as *species inquirenda*, but after checking morphological and morphometrical data we consider it to be a valid species.

DISTRIBUTION

It has only been recorded from the type locality at Humayunnagar, Hyderabad (A.P.), India, from soil around roots of *Coccinia indica* W&A.

15. *Pratylenchus crenatus* Loof, 1960

(Fig. 36)

MEASUREMENTS

- Female holotype (after Loof, 1960): L = 0.45 mm; a = 25; b = 6; c = 21; V = 81; stylet = 17 μm.
- 131 females (after Loof, 1960): L = 0.32-0.60 mm; a = 19.7-29.9; b = 4.9-7.9; c = 16.4-26.8; V = 78-86; stylet = 14-18 μm.
- 10 females (after Ryss, 1988): L = 0.50 (0.43-0.56) mm; a = 21 (18-27); b = 6.1 (5.1-6.9); b′ = 5.2 (4.9-5.5); c = 20 (16-27); c′ = 2.0-3.0; V = 83 (79-84); stylet = 16.5 (16.0-17.5) μm.

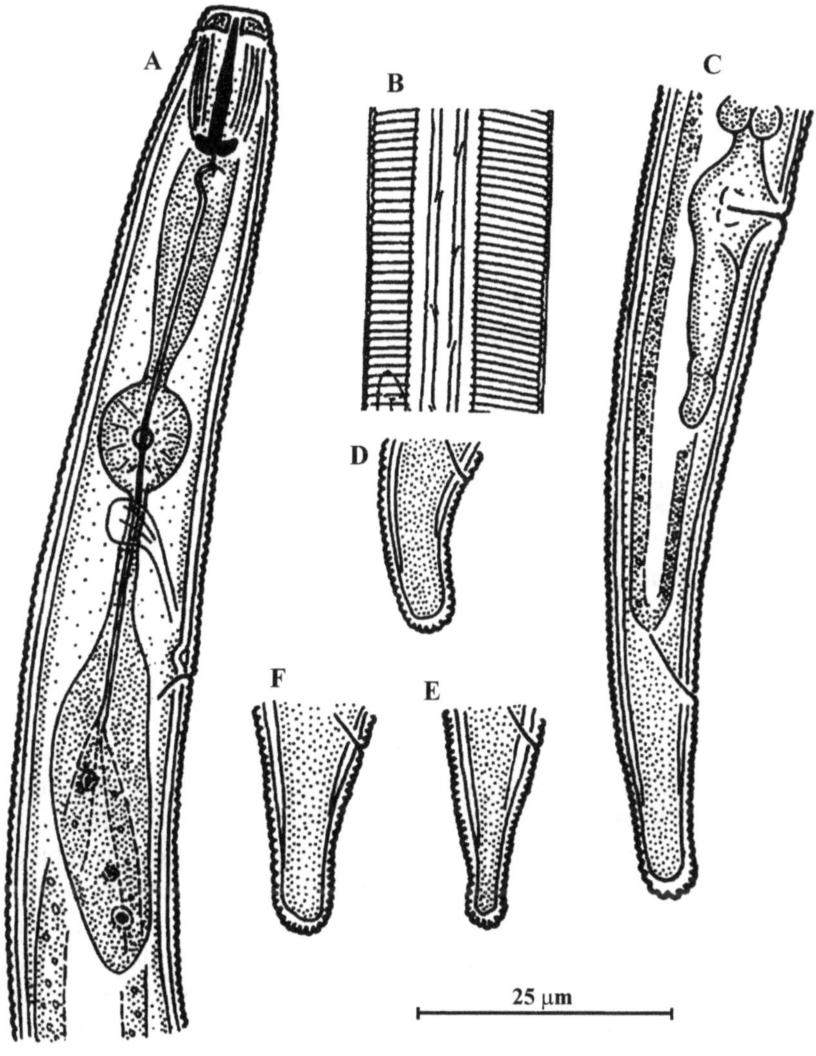

Fig. 36. Pratylenchus crenatus *Loof, 1960. A: Female pharyngeal region; B: Lateral field at mid-body; C: Female posterior region; D-F: Female tails. After Loof (1960).*

- 1 male (after Karssen & Brinkman, 2003): L = 0.43 mm; a = 28.4; c = 21.4; stylet = 16 μm; spicules = 17 μm; gubernaculum = 4.5 μm.

DESCRIPTION

Female

Body usually coiled when killed by gentle heat. Labial region with three annuli. Body annulation prominent. Lateral fields with six lines, outer ones strongly crenate; inner ones interrupted. Vulva usually at 81-84%; only very exceptionally at 78-80%. Ovary with oocytes in single row, except for a short zone near anterior end. Spermatheca small, empty, situated anterior to vulva *ca* 50-90% of vulva to anus distance. Post-vulval uterine sac short, occupying 40-50% of vulva-anus distance (*ca* two body diam. long). Tail broadly rounded, bearing 20-24 annuli; tip often spatulate, coarsely and distinctly annulated.

Karssen and Bolk (2000) observed that an additional character to identify this species was the swelling of the secretory-excretory duct which was clearly seen in fresh females and juveniles, irrespective of geographical origin of the population. Nevertheless, in fixed and permanent mounted specimens this swelling is not noticeable.

Male

Karssen and Brinkman (2003) described the only known specimen of this species. Body vermiform, distinctly annulated, 15 μm in diam. at mid-body, 14 μm at metacorpus, slender and tapering towards both ends. Lateral fields with six lines, inner lines interrupted, not clearly areolated. Labial region slightly offset and rounded, 7.6 $\times$ 2.5 μm in dimension and three times as wide as high. Labial sclerotisation clear, vestibule extension weak. Three lip annuli present, *i.e.*, there are two clear transverse lines. Stylet robust, with small rounded knobs, 3.2 μm wide and 1.9 μm high, their anterior margins directed slightly forward. Dorsal pharyngeal gland orifice 2.5 μm posterior to stylet knobs, metacorpus 49 μm from anterior end. Hemizonid near pharyngo-intestinal junction and anterior to secretory-excretory pore which is located 76 μm from anterior end. Secretory-excretory duct with clear swelling near opening. Pharyngeal gland lobe 25 μm long, relatively small. Testis single, 245 μm long, completely developed, anteriorly outstretched and occupying nearly half of body length. Spicules and gubernaculum characteristic for genus, both ventrally curved. Tail 20 μm long, tip conical and pointed. Bursa arising slightly anterior to spicule head and enveloping tail. Phasmids distinct, extending into bursa, located *ca* 8 μm from tail tip. Important diagnostic features confirming that male belongs to

P. crenatus are presence of three lip annuli, short pharyngeal gland lobe, lateral fields with six lines and secretory-excretory duct swelling (Karssen & Bolk, 2000).

DIAGNOSIS AND RELATIONSHIPS

Pratylenchus crenatus is characterised by: crenate tails and three annuli on the labial region, the males usually absent and the long post-vulval uterine sac.

The matrix code is: A2, B2, C3, D1, E3, F3, G2, H2, I1, J3, K1.

It is very close to *P. pratensis* and *P. convallariae* from which it differs in the lateral fields, position of the vulva and absence of males (see the corresponding descriptions).

DISTRIBUTION

It is widely distributed in the northern temperate zone, predominantly on light sandy soils, where it particularly parasitises cereals and grasses (Loof, 1991). Recorded, apart from the type locality, in many parts of the world. It is widespread in many countries of Europe: Belgium on potato (Pelsmaeker & Coomans, 1987) and rose (Coolen & Hendrickx, 1972a); Bosnia-Hercegovina on corn (Klindic & Petrovic, 1975); Croatia on tobacco (Ostrec & Grubisic, 2003); Denmark on barley (Andersen, 1979) and legumes (Andersen, 1972); England and Wales on cereals (Corbett, 1970); Finland on various crops (Kurppa, 1988); France on wheat (Esmenjaud *et al.*, 1990); Germany on corn (Knuth, 2002), barley (Dowe *et al.*, 1990); Italy on olive (Inserra *et al.*, 1979); The Netherlands on potato (Scholte, 2000); Norway on rose (Stoeen, 1974); Poland in rose nurseries (Wojtowicz & Sobio, 1994), red clover (Kornobis, 1987), corn (Kornobis, 1983), strawberry (Szczygiel, 1974), fruit tree nurseries (Kozlowska & Wasilewska, 1972); Portugal on various crops (Abrantes *et al.*, 1987); Russia on strawberry (Belozerova & Metlitskii, 1972) and various crops (Ryss, 1988); Slovakia on cereals (Liskova *et al.*, 1988), grapevine (Liskova, 1980); Slovenia on corn (Urek *et al.*, 2003); Spain on horticultural crops (Espárrago & Navas, 1995); Switzerland on strawberry (Vallotton, 1989); United Kingdom on oilseed rape (Evans & Webb, 1989), wind toppled trees (Hooper *et al.*, 1990), raspberry (Cotten & Roberts, 1981); Yugoslavia on corn (Grujicic, 1974) and wheat (Grujicic, 1969). It has also been reported from several states of the USA: in Maine on potato (Huettel *et al.*, 1991); New York on potato (Florini *et al.*, 1987); Ohio on potato (Wheeler *et al.*, 1994; Brown *et al.*, 1980); and in Canada on turf grass (Yu *et al.*, 1998), oats (Townshend, 1989a), potato (Olthof *et al.*, 1982; Kimpinski & Smith, 1988), forage grass and legumes (Kimpinski *et al.*, 1984), apple orchards (Vrain & Rousselle, 1980), various crops (Townshend *et al.*, 1978a), tobacco (Olthof & Hopper, 1973). It has also been recorded from several countries in Asia: Azerbaidzhan on *Crocus sativus* L. (Kasimova & Atakishieva, 1980); India on soybean (Raut & Sethi, 1986), various crops (Sethi & Swarup, 1971); and Japan on tobacco (Orui & Mizukubo, 1999a), various crops (Gotoh, 1974). Recorded also in Argentina on various crops (Torres & Chaves, 1999); South Africa on soybean (Fourie *et al.*, 2001), wheat (Jordaan *et al.*, 1992), corn (Jordaan *et al.*, 1989), sunflower (Bolton *et al.*, 1989), sorghum (De Waele & Jordaan, 1988b); Sudan (Decker & El-Amin, 1980); and Australia on subtropical crops (Knight, 2001), carrots (Hay & Pethybridge, 2005) and various crops (Riley & Kelly, 2001).

16. *Pratylenchus cruciferus* Bajaj & Bhatti, 1984
(Fig. 37)

MEASUREMENTS

- Female holotype (after Bajaj & Bhatti, 1984): L = 0.76 mm; a = 34; b = 8.3; b′ = 5.8; c = 24; c′ = 2.5; V = 78; stylet = 16 μm.
- 20 females (after Bajaj & Bhatti, 1984): L = 0.73 (0.65-0.79) mm; a = 33 (26-40); b = 8.2 (7.3-9.2); b′ = 5.8 (5.0-7.0); c = 22 (19-28); V = 77 (76-81); stylet = 15.5 μm.

DESCRIPTION

Female

Body long and cylindrical, C-shaped on fixation. Lateral fields occupying one-quarter or one-fifth of mid-body diam. with four lines, central core of lateral fields with oblique striations. Labial region flat, continuous with body, with three annuli. Labial framework extending up to 2-3 body annuli. Stylet 15-16 μm long, anterior surfaces of stylet knobs concave. Procorpus narrower at junction with rounded to oval median bulb. Pharyngeal gland lobe 31-51 μm long, overlapping intestine for *ca* two corresponding body diam. Opening of dorsal pharyngeal gland 3-4 μm from stylet base. Nerve ring at mid-isthmus.

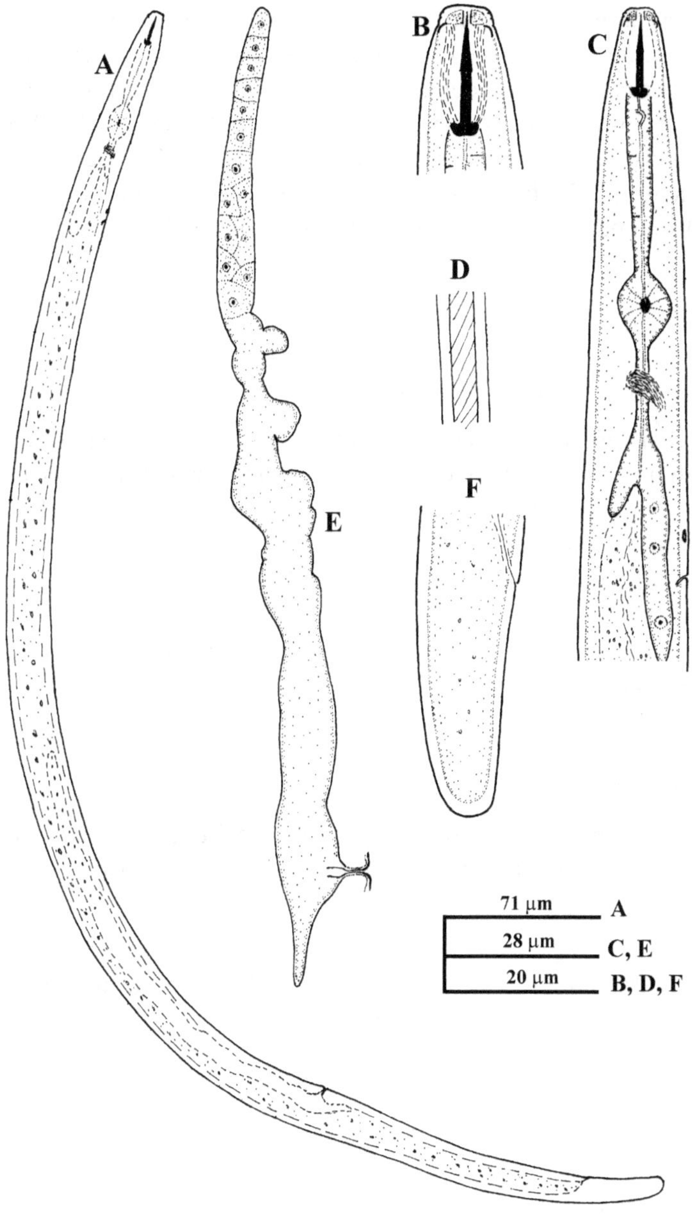

Fig. 37. Pratylenchus cruciferus *Bajaj & Bhatti, 1984. A: Entire female; B: Female labial region; C: Female pharyngeal region; D: Lateral field at mid-body; E: Female reproductive system; F: Female tail. After Bajaj and Bhatti (1984).*

Excretory pore 90-104 μm from anterior end or 3-19 μm posterior to pharyngo-intestinal valve. Hemizonid 2-8 annuli anterior to excretory pore. Spermatheca and columella indistinct. Vulva a transverse slit located at 76-81% of body length. Post-vulval uterine sac 18-30 μm or 0.7-1.4 vulval body diam. long. Anal body diam. = 10-14 μm. Tail subcylindrical to cylindrical with smoothly rounded to truncate terminus, 2.1-3.0 anal body diam. long. Phasmids near mid-tail.

Male

Not found.

Diagnosis and relationships

Pratylenchus cruciferus is characterised by: labial region flat, continuous with body and with three annuli, lateral fields with four lines, central core of lateral fields with oblique striations and hemizonid 2-8 annuli anterior to excretory pore.

The matrix code is: A2, B1, C2, D1, E2, F3, G2, H1, I3, J1, K1.

It is close to *P. thornei* from which it differs in the shape of labial region, body length, presence of oblique lines in the central zone of lateral fields and position of hemizonid (see the corresponding description).

Distribution

It has only been recorded from the type locality at Patanwas village, Bhiwani district, India, from soil around the roots of mustard, *Brassica campestris* L.

17. *Pratylenchus curvicauda* Siddiqi, Dabur & Bajaj, 1991
(Fig. 38)

Measurements

- Female holotype (after Siddiqi, Dabur & Bajaj, 1991): L = 0.47 mm; a = 27.6; b = 5.4; b′ = 3.5; c = 16.2; c′ = 2.9; V = 72; stylet = 16 μm.
- 20 females (after Siddiqi, Dabur & Bajaj, 1991): L = 0.50 (0.45-0.55) mm; a = 24 (21-28); b = 5.9 (5.2-6.8); b′ = 3.4 (3.0-3.9); c = 14.5 (13-18); c′ = 2.7 (2.1-2.9); V = 73 (69-77); stylet = 16 (15-16.5) μm.

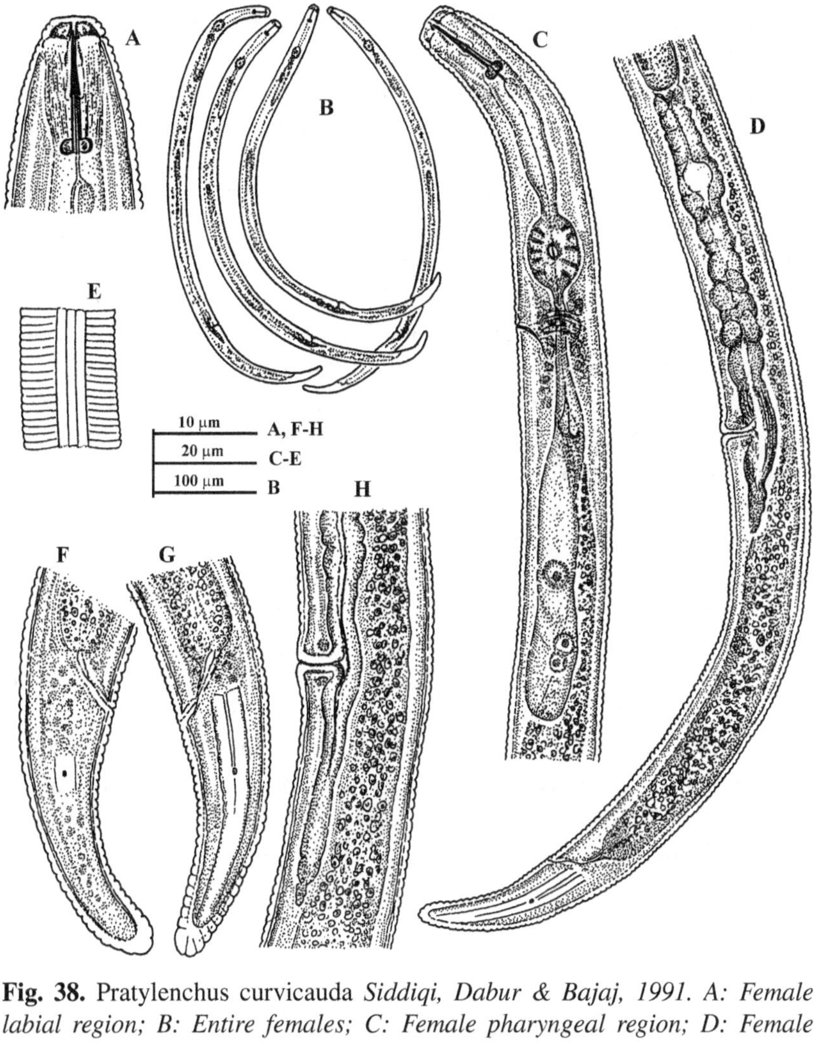

Fig. 38. Pratylenchus curvicauda *Siddiqi, Dabur & Bajaj, 1991. A: Female labial region; B: Entire females; C: Female pharyngeal region; D: Female posterior region; E: Lateral field at mid-body; F, G: Female tails; H: Vulval region. After Siddiqi* et al. *(1991).*

DESCRIPTION

Female

Body strongly ventrally arcuate to C-shaped or irregularly curved, never straight in relaxed condition; max. body diam. 17-25 (21) μm. Body annuli *ca* 1 μm wide at mid-body. Lateral fields occupying *ca* one-third body diam., with four lines, central band slightly narrower than lateral ones and occasionally with one or two additional lines, not areolated posterior to pharynx. Labial region broadly rounded to truncate, continuous with body or slightly offset having three annuli (occasional specimen with two lip annuli). Outer-margins of labial framework extending up to one body annulus. Stylet conus 7.2-8.5 μm long; basal knobs rounded, anteriorly flattened, 4.2 (3.9-4.8) μm across and *ca* 2 μm high. Opening of dorsal pharyngeal gland at 2.5-3.7 μm from stylet base. Median bulb rounded to oval, large and muscular, 15-18 $\times$ 10-12 μm; valvular apparatus 3.4-3.6 μm long $\times$ 2.5-3.0 μm wide. Excretory pore slightly anterior to pharyngo-intestinal junction, 82 (70-89) μm from anterior extremity. Hemizonid 2-4 annuli long, just anterior to excretory pore. Pharyngeal glands extending to 62 (47-80) μm over intestine; nuclei of subventral glands close together. Ovary outstretched, sometimes extending to pharyngeal glands; oocytes arranged in single row except for anterior most part of multiplication zone. Spermatheca small, empty. Uterine egg = 65 $\times$ 21 μm. Post-vulval uterine sac 25 (19-37) μm long, with rudimentary ovarian cells. Vulva on a ventral protuberance of body. Distance between vulva and tail terminus = 129 (111-150) μm; VL/VB usually = 6.5 (5.8-8.5). Annuli between vulva and anus = 77 (60-88). Tail ventrally arcuate, conoid-rounded 32 (27-38) μm long, with 26 (24-31) annuli ventrally; tail terminus narrowly rounded, usually indented. Phasmids located anterior to mid-tail, 16-22 annuli, or 21-29 μm from terminus; lateral fields posterior to phasmids forming a plain band which may bear a central line.

Male

Not found.

DIAGNOSIS AND RELATIONSHIPS

Pratylenchus curvicauda is characterised by: strongly ventrally arcuate to C-shaped body posture, labial region with three annuli, a conoid, ventrally arcuate female tail with phasmids located anterior to its middle being 16-22 annuli or 21-29 μm from tail terminus and by the absence of males.

The matrix code is: A2, B1, C3, D1, E1, F4, G3, H2, I4, J1, K1.

It is close to *P. mulchandi*, *P. vulnus* and *P. zeae*. It can be distinguished from *P. mulchandi* in having a more tapering and ventrally arcuate female tail; phasmids located 16-22 annuli or 21-29 μm from tail terminus and vulva located more anteriorly; and from *P. vulnus* in having a more anteriorly placed vulva, spermatheca without sperm and phasmids located anterior to the middle of the tail. In addition, it differs from *P. zeae* in body habitus after relaxation, length of pharyngeal gland, female tail, female tail tip and position of phasmids (see the corresponding descriptions).

DISTRIBUTION

It has only been recorded from the type locality at Perth, Western Australia, where it was found in soil around the roots of *Trifolium* sp.

18. *Pratylenchus delattrei* Luc, 1958
(Fig. 39)

MEASUREMENTS

- 13 females (after Luc, 1958): L = 0.39-0.47 mm; a = 20.4-25.8; b = 3.7-4.8; c = 18.0-22.3; V = 73-81; stylet = 16.5-18 μm.
- 7 females (after Das & Sultana, 1979): L = 0.44-0.49 mm; a = 20-25; b = 8.1-8.8; c = 18.0-22.7; V = 75-77; stylet = 17-18 μm.
- 14 females (Subramaniyan & Sivakumar, 1991): L = 0.54 (0.52-0.57) mm; a = 24.0 (22.0-27.4); b = 4.3 (4.0-4.9); c = 19.4 (18.3-20.7); c′ = 1.4 (1.2-1.9); V = 77 (76-79); stylet = 16.5 (16-17) μm.
- 48 females (after Subramaniyan & Sivakumar, 1988): L = 0.43 (0.36-0.49) mm; a = 23.7 (20.6-26.1); b = 4.8 (3.6-5.4); b′ = 3.5 (3.0-4.1); c = 20.5 (17.3-23.3); c′ = 1.6 (1.3-2.1); V = 77 (72-79); stylet = 17 (16-18) μm.
- 13 females (after Zarina & Maqbool, 1998): L = 0.40 (0.34-0.46) mm; a = 27.2 (21.6-30.7); b = 4.0 (3.4-5.0); c = 18.1 (14.4-26.3); c′ = 2.5 (1.3-3.1); V = 77 (73-86); stylet = 15 (13-16) μm.

DESCRIPTION

Female

Body slightly ventrally arcuate after heat relaxation and fixation. Cuticular annuli *ca* 1.3-1.5 μm wide at mid-body. Lateral fields one-third of body diam., with four lines. Labial region continuous with body

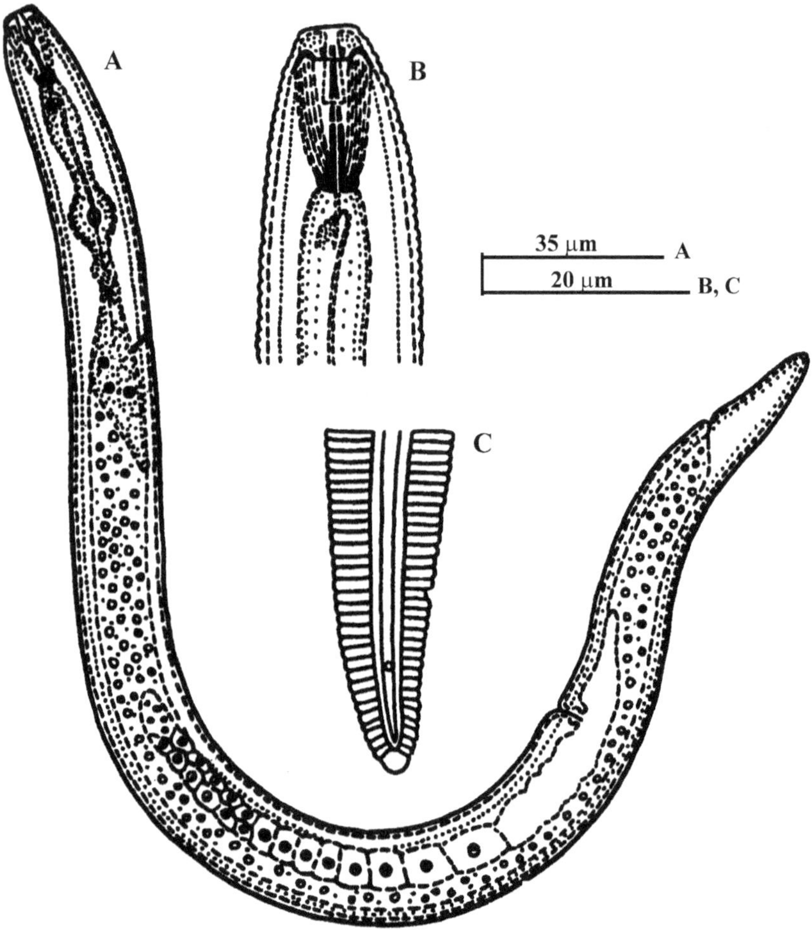

Fig. 39. Pratylenchus delattrei *Luc, 1958. A: Entire female; B: Female labial region; C: Female tail. After Luc (1958).*

contour slightly off set with three distinct annuli. Labial framework well developed, 8.8 μm high and 4.0 μm wide. Stylet robust, basal stylet knobs flattened 1.6 × 4.0 μm diam. Orifice of dorsal pharyngeal gland 1.6-2.4 μm posterior to stylet base. Pharynx pratylenchoid, pharyngeal lobe overlapping intestine ventrally; rounded to oval, median bulb 7.2 × 12.0 μm, with distinct valvular apparatus. Isthmus crossed by nerve ring in posterior half. Nerve ring anterior to excretory pore. Excretory pore 66.4-76.8 μm long from anterior end of body. Hemizonid *ca* three annuli long situated just anterior to excretory pore. Ovary single, prodelphic, outstretched anteriorly, oocytes arranged in single row, spermatheca rounded, without sperm. Vulva a transverse slit, post-vulval uterine sac larger than vulval body diam. Tail subcylindrical with smooth conical to rounded terminus.

Male

Not found.

DIAGNOSIS AND RELATIONSHIPS

Pratylenchus delattrei is characterised by: labial region with three annuli, lateral fields with four incisures, tail subcylindrical with smooth conical to rounded terminus and absence of males.

The matrix code is: A2, B1, C3, D1, E2, F3, G3, H1, I1, J1, K1.

It is very close to *P. andinus* and *P. bolivianus* from which it differs by body and stylet length, lip annulation, position of vulva and tail terminus (see the corresponding descriptions).

Frederick and Tarjan (1989) pointed out that *P. singhi* is almost identical with *P. delattrei* except for the presence of a spermatheca filled with sperm. However, the description of *P. singhi* was based on only seven specimens and thus it was concluded that *P. singhi* is a synonym of *P. delattrei*. Similarly, *Pratylenchus portulacus* Zarina & Maqbool, 1998 was not specifically compared with *P. delattrei*, despite being almost identical in morphology and morphometry. Siddiqi (2000) considered *P. portulacus* to be a junior synonym of *P. delattrei*, an action that is maintained here.

Subramaniyan and Sivakumar (1991) described *Pratylenchus graminis* as a new species differentiated from *P. delattrei* on absence of spermatheca, length of pharyngeal lobe, position of excretory pore and hemizonion and stylet knob shape. These characters are of minor importance and therefore *P. graminis* is considered here as a new junior synonym of *P. delattrei*.

DISTRIBUTION

It has been recorded from the type locality at Southern Madagascar on cotton. It has also been recorded in Sudan on sugarcane (Saadabi, 1988); in Korea on rose and rhapis (Kim & Minagawa, 1996); in Pakistan on purslane, *Portulaca grandiflora* Hk. f. (Zarina & Maqbool, 1998); in Oman on date palm (*P. dactylifera*) (Mani *et al.*, 2005);

and on several hosts and localities in India: crossandra (Vadivelu & Muthukrishnan, 1979; Jothi *et al.*, 2004), turmeric (*Curcuma longa* L.) (Poornima & Vadicelu, 1999), on pigeon pea and peanut (Sharma *et al.*, 1992b), crossandra, corn and tomato (Subramanian & Sivakumar, 1988), sugarcane (Mehta & Sundararaj, 1990), bean (Sundaram *et al.*, 1987).

19. *Pratylenchus dunensis* de la Peña, Moens, van Aelst & Karssen, 2006

(Figs 40, 41)

Measurements

- Female holotype (after de la Peña *et al.*, 2006a): L = 0.49 mm; a = 30.6; b = 6.6; b′ = 4.7; c = 14.4; c′ = 3.4; V = 76; stylet = 16.5 μm.
- 15 females (after de la Peña *et al.*, 2006a): L = 0.51 (0.45-0.58) mm; a = 28.3 (25-32); b = 6.8 (5.8-8.3); b′ = 4.3 (3.8-5.7); c = 15 (13-17); V = 78 (76-79); stylet = 16.5 (16-18) μm.
- 15 males (after de la Peña *et al.*, 2006a): L = 0.45 (0.39-0.48) mm; a = 28.7 (23-31); b = 6.2 (5.4-7.5); b′ = 4.1 (3.8-4.8); c = 15 (12-17); c′ = 2.6 (2.0-3.1); stylet = 14.5 (14-15) μm; spicules = 15.1 (14.5-16.5) μm; gubernaculum = 5.0 (4.5-6) μm.

Description

Female

Body vermiform, slender, tapering towards both ends. Cuticle finely annulated with annuli 1-1.3 μm wide at mid-body. Labial region slightly offset, three times height in diam., rounded with prominent labial sclerotisation and vestibule extension. Two lip annuli, often with one or two incomplete, transverse incisures not visible with light microscope. Labial disc oval, elevated, fused with four submedian lips forming

---

Fig. 40. Pratylenchus dunensis *de la Peña, Moens, van Aelst & Karssen, 2006. A: Entire female body, lateral view; B: Entire male body, lateral view; C: Male labial region; D: Female labial region; E: Female anterior end including pharyngeal gland; F: Female vulval region, ovary and spermatheca; G: Lateral field at mid-body; H: Female tail tips; I: Female posterior end; J: Male posterior end. After de la Peña* et al. *(2006a).*

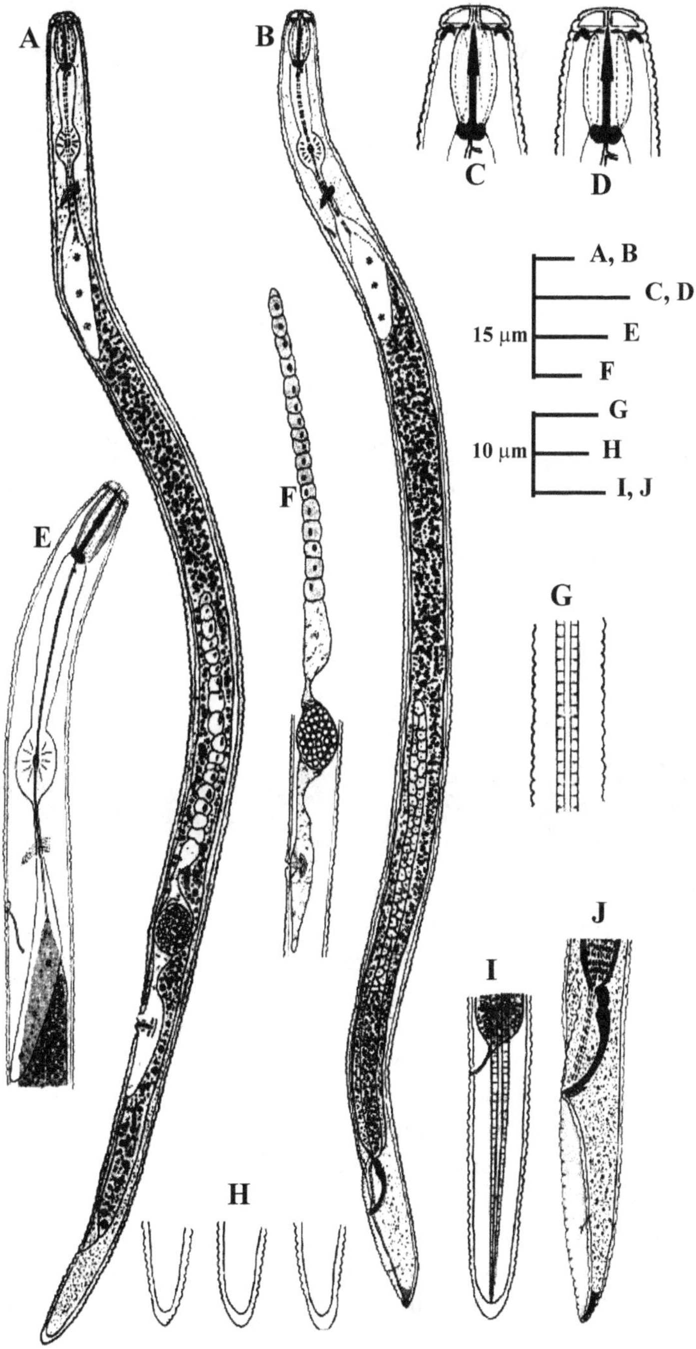
A
B
C
D
A, B
C, D
15 μm
E
F
G
10 μm
H
I, J
E
F
G
J
I
H

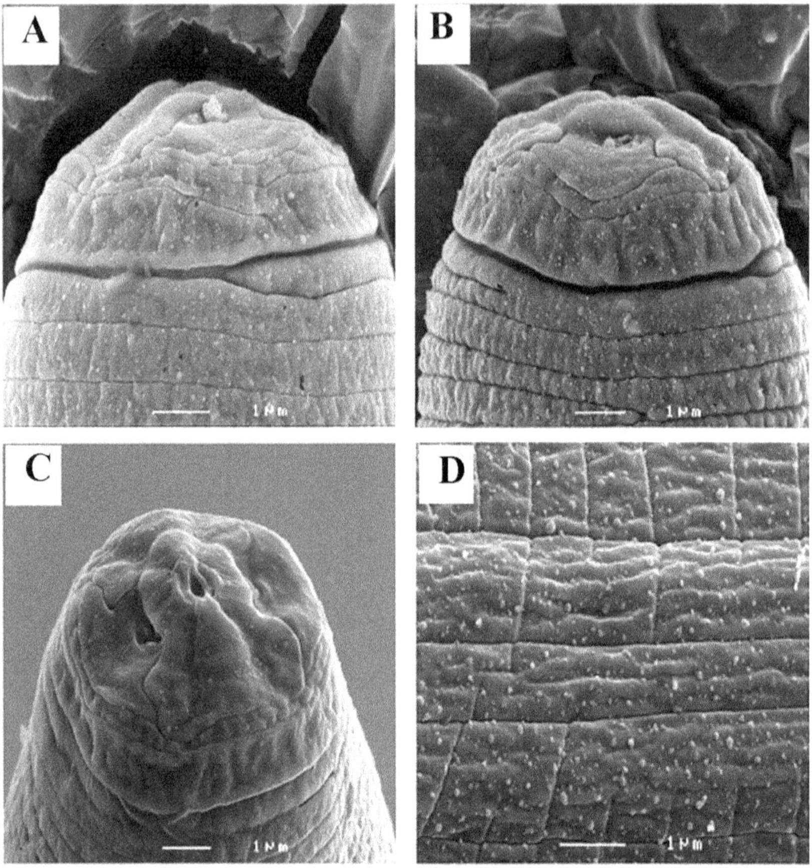

Fig. 41. Pratylenchus dunensis *de la Peña, Moens, van Aelst & Karssen, 2006. SEM studies. A: Lateral view of female labial region; B: Lateral view of male labial region; C:* En face *view of female labial region; D: Female lateral field view at mid-body. After de la Peña* et al. *(2006a).*

smooth, dumbbell-shaped, cephalic cap. Labial sensilla marked by subtle depression on each submedian lip. SEM *en face* view showing oval prestoma. Lateral field with four lines starting posterior to stylet; field occupying *ca* one-third of body diam. at mid-body. At pharyngeal-vulva region, middle ridge narrower than outer ones; at mid-body, width of inner ridge 60-80% of distance between outer lines. Outer incisures partially areolated, with striae crossing outer lateral field ridges every 1-1.5 μm. Lateral lines converging posterior to phasmid. Stylet relatively short and slender, with robust, anteriorly indented, offset knobs. Stylet cone length equal or shorter than length of shaft plus knobs. Hemizonid near isthmus level, 2-3 μm in length, anterior to oval shaped secretory-excretory pore located between nerve ring and pharyngeal junction. Pharyngeal gland lobe three body diam. long. Vulva with well-developed lips, often protruding. Genital tract monoprodelphic with single row of oocytes. Spermatheca round to slightly oval, filled with rounded sperm. Post-vulval uterine sac undifferentiated, short (25 μm). Tail cylindrical with distinct annulations, narrowing in posterior third with smooth, dome-shaped terminus; some specimens (one in five of adult females) with one or two indentations adjacent to smooth tail tip. Hyaline part distinct. Anus round to oval shaped. Small, rounded, phasmid located between inner lateral field lines, posterior to mid-tail.

Male

Occurring abundantly (nearly 50% of adults). Morphologically similar to females, but smaller for all non-sexual characters. Labial region characters as in females, but more truncated in outline. Apical annuli slightly raised. Incomplete transverse incisures and variation in lip annuli as described for females. Single testis anteriorly outstretched. Spicules and gubernaculum ventrally curved. Bursa enclosing tail. Ventral surface of bursa coarsely crenate; phasmidial orifice on bursa located at almost midway between cloaca and tail tip.

Diagnosis and relationships

Pratylenchus dunensis is characterised by two lip annuli with incomplete transverse incisures, a stylet (*ca* 16 μm) with robust, anteriorly indented and offset knobs, vulva at 78% of body length, lateral field with four parallel lines unequally spaced from pharynx to vulva, the inner ridge occupying 60-80% of the distance between outer incisures, lateral field partially areolated through entire body, rounded to oval spermatheca filled with round sperm, conical tail with 31-39 annules, tail tip rounded and smooth with short, conspicuous, hyaline part (*ca* 2 μm), phasmid located between inner lateral field incisures in posterior half of tail. *Pratylenchus dunensis* is further characterised by the abundant presence of males (nearly 50% of adults).

The matrix code is: A1, B2, C3, D2, E2, F4, G1, H1, I4, J1, K2.

It is very close to *P. acuticaudatus*, *P. agilis*, *P. angulatus*, *P. brachyurus*, *P. hexincisus*, *P. macrostylus*, *P. neglectus* and *P. scribneri*, from which it differs by the presence of males *vs* males absent or very rare; from *P. alleni*, *P. brzeskii*, *P. coffeae*, *P. estoniensis*, *P. loosi*, *P. neobrachyurus*, *P. penetrans*, *P. pseudocoffeae* and *P. silvaticus* it differs in body and stylet length, labial region shape, lateral field morphology, position of the vulva and tail length and shape (see the corresponding descriptions).

Although the ITS-RFLP of *P. dunensis* is identical to those of *P. brzeskii* and *P. penetrans*, the D2D3 LSU sequences revealed clear differences between *P. dunensis* and *P. brzeskii* and *P. penetrans* (de la Peña *et al.*, 2006a).

Distribution

It has been recorded only from the type locality in the fore dunes of Groote Keeten, Province Noord Holland, The Netherlands, in soil and roots of *A. arenaria*. The species was also detected in the roots of *Elymus farctus* Viv. at the same area (de la Peña *et al.*, 2006a).

20. *Pratylenchus ekrami* Bajaj & Bhatti, 1984
(Fig. 42)

Measurements

- Female holotype (after Bajaj & Bhatti, 1984): L = 0.60 mm; a = 39; b = 6.9; b′ = 4.0; c = 25; c′ = 2.9; V = 83; stylet = 12 μm.
- 22 females (after Bajaj & Bhatti, 1984): L = 0.53 (0.43-0.63) mm; a = 29 (20-39); b = 6.5 (5.0-7.8); b′ = 4.5 (3.6-5.0); c = 21 (17-30); V = 80 (79-83); stylet = 12 (11-13) μm.
- 8 males (after Bajaj & Bhatti, 1984): L = 0.40-0.54 mm; a = 29-34; b = 5.2; b′ = 4.0; c = 19-26; c′ = 1.9-2.2; stylet = 12 μm.

Description

Female

Body open C-shaped upon fixation, tapering abruptly posterior to vulva. Lateral fields with four lines, extending to tail terminus, outer lines crenate. Lateral occupying *ca* one-half mid-body diam. Labial region continuous with rest of body, anteriorly truncate, with three annuli. Labial framework sclerotised, extending posteriorly up to two

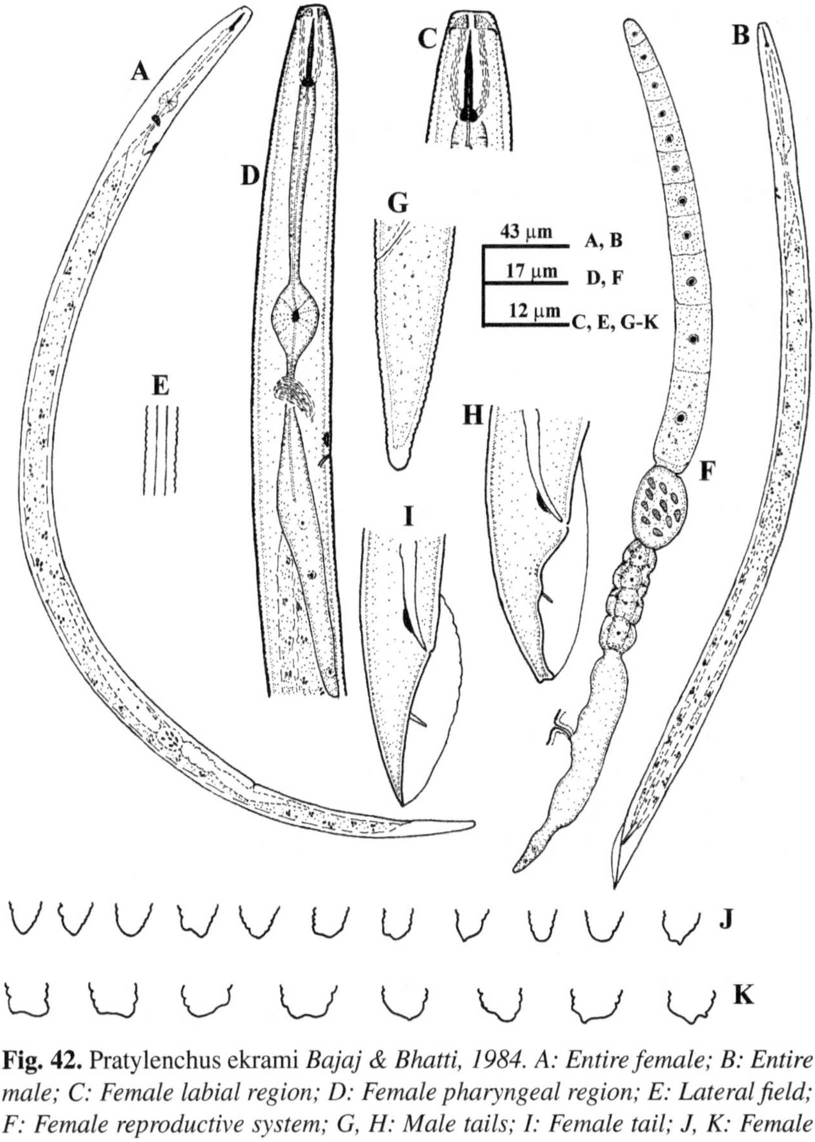

Fig. 42. Pratylenchus ekrami *Bajaj & Bhatti, 1984. A: Entire female; B: Entire male; C: Female labial region; D: Female pharyngeal region; E: Lateral field; F: Female reproductive system; G, H: Male tails; I: Female tail; J, K: Female tail tips. After Bajaj and Bhatti (1984).*

body annuli. Stylet 11-13 μm long with rounded, sloping knobs. Corpus long, almost cylindrical. Median bulb usually post equatorial, shape variable from narrow oval, large rounded to bilobed. Pharyngeal gland lobe 36-55 μm long, overlapping intestine up to 2-3 body diam.

posterior to cardia. Opening of dorsal pharyngeal gland 3-4 μm from stylet base. Nerve ring at mid-isthmus. Excretory pore 74-97 μm from anterior extremity or up to 8 μm anterior or 8 μm posterior to pharyngo-intestinal junction. Hemizonid just anterior to excretory pore. Spermatheca elongate oval, 18-37 × 10-15 μm, usually filled with sperm. Tricolumella well developed. Vulva a transverse slit located at 77-83% of body length. Post-vulval uterine sac 30-37 μm long, occupying 27-36% of vulva-anus distance, usually with 2-3 differentiated cells. Anal body diam. 9-14 μm. Tail elongated conoid to subcylindrical, 1.9-3.0 anal body diam. long with 26-40 tail annuli. Tail tip narrowly to broadly rounded, may be truncate or with a mucron-like process, or bifid. Phasmids anterior to mid-tail.

Male

Spicules 15 μm long. Gubernaculum trough-shaped, 7-8 μm long. Phasmids extending into smooth or crenate caudal alae. Tail tip pointed to bifid.

Diagnosis and relationships

Pratylenchus ekrami is characterised by: labial region continuous with rest of body, anteriorly truncate, with three annuli, stylet 11-13 μm long with rounded, sloping knobs, tail elongated conoid to subcylindrical, tail tip narrowly to broadly rounded.

The matrix code is: A2, B2, C1, D3, E3, F5, G2, H3, I2, J1, K1.

It is very close to *P. vulnus* from which it differs in shape of labial region, number of lip annuli, stylet size, shape of stylet knobs and median bulb, details of lateral lines and number of tail annuli (labial region high, lip annuli usually three or four on one side and four on the other, stylet 14-18 μm long, stylet knobs rounded, central band of lateral lines narrower than outer ones, median bulb oval, relatively narrow in *P. vulnus*).

Distribution

It has been recorded only from the type locality at horticultural farm, Dept of Horticulture, H.A.U., Hissar, India, in soil and roots of *Pyrus malus* L.

21. *Pratylenchus elamini* Zeidan & Geraert, 1991
(Fig. 43)

MEASUREMENTS

- Female holotype (after Zeidan & Geraert, 1991): L = 0.39 mm; a = 28; b = 4.8; b′ = 3.4; c = 17; c′ = 2.6; V = 75; stylet = 13.5 μm; o = 19.
- 22 females (after Zeidan & Geraert, 1991): L = 0.41 (0.34-0.51) mm; a = 28 (24-32); b = 5.8 (4.7-6.8); b′ = 3.8 (3.3-4.5); c = 18 (16-20); c′ = 2.6 (2.0-3.7); V = 75 (72-77); stylet = 14 (13-14.5) μm; o = 18 (15-21).
- 3 females (after Zeidan & Geraert, 1991): L = 0.44 (0.39-0.48) mm; a = 27 (24-30); b = 5.7 (5.4-6.0); b′ = 4.1 (3.9-4.2); c = 19 (17-21); c′ = 2.9 (2.4-3.4); V = 75-76; stylet = 13.5-14 μm; o = 19-21.

DESCRIPTION

Female

Short nematodes with slender bodies, showing various body postures upon relaxation varying from rather straight (three females), sharply curved anteriorly (two), C-shaped (nine) and almost U-shaped (one) or spiral (two). Cuticle transversely striated with annuli less than 1 μm wide. Lateral fields starting *ca* four annuli posterior to labial region with two lines, increasing to four in ten females and five in seven specimens; outer lines areolated; on tail lateral lines decreasing to three or four. Labial region with three annuli and heavily sclerotised framework with outer margins extending posteriorly between first and second body annuli. SEM studies show a dome-shaped labial region, distinctly offset from rest of body with three annuli (sometimes partly subdivided) and an undivided apical region with six labial sensilla appearing as pits close to mouth opening and two slit-like amphidial apertures in dorso-ventral direction. Stylet well developed, with conus almost equal to or slightly shorter than shaft; basal knobs with rounded and anteriorly slightly indented margins. Pharynx with elliptical median bulb and rather long glandular lobe overlapping intestine over poorly developed pharyngo-intestinal junction. Hemizonid *ca* two annuli wide and situated one or two annuli anterior to excretory pore. Female genital tract anteriorly outstretched with long post-vulval uterine branch. Oocytes arranged in

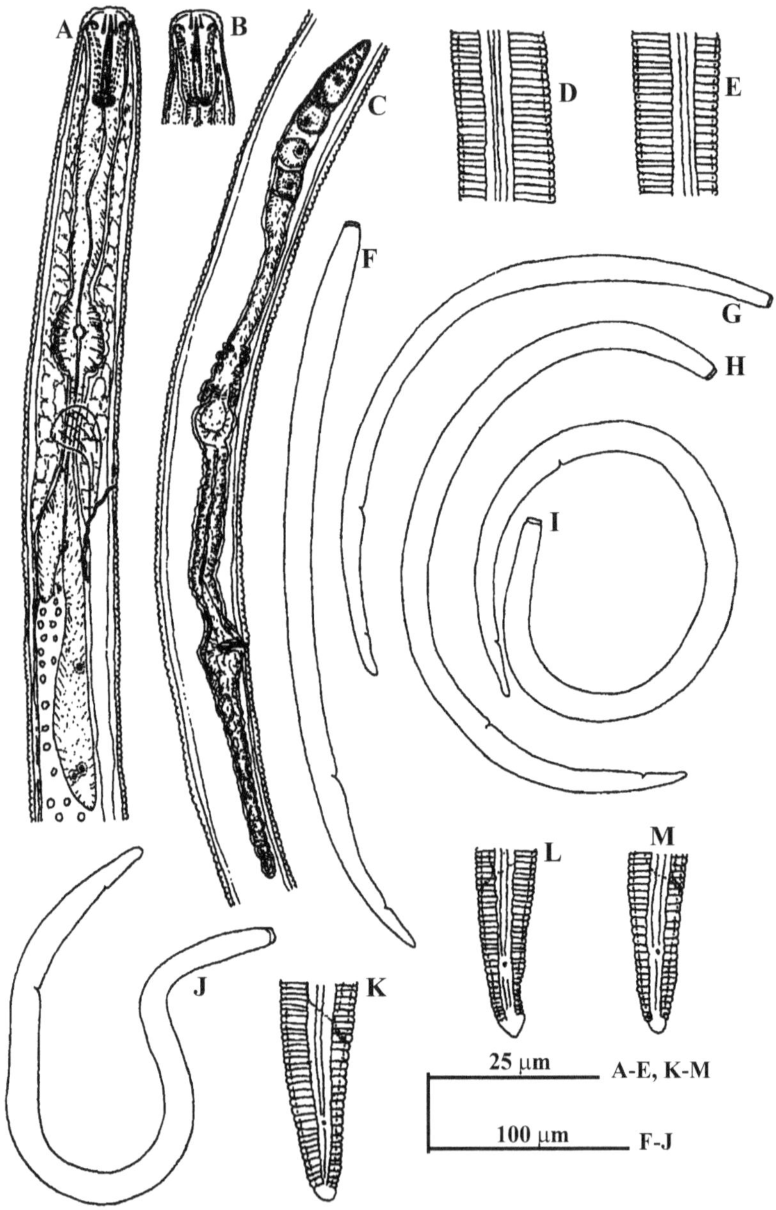

Fig. 43. Pratylenchus elamini *Zeidan & Geraert, 1991. A: Female pharyngeal region; B: Female labial region; C: Female posterior region; D, E: Lateral field at mid-body; F-J: Entire females; K-M: Female tails. After Zeidan and Geraert (1991).*

single file, spermatheca oval and devoid of sperm. Vulva a transverse slit, slightly raised from body contour in most females. Tail conical in shape with an almost conical smooth terminus. Phasmids pore-like.

Male

Not found.

DIAGNOSIS AND RELATIONSHIPS

Pratylenchus elamini is characterised by: short body with wide variation of body shape upon relaxation, variable anterior body shape, three labial annuli, short stylet (13.0-14.5 μm), anterior vulva position (72-77%), a long post-vulval uterine sac (1.5-2.9 times corresponding vulval body diam.), smooth conical tail terminus, males absent and with spermatheca empty.

The matrix code is: A2, B1, C2, D3, E2, F5, G3, H1, I2, J1-2, K2.

It is close to *P. delattrei*, *P. kralli*, *P. microstylus*, *P. pseudopratensis* and *P. sudanensis* from which can be differentiated by stylet length, post-vulval uterine sac, tail shape and tail tip (see the corresponding descriptions).

DISTRIBUTION

It has been recorded only from fallow soil at the type locality at Kassala City, Eastern Sudan from the rhizosphere of *Psidium guajava* L. in an orchard.

22. *Pratylenchus estoniensis* Ryss, 1982
(Fig. 44)

MEASUREMENTS

- Female holotype (after Ryss, 1982): L = 0.44 mm; a = 20; b = 5.6; b′ = 4.1; c = 17; c′ = 1.0; V = 82; stylet = 16 μm.
- 11 females (after Ryss, 1982): L = 0.35-0.50 mm; a = 17-38; b = 4.6-7.6; b′ = 3.5-5.5; c = 16-22; c′ = 0.8-1.1; V = 79-86; stylet = 15.5-17 μm.

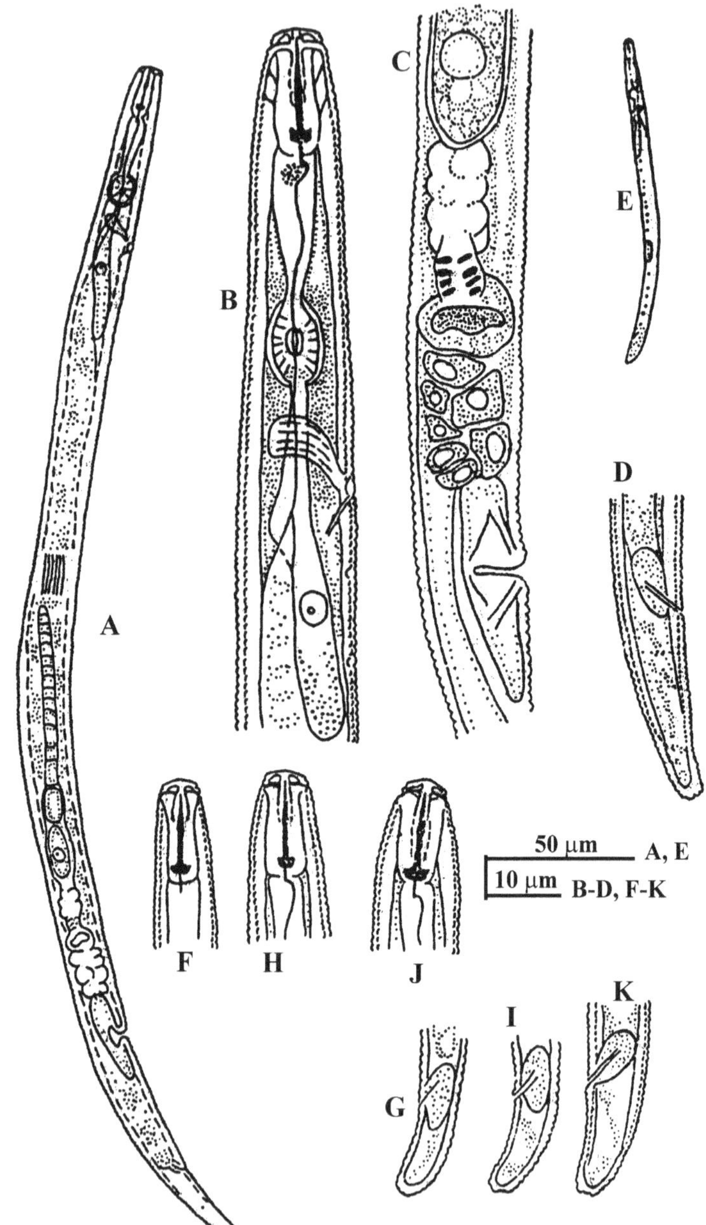

Fig. 44. Pratylenchus estoniensis *Ryss, 1982. A, E: Entire females; B: Female pharyngeal region; C: Female posterior region; D, G, I, K: Female tails; F, H, J: Female labial regions. After Ryss (1982).*

Description

Female

Body short and thick. Labial region with two annuli. Stylet knobs slightly pointed and directed forward. Anterior cephalid located at four annuli, posterior cephalid, at eight annuli posterior to base of labial region. Distance from posterior edge of metacorpal bulb to end of pharyngeal gland lobe usually equal to distance from posterior edge of bulb to oral opening. Lateral field with six lines, two central ones sometimes represented by slanting lines. Vulva located at 77-83% of body length; post-vulval uterine sac lacking rudimentary parts of posterior genital branch in the form of oogonia or somatic nuclei; its length not exceeding 1.5 body diam. at vulval level. Spermatheca round, with an internal cavity, not containing sperm. Tail conoid-rounded, tail tip serrated, 1.5-2 times as long as anal diam.; with 22-30 annuli on ventral surface.

Male

Not found.

Diagnosis and relationships

Pratylenchus estoniensis is characterised by: labial region with two annuli, lateral fields with six lines, stylet knobs slightly pointed and directed forward, spermatheca round and lacking sperm and tail conoid-rounded with serrated tip.

The matrix code is: A1, B1, C3, D2, E3, F1, G3, H2, I2, J3, K1.

It can be distinguished from the closely related species *P. crenatus* and *P. neglectus* by the number of annuli in the labial region, shape of spermatheca and shape of tail tip (see the corresponding description).

Distribution

It has only been recorded from the type locality at Vyiste in the Parnu region, Estonia, from roots of *Rubus saxatilis* L., *Phleum pratense* L. and *Calamagrostis epigeios* (L.) Roth.

23. *Pratylenchus fallax* Seinhorst, 1968
(Fig. 45)

MEASUREMENTS

- Female holotype (after Seinhorst, 1968): L = 0.52 mm; a = 29; b = 5.2-6.7; c = 18; V = 80; stylet = 17 μm.
- 10 females (after Seinhorst, 1968): L = 0.42-0.56 mm; a = 23-33; b = 5.2-6.7; c = 18-24; V = 77-81; stylet = 16-17 μm.
- 10 males (after Seinhorst, 1968): L = 0.40-0.50 mm; a = 26-33; b = 5.0-6.2; c = 16-25; stylet = 15 μm; spicules = 14-16 μm.
- 10 females (after Ryss, 1988): L = 0.49 (0.44-0.53) mm; a = 29 (24-32); b = 5.9 (5.4-6.3); b′ = 4.0 (3.5-4.5); c = 22 (17-25); c′ = 2.5 (2.0-3.0); V = 79 (77-82); stylet = 16 (15-17) μm.
- 10 males (after Ryss, 1988): L = 0.44 (0.38-0.46) mm; a = 27 (23-29); b = 5.6 (5.0-6.0); b′ = 4.0 (3.7-4.6); c = 19 (15-23); c′ = 2.4 (1.9-3.1); stylet = 15 (15-16) μm; spicules = 15 (14-16) μm; gubernaculum = 4 μm.

DESCRIPTION

Female

Body straight or slightly curved in specimens killed by heat. Labial region with three often rather flat and obscure annuli. Posterior edges of labial framework projecting two body annuli into body. Lateral fields broad, with four lines up to a short distance anterior to phasmids and three from there to tail tip. Some additional lines usually present running obliquely between inner lines of lateral fields. Stylet knobs anteriorly flattened or pointing forward. Excretory pore at, or posterior to, level of nerve ring. Ovary outstretched, oocytes in single file except for short region near anterior end-of ovary. Spermatheca round, sometimes empty and then narrower and longer than when filled with sperm. Distance between spermatheca and vulva 42-71% of vulva-anus distance. Post-vulval uterine sac *ca* one-fourth to one-third of the vulva-anus distance; its posterior part often consisting of two or three rudimentary elements. Tail conical, with 16-26 rather narrow annuli. Tip rounded or with slightly irregular contour, distinctly crenate (generally) to almost smooth (rarely). However, this is only clearly visible in lateral views of tail. In dorsoventral positions (more common in mounted specimens) the

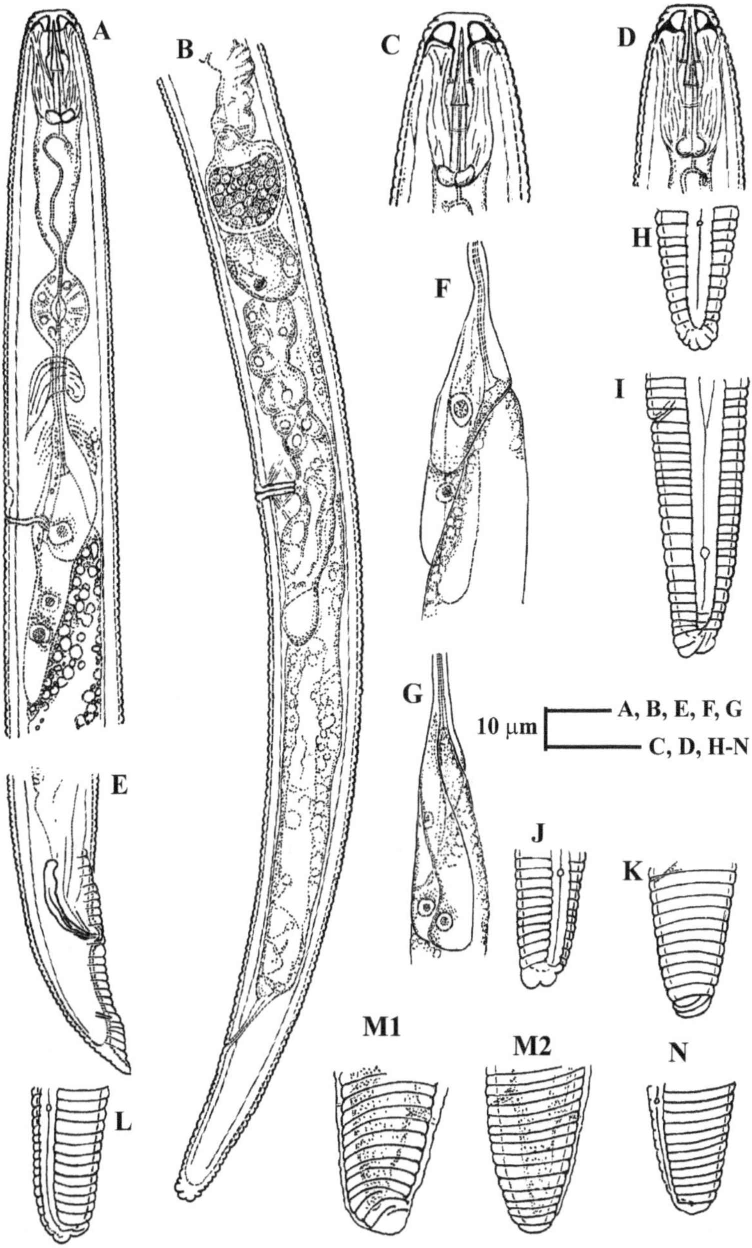

Fig. 45. Pratylenchus fallax *Seinhorst, 1968. A: Female pharyngeal region; B: Female posterior region; C: Female labial region; D: Male labial region; E: Male tail; F, G: Pharyngeal glands; H-N: Female tails. After Seinhorst (1968).*

annulation is only detected by careful focusing on successive lines. Phasmids at 9-13 annulus from tip.

Male

About one male found per five females. Testis with double row of spermatocytes. Spicules slender, with well marked manubria and ventrally arcuate shaft. Gubernaculum *ca* 4 μm long; bursal margins distinctly crenate.

DIAGNOSIS AND RELATIONSHIPS

Pratylenchus fallax is characterised by: labial region with three often rather flat and obscure annuli, tail conical with rounded distinctly crenate tip, spermatheca round and males not rare (1 : 5 females).

The matrix code is: A2, B2, C3, D2, E2, F3, G3, H2, I2, J1, K1.

It is close to *P. convallariae* from which it differs by its shorter body length, narrower, more numerous tail annuli and lower male to female ratio. It is also close to *P. penetrans* from which it differs by fewer males, a generally longer pharyngeal overlap and occasional populations with crenate tails. It is easily distinguished from *P. pratensis* by the round spermatheca.

Pratylenchus cerealis Haque, 1966 was described from flattened specimens with anterior and abnormal outline of posterior part of neck and anterior position of excretory pore and was considered to be a species of uncertain status by Loof (1978, 1991). Frederick and Tarjan (1989) examined glycerin-mounted specimens of *P. cerealis* and did not find significant morphological or biometric differences between this species and *P. fallax*, except in the much lower a ratio for *P. cerealis*. As pointed out by Frederick and Tarjan (1989) the principle of priority demands that the oldest named species becomes the senior synonym; even though the description by Seinhorst is more precise than that of Haque (1966); however, several authors, *e.g.*, Siddiqi (2000) and Ryss (2002a), have maintained *P. cerealis* as an uncertain species, an opinion with which we agree.

Perry *et al.* (1980) demonstrated that *P. penetrans* and *P. fallax* could be separated on the basis of the biological species concept as only a small proportion of interspecific crosses were successful. On this basis, the specific status of *P. fallax* and *P. penetrans* is endorsed.

DISTRIBUTION

Pratylenchus fallax was described from an apple orchard with grass cover on loam soil, at Doornenburg, The Netherlands. It is widespread in sandy or sandy-peat soils in The Netherlands (Seinhorst, 1968, 1977b), occurring especially in meadows and orchards on loam and sandy loam soils and in places with a more or less natural vegetation, and is associated with grass and ornamentals in many European countries (Webb, 1990). It has been also reported on other crops in Europe: Belgium on rose (Coolen & Hendrickx, 1972a); Croatia on tobacco (Ostrec & Grubisic, 2003); England on raspberry (Cotten & Roberts, 1981); England and Wales on cereals (Corbett, 1970); Estonia in the roots and rhizosphere of mandarin, orange and lemon trees (Ryss, 1992); Germany on various crops (Decker & Dowe, 1974); Poland (Brzeski & Szczygiel, 1977); Russia on citrus and other crops (Ryss, 1988; Tskitishvili, 1983); Spain in natural environments (Gómez Barcina *et al.*, 1991) and olive nurseries (Nico *et al.*, 2002). It has also been recorded in the United States in Iowa on strawberries (Norton, 1984) and North Dakota on various crops (Donald & Hosford, 1980), as well as the Canadian provinces of Quebec and Ontario on turf grass (Yu *et al.*, 1997, 1998), various crops (Townshend *et al.*, 1978a). It is also recorded from various crops in India (Khan & Singh, 1974) and Japan (Gotoh, 1974).

24. *Pratylenchus flakkensis* Seinhorst, 1968
(Fig. 46)

MEASUREMENTS

- Female holotype (after Seinhorst, 1968): L = 0.48 mm; a = 25; b = 5.5; c = 19; V = 76; stylet = 17 μm.
- 10 females (after Seinhorst, 1968): L = 0.42-0.57 mm; a = 20-27; b = 5.2-7.1; c = 12-18; V = 73-77; stylet = 17 μm.
- 7 males (after Seinhorst, 1968): L = 0.42-0.49 mm; a = 27-33; b = 5.1-6.5; c = 18-21; stylet = 16 μm.
- 10 females (after Ryss, 1988): L = 0.51 (0.48-0.60) mm; a = 25 (18-28); b = 6.1 (5.0-6.8); c = 17 (15-20); c′ = 2.5 (2.0-3.0); V = 76 (73-80); stylet = 16 (15.5-17) μm.
- 10 males (after Ryss, 1988): L = 0.45 (0.44-0.51) mm; a = 31 (25-35); b = 5.5 (5.3-6.2); c = 21 (17-24); c′ = 2.5 (2.0-2.6); stylet = 15 (15-16) μm; spicules = 15 μm; gubernaculum = 4-6 μm.

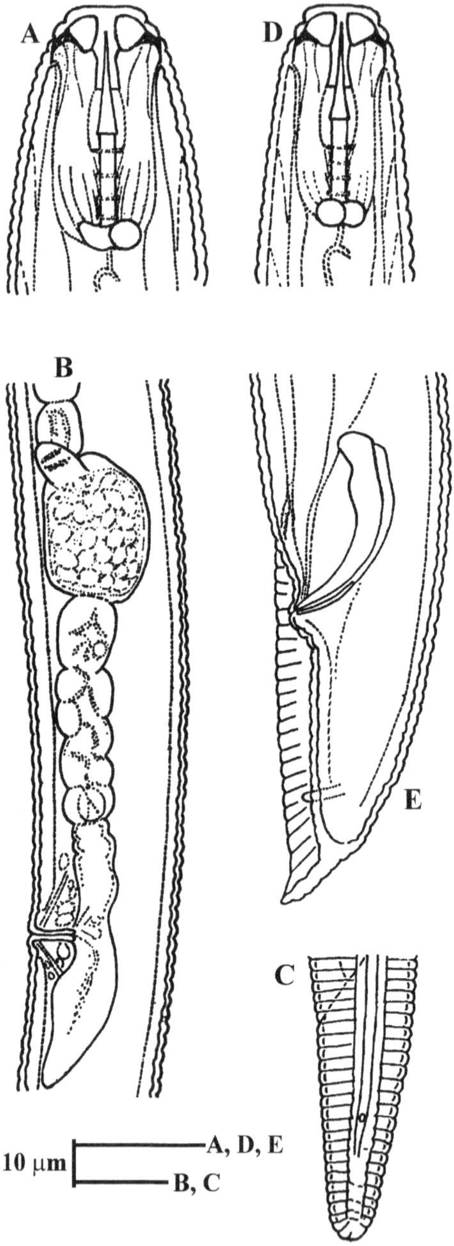

Fig. 46. Pratylenchus flakkensis *Seinhorst, 1968. A: Female labial region; B: Female posterior region; C: Female tail; D: Male labial region; E: Male tail. After Seinhorst (1968).*

- 8 females (after Van den Berg & Quénéhervé, 2000): L = 0.53 (0.51-0.55) mm; a = 29 (27-32); b = 3.5; c = 17 (16-18); c′ = 2.5-3.0; V = 77 (75-79); stylet = 15 (14.5-16) μm.
- 14 females (after Zarina & Maqbool, 1998): L = 0.50 (0.42-0.57) mm; a = 29 (27-33); b = 4.8 (3.5-6.9); c = 20.4 (17.6-34.7); c′ = 2.5 (1.4-2.6); V = 81 (79-83); stylet = 15 (13-16) μm.
- 15 males (after Zarina & Maqbool, 1998): L = 0.45 (0.37-0.53) mm; a = 32 (27-37); b = 5.5 (3.5-7.5); c = 20.8 (17.6-24.0); c′ = 1.8 (1.4-2.7); T = 38 (30-47); stylet = 14 (13-16) μm; spicules = 16 (14-18) μm; gubernaculum = 5 (3.5-6.5) μm.

DESCRIPTION

Female

Labial region with two annuli. Posterior edge of labial framework strongly sclerotised, extending posteriorly into body for *ca* two annuli. Lateral fields with four lines, band between inner lines plain. Stylet knobs with forward pointing anterior margins. Excretory pore at, or posterior to, level of nerve ring. Lateral fields areolated on posterior third of tail. Ovary outstretched, oocytes in single file. Spermatheca round to angular, located at 41-66% of vulva-anus distance anterior to vulva. Post-vulval uterine sac extending for 25-30% of distance between vulva and anus. Tail conical with 18-24 annuli and (sometimes faint) annulation around tip. Phasmids at 12-18 annuli from tip.

Male

First annulus on labial region distinctly wider than second. Testis outstretched, spermatocytes in double row. Bursa with coarsely crenate margin. Phasmids located at 12-18 annuli from tip or at one-third of tail length.

DIAGNOSIS AND RELATIONSHIPS

Pratylenchus flakkensis is characterised by: labial region with two annuli, stylet knobs with forward pointing anterior margins, annulated tail tip and presence of males.

The matrix code is: A1, B2, C3, D2, E2, F2, G3, H2, I1, J1, K1.

It can be distinguished from closely related species (*P. alleni*, *P. loosi* and *P. zeae*) by stylet length, the length of the post-vulval uterine sac, and number of tail annuli (see the corresponding descriptions).

DISTRIBUTION

It has been recorded from the type locality in a pasture at Middelharnis, The Netherlands. It has been also found in Guadeloupe from a humid forest (Van den Berg & Quénéhervé, 2000); in Pakistan on rose and fern (*Cheilantes* sp.) (Zarina & Maqbool, 1998); in Estonia (Ryss, 1992); in Peru on sweet potato (Anguiz & Canto-Saenz, 1991); in Russia on *Poa pratensis* (Ryss, 1986, 1988) and citrus crops (Tskitishvili, 1983); in Iowa, USA, on *Helianthus*, *Plantago* and corn (Williams, 1982); in the Great Lakes basin of North America (Townshend *et al.*, 1978a); in South Africa (Van den Berg, 1976); and in arable land at Ouddorp and in soil at Beckenham, England (Seinhorst, 1968).

25. *Pratylenchus gibbicaudatus* Minagawa, 1982
(Fig. 47)

MEASUREMENTS

- Female holotype (after Minagawa, 1982): L = 0.51 mm; a = 17.9; b = 6.6; c = 16.0; c′ = 2.1; V = 72; stylet = 15 μm.
- 30 females (after Minagawa, 1982): L = 0.48 (0.41-0.53) mm; a = 20 (16.0-29.7); b = 6.5 (5.3-9.5); c = 15.5 (12.8-19.0); c′ = 2.4 (1.6-3.9); V = 73 (70-77); stylet = 15.5 (14-16.5) μm.
- 1 male (after Minagawa, 1982): L = 0.47 mm; a = 32.2; b = 7.9; c = 18.6; c′ = 2.0; stylet = 15 μm; spicules = 18 μm; gubernaculum = 5 μm.

DESCRIPTION

Female

Body stout, straight to open C-shaped after fixation. Body diam. slightly increasing from near median pharyngeal bulb to vulval region, decreasing regularly from vulva to near tail terminus. Body annuli faint, 1.0 $\pm$ 0.1 (0.7-1.2) μm apart at mid-body. Labial region continuous with body contour, flattened in front, with two annuli, 2.0-2.5 (2.4 $\pm$ 0.2) μm high, 6.5-8.5 μm wide. Labial framework well developed, strongly sclerotised, arch-shaped, margin extending by two annuli into body. Lateral fields *ca* 6.5-10.0 μm wide with four lines, inner two lines fused around phasmid. Stylet stout; knobs round or slightly flattened at anterior

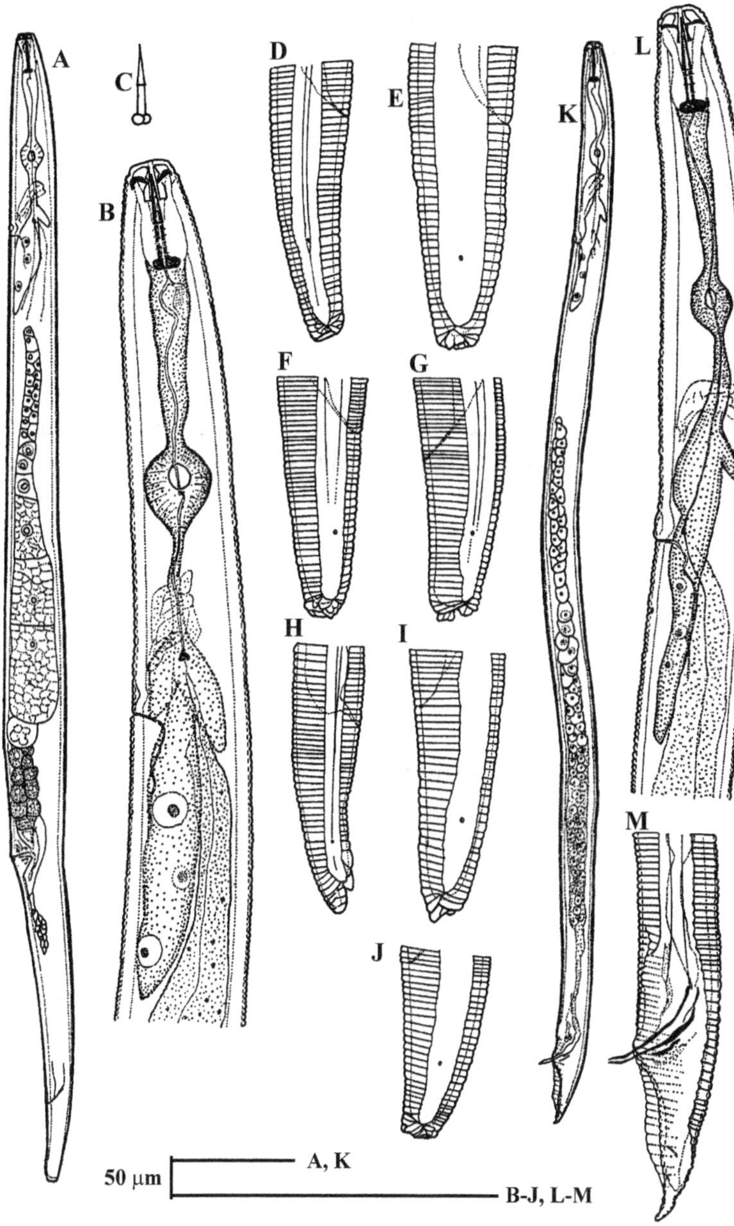

Fig. 47. Pratylenchus gibbicaudatus *Minagawa, 1982. A: Entire female; B: Female pharyngeal region; C: Stylet; D-J: Female tails; K: Entire male; L: Male pharyngeal region; M: Male tail. After Minagawa (1982).*

surface, 2.0-2.5 (2.1±0.2) μm high, 3.5-4.5 (4.0±0.3) μm wide. Dorsal pharyngeal gland orifice at 2.5-4.0 (3.2±0.3) μm posterior to stylet base. Metacorpus round, occupying nearly entire body diam., 35-62 (52±5.1) μm from anterior end. Pharynx 62-99 (77 ± 9.4) μm long. Glandular pharyngeal basal lobe terminating 107-218 (132 ± 25.6) μm from lips. Excretory pore located at 71-105 (84±6.8) μm, 15.6-20.3 (17.5±1.1)% of body from anterior end. Hemizonid located immediately anterior to excretory pore, two-annuli long; hemizonion one annulus long, *ca* 15 μm posterior to hemizonid in some specimens. Reproductive system well developed. Spermatheca round without sperm. Post-vulval uterine sac 21.5-47 (34 ± 8.4) μm long, 0.7-2.6 times (1.5 ± 0.5) body diam. at vulva. Distance between vulva and anus 2.8-4.5 times (3.3 ± 0.5) tail length. Tail 25.3-36.6 (31.2±2.9) μm long with 24-39 (30±2.5) annuli. Phasmids in posterior half of tail, 7.5-16 (12.5 ± 2.0) μm from tail tip. Tail terminus annulated, truncate, or broadly rounded with shallow notch at the tip and rarely bifurcate.

Male

General appearance similar to female but body more slender and pharyngeal gland less developed. Spicules arch-shaped; gubernaculum small.

DIAGNOSIS AND RELATIONSHIPS

Pratylenchus gibbicaudatus is characterised by: two lip annuli, crenate tail terminus and presence of males.

The matrix code is: A1, B2, C2, D2, E1, F5, G2, H2, I3, J1, K1.

It can be distinguished from the closely related species (*P. flakkensis*) by flattened anterior surface of stylet knobs, more posteriorly located phasmid and greater number of tail annuli (24-39 *vs* 18-24) (see the corresponding description).

DISTRIBUTION

It has been recorded from the type locality from rhizosphere of *Scirpus wichurai* Böcklr. f. *concolor* (Maxim.) T. Koyama grown in a swamp adjacent to pastures in the Somma area of Mt. Aso, Kumamoto Prefecture, Japan. This species parasitises banana and causes a reddish-black cortical necrosis typical of lesion nematodes in American Samoa (Brooks, 2004).

26. *Pratylenchus goodeyi* Sher & Allen, 1953
(Figs 48, 49)

MEASUREMENTS

- Female holotype (after Sher & Allen, 1953): L = 0.59 mm; a = 27; b = 5.8; c = 18; V = 74.
- Females (after Sher & Allen, 1953): L = 0.64-0.68 mm; a = 27-37; b = 5.5-6.1; c = 16-18; V = 73-75; stylet = 17 μm.
- Males (after Sher & Allen, 1953): L = 0.55-0.57 mm; a = 26; b = 5.4-5.8; c = 17-18; T = 54; stylet = 16 μm.
- 10 females (after Ryss, 1988): L = 0.40-0.56 mm; a = 31 (24-34); b = 5.8 (5.5-6.5); b′ = 4.4 (3.8-4.7); c = 16 (14-18); c′ = 2.8 (2.5-3.0); V = 74 (73-75); stylet = 15 (14-15) μm.
- 10 males (after Ryss, 1988): L = 0.46 (0.40-0.51) mm; a = 35 (25-37); b = 6.1 (5.8-6.6); b′ = 4.6 (3.9-4.9); c = 23 (20-25); stylet = 14 μm; spicules = 15-16 μm; gubernaculum = 5-6 μm.
- 13 females (after Sakwe & Geraert, 1994): L = 0.52 (0.43-0.61) mm; a = 25.4 (21.7-28.8); b = 6.4 (5.4-7.7); b′ = 4.1 (3.7-4.8); c = 16.8 (14.7-18.5); c′ = 2.4 (1.7-2.8); V = 75 (72-78); stylet = 15.5 (14.5-16.5) μm.
- 4 males (after Sakwe & Geraert, 1994): L = 0.54 (0.47-0.65) mm; a = 26.5 (22.8-29.2); b = 6.6 (5.9-7.1); b′ = 4.6 (3.8-5.1); c = 17.8 (15.2-19.4); c′ = 2.2 (1.9-2.4); T = 55 (45-64); stylet = 14.5 (13.0-15.5) μm.
- 15 females (after Troccoli *et al.*, 1996b): L = 0.56 (0.49-0.71) mm; a = 23.9 (19.7-29.5); b = 6.3 (4.1-8.1); b′ = 4.1 (2.9-5.3); c = 16.5 (13.8-20.0); c′ = 2.5 (1.9-3.2); V = 75 (73-79); stylet = 16.0 (15.5-17.5) μm.
- 10 males (after Troccoli *et al.*, 1996b): L = 0.49 (0.38-0.53) mm; a = 27.8 (24.0-31.7); b = 6.4 (5.8-7.3); b′ = 4.2 (3.9-4.6); c = 15.7 (12.5-17.2); c′ = 2.7 (2.4-3.3); T = 45 (35-50); stylet = 14.5 (13.5-15.5) μm.

DESCRIPTION

Female

Body slender, heat-relaxed habitus almost straight but slightly ventrally arcuate posteriorly. Lateral fields with four inconspicuous lines, two outer bands partially areolated, clearly observed with SEM (Corbett

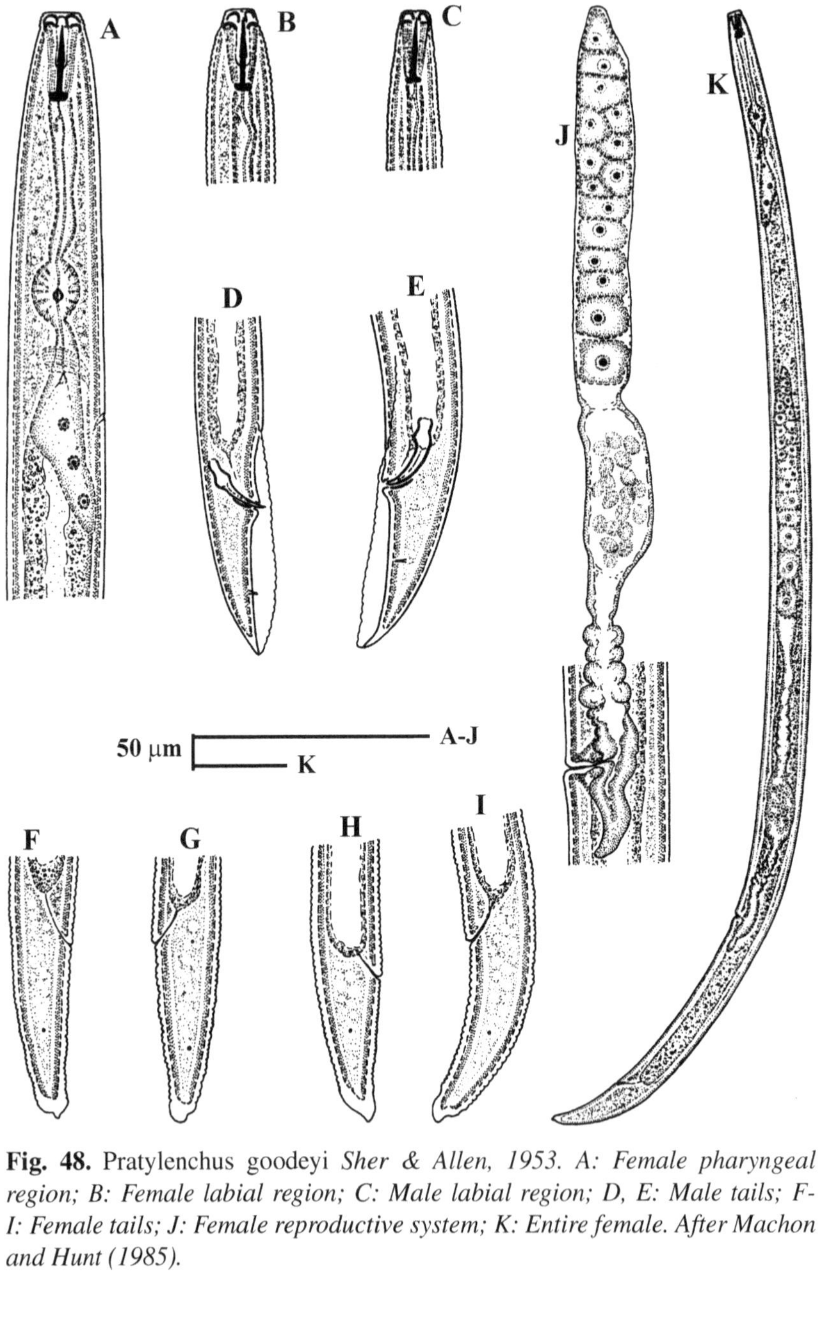

Fig. 48. Pratylenchus goodeyi *Sher & Allen, 1953. A: Female pharyngeal region; B: Female labial region; C: Male labial region; D, E: Male tails; F-I: Female tails; J: Female reproductive system; K: Entire female. After Machon and Hunt (1985).*

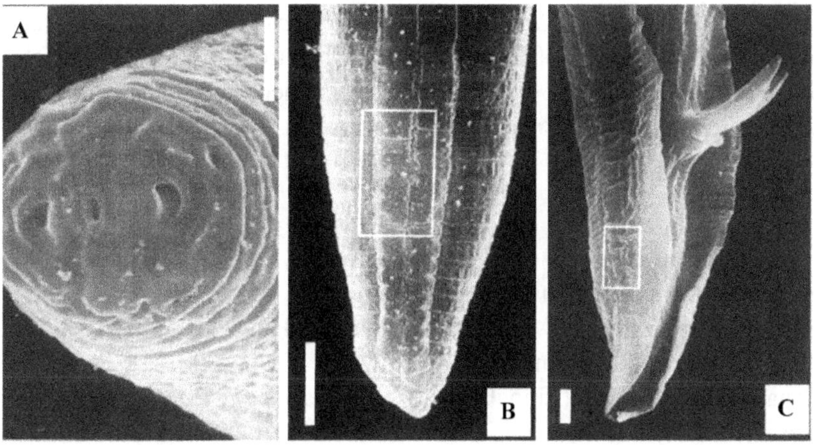

Fig. 49. *SEM micrographs of* Pratylenchus goodeyi *Sher & Allen, 1953. A: Female* en face *view; B: Female tail; C: Male tail. Rectangular boxes indicate phasmid position. (Scale bars: A = 2 μm; B = 5 μm; C = 3 μm.) After Hernández* et al. *(2000).*

& Clark, 1983). Labial region with four annuli. *En face* view characterised by fusion of subdorsal, subventral and lateral lips with oral disc (Hernández *et al.*, 2000). Stylet 16-18 μm long with pronounced, anteriorly flattened, knobs. Ovary with a single row of oocytes, except for double row near anterior end. Post-vulval uterine sac *ca* vulval body diam. long. Spermatheca large, sub-rectangular, filled with sperm. Tail conoid, ventrally concave with dorsal contour sinuate just prior to tail tip. Tail usually with 22-24 ventral annuli (range: 19-27). Phasmid usually conspicuous, 0-14 annuli (Corbett & Clark, 1983) from tail tip.

Male

Common. Similar in general form to female. Single outstretched testis, spermatocytes in a double row. Spicules slender, arcuate. Gubernaculum simple. Bursa enveloping tail tip.

DIAGNOSIS AND RELATIONSHIPS

Pratylenchus goodeyi is characterised by: labial region with four annuli, lateral fields with four inconspicuous lines, the two outer bands partially areolated, spermatheca large, sub-rectangular in shape and tail conoid, ventrally concave.

The matrix code is: A3, B2, C3, D4, E1, F1, G3, H3, I3, J1, K2.

It is closely related to *P. morettoi* and *P. typicus* from which it differs by stylet length, shape of the stylet knobs, position of the vulva, length of the post-vulval uterine sac, shape of spermatheca and shape of tail and tail terminus (see the corresponding descriptions).

DISTRIBUTION

It has been recorded from the type locality at Kew Gardens, UK, on roots of banana. It has also been recorded in numerous localities, especially in banana growing areas (Machon & Hunt, 1985): in Australia (Stanton *et al.*, 2001); in Uganda (Talwana *et al.*, 2000, 2003); in the Canary Islands, Spain (Pinochet, 1998; Iglesias *et al.*, 1999); in Madeira Island, Portugal (Nico *et al.*, 1999); in Cameroon on several crops (Sakwe & Geraert, 1994; Bridge *et al.*, 1995; Fogain, 2001); in Ethiopia (Peregrine & Bridge, 1992); Ivory Coast (Sarah, 1989); on banana in Kenya (Waudo *et al.*, 1990, 1998) and Uganda (Speijer & De Waele, 2001) and several crops and weeds (Prasad *et al.*, 1995); in Tanzania (Mbwana *et al.*, 1995); in Greece from soil around boxwood (Koliopanos & Kalyviotis-Gazelas, 1969) and Crete (Vovlas *et al.*, 1994); in Ethiopia (Peregrine & Bridge, 1992); on grapevine in Australia (McLeod, 1978); and in Russia on *Musa sapientum* L. (Ryss, 1988) and strawberry (Vladimirova, 1972). In several studies *P. goodeyi* was found in a number of African countries associated with higher and cooler altitudes and therefore is considered an indigenous African nematode with an 'afro-montane' distribution (Price & Bridge 1995).

27. *Pratylenchus hexincisus* Taylor & Jenkins, 1957
(Figs 50, 51)

MEASUREMENTS

- 95 females (after Taylor & Jenkins, 1957): L = 0.44 (0.34-0.54) mm; a = 26 (19-31); b = 5.0 (4.2-5.8); c = 19 (17-24); V = 78 (75-82); stylet = 15 (14.5-15.5) μm.
- 35 females (after Loof, 1964): L = 0.43 (0.37-0.51) mm; a = 22.6 (18.2-28.8); b = 7.2 (5.9-8.4); c = 18.6 (16.1-22.7); V = 79 (74-81); stylet = 15 (14-16) μm.
- 2 males (after Loof, 1964): L = 0.38-0.42 mm; a = 29-33; b = 4.2-4.5; c = 18-19; T = 41-44; stylet = 14 μm; spicules = 16-17 μm.

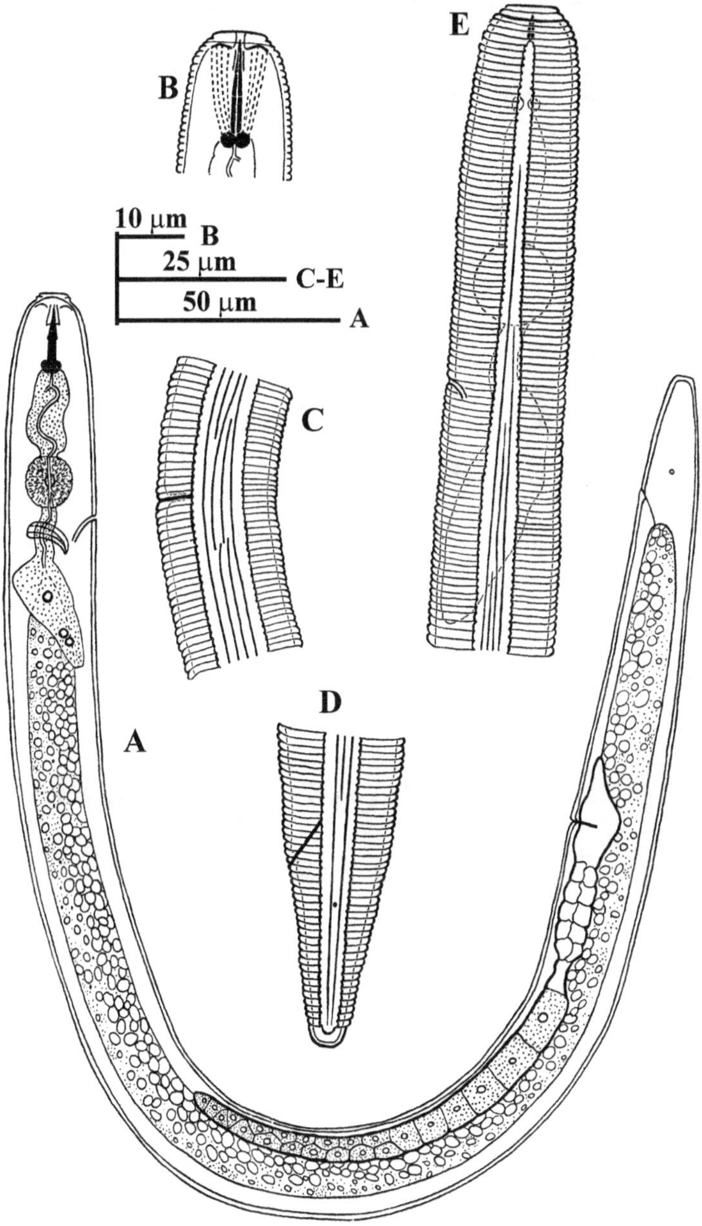

Fig. 50. Pratylenchus hexincisus *Taylor & Jenkins, 1957. A: Entire female; B: Female labial region; C: Lateral field at mid-body; D: Female tail; E: Female pharyngeal region. After Taylor and Jenkins (1957).*

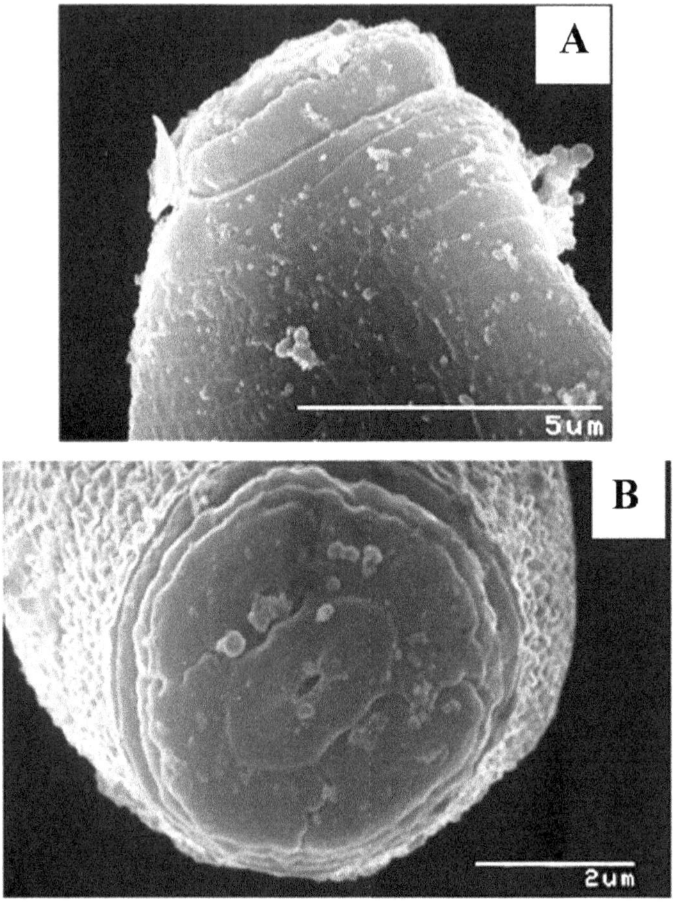

Fig. 51. *SEM micrographs of* Pratylenchus hexincisus *Taylor & Jenkins, 1957. A: Female labial region; B: Female* en face *view. After Inserra* et al. *(2007).*

- 11 females (after Inserra *et al.*, 2007): L = 0.49 (0.44-0.53) mm; a = 21.8 (19.8-24.2); b = 6.0 (5.2-6.8); c = 17.7 (16.3-19.2); c′ = 2.3 (2.2-2.5); V = 78 (76-79); stylet = 15.1 (14.5-15.5) μm.

DESCRIPTION

Female

Body short and somewhat stout, tapering towards both ends. Cuticle marked with transverse striations *ca* 1 μm apart, occurring slightly further apart on tail of some specimens. Labial region low, flattened, with two annuli. Lateral margins of labial framework extending posteriorly *ca* 1 μm into body. SEM observations by Inserra *et al.* (2007) indicated that *en face* view is divided showing dumbbell-shaped submedian lip sectors fused with an oval oral disc (Fig. 51B) and a lateral view of labial region showed second lip annulus to be thicker than first (Fig. 51A). Lateral fields marked by six lines except toward extremities where they are reduced in number, occupying *ca* one-third of body diam. in vulval region. Near mid-body these lines may be broken into irregular lines. Outer lines crenate, inner lines somewhat irregular, particularly in vulval region where breaks and oblique lines occur frequently. Only four lines occur on tail. Stylet 15 μm long with somewhat rounded knobs. Stylet guiding apparatus 3.2-3.6 μm long. Pharyngeal gland orifice 2.2 (1.7-2.5) μm from stylet base. Valve of median bulb anterior to centre. Nerve ring surrounding base of isthmus. Basal bulb with overlap *ca* as long as body diam. Hemizonid just anterior to excretory pore. Excretory pore located in isthmus region; nerve ring surrounding isthmus. Pharyngeal glands ventrally overlapping intestine, which contains granules that are larger and denser towards ventral surface. Dorsal pharyngeal gland nucleus larger than subventral nuclei. Intestine terminating in an oblique rectum opening through cuticle by means of a conspicuous anus. Cuticularised vagina extending transversely *ca* one-half body diam. and opening by means of a prominent vulva. Anterior ovary outstretched with oocytes arranged in single row except for short region of multiplication. Post-vulval uterine sac *ca* as long as body diam. No spermatheca observed. Tail slightly convex-conoid to broad, with *ca* 20 obscure annuli. Phasmids in anterior one-quarter to two-fifths of tail. Tail tip rounded and smooth.

Male

Not found.

DIAGNOSIS AND RELATIONSHIPS

Pratylenchus hexincisus is characterised by: labial region low, flattened, bearing two annuli, lateral fields with six lines except toward extremities where they are reduced in number, somewhat rounded stylet knobs and broad rounded terminus of the convex-conoid tail.

The matrix code is: A1, B2, C2, D1, E2, F2, G3, H1, I1, J3, K1.

It is close to *P. brachyurus*, *P. brzeskii* and *P. coffee* from which it can be differentiated by body and stylet length, position of excretory pore, position of vulva, post-vulval uterine sac length, absence of males (see the corresponding descriptions).

DISTRIBUTION

It has been recorded at the type locality in Kent County, Maryland, USA, from roots of corn. It has been recorded in several states of USA: Iowa on corn (Williams, 1982); Missouri on soybean (Niblack, 1992); California on peach (Duncan *et al.*, 1992); South Dakota on corn (Smolik & Evenson, 1987); and Indiana on corn (Bergeson & Ferris, 1986). It has also been recorded in Hungary on apricot (Dabaj & Jenser, 1991); in China (Yin, 1991); in Slovenia on corn (Urek *et al.*, 2003); in several localities at northern and central Spain (Arias & Romero, 1979); in Venezuela on *Pennisetum purpureum* Schumach. and *Citrus* sp. (Loof, 1964); and in Argentina on soybean (Doucet, 1988).

28. *Pratylenchus hippeastri* Inserra, Troccoli, Gozel, Bernard, Dunn & Duncan, 2007

(Figs 52, 53)

MEASUREMENTS

- Female holotype (after Inserra *et al.*, 2007): L = 0.56 mm; a = 26.7; b = 5.7; c = 15.6; c′ = 2.8; V = 78; stylet = 15.5 μm.
- 22 females (after Inserra *et al.*, 2007): L = 0.59 (0.55-0.63) mm; a = 25.5 (23.2-27.9); b = 6.5 (5.7-7.1); c = 16.1 (14.6-18.7); c′ = 2.6 (2.2-2.9); V = 77 (75-78); stylet = 15.5 (15-16) μm.

DESCRIPTION

Female

Body slender with mean a ratio > 25. Labial region flat, consisting of two annuli, 2.0 μm high and 7.5 μm wide on average; first annulus slightly narrower and lower than second. Facial pattern plain and smooth with all labial sectors fused together and with oral disc. Minute and incomplete initiation of a third lip annulus visible on one side of labial region in some specimens. Stylet robust with ellipsoidal knobs. Hemizonid just anterior to excretory pore and two body annuli long. Pharyngeal glands overlapping intestine ventrally. Dorsal gland

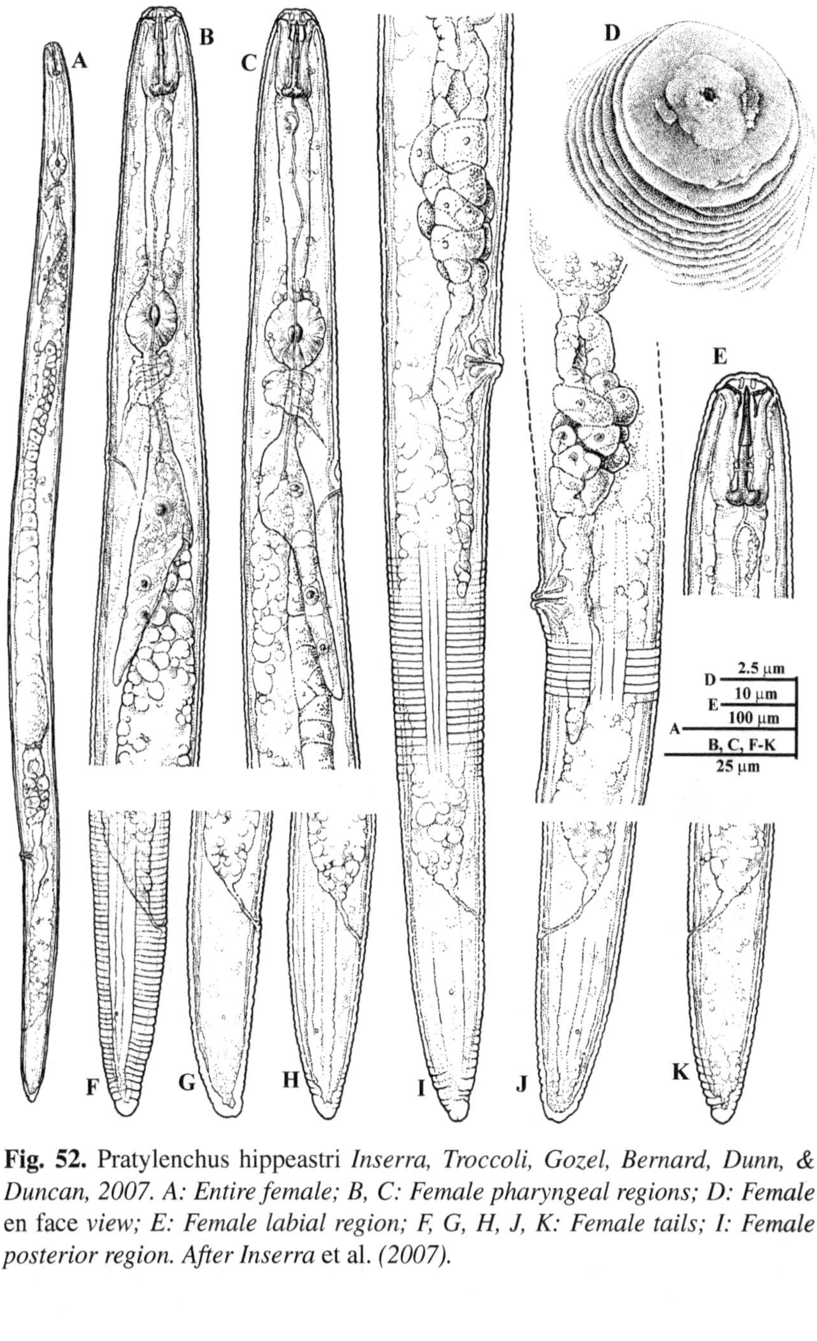

Fig. 52. Pratylenchus hippeastri *Inserra, Troccoli, Gozel, Bernard, Dunn, & Duncan, 2007. A: Entire female; B, C: Female pharyngeal regions; D: Female* en face *view; E: Female labial region; F, G, H, J, K: Female tails; I: Female posterior region. After Inserra* et al. *(2007).*

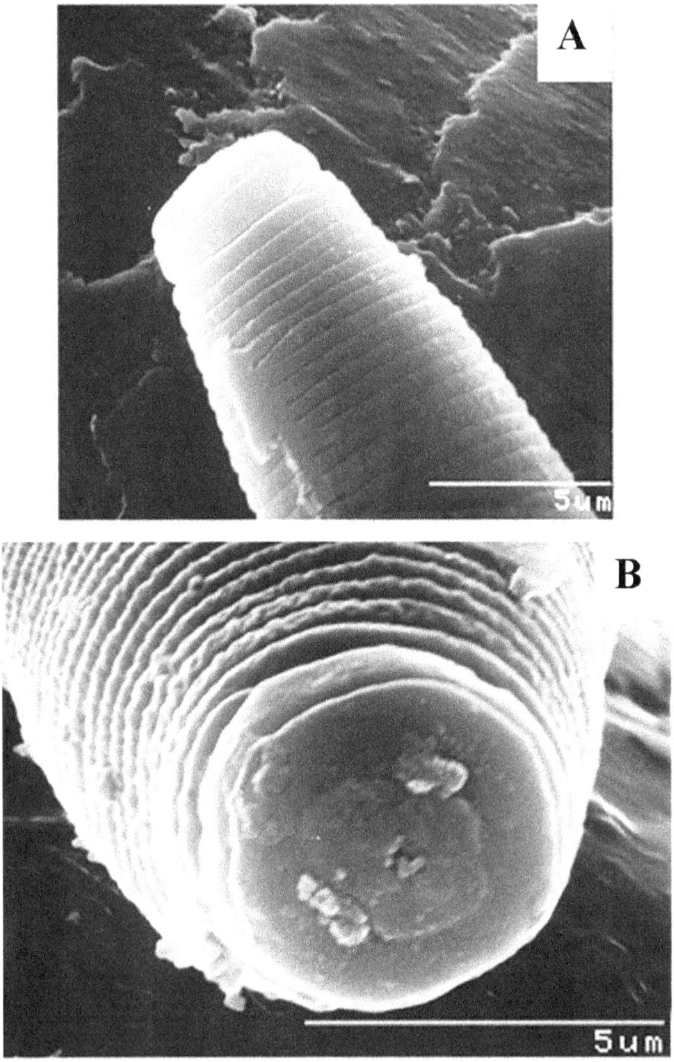

Fig. 53. *SEM micrographs of* Pratylenchus hippeastri *Inserra, Troccoli, Gozel, Bernard, Dunn, & Duncan, 2007. A: Female labial region; B: Female* en face *view. After Inserra* et al. *(2001).*

nucleus at level of or just posterior to cardia. Nuclei of ventrosublateral glands in distal third of pharyngeal lobe. Lateral field marked by four lines; occasionally five observed in some live specimens. Longitudinal striae present in central band of some specimens. Ovary extending to pharyngeal glands in a few specimens. Spermatheca rectangular in shape with a round central cavity, sperm absent. In some specimens spermatheca cavity almost 7 μm in diam. Columnar cells of uterus distinct and disposed in three rows. Vulval lips usually slightly raised or protruding from body surface in some specimens. Post-uterine branch tapered with distinct cells occasionally visible in distal portion. Phasmids located in distal third of tail. Tail conoid, bluntly pointed, usually with a ventral constriction or subhemispherical smooth terminus. A subhemispherical and slightly indented tail terminus was observed in some specimens. Hyaline portion of tail terminus 2-4 μm long.

Male

Not found.

DIAGNOSIS AND RELATIONSHIPS

Pratylenchus hippeastri is characterised by: a slender body, a flat, plain and smooth face, labial region with two lip annuli (and an incomplete third annulus), with the second lip annulus thicker than the first, ellipsoidal stylet knobs, a rectangular empty spermatheca with a large round cavity and a conoid tail with bluntly pointed terminus usually with a ventral constriction or subhemispherical and smooth.

The matrix code is: A1, B1, C2, D4, E2, F5, G3, H1, I3, J1, K1.

It is close to *P. scribneri* with which it was considered conspecific since 1958 when Birchfield and Christie reported this root-lesion nematode from amaryllis in Florida, but differs in stylet and tail length, *en face* view, tail and tail terminus shape (see the corresponding description). It is also close to *P. hexincisus* because both species share similar morphometric parameters, but differs in *en face* view and lateral field (see the corresponding description). Additionally, molecular studies based on the D2/D3 and ITS sequences of *P. hippeastri* clearly separated this species from two populations of *P. hexincisus* and a population of *P. scribneri* (Inserra *et al.*, 2007).

DISTRIBUTION

It has been recorded only from the type locality at the Nematode Assay Laboratory, University of Florida, Gainesville, Florida, USA, from roots of amaryllis (*Hippeastrum* sp.).

29. *Pratylenchus jaehni* Inserra, Duncan, Troccoli, Dunn, dos Santos, Kaplan & Vovlas, 2001

(Figs 54, 55)

MEASUREMENTS

- Female holotype (after Inserra *et al.*, 2001): L = 0.48 mm; a = 20.4; b = 6.1; c = 19.2; c′ = 0.94; V = 78; stylet = 15 μm.
- 20 females (after Inserra *et al.*, 2001): L = 0.49 (0.43-0.54) mm; a = 20.7 (16.6-25.8); b = 6.0 (5.4-6.7); c = 19.2 (17.1-23.7); V = 78 (77-80); stylet = 14.5 (14-15) μm.
- 20 males (after Inserra *et al.*, 2001): L = 0.50 (0.44-0.54) mm; a = 27.8 (25.3-30.9); b = 5.9 (5.4-6.4); c = 21.5 (18.8-23.9); T = 38 (37-40); stylet = 14.5 (14-15) μm; spicules = 16.0 (15.0-17.5) μm; gubernaculum = 5.5 (4.0-6.5) μm.

DESCRIPTION

Female

Body stout with a mean a ratio < 23 (max. 26.0 μm). Labial region with two lip annuli and a plain and smooth face with all labial sectors fused together and with oral disc. Lateral fields marked by four lines. Longitudinal and interrupted striae present in central band of some specimens. Lateral fields width at mid-body 5.5-7.5 μm. Stylet robust with mean length < 15 μm; however, specimens with stylet 16 μm long occur occasionally. Stylet knobs < 2.7 μm high. Dorsal pharyngeal gland opening *ca* 2.5 μm from stylet knobs. Hemizonid just anterior to excretory pore at level of nerve ring and two body annuli long. Hemizonion distinct, located 7-8 annuli posterior to hemizonid. Pharyngeal glands overlapping intestine ventrally for *ca* 33 μm from pharyngo-intestinal valve. Maximum length of pharyngeal overlap 48 μm in live specimens. Ovary not extending to pharyngeal glands. Spermatheca subspherical, filled with sperm. Columnar cells of uterus distinct and disposed in four rows of four cells each. Post-vulval uterine sac 15-39 μm long. Vulva located at 77-80% of body length. Phasmids located at *ca* middle tail. Tail 21-31 μm long. Tail terminus usually hemispherical or subhemispherical and smooth. Specimens with tail terminus truncate and smooth or slightly indented occur occasionally.

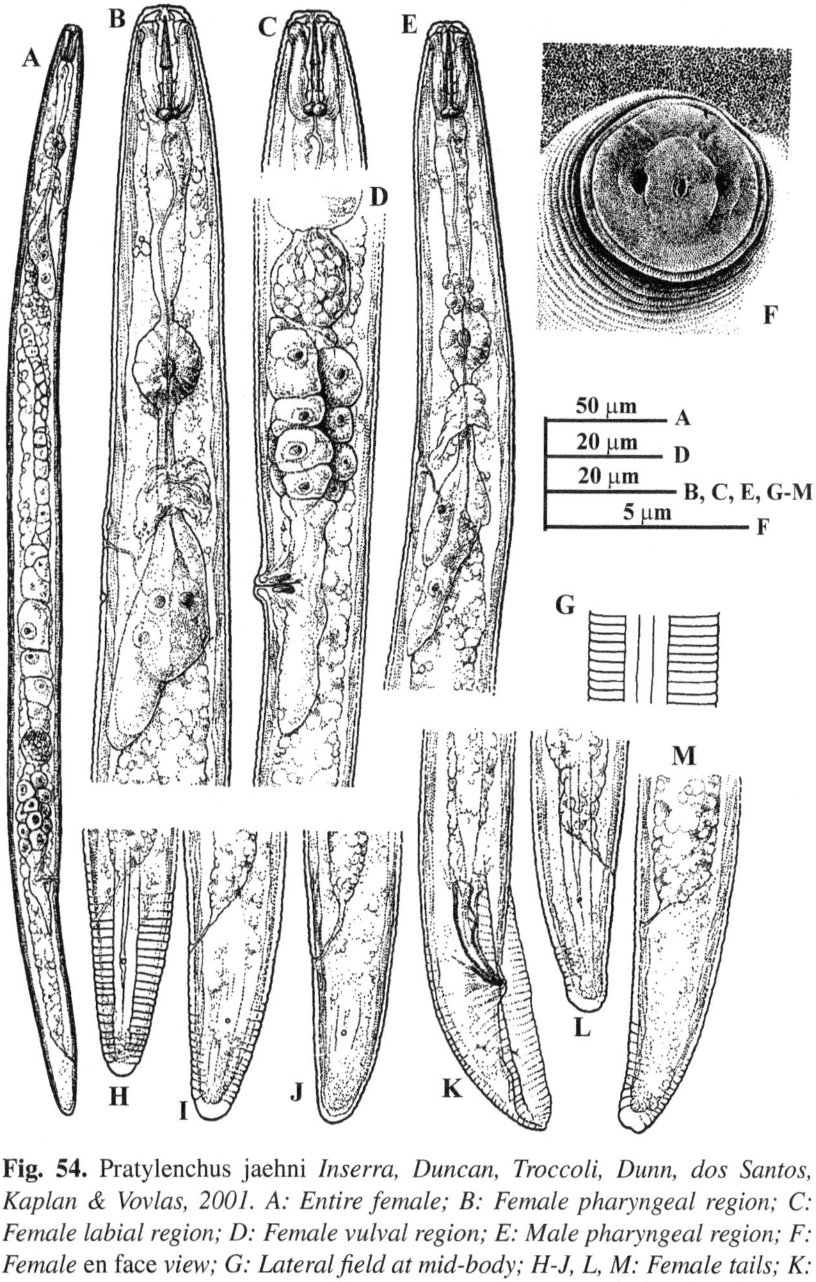

Fig. 54. Pratylenchus jaehni *Inserra, Duncan, Troccoli, Dunn, dos Santos, Kaplan & Vovlas, 2001. A: Entire female; B: Female pharyngeal region; C: Female labial region; D: Female vulval region; E: Male pharyngeal region; F: Female* en face *view; G: Lateral field at mid-body; H-J, L, M: Female tails; K: Male tail. After Inserra* et al. *(2001).*

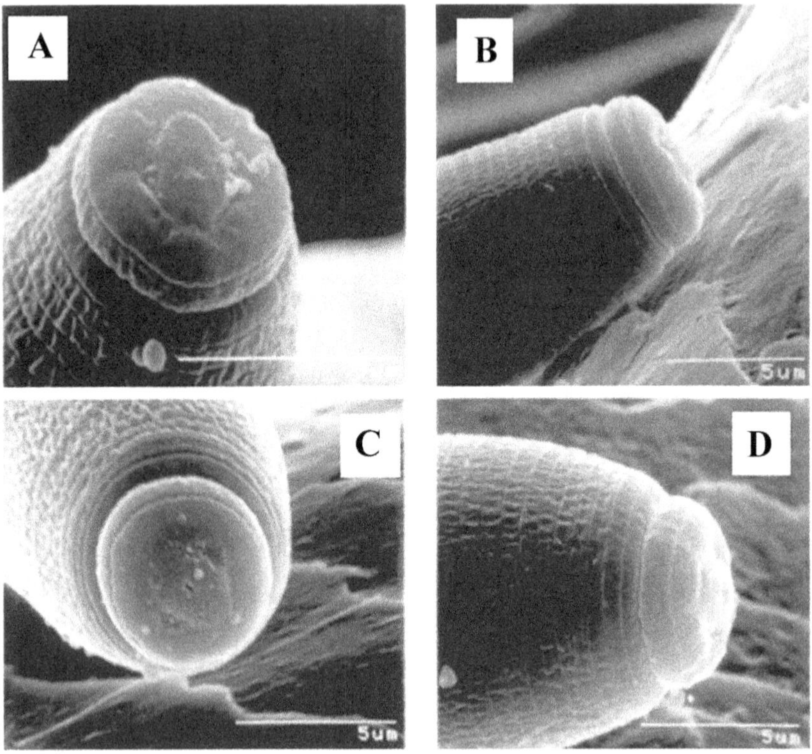

Fig. 55. *SEM micrographs of* Pratylenchus jaehni *Inserra, Duncan, Troccoli, Dunn, dos Santos, Kaplan & Vovlas, 2001. A: Female* en face *view; B: Female labial region; C: Male* en face *view; D: Male labial region. After Inserra* et al. *(2001).*

NOTE

Morphometric values of fixed specimens of *P. jaehni* and *P. coffeae* were consistently smaller than those of live specimens, except for ratios a and V. The variability was not significantly different for the two species and, except for an aberrant value for dorsal pharyngeal gland opening, the range of shrinkage of characters in fixed specimens was 0-13% with a mean of 6.4%. Variability in the range of 0.5 μm of the stylet length measurements was observed when fixed specimens were measured by different operators.

Male

Morphology similar to that of female, including *en face* morphology. However, males have shorter stylet than females (14.0-14.9 *vs* 14.5-15.5 μm) and smaller stylet knobs (1.8-2.0 μm high and 3.0-4.0 μm wide *vs* 1.9-2.7 and 4.0-5.0 μm). Lateral fields marked by four lines and no striae were observed within bands of examined specimens. Reproductive system characterised by single testis not extending to pharyngeal glands. Bursa margin smooth or finely crenate proximally.

DIAGNOSIS AND RELATIONSHIPS

Pratylenchus jaehni is characterised by: labial region with two annuli, stylet ⩽ 15 μm, female tail terminus usually hemispherical or subhemispherical and smooth and presence of males.

The matrix code is: A1, B2, C2, D2, E2, F4, G2, H1, I2, J1, K1.

It is close to *P. coffeae* and *P. loosi*. It differs from *P. coffeae* and *P. loosi* by only a few morphological characters of the females. The mean values of stylet length, stylet knob height and vulva position are smaller (⩽15 *vs* ⩾15 μm, ⩽2.7 *vs* ⩾2.7 and ⩽79 *vs* ⩾79%) than those in *P. coffeae* and *P. loosi*, respectively. The mean value of ratio a is also smaller in the new lesion nematode from Brazil (⩽23.0 *vs* ⩾28.0 and 24.7) than in *P. coffeae* and *P. loosi*, respectively. No morphological differences were observed among *P. jaehni* males and those of *P. coffeae* and *P. loosi*. Furthermore, the high range values of the stylet length, vulva position and ratio a of females of *P. jaehni* overlap the low range values of *P. coffeae* and *loosi*. The tail terminus is usually subhemispherical and smooth in *P. jaehni*, whereas it is commonly truncate and indented in the standard populations of *P. coffeae* and bluntly or finely pointed in *P. loosi*. Specimens of *P. coffeae* with a subhemispherical and smooth tail terminus occur in all the eight populations studied and also in *P. loosi*, further complicating the separation of *P. jaehni* from *P. coffeae* and *P. loosi* when only a few specimens are available for diagnosis. An almost truncate tail terminus, similar to *P. coffeae*, was observed in some specimens of *P. jaehni*. These morphological similarities require examination of at least ten specimens to obtain a reliable differentiation of this new lesion nematode from *P. coffeae* and *P. loosi*.

Pratylenchus jaehni appears to be reproductively isolated from *P. coffeae* and molecular distinctions between *P. jaehni*, *P. coffeae* and

P. loosi are more clear-cut than morphological differences (Duncan *et al.*, 1999). Nevertheless, the results of this study emphasise the importance of the use of behavioural and morphological information in combination with molecular analyses for the separation of closely related species. Indeed, differences in DNA sequences also occur among *P. jaehni* and the *P. coffeae* population from São Paulo (Duncan *et al.*, 1999), despite the ability of the latter population to mate with *P. jaehni*. Further study of these Brazilian populations (all of which formed the PCA-group V in Duncan *et al.*, 1999) is needed to resolve their evolutionary relationships. Thus, while it is important to be aware of the genetic diversity among these populations, determining a basis for their morphological discrimination as separate species should be attempted only for those that exhibit major biological differences of agricultural importance. Renaming those populations, in the absence of behavioural or unambiguous morphological differences, could reduce rather than strengthen the usefulness of our current classification system.

DISTRIBUTION

It has been recorded only from the type locality at the Sitio das Antas farm on road SP133 near Bairro São João, county of Itápolis, in São Paulo State, Brazil, from roots of citrus (*Citrus uranium* L.).

30. *Pratylenchus japonicus* Ryss, 1988

(Figs 56, 57)

MEASUREMENTS

- Female lectotype (after Mizukubo *et al.*, 1997): L = 0.42 mm; a = 24; b = 5.3; b′ = 3.3; c = 17.0; c′ = 2.5; V = 87; stylet = 19 μm.
- 18 females (after Mizukubo *et al.*, 1997): L = 0.52 (0.40-0.62) mm; a = 31.7 (24.7-36.0); b = 6.4 (5.6-7.1); b′ = 4.0 (3.1-5.1); c = 17.6 (13.3-23.5); c′ = 3.2 (2.6-3.8); V = 86 (84-88); stylet = 19.8 (18.5-21.5) μm.
- 3 females (after Mizukubo *et al.*, 1997): L = 0.52 (0.45-0.55) mm; a = 29.8 (28.2-32.4); b = 6.1 (5.5-6.6); b′ = 3.9 (3.5-4.3); c = 18.5 (16.5-20.6); c′ = 2.7 (2.4-2.9); V = 87 (86-88); stylet = 19.5 (18.5-21.0) μm.

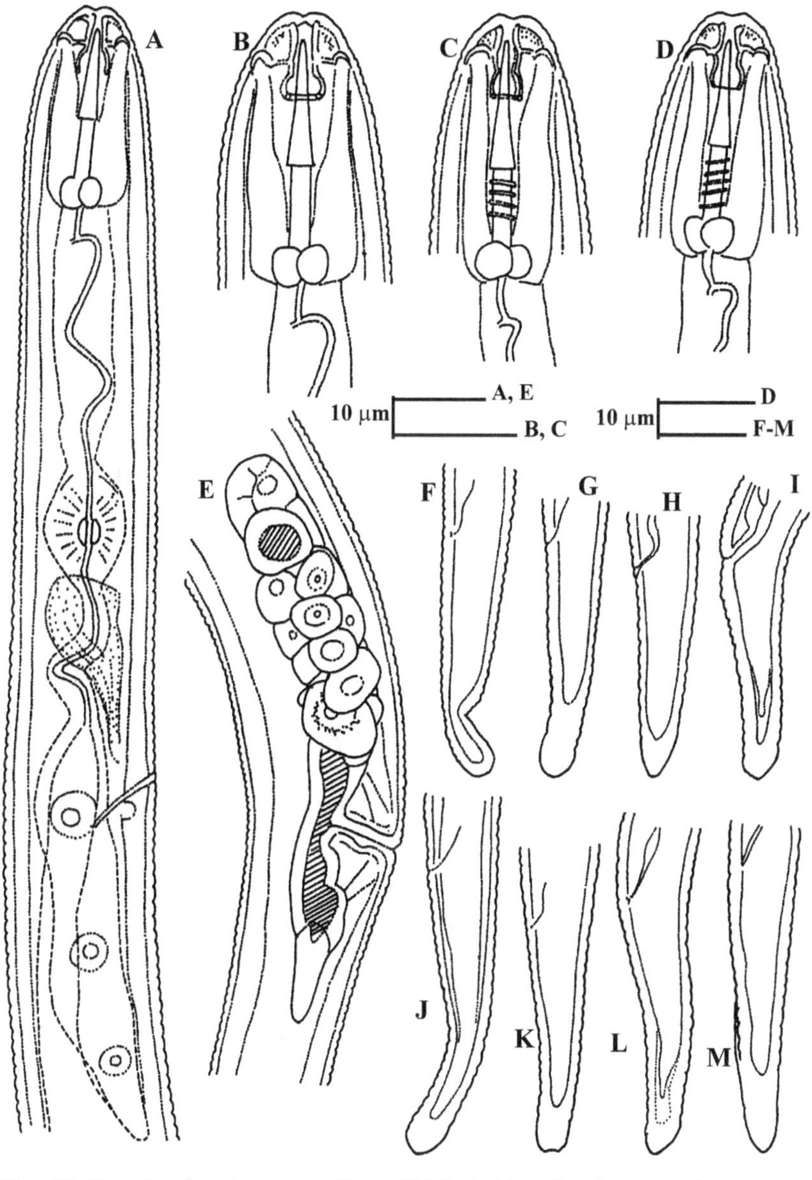

Fig. 56. Pratylenchus japonicus *Ryss, 1988. A: Female pharyngeal region; B-D: Female labial regions; E: Vulval region; F-M: Female tails. After Mizukubo* et al. *(1997).*

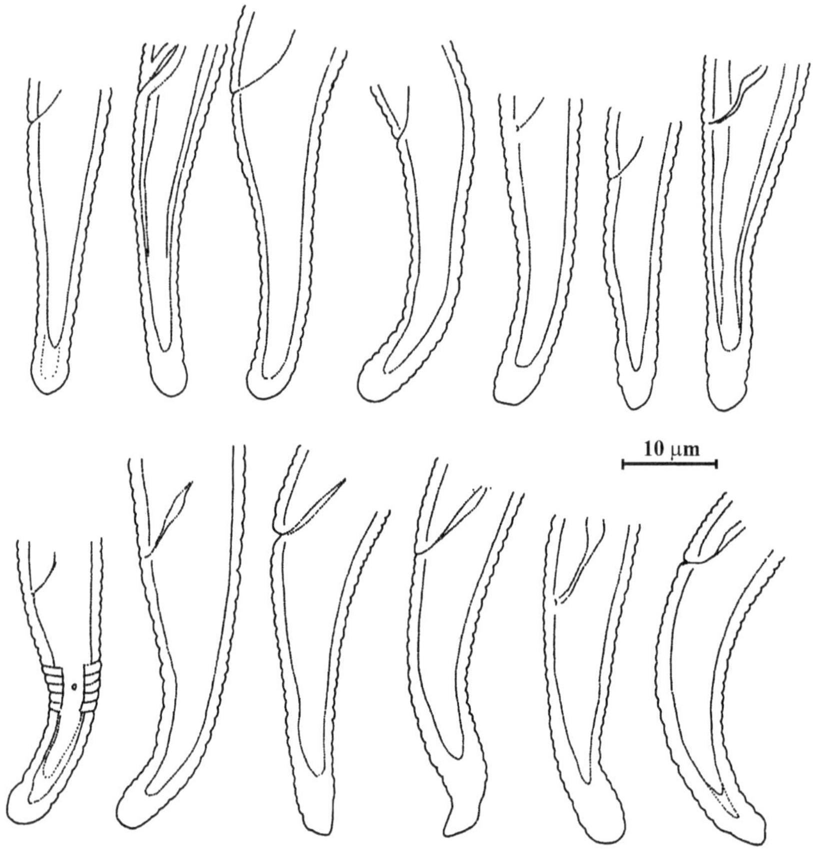

Fig. 57. Pratylenchus japonicus *Ryss, 1988. Morphological variation in female tails. After Mizukubo* et al. *(1997).*

- 50 females (after Minagawa, 1982): L = 0.52 (0.44-0.70) mm; a = 25.3 (16.0-36.6); b = 7.0 (5.3-10.3); c = 17.4 (12.9-20.7); c′ = 2.9 (1.4-4.5); V = 86 (83-89); stylet = 19.3 (18.3-20.8) μm.
- 22 females (after Minagawa, 1982): L = 0.54 (0.49-0.62) mm; a = 27 (23-30); b = 7.4 (6.5-7.9); c = 17 (14-21); c′ = 2.8 (2.3-3.6); V = 86 (82-89); stylet = 19 (18.5-19.5) μm.

DESCRIPTION

Female

Body rather slender, curved ventrally when heat-relaxed. Annuli 0.9 (0.8-1.0) μm wide at mid-body. Labial region tapering anteriorly, rounded, elevated, 3.1 (2.8-3.3) μm high, 7.7 (7.2-8.2) μm diam., continuous with body contour. Labial framework moderately developed, laterally extending into body for 1-2 annuli; vestibule extension well developed, cask-shaped. Labial region with three shallow annuli on both sides, often indiscernible by light microscope but distinct in SEM image. In *en face* view of first annulus, oral aperture oval and widened, with six pore-like inner sensilla at margins, median sensilla indiscernible; amphidial apertures slit-like, oblique; median segment fused to lateral segment (Group I *sensu* Corbett & Clark, 1983). Cephalids not seen. Lateral fields 5.8 (5.6-5.9) μm wide or 35 (31-39)% of max. body diam., consisting of three equally spaced bands separated by two narrow but deep inner grooves; almost plain in mid-body region; areolated in tail region. Lateral fields never extending to tail tip; inner band fused to single groove shortly before phasmid. Stylet 19.8 (18.5-21.5) μm long, 2.6 (2.3-2.8) times as long as labial region diam. Stylet knobs massive, 3.7 (3.3-4.6) μm high, 4.7 (4.3-5.6) μm across; in width/height ratio = 1.3 (0.9-1.5); shape rather constant, mostly flattened anteriorly or rounded, never indented or sloping posteriorly. Dorsal pharyngeal gland opening 3.3 (2.9-3.9) μm from stylet base. Metacorpus oval, occupying *ca* half of corresponding body diam.; valve conspicuous, 60 (51-66) μm from anterior body end, or 63 (42-74)% of pharynx length. Pharyngeal basal lobe extending 132 (108-156) μm from anterior body end, ventrally overlapping intestine 46.3 (29.5-66.8) μm or a distance of 2.8 (1.6-3.8) times corresponding body diam. Pharyngeal nuclei usually in tandem. Excretory pore 83 (65-92) μm from anterior end, 95.6 (62-108)% of pharyngeal length and 74 (71-77) body annuli from anterior end; annuli between anterior body end and excretory pore 1.1 μm across. Hemizonion lenticular, 2-3 annuli long, level with or immediately anterior to excretory pore. Gonad outstretched, 126 (87-162) μm long; ovary with oocytes generally in a row (rarely partially double); spermatheca indistinct, empty; when observed, rounded, at 44 μm from vulva, 9.2 μm long, 7.9 μm diam. Uterus with tricolumella of 15 cells. Post-vulval uterine sac *ca* as long as vulval body diam. Vagina narrow walled, anteriorly inclined to body axis, 8.8 (6.6-10) μm long, or 56 (48-63)% of vulval body diam. Tail 30 (24-34) μm long, with 22 (19-26) annuli, distinctly conoid; tail tip rounded; terminal cuticular portion moderately broad and elongate, 3.2 (2.6-3.8) times as long as anal body diam.; posterior half distinctly more slender than anterior half due to rather extensive tapering from anus to middle level; tip with some variation in shape: mostly rounded (52%), often bluntly pointed (33%), rarely flattened or concave (9%) and very rarely clavate; terminus smooth, without evident annuli. Tail terminal cuticle rather thick, 5.5 (2.9-9.2) μm thick; phasmids pore-like, in anterior half of tail, centred in lateral fields, 11 (9.2-12.4) μm posterior to anus or 62.9 (60-69)% from tail tip.

NOTE

Amplified product of ITS regions in rDNA is a single fragment of *ca* 0.8 kb. The restriction patterns and the DNA fragment sizes obtained after digestion of ten endonucleases showed that DNA fragment sizes obtained were unique among the *Pratylenchus* species (Mizukubo *et al.*, 1997).

Male

Not found.

DIAGNOSIS AND RELATIONSHIPS

Pratylenchus japonicus is characterised by: the first lip annulus with distinctly convex anterior margin, long and stout stylet (18-21 μm), posterior vulva (V = 84-88), empty spermatheca, shorter post-vulval uterine sac and slender tail with smooth terminus.

The matrix code is: A2, B1, C4, D1, E4, F2, G3, H1, I3, J1, K1.

It is close to *P. brachyurus* and *P. macrostylus* from which it can be distinguished by number of lip annuli, shape of the first lip annulus, the elongate stylet, posterior vulva and tail terminus (see the corresponding descriptions).

Ryss (1988) distinguished the Japanese *P. macrostylus* Wu, 1971 from North American populations because of the differences in stylet length and proposed a new subspecies (*P. macrostylus japonicus* Ryss, 1988). Similar differences were observed by Hartman and Eisenback (1991); Mizukubo *et al.* (1997) elevated it to the specific rank based on the morphological, morphometrical and molecular data, which clearly differentiated *P. macrostylus* and *P. japonicus*.

DISTRIBUTION

It has been recorded only from the type locality at Japan from *Quercus acutissimus* Carruth. *Q. dentata* Thumb., *Q. serrata* Murray and *Prunus jamasakura* Siebold *ex* Koidzumi.

31. *Pratylenchus kasari* Ryss, 1982
(Fig. 58)

MEASUREMENTS

- Female holotype (after Ryss, 1982): L = 0.66 mm; a = 36; b = 6.0; b′ = 3.8; c = 19; c′ = 1.3; V = 75; stylet = 17 μm.
- 14 females (after Ryss, 1982): L = 0.56-0.77 mm; a = 32-44; b = 5.6-8.4; b′ = 3.3-4.2; c = 16-20; c′ = 1.2-1.6; V = 75-81; stylet = 16-17.5 μm.
- Females (after Brzeski, 1998): L = 0.39-0.77 mm; a = 25-44; b = 4.4-8.4; b′ = 3.3-4.2; c = 2.8-4.0; V = 75-81; stylet = 15-17.5 μm.
- 7 males (after Ryss, 1982): L = 0.57-0.70 mm; a = 34-44; b = 5.7-7.3; b′ = 3.5-4.3; c = 18-22; c′ = 1.2-1.4; T = 47-67; stylet = 16-17 μm; spicules = 20-21 μm.

DESCRIPTION

Female

Body narrow and slender. Lateral fields with four lines; in some specimens a broken fifth line visible posterior to trophic-genital section, four (sometimes two) lines on tail. Stylet long, with rounded, anteriorly directed knobs. Opening of dorsal pharyngeal gland 3.5-4 μm posterior to stylet. Labial area dome-shaped, with three annuli. Anterior cephalid at 3-4 annuli from base of cephalic capsule; posterior cephalid at eight annuli posterior. Pharyngeal gland lobe long. Hemizonid immediately anterior to excretory pore, its length equal to width of 3-4 annuli. Hemizonion situated at 20-24 annuli posterior to hemizonid, its length equal to width of one annulus. Spermatheca oval, its length (28-77 μm) usually exceeding its diam. by 3-5 (sometimes two) times, containing sperm 5 μm in diam. Rudiment of posterior gonad with group of 10-20 small nuclei located posterior to post-vulval uterine sac. Tail length exceeding anal diam. by 3-4 times. Phasmids generally located

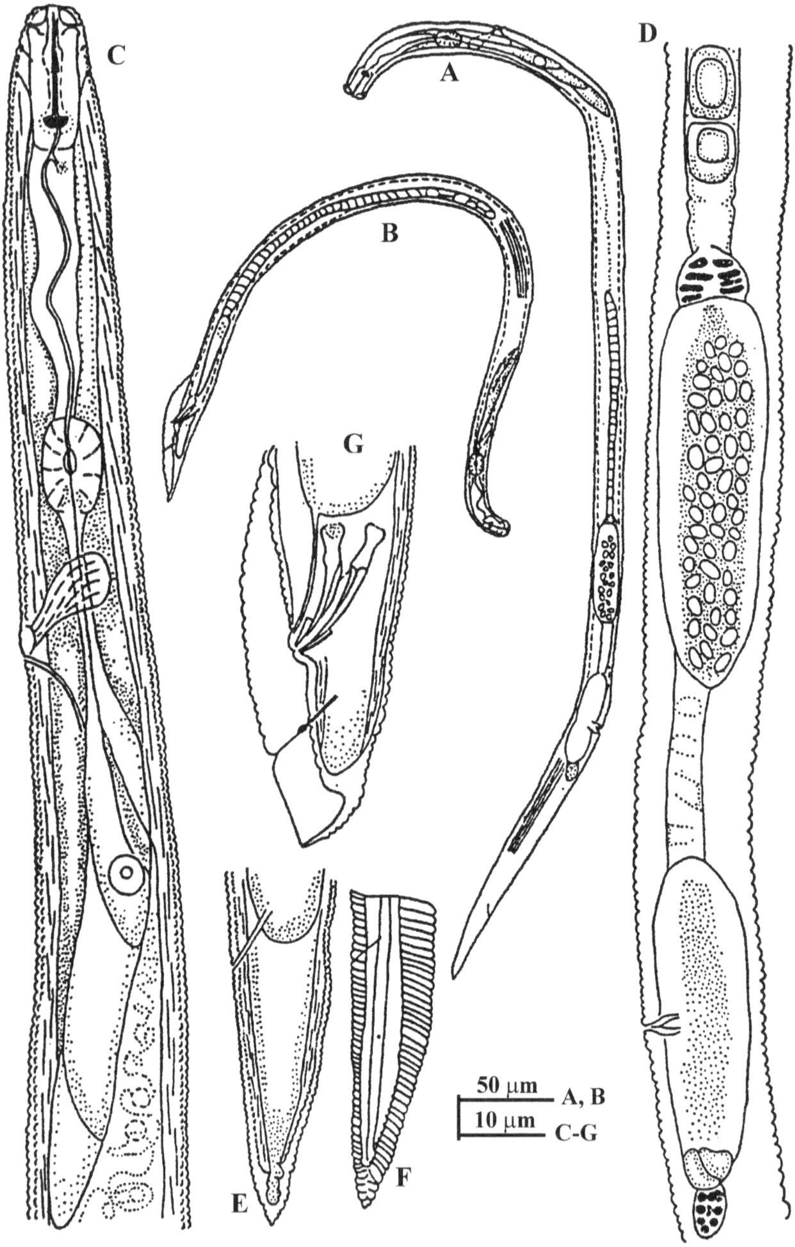

Fig. 58. Pratylenchus kasari *Ryss, 1982. A: Entire female; B: Entire male; C: Female pharyngeal region; D: Vulval region; E, F: Female tails; G: Male tail. After Ryss (1982).*

in anterior half of tail. Tail conoid pointed with 32-44 annuli on ventral surface of tail, not counting terminal annuli. Tip sharply conical, serrated.

Male

Stylet somewhat shorter than that of female. Sexual dimorphism not expressed. Spicule shape typical for pratylenchids but very large.

DIAGNOSIS AND RELATIONSHIPS

Pratylenchus kasari is characterised by: labial region with three annuli, elongate-oval spermatheca and tail conoid pointed with sharply conical and serrated terminus.

The matrix code is: A2, B2, C3, D3, E2, F3, G3, H2, I3, J1, K1.

It can be distinguished from closely related species (*P. pratensis*) from which it differs by the longer stylet, elongate spermatheca, the presence of a rudimentary posterior branch of the gonad posterior to the post-vulval uterine sac, the number of annuli on tail, the tail shape and length of spicules (see the corresponding description).

DISTRIBUTION

It has been recorded from the type locality at Pechkovo, Leningrad, Russia, in soil around the roots of tufted hair-grass (*Deschampsia caespitose* L.); and in forest at Poland (Brzeski, 1998).

32. *Pratylenchus kralli* Ryss, 1982
(Fig. 59)

MEASUREMENTS

- Female holotype (after Ryss, 1982): L = 0.41 mm; a = 24; b = 5.8; b′ = 4.4; c = 20; c′ = 0.9; V = 76; stylet = 14 μm.
- 18 females (after Ryss, 1982): L = 0.40-0.50 mm; a = 20-33; b = 4.8-6.5; b′ = 3.9-4.5; c = 17-23; V = 74-80; stylet = 14-15 μm.
- 10 males (after Ryss, 1982): L = 0.38-0.45 mm; a = 21-34; b = 5.2-6.4; b′ = 3.9-6.5; c = 16-23; T = 41-54; stylet = 12.5-14 μm; spicules = 14-15.5 μm.

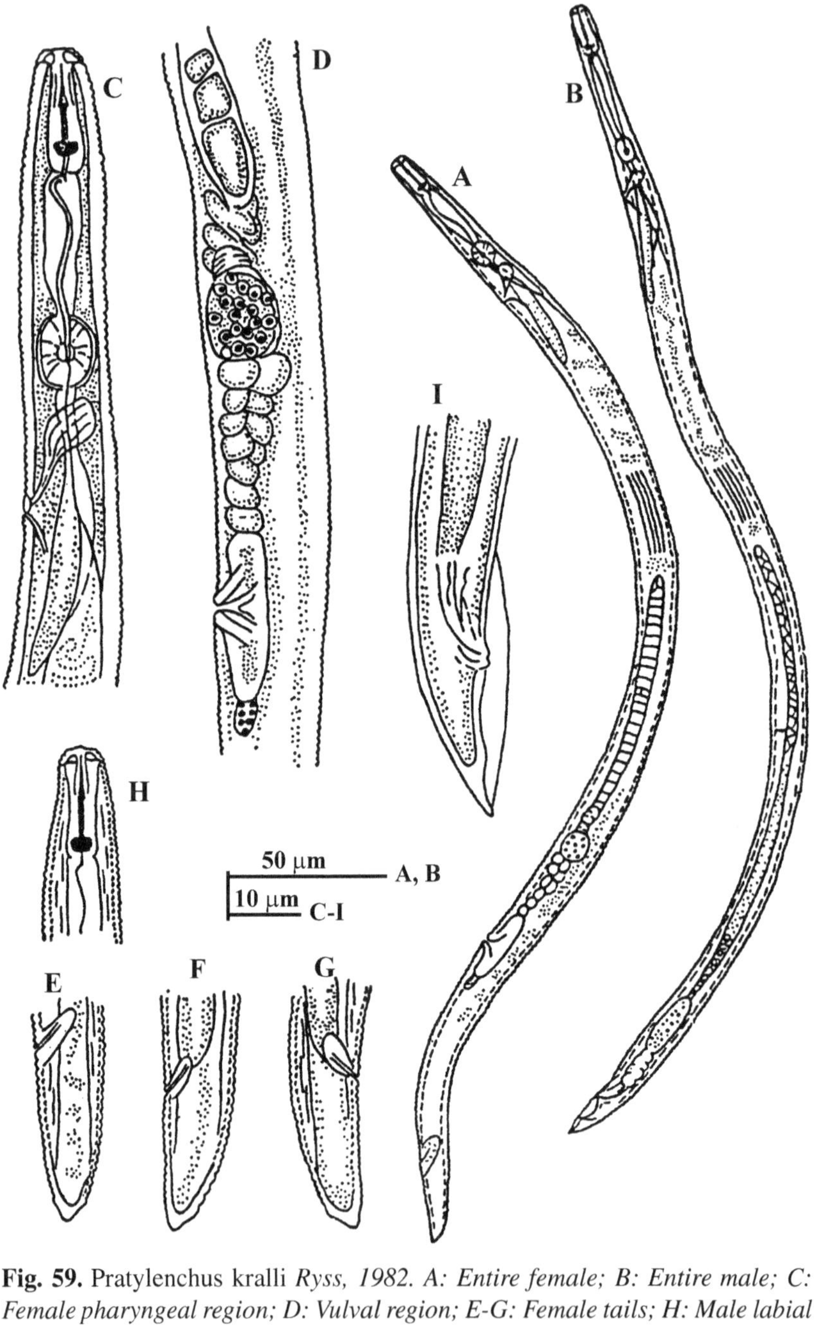

Fig. 59. Pratylenchus kralli *Ryss, 1982. A: Entire female; B: Entire male; C: Female pharyngeal region; D: Vulval region; E-G: Female tails; H: Male labial region; I: Male tail. After Ryss (1982).*

DESCRIPTION

Female

Body narrow, slender. Labial area dome-shaped, high, with three annuli. Lateral fields with four lines. Stylet small, knobs wide, forwardly directed. Opening of dorsal pharyngeal gland 2 μm posterior to stylet. Metacorpal bulb round. Pharyngeal gland lobe narrow. Spermatheca round, sometimes slightly oval; 14-17 μm diam., filled with small spermatozoids 2 μm in diam. Length of posterior branch of uterus exceeding vulval body diam. by more than 1.5 times, not containing oogonia. Posterior to post-vulval uterine sac a rudiment of posterior branch of gonad occasionally observed with several small nuclei which, however, are not nuclei of oogonia but are somatic in origin. Tail more than twice anal diam. long. Tail conoid with 16-23 annuli on ventral surface. Tip smooth, pointed and showing a slight groove in direction of ventral surface of body (Fig. 59E-F). Serration observed in rare instances in tip region.

Male

Stylet somewhat shorter than that of female. Spicule shape typical for pratylenchids, small.

DIAGNOSIS AND RELATIONSHIPS

Pratylenchus kralli is characterised by: labial region with three annuli, lateral fields with four lines, spermatheca round, tail conoid with smooth terminus.

The matrix code is: A2, B2, C2, D2, E2, F1, G3, H1, I1, J1, K1.

It can be distinguished from closely related species (*P. penetrans* and *P. vulnus*) from which it differs by body and stylet length, position of the vulva, shape of the spermatheca and shape of the tail terminus (see the corresponding descriptions).

Frederick and Tarjan (1989) synonymised *P. ventroprojectus* with *P. kralli*. However, the tail shape in *P. ventroprojectus* is described as "short, broadly conoid, with terminus truncate, coarsely or not annulated truncate with a distinct subventral projection"; whereas in *P. kralli* the tail is described as "conoid, with pointed and smooth tip". Therefore, both species are herein regarded as valid.

DISTRIBUTION

It has only been recorded from the type locality at Parnu (Papiniidus region), Estonia, in soil around roots of blackcurrant.

33. *Pratylenchus kumaoensis* Lal & Khan, 1990
(Fig. 60)

MEASUREMENTS

- Female holotype (after Lal & Khan, 1990): L = 0.42 mm; a = 25.8; b = 6.0; b′ = 4.5; c = 20.0; c′ = 2.3; V = 82; stylet = 14.5 μm.
- 10 females (after Lal & Khan, 1990): L = 0.44 (0.39-0.48) mm; a = 27.4 (25.0-29.0); b = 6.1 (5.8-6.5); b′ = 4.5 (3.9-4.8); c = 21.2 (20.0-24.0); c′ = 2.3 (2.0-2.5); V = 82 (81-83); stylet = 14.5 (14-15) μm.

DESCRIPTION

Female

Body straight or slightly arcuate when heat-relaxed. Labial region almost continuous with body contour with three annuli, heavily sclerotised skeleton extending for two annuli. Lateral fields with four lines arising at metacorpus level, continuous up to posterior third of tail. Stylet 14-15 μm long with rounded basal knobs. Opening of dorsal pharyngeal gland 2.8-3.5 μm posterior to stylet-knobs. Pharynx typical of genus. Excretory pore located 64-80 μm from anterior end, opposite to pharyngo-intestinal junction. Hemizonid distinct, anteriorly adjacent to excretory pore, extending 2-3 body annuli. Nerve ring encircling isthmus at *ca* 56-63 μm from anterior end. Vulva a transverse slit, 322-392 μm from anterior end. Spermatheca inconspicuous, without sperm. Post-vulval uterine sac *ca* 15-22 μm long. Tail conoid-rounded 20-23 μm long, or 2.0-2.5 anal body diam., bearing 21-25 annuli. Tail terminus crenate. Phasmids small, located at mid-tail.

Male

Not found.

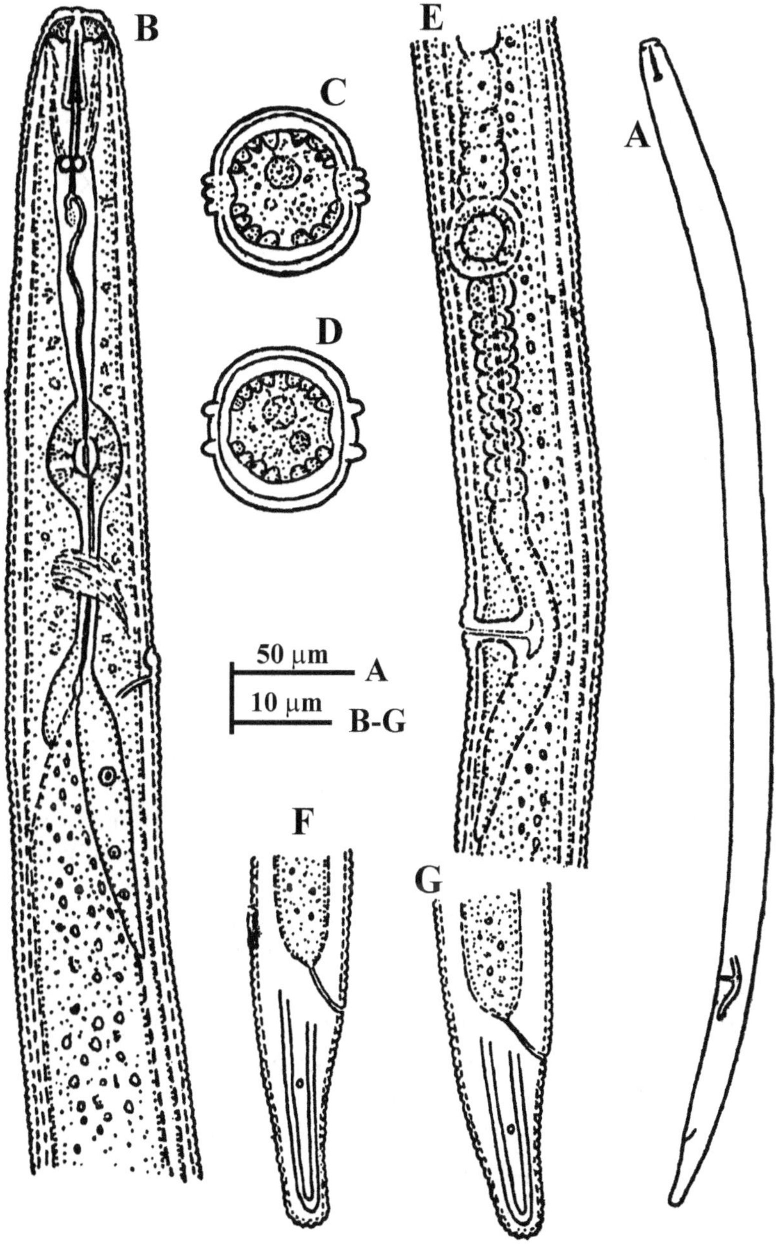

Fig. 60. Pratylenchus kumaoensis *Lal & Khan, 1990. A: Entire female; B: Female pharyngeal region; C, D: Female cross-sections at mid-body; E: Vulval region; F, G: Female tails. After Lal and Khan (1990).*

DIAGNOSIS AND RELATIONSHIPS

Pratylenchus kumaoensis is characterised by: labial region with three annuli, a relatively small stylet (14-15 μm), lateral fields with four lines, tail conoid-rounded and tail terminus crenate.

The matrix code is: A2, B1, C2, D1, E3, F3, G3, H2, I1, J1, K1.

It can be distinguished from closely related species (*P. crenatus* and *P. teres*) by stylet length, lateral fields and position of the vulva (see the corresponding descriptions).

DISTRIBUTION

It has only been recorded from the type locality at Forest Nursery, IVRI Substation, Mukteshwar, Uttar Pradesh, India, from soil around roots of *Populus* spp.

34. *Pratylenchus loosi* Loof, 1960
(Fig. 61)

MEASUREMENTS

- Female holotype (after Loof, 1960): L = 0.60 mm; a = 27.7; b = 6.8; c = 17.8; V = 82; stylet = 16 μm.
- 34 females (after Loof, 1960): L = 0.58 (0.48-0.64) mm; a = 32 (28-36); b = 6.4 (5.7-7.1); c = 20 (18-25); V = 83 (79-85); stylet = 14-18 μm.
- 15 males (after Loof, 1960): L = 0.46 (0.38-0.58) mm; a = 35 (28-41); b = 5.9 (5.4-6.7); c = 20.8 (18.5-23.2); T = 42 (36-52); stylet = 12-16 μm.
- 10 females (after Ryss, 1988): L = 0.65 (0.46-0.70) mm; a = 33 (27-35); b = 6.5 (5.6-6.8); b′ = 4.1 (3.5-5.0); c = 21 (17-25); c′ = 2.8 (2.5-3.5); V = 83 (79-85); stylet = 16 (15-18) μm.
- 10 males (after Ryss, 1988): L = 0.53 (0.49-0.56) mm; a = 36 (31-38); b = 5.7 (5.2-6.5); b′ = 4.0 (3.4-4.5); c = 20 (17-24); c′ = 2.7 (2.4-3.0); stylet = 15 (13-16) μm; spicules = 18 (16-20) μm; gubernaculum = 4-7 μm.
- 20 females (after Inserra *et al.*, 2001): L = 0.52 (0.43-0.62) mm; a = 24.7 (19.5-29.9); b = 5.7 (4.7-7.1); c = 18.3 (16.2-20.5); V = 80 (79-83); stylet = 16 (15.0-17.5) μm.

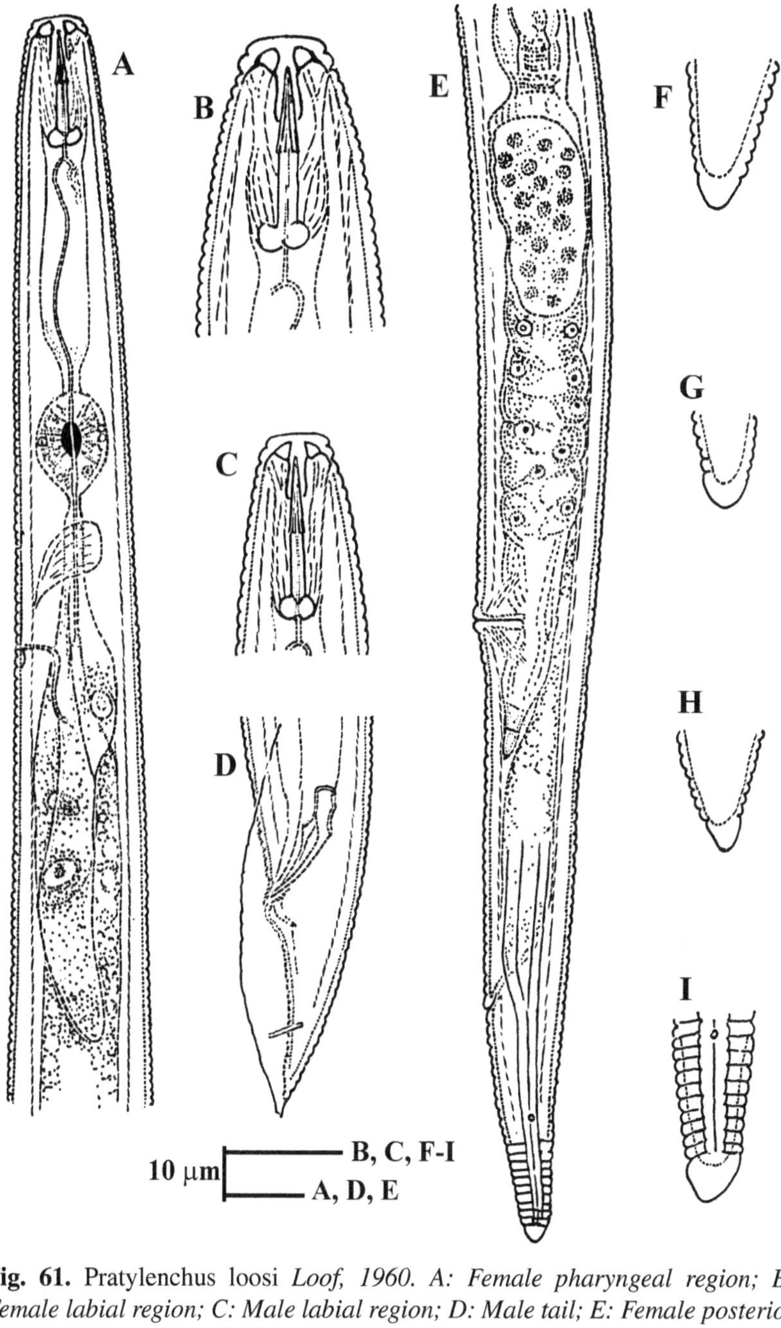

Fig. 61. Pratylenchus loosi *Loof, 1960. A: Female pharyngeal region; B: Female labial region; C: Male labial region; D: Male tail; E: Female posterior region; F-I: Female tails. After Seinhorst (1977).*

– 10 males (after Inserra *et al.*, 2001): L = 0.56 (0.50-0.61) mm; a = 28.3 (25.7-32.1); b = 6.2 (5.8-6.7); c = 21.4 (19.7-24.3); T = 45 (32-59); stylet = 15.0 (15.0-15.5) μm.

DESCRIPTION

Female

Body very slender, almost straight after heat-relaxing, narrowing fairly suddenly just posterior to vulva to *ca* 0.8 times diam. anterior to vulva, then cylindrical to anus. Lateral field broad with four (occasionally five or even six) lines anterior and three posterior to anus. Labial region rounded, with two annuli. Under SEM most specimens with two lip annuli on one side and three on other, some with three annuli and rare specimens with two annuli or three annuli on one side and four annuli on other (Baujard *et al.*, 1990). *En face* view by SEM shows a distinct labial disc with lateral labial sectors fused, no separation between labial disc and submedian lips (Baujard *et al.*, 1990). Posterior edges of labial framework extending one or two body annuli into body. Stylet knobs rounded, broadly fused to shaft. Spermatheca rectangular, up to 20 μm long, rarely more rounded; located 0.5-0.9 times vulva-anus distance anterior to vulva. Post-vulval uterine sac 30-50% of vulva-anus distance. Tail *ca* 2.5 anal body diam. long, tapering gradually to smooth conical tip, usually with narrowly rounded terminus.

Male

Abundant. Spicules slender with well marked manubria and ventrally arcuate shaft, 16-20 μm long; gubernaculum 4-7 μm long, bursal margins faintly crenate.

DIAGNOSIS AND RELATIONSHIPS

Pratylenchus loosi is characterised by: labial region rounded, with two annuli, spermatheca oval, rarely more rounded, tail tapering gradually to smooth conical tip, usually with narrowly rounded terminus and abundant males.

The matrix code is: A1, B2, C3, D4, E3, F2, G3, H3, I2, J1, K1.

It is close to *P. coffeae*, *P. neobrachyurus*, *P. panamaensis* and *P. pseudocoffeae* from which it can be distinguished by shape of the labial region, body and stylet length, position of the vulva and female tail shape (see the corresponding descriptions).

DISTRIBUTION

It has been recorded from the type locality in Sri Lanka from roots of tea, *Camellia sinensis* (L.) Kuntze. It is a serious pathogen of tea in Sri Lanka and other parts of the world (Seinhorst, 1977a). It has also been reported from citrus in India (Sethi & Swarup, 1971) and from apple and citrus in Japan (Gotoh & Oshima, 1963). It has also been recorded in Java on coffee (Inserra *et al.*, 2001); in Korea on tea (Park-Byeong *et al.*, 2002); in Taiwan on tea (Wu *et al.*, 2002); in Guadeloupe on breadfruit *Artocarpus altilis* (Parkins.) Fosb. (Van den Berg & Quénéhervé, 2000); in Iran (Pourjam *et al.*, 1997, 1999b); in Florida on pasture grasses (Inserra *et al.*, 1996); in American Samoa on banana (Brooks, 2004); in China on fruit trees (Wang, 1993); and in Russia on tea (Ryss, 1988).

35. *Pratylenchus macrostylus* Wu, 1971

(Figs 62, 63)

MEASUREMENTS

- Female holotype (after Wu, 1971): L = 0.66 mm; a = 29; b = 6.5; c = 21; V = 87; stylet = 23.5 μm.
- 21 females (after Wu, 1971): L = 0.51-0.68 mm; a = 22-33; b = 5.0-7.4; c = 16-24; V = 85-89; stylet = 22-24 μm.
- 1 male (after Wu, 1971): L = 0.43 mm; a = 30; b = 5; c = 21; T = 40; stylet = 19 μm; spicules = 15 μm; gubernaculum = 2.8 μm.
- 25 females (after Hartman & Eisenback, 1991): L = 0.68 (0.58-0.83) mm; a = 24.5 (17.5-34.3); b = 6.5 (5.6-7.9); b′ = 14.5 (8.6-25.3); c = 23.0 (18.4-29.1); V = 86 (84-88); stylet = 25 (22-28) μm.

DESCRIPTION

Female

Body tapering at both ends. Cuticle marked by fine annulation. Lateral fields with four lines, often faint with broken diagonal striae in middle band. Labial region cap arch-shaped, angular; height 3.1 ± 0.3 μm, diam. 8.9 ± 0.4 μm. Labial region slightly offset from body, marked with two annuli in lateral view, in some cases two annuli on one side and three on other. In SEM, stoma slit-like,

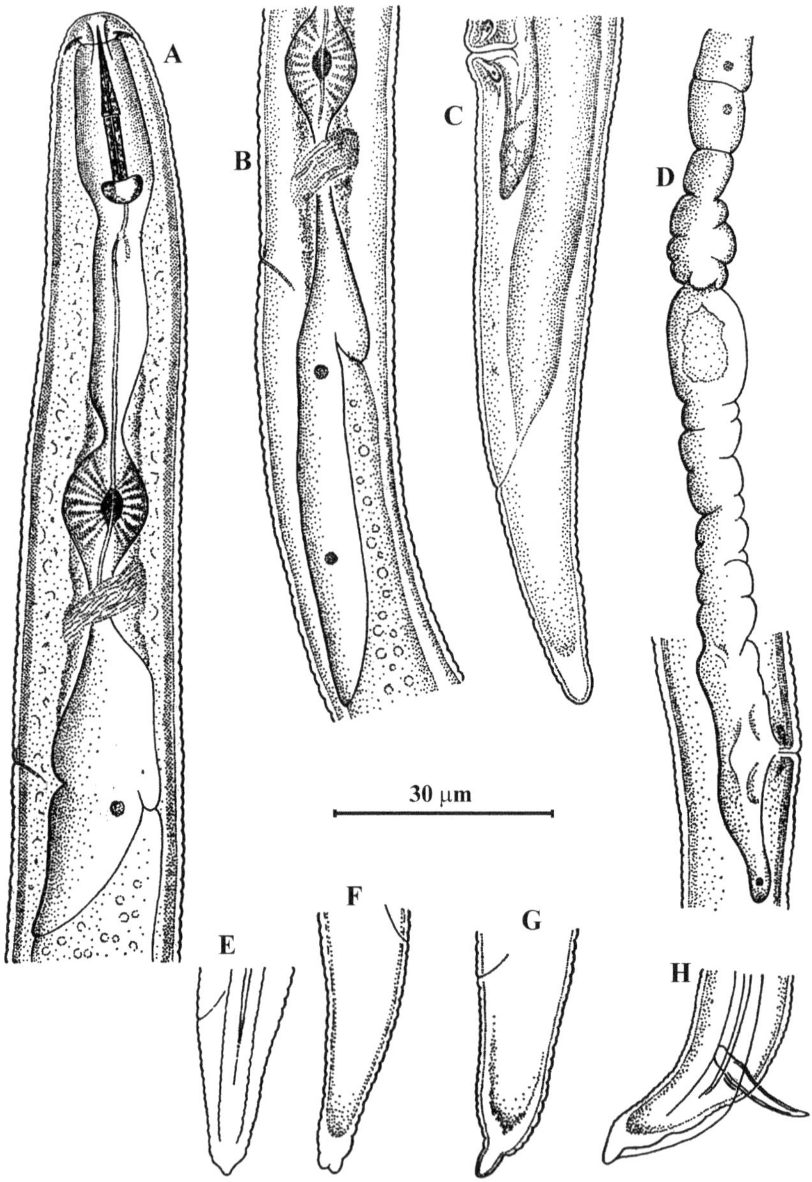

Fig. 62. Pratylenchus macrostylus *Wu, 1971. A: Female pharyngeal region; B: Female pharyngeal region; C: Female posterior region; D: Detail of female reproductive system; E-G: Female tails; H: Male tail. After Wu (1971).*

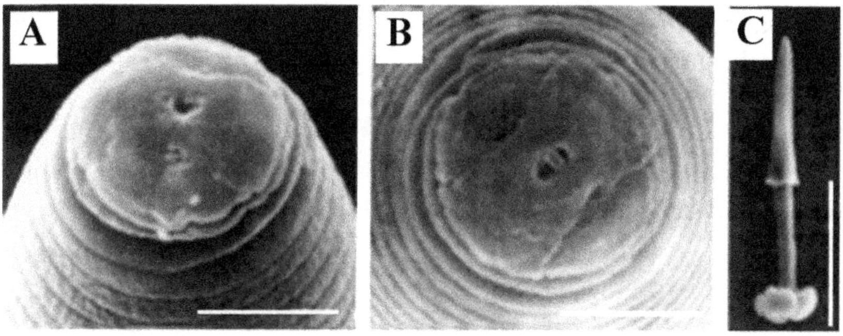

Fig. 63. *SEM micrographs of* Pratylenchus macrostylus *Wu, 1971. A, B: Female* en face *views; C: Female stylet. (Scale bars = 10 μm.) After Hartman and Eisenback (1991).*

labial sensilla pore-like, often obscure; prestoma ovoid. Amphidial openings small, slit-like. Medial lips angular, narrowing near amphidial openings, similar to those of Group 2 (Corbett & Clark, 1983). Labial framework strongly developed, hexaradiate; vestibule and vestibule extension prominent. Stylet robust; conus equal to length of shaft and knobs; shaft cylindrical, fluted and slightly constricted near junction with knobs; knobs prominent, flattened, often cupped anteriorly with anterior projections tapered, rounded, smooth posteriorly. Procorpus wide, cylindrical, narrowing posteriorly near junction with metacorpus. Distance of dorsal pharyngeal gland orifice to stylet base 4.1-8.4 μm; dorsal pharyngeal gland orifice ampulla distinct. Metacorpus muscular; cuticularised valve plates prominent. Nerve ring distinct, hemizonid usually conspicuous, immediately anterior to excretory pore. Basal gland lobe overlapping intestine ventrally, length of overlap 50.0 (28.3-78.3) μm; slight dorso-lateral overlap near pharyngo-intestinal valve. Dorsal pharyngeal gland nucleus conspicuous, anterior to subventral gland nuclei. Subventral pharyngeal gland nuclei near excretory pore, parallel or perpendicular to body axis. Pharyngo-intestinal valve faint. Excretory pore in vicinity of pharyngo-intestinal valve. Intestine filled with globules; rectum cuticularised. Anus small, pore-like. Ovary one, outstretched, rarely flexed; spermatheca round to ovoid, 20.5 (16.5-25.0) μm long, 14.5 (11.0-18.8) μm diam., distance from spermatheca to vulva variable, 58 (41-98) μm. Uterus with tricolumella of 15 cells, occasionally 12. Post-vulval uterine sac short, rounded, occasionally conical posteriorly. Phasmids pore-like. Tail typically conoid; terminus shape variable, usually bluntly rounded with no striations, rarely crenate; 17-26 tail annuli.

Male

General appearance similar to that of female. Testis 171 μm long, reflexed at distal end. Tail length 21 μm.

DIAGNOSIS AND RELATIONSHIPS

Pratylenchus macrostylus is characterised by: stout body, labial region arch-shaped, angular, long and robust stylet and posterior located vulva.

The matrix code is: A1, B2, C5, D2, E4, F2, G3, H1, I3, J1, K1.

It can be distinguished from closely related species *P. brachyurus* and *P. japonicus* by number of lip annuli, shape of the first lip annulus, the elongate stylet, posterior vulva and tail terminus (see the corresponding descriptions).

DISTRIBUTION

Recorded, apart from type locality at Algonquin Park, Ontario, Canada, in soil around the roots of *Betura papyrifera* Marsh., *Picea glauca* (Moench) Voss and *Pseudotsuga menziesii* (Mirb.) Franco; in North Carolina on Fraser fir, *Abies fraseri* (Pursh) Poir. and red spruce, *Picea rubens* Sarg. (Hartman & Eisenback, 1991); and Uzbekistan on cabbage (Rizaeva, 1980).

36. *Pratylenchus manaliensis* Khan & Sharma, 1992
(Fig. 64)

MEASUREMENTS

- Female holotype (after Khan & Sharma, 1992): L = 0.50 mm; a = 26.4; b = 6.2; c = 17; V = 80; stylet = 15.5 μm.
- 21 females (after Khan & Sharma, 1992a): L = 0.57 (0.43-0.64) mm; a = 26 (22-32); b = 5.5 (5-7); c = 22 (16-25); V = 81 (78-83); stylet = 14-16 μm.
- Male (after Khan & Sharma, 1992): L = 0.49 mm; a = 29; b = 6.5; c = 20; T = 49; stylet = 14.5 μm.

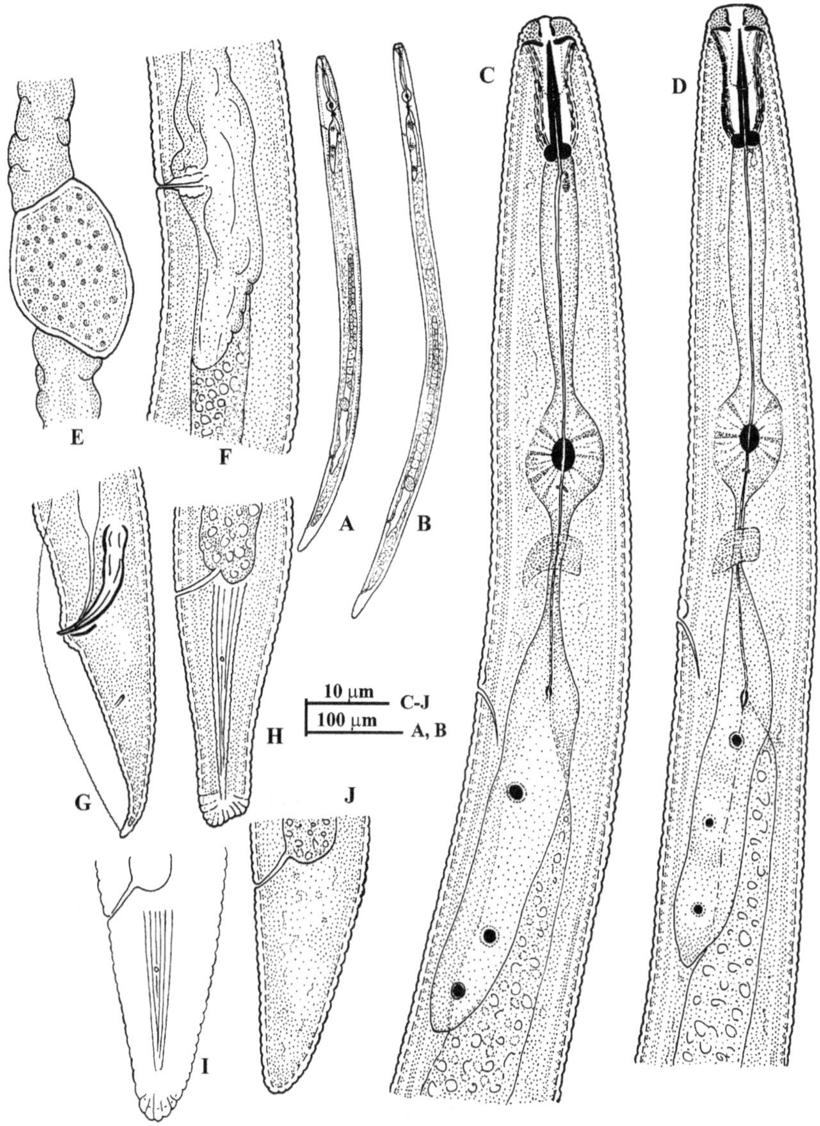

Fig. 64. Pratylenchus manaliensis *Khan & Sharma, 1992. A, B: Entire females; C: Female pharyngeal region; D: Male pharyngeal region; E: Spermatheca; F: Vulval region; G: Male tail; H-J: Female tails. After Khan and Sharma (1992b).*

DESCRIPTION

Female

Female body almost straight anteriorly and slightly curved ventrally in posterior half when fixed. Labial region with three annuli, first one convex. Body cuticle finely striated, 1.2-1.5 μm apart at mid-body. Labial framework well developed, strongly sclerotised, anchor-shaped, its outer margins extending into second body annulus. Lateral fields with six lines, outer two crenate, inner lines smooth and clear in posterior half of body but oblique broken lines often present in pharyngeal region. Stylet robust, metenchium slightly longer than telenchium. Basal stylet knobs rounded. Orifice of dorsal pharyngeal gland 3 μm posterior to stylet base. Excretory pore 75-90 μm from anterior end, slightly anterior to pharyngeal lobe. Hemizonid just anterior to excretory pore. Pharynx typically pratylenchoid, pharyngeal lobe 50-60 μm long, overlapping intestine ventrally for *ca* 2-2.5 body diam. Vulva a transverse slit, reproductive system, prodelphic, outstretched anteriorly. Oocytes arranged in single row. Spermatheca oval to somewhat rectangular with few sperm. Post-vulval uterine sac 18-23 μm long, 1.0-1.5 vulval body diam. long. Tail conoid-rounded, 25-30 μm long, consisting of 19-23 annuli, gradually and slightly depressed on both sides and terminating in a rounded, crenate, terminus.

Male

Similar to female in general morphology. Labial region with slight sexual dimorphism. Spicules paired, 19-23 μm long, slightly arcuate ventrally. Gubernaculum simple crescent shape, 4-5 μm long. Bursa extending to terminus, outer margin finely striated. Single testis with spermatocytes, leading to *vas deferens* and opening through cloaca. Phasmids at about mid-tail.

DIAGNOSIS AND RELATIONSHIPS

Pratylenchus manaliensis is characterised by: labial region with three annuli, lateral fields with six lines, intestine overlapping the rectum, an oblong spermatheca and the presence of males.

The matrix code is: A2, B2, C2, D4, E3, F3, G3, H2, I4, J3, K1.

It can be distinguished from closely related species (*P. crenatus* and *P. teres*) by the number of lip annuli, the shape of the stylet knobs, shape of the spermatheca, length of the post-vulval uterine sac and the presence of males (see the corresponding descriptions).

DISTRIBUTION

It has only been recorded from the type locality at Manali (HP), India, in soil around the roots of apple (*Malus domestica* B.).

37. *Pratylenchus mediterraneus* Corbett, 1983
(Fig. 65)

MEASUREMENTS

- Female holotype (after Corbett, 1983): L = 0.48 mm; a = 28; b = 5.8; b′ = 4.5; c = 21; V = 78; stylet = 15 μm.
- 10 females (after Corbett, 1983): L = 0.51 (0.43-0.58) mm; a = 27 (24-31); b = 6.4 (5.2-7.6); b′ = 4.3 (3.9-4.7); c = 21 (17-25); V = 79 (77-80); stylet = 15 (14-16) μm.
- 5 males (after Corbett, 1983): L = 0.48 (0.43-0.54) mm; a = 31 (27-35); b = 6.0 (5.6-6.3); b′ = 4.4 (4.1-4.6); c = 20 (18-22); T = 40 (34-42); stylet = 14 (13-15) μm; spicules = 17 (16-18) μm; gubernaculum = 5 (4.5-5.5) μm.
- 9 females (after Choi *et al.*, 2006): L = 0.56 (0.48-0.63) mm; a = 24.7 (22-27.1); b = 6.7 (6.0-7.6); c = 17.7 (16.4-19.5); V = 80 (78-84); stylet = 15.4 (14.7-16.1) μm.

DESCRIPTION

Female

Body slender, slightly curved ventrally when heat-relaxed, often narrowing abruptly posterior to vulva. Lateral fields with four lines, outer ones crenate, occasionally areolated in outer bands. Middle band variously ornamented in mid-body, oblique striae becoming double lines near vulva giving appearance of six lines in lateral fields. Lateral fields occasionally widening in mid-tail, two middle lines continuing posterior to phasmid, which lies between them, approximately mid-tail (7-12 annuli from tail tip). Labial region with three annuli, occasionally with first annulus indistinct to give appearance of two wide annuli, high, barely offset from the body with skeleton extending into body *ca* two

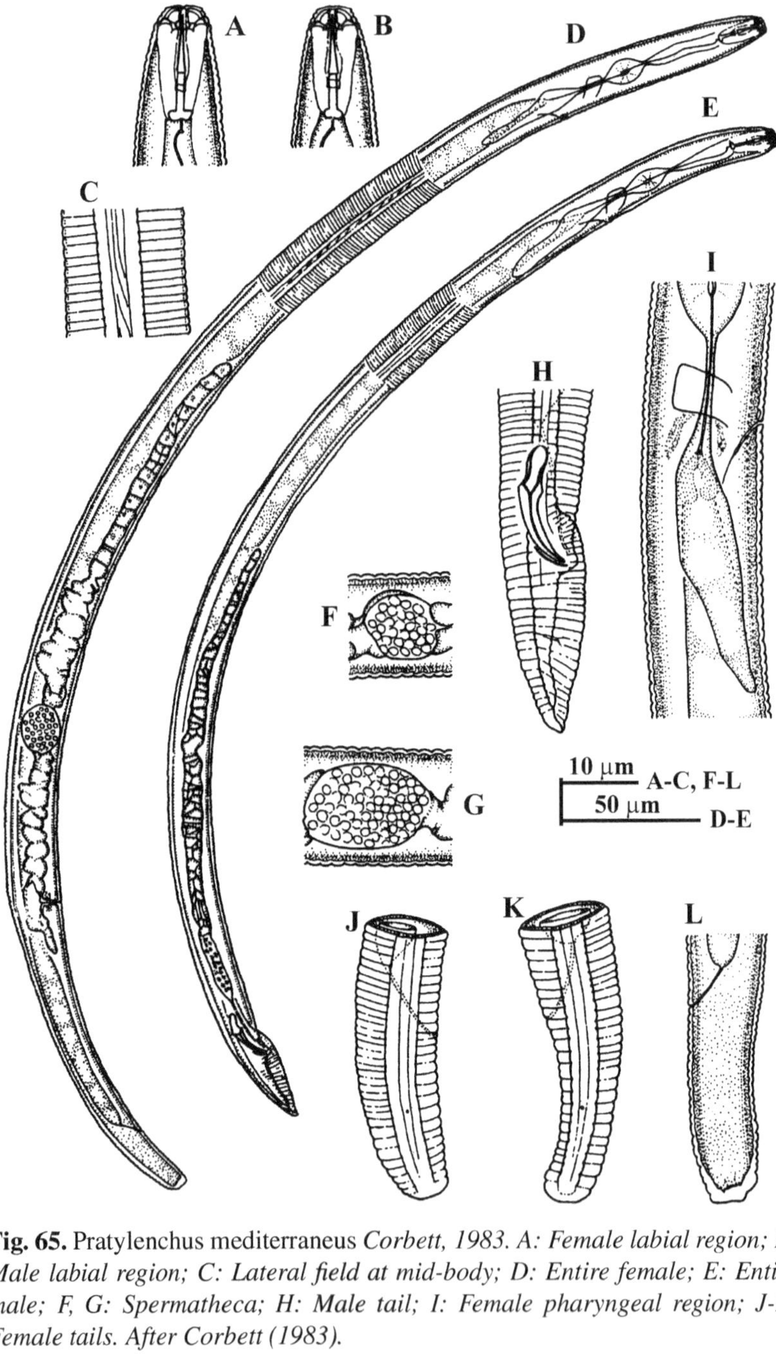

Fig. 65. Pratylenchus mediterraneus *Corbett, 1983. A: Female labial region; B: Male labial region; C: Lateral field at mid-body; D: Entire female; E: Entire male; F, G: Spermatheca; H: Male tail; I: Female pharyngeal region; J-L: Female tails. After Corbett (1983).*

annuli. *En face* view by SEM shows subdorsal and subventral segments of first lip annulus fused to produce a pandurate outline separated from lateral segments each bearing an amphidial aperture on inner edge. Oral aperture oval surrounded by six small pores. Stylet long with rounded basal knobs, dorsal gland opening 1.0-3.3 μm posterior to stylet base. Median bulb ovate, well developed; nerve ring encircling isthmus just anterior to posterior glandular part of pharynx. Pharyngeal lobe overlapping intestine ventrally and laterally 25-55 μm, approximately 1.5-2.5 body diam. long; occasionally pharyngeal glands appearing to be in separate lobes rather than enclosed in single lobe. Excretory pore opposite pharyngo-intestinal junction to one body diam. anterior. Single anterior gonad with spherical to sub-spherical spermatheca, sometimes oval, when full of sperm, located 38-87 μm anterior to vulva. Anterior reproductive tract occupying 27-40% of body with a short post-vulval uterine sac 18-25 μm (approximately one body diam.) long. Tail sometimes bent strongly ventrally, coarsely annulated with 15-22 annuli, smooth tip broadly rounded to truncate, sometimes notched where lateral fields apparently pass around tip, or with cuticular thickening resembling a pad.

Male

Common, body slender, slightly curved ventrally when dead, tail enveloped by crenate-edged bursa. Anterior part of body similar to female but with more slender median pharyngeal bulb. Lateral fields with four lines expanding onto and terminating on bursa as a small pore (the phasmidial tube). Gonad a single outstretched testis (reflexed in one specimen) leading into *vas deferens* opening into cloaca through which the spicules emerge. Spicules paired, cephalated, ventrally arcuate to a blunt point, resting on a simple curved gubernaculum.

Diagnosis and relationships

Pratylenchus mediterraneus is characterised by: labial region with three annuli, vulva at 77-80%, spherical to sub-spherical spermatheca, tail broadly rounded to truncate with smooth tip, males present.

The matrix code is: A2, B2, C2, D2, E2, F3, G2, H1, I3, J1, K1.

It can be separated from the closely related species *P. thornei* in labial region shape and *en face* pattern but differs in having a shorter stylet and by reproducing sexually, with males being common. These differences have been supported by recent molecular analysis of the D3 expansion of the 26S ribosomal genes, which indicated that *P. mediterraneus* and *P. thornei* sequences showed differences in some crucial nucleotide positions, strongly indicating that *P. mediterraneus* should be maintained as a valid species (De Luca *et al.*, 2004). In addition, it differs further from *P. sudanensis* and *P. pseudopratensis* in spermatheca shape and from *P. fallax* and *P. penetrans* in *en face* pattern (Corbett & Clark, 1983) and in labial region shape; and from *P. fallax* in having a smooth tail tip and from *P. penetrans* in having a broadly rounded to truncate tail.

DISTRIBUTION

It has been recorded from the type locality at Sa'ad, Israel, around the roots of *Trifolium alexandrinum* L. It has also been recorded on chickpea in Turkey (Di Vito *et al.*, 1994); chickpea and lentil in Syria (Greco *et al.*, 1992); on wheat at Northern Negev region, Israel (Orion & Shlevin, 1989); and on chrysanthemum in Korea (Choi *et al.*, 2006).

38. *Pratylenchus microstylus* Bajaj & Bhatti, 1984
(Fig. 66)

MEASUREMENTS

- Female holotype (after Bajaj & Bhatti, 1984): L = 0.42 mm; a = 23; b = 6.4; b′ = 3.9; c = 18; c′ = 2.0; V = 76; stylet = 11 μm.
- 11 females (after Bajaj & Bhatti, 1984): L = 0.39 (0.33-0.46) mm; a = 22 (19-26); b = 6.0 (5.3-6.4); b′ = 3.8 (3.1-4.4); c = 18 (16-22); c′ = 2.1 (1.9-2.3); V = 76 (75-77); stylet = 11-12 μm.

DESCRIPTION

Female

Body short and stout, assuming C-shape upon fixation. Lateral fields with four lines, occupying 25% of mid-body diam. and extending to tail tip. Labial region low, almost continuous with rest of body, with three annuli. Labial framework sclerotised. Stylet 11-12 μm long with anteriorly flattened knobs. Corpus constricted near junction with oval median bulb located at 40-42% of pharyngeal length. Pharyngeal gland lobe 2-3 corresponding body diam. or 28-46 μm long. Opening of

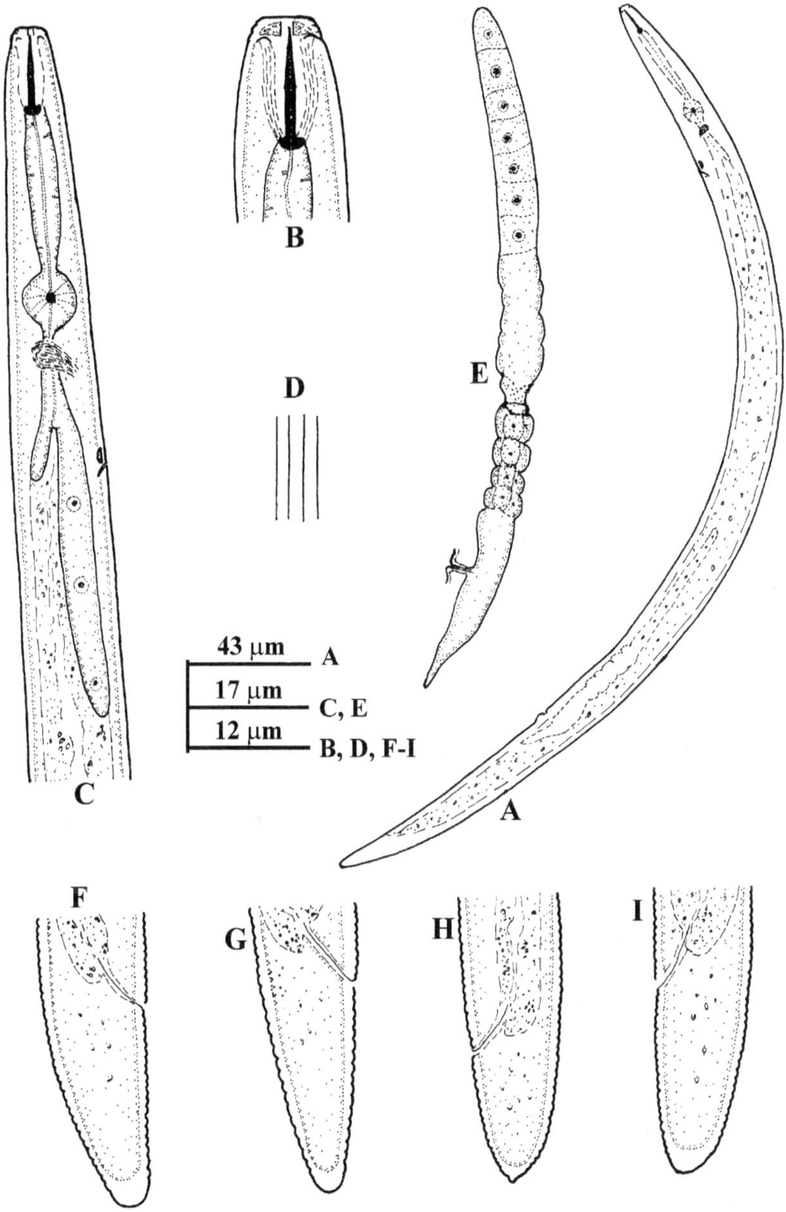

Fig. 66. Pratylenchus microstylus *Bajaj & Bhatti, 1984. A: Entire female; B: Female labial region; C: Female pharyngeal region; D: Lateral field at mid-body; E: Female reproductive system; F-I: Female tails. After Bajaj and Bhatti (1984).*

dorsal pharyngeal gland 3-4 μm from stylet base. Nerve ring encircling isthmus just posterior to median bulb. Excretory pore 59-64 μm from anterior end, or 3 μm anterior or 7 μm posterior to pharyngo-intestinal valve. Hemizonid just anterior to excretory pore. Spermatheca indistinct. Tricolumella well developed. Post-vulval uterine sac length less than vulval body diam. long. Vulva a transverse slit located at 75-77% of body length. Uterine egg in one specimen 53 × 17 μm in size. Intestine overlapping rectum dorsally. Anal body diam. 7-12 μm. Tail usually conoid, striated, with rounded smooth terminus. Tail in one specimen with a terminal process. Phasmids located near mid-tail.

Male

Not found.

DIAGNOSIS AND RELATIONSHIPS

Pratylenchus microstylus is characterised by: labial region low with three annuli, short stylet, spermatheca indistinct, tail usually conoid, striated, with rounded smooth terminus.

The matrix code is: A2, B1, C1, D1, E2, F2, G3, H1, I2, J1, K1.

It can be distinguished from closely related species (*P. mulchandi*) by smaller body length, a shorter stylet and post-vulval uterine sac and overlapping rectum (L = 440-580 μm; stylet = 16-20 μm; post-vulval uterine sac more than 1.5 vulval body diam. long; intestine not overlapping rectum in *P. mulchandi* – see the corresponding description).

DISTRIBUTION

It has only been recorded from the type locality at Hissar, India, from soil and roots of *Sorghum halepense* (L.) Pers.

39. *Pratylenchus morettoi* Luc, Baldwin & Bell, 1986
(Figs 67, 68)

MEASUREMENTS

- Female holotype (after Luc *et al.*, 1986): L = 0.72 mm; a = 35.1; b = 6.2; b′ = 5.2; c = 15.6; c′ = 3.4; V = 76; stylet = 16 μm.
- 36 females (after Luc *et al.*, 1986): L = 0.74 (0.56-0.93) mm; a = 34 (26-40); b = 6.6 (5.3-7.4); c = 15.2 (13-19); c′ = 3.6 (2.8-4.5); V = 76 (73-80); stylet = 16.5 (14-19) μm.

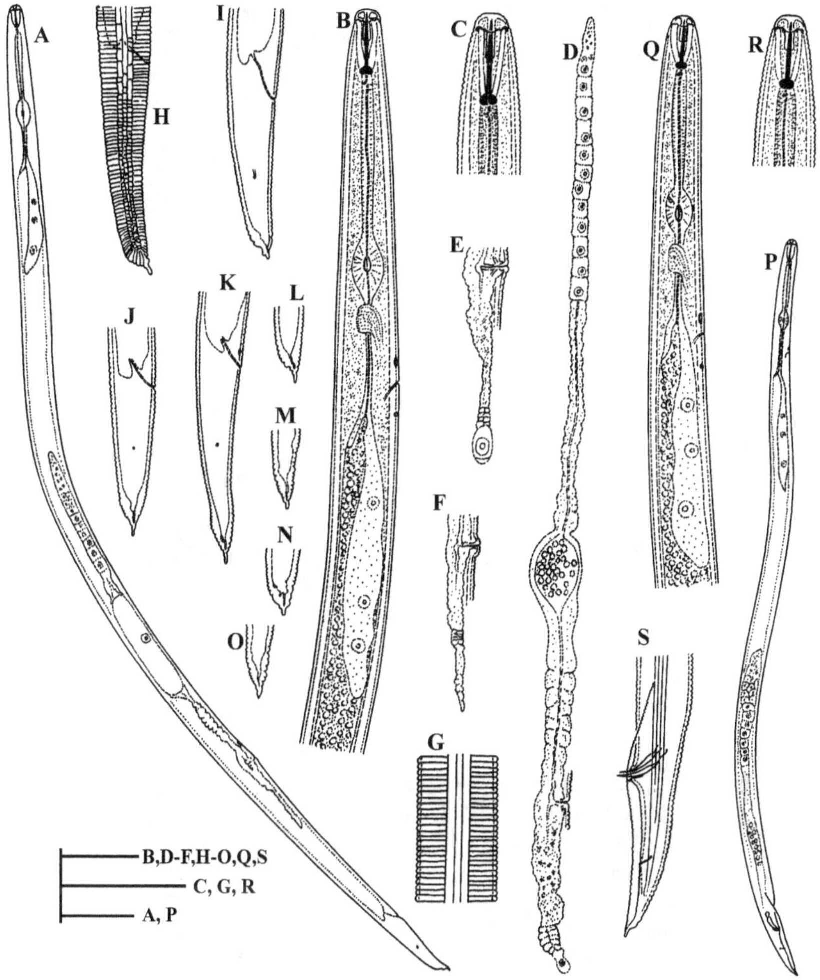

Fig. 67. Pratylenchus morettoi *Luc, Baldwin & Bell, 1986. A: Entire female; B: Female pharyngeal region; C: Female labial region; D: Female reproductive system; E, F: Post-vulval uterine sac; G: Lateral field at mid-body; H-O: Female tails; P: Entire male; Q: Male pharyngeal region; R: Male labial region; S: Male tail. After Luc* et al. *(1986).*

- 22 males (after Luc *et al.*, 1986): L = 0.60 (0.51-0.68) mm; a = 34.7 (30-38); b = 5.8 (5.1-6.5); c = 15.8 (14-17); c′ = 3.1 (2.7-3.6); stylet = 14.5 (13-17.5) μm; spicules = 18.5 (15-21) μm; gubernaculum = 5 (4.0-6.5) μm.

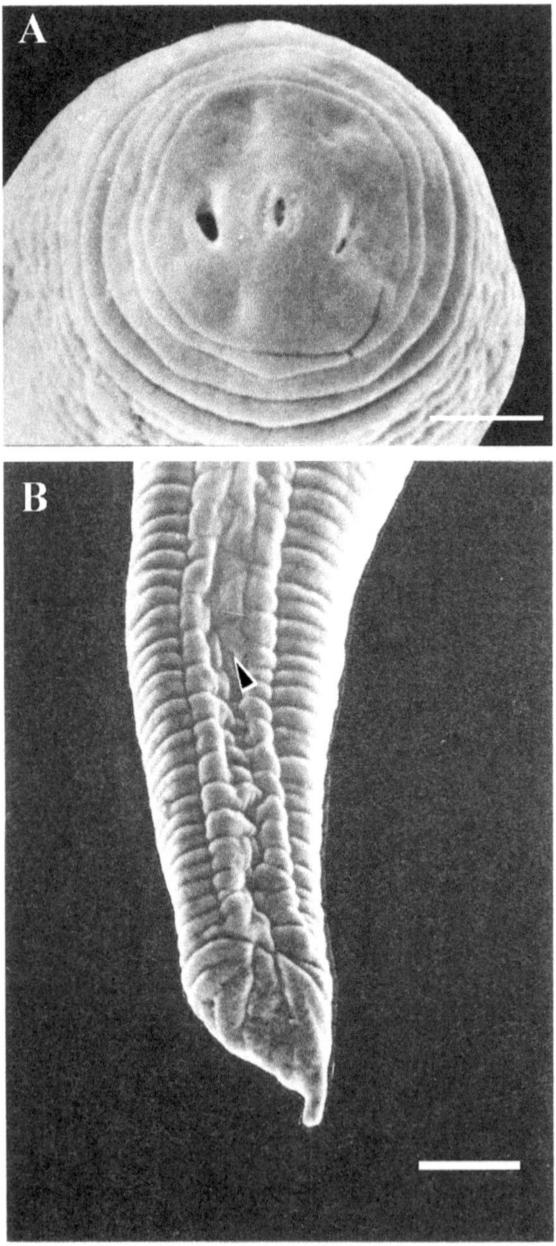

Fig. 68. *SEM micrographs of* Pratylenchus morettoi *Luc, Baldwin & Bell, 1986. A: Female* en face *view; B: Female tail (arrowhead = phasmid). (Scale bars: A = 5 μm; B = 10 μm.) After Luc* et al. *(1986).*

DESCRIPTION

Female

Habitus slightly and regularly ventrally curved after heat relaxation. Body thin, elongate, slightly tapering at anterior extremity; posterior extremity long conical. Cuticle thin (0.5 μm); annuli narrow (1.1 μm at mid-body). Lateral fields composed of four equidistant lines, 6-9 μm wide at mid-body, not areolated except in anterior part and on tail. Labial area dome-shaped, with three or four annuli, not separated from rest of body. SEM *en face* plain; labial sectors fused together and with labial disc; amphidial aperture a large slit; oral aperture ovate, surrounded by six pits (apertures of inner labial sensillae). Cephalids not seen. Stylet straight; basal knobs rounded or slightly sloping posteriorly. Dorsal pharyngeal gland opening at 3-4 μm from base of stylet. Procorpus long and thin; median bulb well marked, ovate, occupying *ca* half of corresponding body diam., with prominent valve; isthmus long and thin; pharyngeal glands in tandem, elongate, with long ventral overlap of intestine; pharyngeal gland nuclei in tandem, posterior one (dorsal gland) somewhat larger. Intestine without fasciculi, forming a very short post-anal sac. Nerve ring encircling anterior half of isthmus. Excretory pore at 95-128 μm from anterior end. Hemizonid flat, 3-4 μm long, 2-5 μm anterior to excretory pore. Hemizonion lenticular, 1.5-2 μm wide, 11-19 μm posterior to excretory pore. Vulva plain; vulval lips slightly protruding; no lateral flaps, no epiptygma. Anterior genital branch straight; ovary with oocytes in one row; spermatheca rounded, axial, thick walled, containing globular spermatozoa 2 μm in diam.; columnar uterus of medium length. Post-vulval uterine sac 46-74 (59) μm long, degenerate with only columnar part of uterus recognisable; in some cases, a cell with a prominent nucleus (remnant of ovary cap cell) present at distal end. Tail long conical; extremity variable, from more or less pointed to stout, but always showing a terminal projection and a terminal thickening of cuticle; lateral fields areolated on posterior two-thirds of tail, with four lines nearly reaching tail extremity. Phasmids punctiform, 17.5-35 μm (26.5) posterior to anus or at 46 (37-54)% of tail length. Caudalids rarely seen, 1.5-2 μm long, 2-3 μm anterior to anus.

Male

Similar to female except for smaller body and stylet length. No secondary sexual dimorphism visible in labial area or pharynx. Spicules curved, cephalated; gubernaculum thin, anteriorly recurved, slightly projecting from cloaca. Caudal alae well developed, crenate, enveloping tail extremity. Tail conical, extremity pointed. Phasmids 12-23 μm (17) posterior to anus, or 43 (30-56)% of tail length, located in notch on edge of caudal alae.

Diagnosis and relationships

Pratylenchus morettoi is characterised by: body thin and elongate, lip area dome shaped, long pharyngeal overlap with the intestine, tail elongate and more conical than cylindrical and the thin, well-marked projection at the tail terminus.

The matrix code is: A2, B2, C3, D3, E2, F6, G3, H3, I4, J1, K1.

It can be distinguished from the closely related species (*P. goodeyi*, *P. pratensis* and *P. ventroprojectus*) by the thin, well-marked ventral projection (see the corresponding descriptions).

Distribution

It has only been recorded from the type locality near Farmington, San Joaquin Co., California, USA, from soil around roots of *Plagiobothrys stipitatus* (E. Greene) I.M. Johnston var. *micranthus* (Piper) I.M. Johnston.

40. *Pratylenchus mulchandi* Nandakumar & Khera, 1970
(Fig. 69)

Measurements

- 17 females (after Zeidan & Geraert, 1991): L = 0.54 (0.47-0.60) mm; a = 30 (26-32); b = 6.0 (5.0-6.8); b′ = 4.1 (3.5-4.7); c = 16 (15-19); c′ = 2.9 (2.1-3.2); V = 75 (72-77); stylet = 17.5 (16.5-18.5) μm.
- 1 male (after Zeidan & Geraert, 1991): L = 0.53 mm; a = 33; b = 7.1; b′ = 4.3; c = 20; c′ = 2.4; T = 34; stylet = 15.5 μm; spicules = 14 μm; gubernaculum = 4.5 μm.
- 8 females (after Zeidan & Geraert, 1991): L = 0.53 (0.49-0.58) mm; a = 29 (28-32); b = 6.4 (5.9-6.6); b′ = 3.9 (3.4-4.6); c = 16 (13-18); c′ = 3.0 (2.5-3.9); V = 73 (71-75); stylet = 17.5 (16.5-18.5) μm.

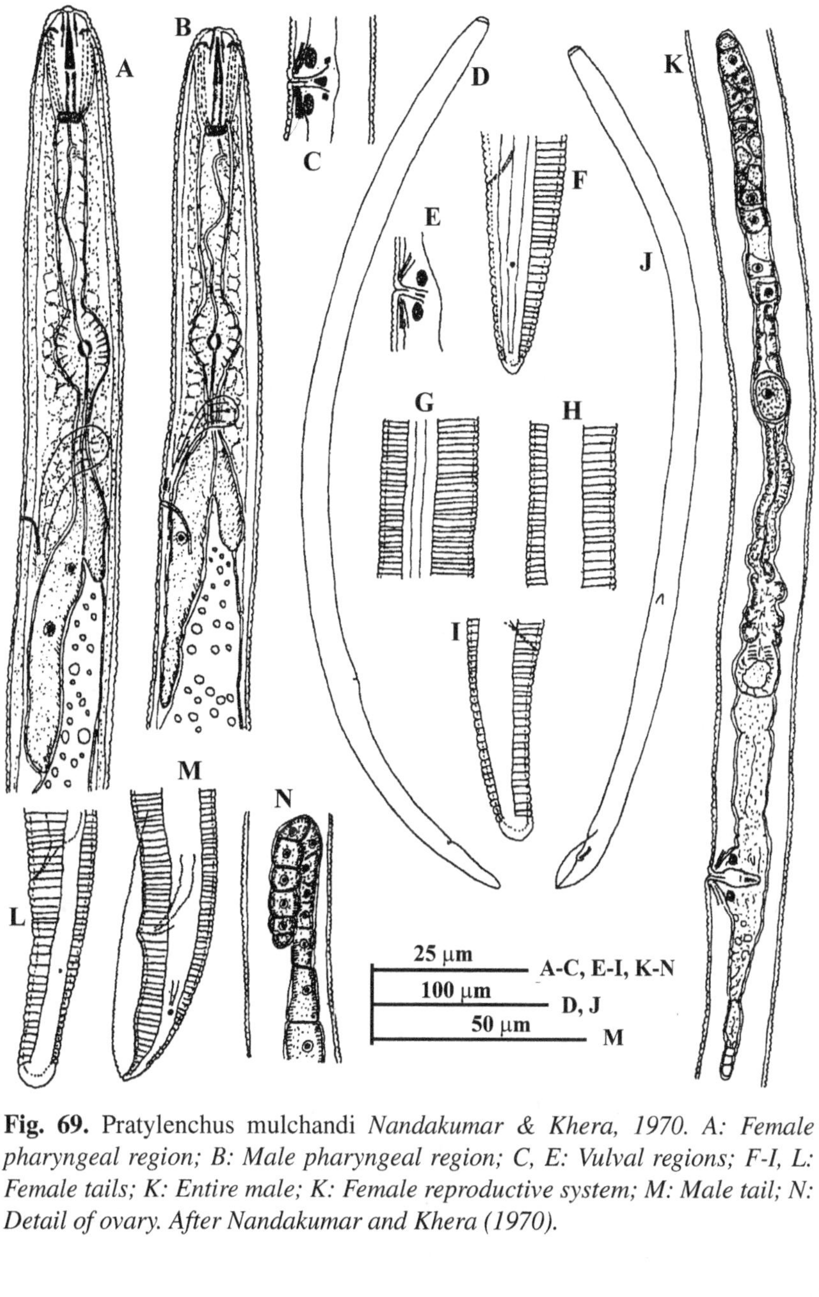

Fig. 69. Pratylenchus mulchandi *Nandakumar & Khera, 1970. A: Female pharyngeal region; B: Male pharyngeal region; C, E: Vulval regions; F-I, L: Female tails; K: Entire male; K: Female reproductive system; M: Male tail; N: Detail of ovary. After Nandakumar and Khera (1970).*

DESCRIPTION

Female

Moderately long nematodes with rather thin bodies showing slightly ventrally arcuate bodies when relaxed. Cuticle transversely striated with fine, less than 1 μm wide, annuli. Lateral fields with four lines; some specimens with two lines by LM, SEM showing disappearance of two or more lines at vulva level. Labial region with three annuli and heavily sclerotised framework. SEM showing a dome-shaped labial region with three annuli, offset from body; with undivided front and oral disc not showing six pits of labial sensilla; amphidial apertures dorso-ventral slits. Stylet very strong, with conical part as long as or shorter than shaft and provided with rounded and slightly anteriorly curved basal knobs. Pharynx with elliptical to oval median bulb and moderately long glandular lobe overlapping intestine ventrally over indistinct pharyngo-intestinal junction. Hemizonid *ca* two annuli wide, situated at *ca* one annulus anterior to excretory pore. Female genital tract anteriorly outstretched in most females, reflexed in a single female. Post-vulval uterine sac long. Oocytes arranged in one or two files; one egg observed in one female measuring 31 $\times$ 23 μm. Spermatheca oval in shape, empty. Vulva a transverse slit. Vagina with a highly refractive triangular peg of variable size and shape. Tail with smooth conical to rounded terminus.

Male

Anterior region similar to female. Lateral fields, as in some females, with only two lines distinct over most of body; two inner lines could just be detected anterior to phasmid. Testis anteriorly outstretched. Bursa enveloping entire tail.

DIAGNOSIS AND RELATIONSHIPS

Pratylenchus mulchandi is characterised by: labial region with three annuli, lateral fields with four lines, oval spermatheca and tail with smooth conical to rounded terminus.

The matrix code is: A2, B2, C3, D3, E2, F5, G3, H1, I2, J1, K1.

It is close to *P. bolivianus*, *P. delattrei* and *P. microstylus* from which it differs by body and stylet length, position of vulva, shape of the spermatheca and female tail shape (see the corresponding descriptions).

DISTRIBUTION

It has been recorded from the type locality at New Halfa City, Eastern Sudan, from the rhizosphere of sugar cane. It has been also recorded in India on pulse crops (Subramaniyan & Sivakumar, 1991; Ali & Askary, 2001); and in Sudan on sugarcane (Zeidan & Geraert, 1991).

41. *Pratylenchus neglectus* (Rensch, 1924) Filipjev & Schuurmans Stekhoven, 1941

(Figs 70-73)

MEASUREMENTS

- Female neotype (after Loof, 1960): L = 0.51 mm; a = 21.2; b = 6.3; c = 22.3; V = 79; stylet = 16 μm.
- 9 females (after Rensch, 1924): L = 0.31-0.58 mm; a = 16.5-32.2; b = 4.9-7.8; c = 20 (13.8-26.8); V = 81 (76-87); stylet = 15-19 μm.
- 2 males (after Rensch, 1924): L = 0.42-0.52 mm; a = 25.5-28.9; b = 6.2-6.3; c = 17.3-21.7; T = 42-56; stylet = 15 μm.
- Females (after Sher & Allen, 1953): L = 0.31-0.55 mm; a = 18-25; b = 4.0-6.3; c = 16-22; V = 80-88; stylet = 16-18 μm.
- 1 male (after Sher & Allen, 1953): L = 0.34 mm; a = 22; b = 4.8; c = 20; stylet = 14 μm.
- 6 females (after Loof, 1960): L = 0.43 (0.33-0.49) mm; a = 22.5 (20.5-26.4); b = 6.0 (5.7-6.1); c = 20.5 (17.7-22.9); V = 83 (82-84); stylet = 16-17 μm.
- 10 females (after Ryss, 1988): L = 0.60 (0.41-0.70) mm; a = 24 (17-31); b = 6.3 (4.9-7.1); c = 21 (13-31); c′ = 1.9 (1.5-2.5); V = 83 (77-85); stylet = 16 (15.5-17.5) μm.
- 2 females (after Zeidan & Geraert, 1991): L = 0.45-0.46 mm; a = 19-24; b = 3.6-3.8; c = 17-18; c′ = 2.4-2.5; V = 83-85; stylet = 16.5-17.5 μm.
- 31 females (after Mizukubo & Minagawa, 1991): L = 0.41 (0.39-0.44) mm; a = 23.1 (19.8-26.0); b = 5.5 (5.0-6.2); b′ = 3.8 (3.3-4.4); c = 18.1 (16.6-20.3); c′ = 2.3 (2.0-2.5); V = 81 (80-82); stylet = 16 (15-16) μm.
- 17 females (after Mizukubo & Minagawa, 1991): L = 0.42 (0.38-0.47) mm; a = 22.4 (16.5-27.0); b = 5.4 (4.5-6.6); b′ = 3.7 (3.3-

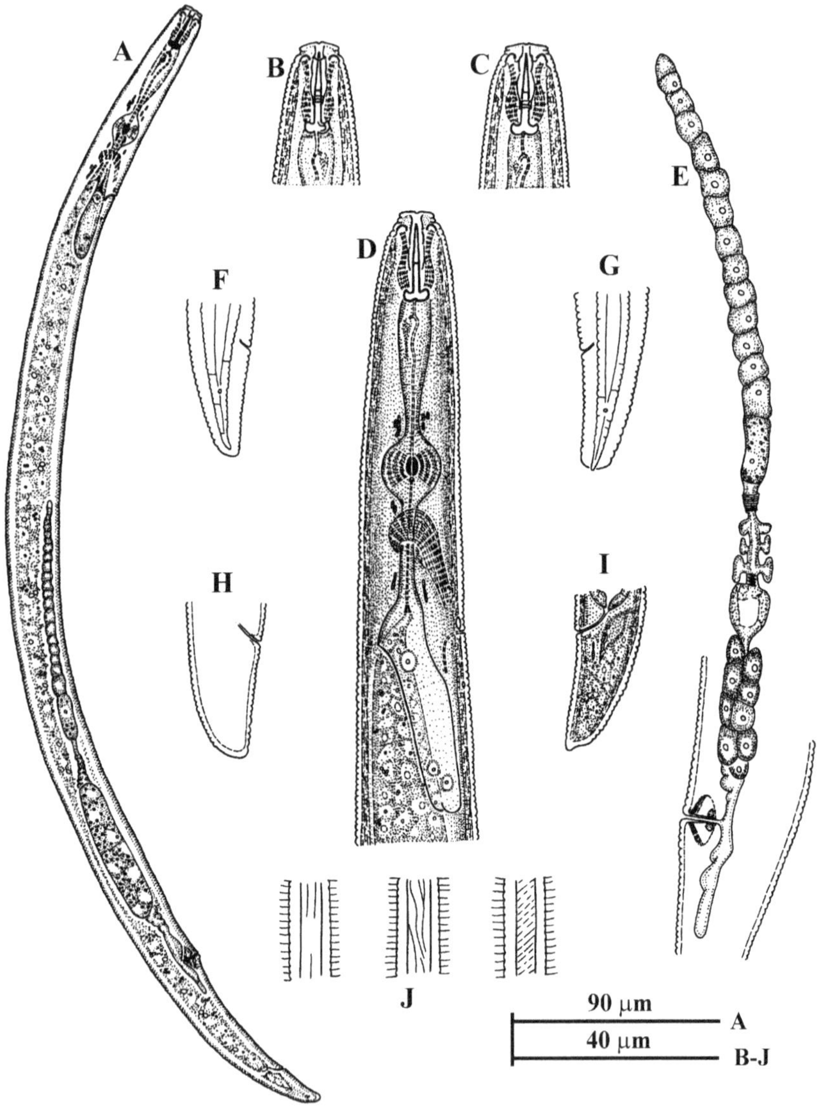

Fig. 70. Pratylenchus neglectus *(Rensch, 1924) Filipjev & Schuurmans Stekhoven, 1941. A: Entire female; B, C: Female labial regions; D: Female pharyngeal region; E: Female reproductive system; F-I: Female tails; J: Variability of lateral field. After Townshend and Anderson (1976).*

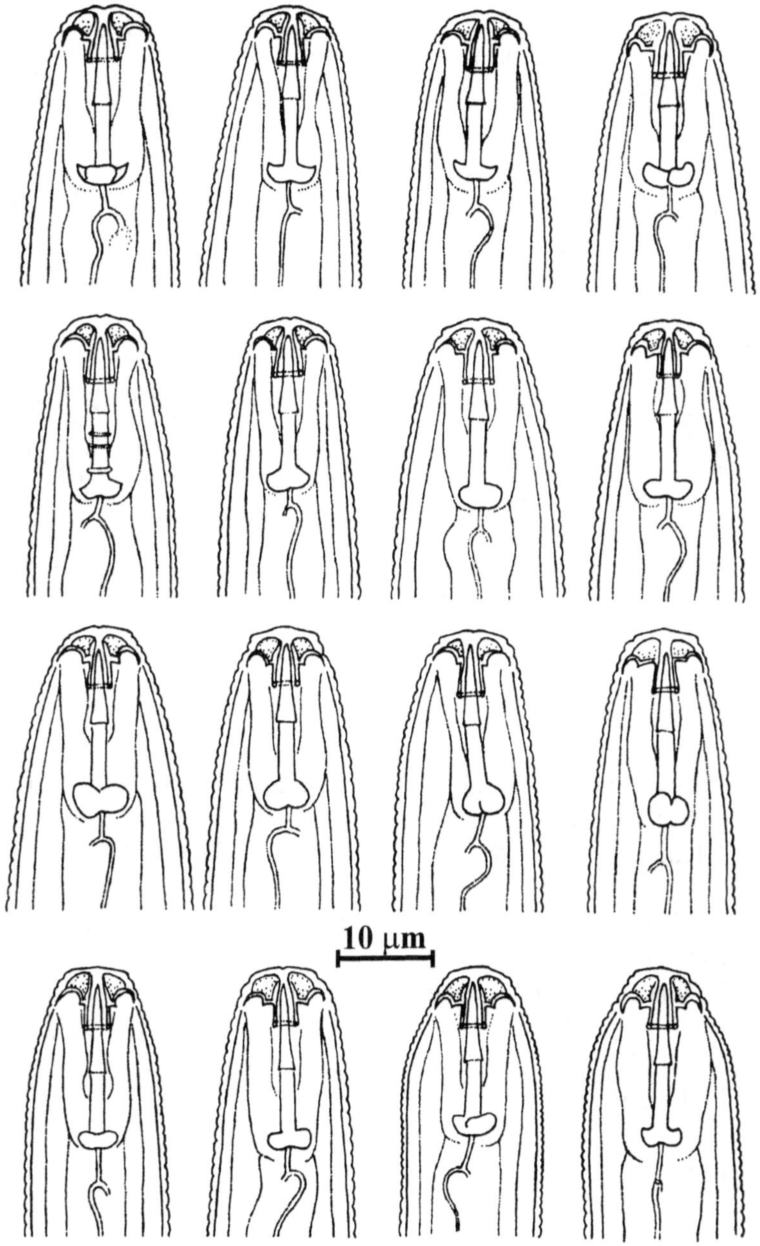

Fig. 71. *Variability in the female labial region of* Pratylenchus neglectus *(Rensch, 1924) Filipjev & Schuurmans Stekhoven, 1941. After Mizukubo and Minagawa (1991).*

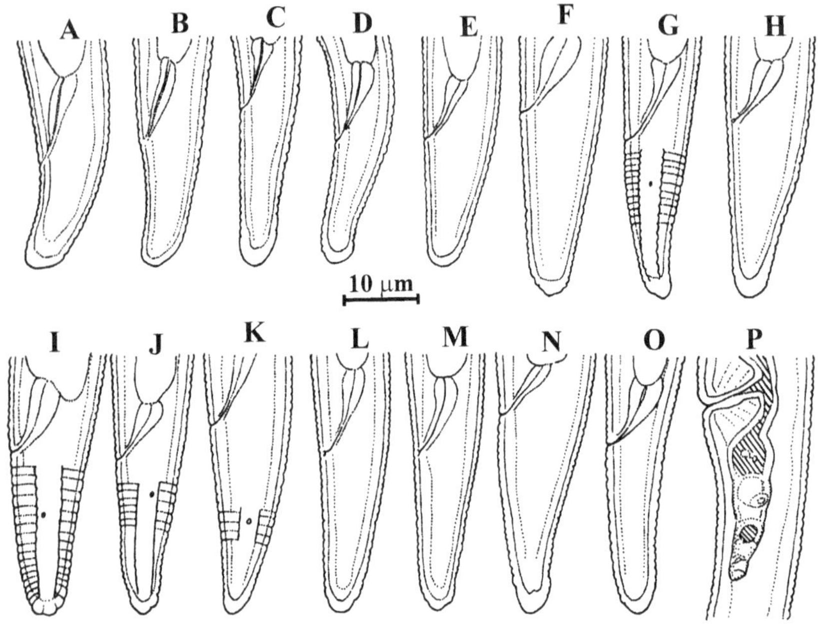

Fig. 72. *Variability in the female tails of* Pratylenchus neglectus *(Rensch, 1924) Filipjev & Schuurmans Stekhoven, 1941. A-O: Female tails; P: Vulval region. After Mizukubo and Minagawa (1991).*

4.5); c = 17.8 (14.6-21.4); c′ = 2.3 (2.0-2.8); V = 81 (78-83); stylet = 16 (15-17) μm.

- 36 females (Doucet & Cagnolo, 1998): L = 0.43 (0.30-0.60) mm; a = 26.4 (20.4-32.8); b = 5.8 (4.4-7.0); b′ = 4.0 (3.1-4.7); c = 22.9 (16.3-27.4); c′ = 2.0 (1.6-2.4); V = 81 (78-84); stylet = 16.6 (16.0-17.5) μm.
- 32 females (after Pourjam *et al.*, 1999a): L = 0.49 (0.34-0.59) mm; a = 26.7 (22.3-32.2); b = 6.0 (4.3-7.6); b′ = 4.3 (3.4-5.2); c = 21.4 (14.8-28.2); c′ = 2.1 (1.5-3.0); V = 82 (80-85); stylet = 16 (14-18) μm.

DESCRIPTION

Female

A parthenogenetic species characterised by great variation in body length and diam., tail shape and thickness and early maturity of adults, which is sometimes completed during final ecdysis. Feeding, mature

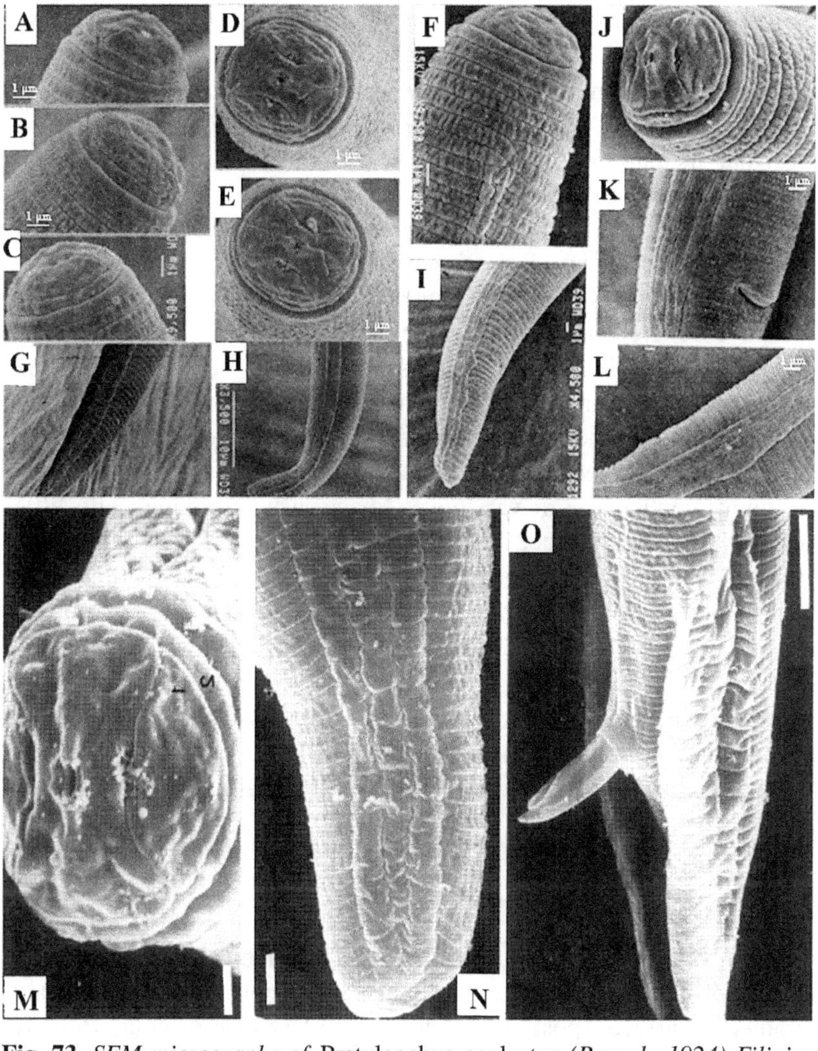

Fig. 73. *SEM micrographs of* Pratylenchus neglectus *(Rensch, 1924) Filipjev & Schuurmans Stekhoven, 1941. A-C, F: Female labial regions; D, E, J, M: Female* en face *views (1, 2 refers to number of lip annuli); G-I, N: Female tails; K, L: Lateral field at mid-body; O: Male tail. (Scale bars: M = 1 μm; N = 2 μm; O = 3 μm.) After Pourjam* et al. *(1999); Hernández* et al. *(2000).*

females extracted from roots are often markedly more robust than those from the rhizosphere, which become more linear when relaxed. Labial region bluntly rounded, with two annuli, second annulus wider than first, apical one comprising the lips. As shown by Corbett and Clark (1983) and Sher and Bell (1975), the dorsal and ventral submedian lips are fused, forming a large cap while lateral lips are delineated, amphidial apertures set at an oblique angle. Lateral fields with four lines, but median zone often marked by one or two longitudinal or several oblique striae. Stylet knobs 4-6 μm across, typically indented on anterior surfaces. Pharyngeal gland overlapping intestine ventrally or ventrolaterally, subventral gland nuclei at end of lobes, not in tandem. Excretory pore 75-87 μm from anterior end. Hemizonid immediately anterior to excretory pore, extending over 2-3 body annuli. Female monodelphic, prodelphic, ovary outstretched with oocytes in tandem, occasionally extending to base of pharynx. Intra-uterine eggs in older adults may be segmented or contain early first-stage juveniles. Post-vulval uterine sac less than or equal to body diam., 12-18 μm long and undifferentiated. Tail variable in shape, usually conoid with little curvature of ventral surface and usually with 15-20 annuli. Tail terminus without annulation, usually rounded, but may be obliquely truncate or slightly digitate. Phasmids in posterior half of tail.

Male

Males seldom found, only three specimens known, of no appreciable diagnostic or reproductive significance. Similar to female. Single out-stretched testis. Phasmids slightly posterior to mid-tail, not appearing to extend into delicate bursa.

DIAGNOSIS AND RELATIONSHIPS

Pratylenchus neglectus is characterised by: labial region with two annuli, second annulus wider than first, stylet knobs typically indented on anterior surfaces, post-vulval uterine sac less than or equal to body diam., tail variable in shape usually conoid with little curvature of ventral surface and tail terminus without annulation, usually rounded, but may be obliquely truncate or slightly digitate.

The matrix code is: A1, B2, C3, D1, E2, F1, G3, H1, I1, J1, K1.

It is close to *P. panamaensis*, *P. roseus* and *P. yamagutii* from which it differs by body and stylet length, position of the vulva, tail shape and tail terminus (see the corresponding descriptions).

As indicated by Loof (1960) and Frederick and Tarjan (1989), there are no reliable morphometric or physical criteria which separate *P.*

neocapitatus Khan & Singh, 1975 from *P. neglectus*, thus the former is maintained as synonym of the latter.

Loof (1978) synonymised *P. similis* to *P. neglectus* on the basis of re-examination of the four specimens from Jadid, finding that the stylet length was 16 μm, not 13-14 μm, as originally reported. This action is maintained here.

Pratylenchus gotohi Mizukubo & Minagawa, 1991 was described from several uncultivated hosts and localities at Japan: *Miscanthus sinensis* Anderss. and *Sasa* sp. at Teradoko, Kokonoe and Oita; and *Pennisetum alopecuroides* Hameln and *Imperata cylindrica* L. at Kuroishi, Nishigoshi and Kumamoto. This species was differentiated from *P. neglectus* by the more posterior excretory pore (77 (72-85) *vs* 67 (63-74) μm); longer posterior uterine branch (5.4 (4.5-8.1) *vs* 3.7 (3.3-4.3) μm); anterior position of the vulva (V = 81 (78.3-82.8) *vs* 82.2 (81.2-82.8)); higher stylet knobs (2.0-3.3 *vs* 1.7-2.0 μm); lower m-value (47.8 (45.8-50.0) *vs* 50.7 (47.8-54.2)%); and number of tail annuli (17 *vs* 20). Pourjam *et al.* (1999a) studied several populations of *P. neglectus* from Iran and found that morphological and morphometrical characters of these populations were similar to *P. gotohi* and therefore they considered both species as synonymous. We concur with this action and therefore both species are maintained here as synonyms. Additionally, Mizukubo and Minagawa (1991) showed great variation in the morphology of the stylet knobs of several populations of *P. neglectus* (Fig. 71) and slight differences in tail terminus between some populations (Fig. 72).

DISTRIBUTION

It has been recorded from the type locality at Theessen near Maagdeburg, Germany, from roots of rye. It has been recorded on several hosts in various localities in North America: California in pistachio orchards (McKenry & Kretsch, 1984) and from soil around roots of pear (Sher & Allen, 1953); Iowa on corn (Williams, 1982); New York (Timper & Brodie, 1997); Ohio on potato (Brown *et al.*, 1980); Washington on wheat (Mojtahedi *et al.*, 1992); in dry land field crops in the semi-arid Pacific Northwest (Smiley *et al.*, 2004); and in Canada on turf grass (Yu *et al.*, 1998) and potato (Olthof & Wolynetz, 1991). It has been widely recorded in several countries of Europe: Bulgaria in fruit tree nurseries (Choleva *et al.*, 1984); Croatia on tobacco (Ostrec & Grubisic, 2003) and soybean (Kelic, 1991); in Estonia (Ryss, 1992); France on colza (Scotto

La Massese *et al.*, 1981); Germany on corn (Knuth, 2002); Italy on cereals (Inserra *et al.*, 1978; Palmisano, 1992; Troccoli, 1995) and corn (Tacconi *et al.*, 1988b); Poland on alfalfa and red clover (Kornobis, 1987, 1990) and cabbage (Brzeski, 1969); Portugal (Abrantes *et al.*, 1987); Russia on cereals (Shagalina *et al.*, 1988); Slovakia on cereals (Valocka & Sabova, 1978); Slovenia on corn (Urek *et al.*, 2003); Spain in horticultural crops (Espárrago & Navas, 1995), on cereals and legumes (Castillo *et al.*, 1996a; Talavera & Tobar Jiménez, 1997), in mountainous pastures (Talavera & Navas, 2002), several hosts and localities (Gómez Barcina *et al.*, 1989; Peña-Santiago, 1990; Hernández *et al.*, 2000; Escuer & Bello, 2003); and Yugoslavia (Grujicic, 1974). It has been recorded in some countries of Africa: Algeria and Tunisia on several crops (Troccoli *et al.*, 1992; Troccoli & Di Vito, 2002); Morocco on wheat (Meskine & Abbad Andalousi, 1993); South Africa on soybean and wheat (Jordaan *et al.*, 1992; Fourie *et al.*, 2001). It also recorded from Asia: India on sugarcane (Mehta & Sundararaj, 1990); Iran (Pourjam *et al.*, 1999a; Kheiri *et al.*, 2002a); in several localities of Japan (Mizukubo & Minagawa, 1991; Orui & Mizukubo, 1999b); Oman (Mani *et al.*, 1997) and Turkey (Menna & Handoo, 2006). It is recorded in Central and South America from several hosts and localities in Argentina (Doucet & Cagnolo, 1998; Torres & Chaves, 1999); Mexico on strawberry (Sandoval Hernandez & Teliz Ortiz, 1990); and from cereals in Western Australia (Riley & Kelly, 2002) and carrots in Tasmania, Australia (Hay & Pethybridge, 2005).

42. *Pratylenchus neobrachyurus* Siddiqi, 1994
(Fig. 74)

MEASUREMENTS

- Female holotype (after Siddiqi, 1994): L = 0.41 mm; a = 31.5; b = 5.5; b′ = 3.3; c = 16; c′ = 2.9; V = 82; stylet = 16.5 μm.
- 10 females (after Siddiqi, 1994): L = 0.36 (0.31-0.41) mm; a = 25 (20-31); b = 4.9 (4.7-5.5); b′ = 3.0 (2.9-3.5); c = 15.2 (14.5-16.8); c′ = 2.8 (2.4-3.0); V = 81 (80-84); stylet = 15.5 (14.5-17.0) μm.
- 2 males (after Siddiqi, 1994): L = 0.40-0.41 mm; a = 24-28; b = 4.6-4.9; b′ = 2.9-3.0; c = 16-17; c′ = 2.2-2.8; T = 40-42; stylet = 13.5-14.0 μm.

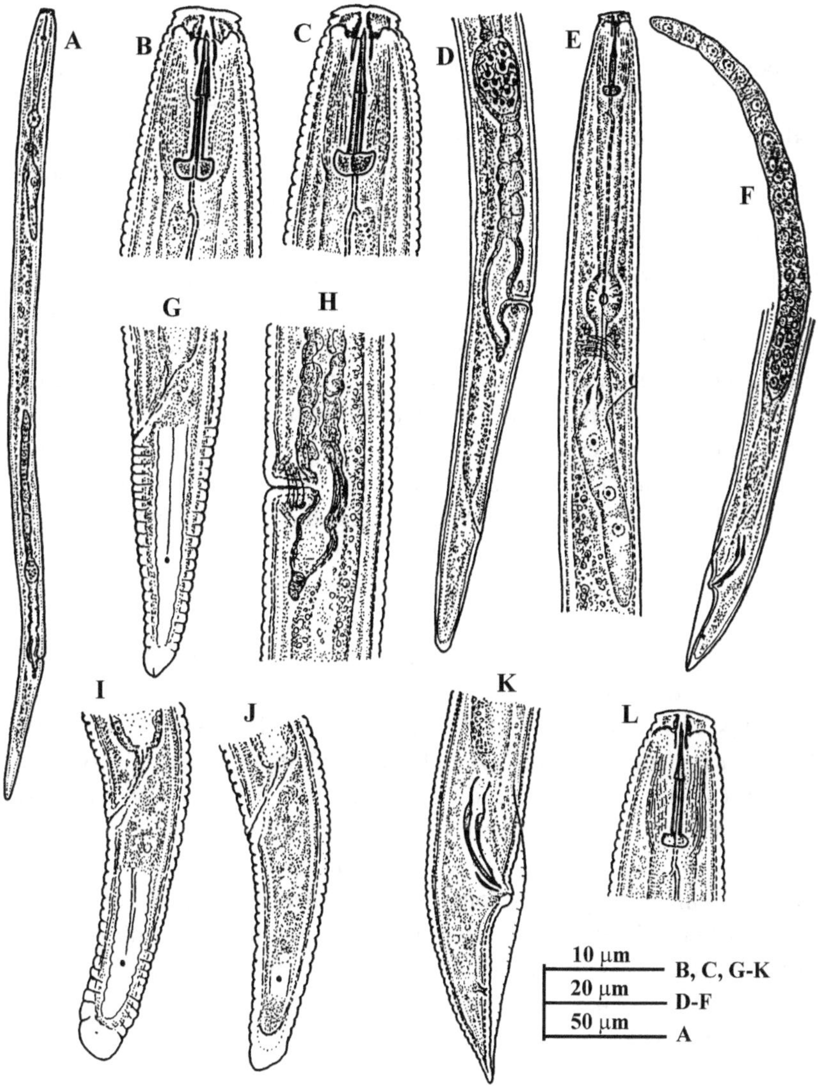

Fig. 74. Pratylenchus neobrachyurus *Siddiqi, 1994. A: Entire female; B, C: Female labial regions; D: Female posterior region; E: Female pharyngeal region; F: Male reproductive system; G, I, J: Female tails; H: Vulval region; K: Male tail; L: Male labial region. After Siddiqi (1994).*

DESCRIPTION

Female

Body straight to slightly arcuate ventrally upon relaxation; max. diam. 14-17 μm. Cuticle finely annulated; annuli 1.0-1.1 μm wide near mid-body. Lateral fields *ca* one-third of body diam., in pharyngeal-vulva region with four lines forming three bands of which middle band slightly narrower than outer ones, in vulva-phasmid region usually with three lines forming two equal bands and in post-phasmid tail region with two lines; not areolated. Labial region low, truncate, almost continuous with body, with two annuli; anterior annulus angular, similar to that of *P. brachyurus* (hence the species name); labial framework heavily sclerotised; outer margins extending one annulus into body. Cephalids not seen. Stylet robust, *ca* twice labial region diam. long, in two almost equal parts; conus tubular, 6.5-8.0 μm long; basal knobs large, rounded with anteriorly flattened surfaces, 4.5-5.0 μm across and 2.4-2.6 μm high. Orifice of dorsal pharyngeal gland 3.0-3.5 μm posterior to stylet base. Median pharyngeal bulb round to oval, 10-12 $\times$ 9-10 μm. Excretory pore opposite junction or slightly posterior to it, 74 (67-79) μm from anterior end of body, 0-1 annulus posterior to hemizonid which is distinct and 2-3 annuli wide. Pharyngeal glands extending mostly on ventral side of intestine; distance from anterior end of body to end of glands 117 (108-126) μm. Vulva a depressed transverse slit, at 65 (58-72) μm from tail tip; lips not raised. Annuli between vulva and anus 40 (38-42). Vagina extending halfway into body. Ovary anteriorly outstretched, short with 7-10 oocytes. Spermatheca oval, 17-23 μm long and 9-10 μm diam., in several females with sperm. Post-vulval uterine sac 1-1.5 vulval body diam. long, with rudiments of gonad. Rectum *ca* one anal body diam. long. Tail conoid to subcylindroid, with rounded terminus which may be smooth or marked by single indentation, 23 (19-26) μm, or 2.7 (2.4-3) anal body diam. long, with 21 (19-23) annuli; phasmids pore-like, a little posterior to mid-tail.

Male

Body straight to arcuate; max. diam. 12-13 μm. Body annuli *ca* 1.1 μm wide near middle. Lateral fields one-fourth of body diam., with four lines. Hemizonid three annuli long, just posterior to excretory pore. Stylet 13.5-14.0 μm long; knobs 3.3-3.5 μm across. Orifice of dorsal pharyngeal gland 4 μm posterior to stylet knobs. Median pharyngeal bulb oval, *ca* 10 × 7 μm. Lips of cloacal aperture rounded, elevated. Spicules ventrally curved near middle, 13-15 μm long. Gubernaculum simple, 4.5 μm long. Tail conoid with round or indented tip, completely enveloped by bursa. Phasmids just posterior to mid-tail.

Diagnosis and relationships

Pratylenchus neobrachyurus is characterised by: labial region with two annuli, first annulus angular, lateral fields with four lines in pharyngeal-vulva region, three lines in vulva-phasmid region and two lines in post-phasmid tail region, spermatheca oval and tail conoid to subcylindroid, with a rounded terminus which may be smooth or marked by single indentation.

The matrix code is: A1, B2, C2, D3, E3, F1, G3, H1, I2, J1, K1.

It is close to *P. alleni*, *P. brachyurus* and *P. loosi* from which it differs by being bisexual and having a smaller body, shape of labial region, a shorter stylet with anteriorly flattened knobs and a more anterior position of vulva and female tail shape (see the corresponding descriptions).

Distribution

It has been recorded from the type locality at Carimagua, Colombia, from soil around roots of grass *Andropogon guyanus* L. and in Venezuela (Crozzoli, 2002).

43. *Pratylenchus okinawaensis* Minagawa, 1991
(Fig. 75)

Measurements

- Female holotype (after Minagawa, 1991): L = 0.46 mm; a = 24.4; b = 5.4; b′ = 3.4; c = 18.2; c′ = 3.4; V = 78.5; stylet = 17 μm.
- 25 females (after Minagawa, 1991): L = 0.46 (0.39-0.51) mm; a = 24.9 (21.0-29.3); b = 5.3 (4.7-6.0); b′ = 3.7 (2.9-5.0); c = 18.3 (15.6-21.4); c′ = 2.3 (1.6-2.9); V = 79 (77-82); stylet = 17 (16-18) μm.
- 24 males (after Minagawa, 1991): L = 0.43 (0.37-0.49) mm; a = 27.3 (23.9-31.1); b = 5.1 (4.2-5.8); b′ = 4.0 (3.5-4.4); c = 16.8 (13.9-19.6); c′ = 2.3 (2.0-2.7); T = 52 (35-63); stylet = 16 (14-17) μm; spicules = 18 (17-19) μm; gubernaculum = 3.9 (3.3-4.7) μm.

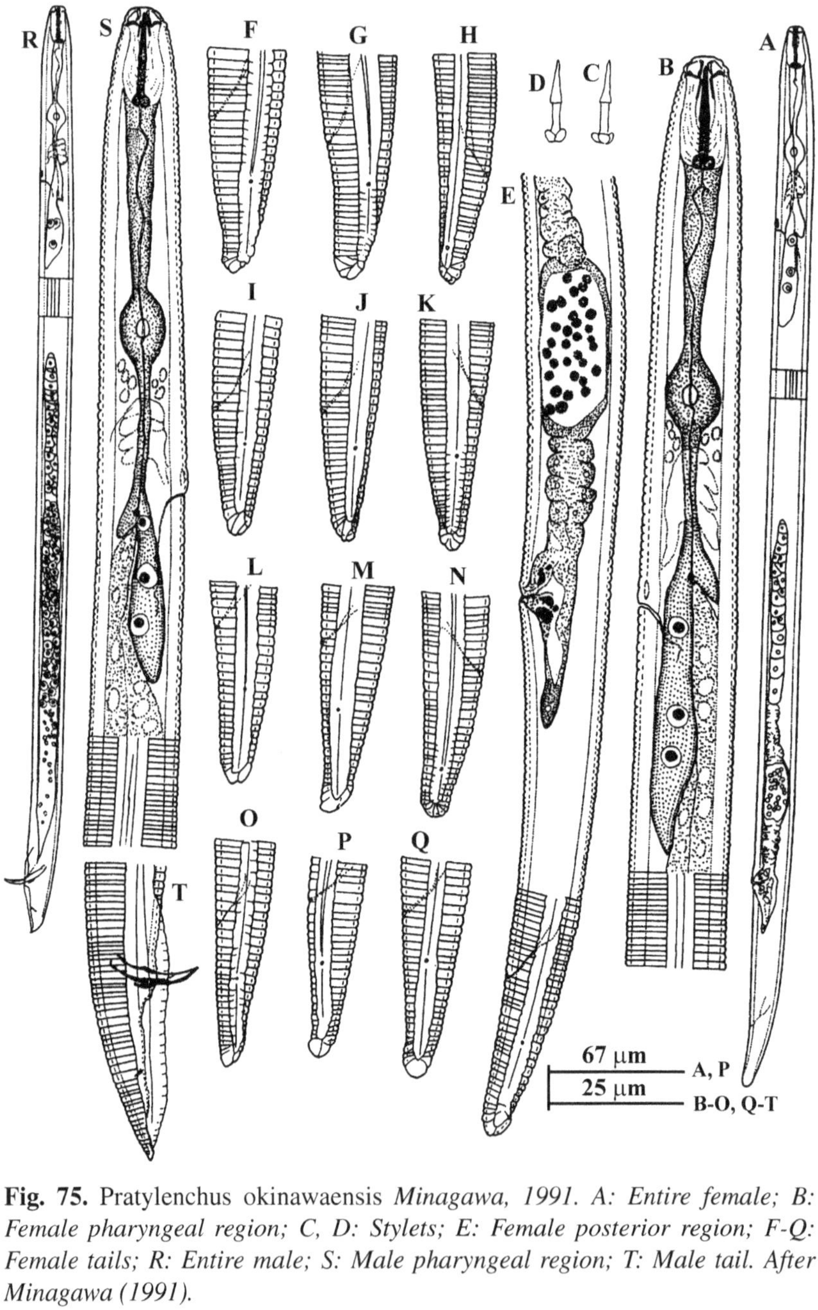

Fig. 75. Pratylenchus okinawaensis *Minagawa, 1991. A: Entire female; B: Female pharyngeal region; C, D: Stylets; E: Female posterior region; F-Q: Female tails; R: Entire male; S: Male pharyngeal region; T: Male tail. After Minagawa (1991).*

Description

Female

Body straight or slightly curved ventrally after heat-relaxation; narrowing posterior to vulva to broad rounded tail terminus and to flattened lip at anterior end, finely annulated. Labial region with two annuli, rarely three (one of 25 females measured), setoff from body by a small constriction. Labial framework moderately sclerotised, extending into body mostly one annulus, occasionally 1.5-2 annuli. *En face* view, subdorsal and subventral sectors fused to oral disc, forming angular and broad dumbbell shape; oral opening oval, surrounded by six small pores on convex edge of opening. Lateral edges of submedian sectors of first lip annulus irregularly shaped, seemingly intermediate between those of Group 2 (rounded dumbbell-shaped patterns and analogies) and Group 3 (angular dumbbell-shaped patterns) *sensu* Corbett and Clark (1983). Lateral fields with four lines, outer lines irregularly crenate in tail region. Stylet moderate; knobs rounded, 2.7 (2.3-3.0) μm high, 3.9 (3.3-6.3) μm across. Dorsal pharyngeal gland orifice 2.7 (2.0-3.7) μm posterior to stylet base. Median bulb oval, well developed, nerve ring around isthmus; pharyngeal lobes overlapping intestine ventrally and laterally; pharyngo-intestinal junction around level of excretory pore. Excretory pore 87 (77-93) μm from labial region, immediately posterior to hemizonid, which is two or three annuli long. Reproductive system single; spermatheca oval to elongate oval 32.6 (16.0-50.0) μm long and 11.2 (8.0-13.3) μm diam., filled with rounded spermatozoa. Post-vulval uterine sac short, 20.4 (13.3-28.3) μm or 1.0 (0.8-1.3) vulval body diam. long. Tail 25.4 (20.0-30.0) μm long, with 21 (17-25) annuli, terminus serrate or irregularly annulated. Phasmids usually in anterior half of tail, 16.1 (13.0-18.7) μm from tail tip.

Male

Body slender, slightly curved ventrally when heat-relaxed. Pharynx and stylet similar to female but less developed. Lateral fields with four lines and oblique striae in central zone in some specimens and terminating on bursa. Gonad single with outstretched testis. Spicules arched, with blunt point; gubernaculum crescent-shaped. Phasmids rodlike, usually at mid-tail region.

DIAGNOSIS AND RELATIONSHIPS

Pratylenchus okinawaensis is characterised by: two lip annuli, crenate and rounded tail terminus and elongate oval spermatheca filled with spermatozoa.

The matrix code is: A1, B2, C3, D3, E2, F4, G3, H2, I4, J1, K1.

It is close to *P. flakkensis*, *P. gibbicaudatus* and *P. yassini* from which it can be distinguished by number of lip annuli, stylet length, position of the vulva and morphology of tail terminus (see the corresponding descriptions).

DISTRIBUTION

It has only been recorded from the type locality at Kunigami, Okinawa, Japan, in the rhizosphere of the following plants: *Kalopanax pictus* (Thunb.) Nakai, *Glochidion triandrum* (Blanco) C.B. Robinson, *Macaranga tanarius* Muel.-Arg., *Daphniphyllum teijsmannii* Zoll., *Pittosporum tobira* (Thunb.) Aiton, *Litsea acuminata* (Bl.) Kurata, *Persea japonica* (Sieb. & Zucc.) Kosterm., *Cryptomeria japonica* (L.f.) D. Don and *Pinus luchuensis* Mayr.

44. *Pratylenchus panamaensis* Siddiqi, Dabur & Bajaj, 1991
(Figs 76, 77)

MEASUREMENTS

- Female holotype (after Siddiqi, Dabur & Bajaj, 1991): L = 0.44 mm; a = 29.3; b = 5.3; b′ = 3.6; c = 19; c′ = 2.2; V = 81; stylet = 16 μm.
- 25 females (after Siddiqi, Dabur & Bajaj, 1991): L = 0.48 (0.42-0.59) mm; a = 25 (21-30); b = 5.7 (4.7-7.3); b′ = 4 (3.0-4.6); c = 21 (17-26); c′ = 1.9 (1.6-2.5); V = 80 (77-83); stylet = 16 (15-17) μm.
- 25 males (after Siddiqi, Dabur & Bajaj, 1991): L = 0.43 (0.38-0.46) mm; a = 27 (22-29); b = 5.5 (4.7-6.4); b′ = 3.9 (3.2-4.3); c = 19 (15-23); c′ = 2.0 (1.4-2.5); T = 48 (36-60); stylet = 15 (14-15.5) μm; spicules = 16 (15-17) μm; gubernaculum = 4 (3-5) μm.
- 30 females (after Golden *et al.*, 1992): L = 0.50 (0.43-0.55) mm; a = 19.8 (15.0-24.9); b = 3.9 (3.5-4.5); c = 19.9 (16.6-24.6); V = 80 (74-84); stylet = 17 (16-18) μm.

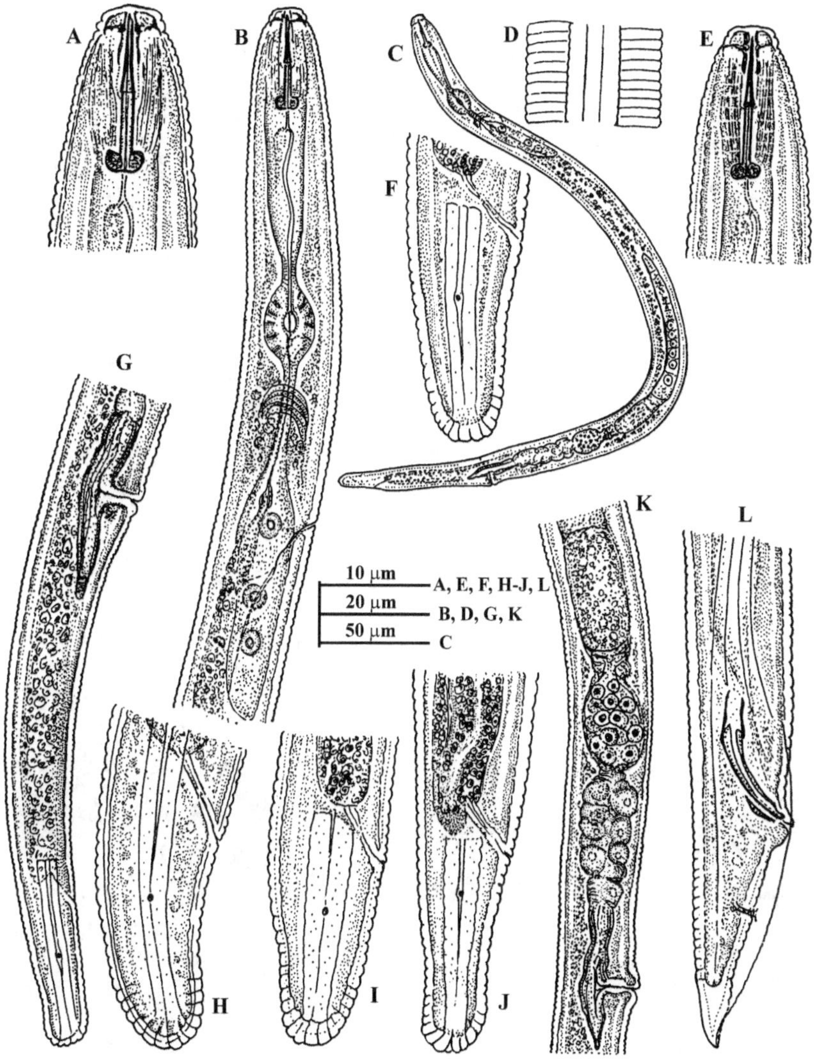

Fig. 76. Pratylenchus panamaensis *Siddiqi, Dabur & Bajaj, 1991. A: Female labial region; B: Female pharyngeal region; C: Entire female; D: Lateral field at mid-body; E: Male labial region; F, H-J: Female tails; G, K: Female posterior regions; L: Male tail. After Siddiqi* et al. *(1991).*

- 20 males (after Golden *et al.*, 1992): L = 0.43 (0.38-0.46) mm; a = 24.5 (17.6-30.7); b = 3.8 (3.4-4.1); c = 21.5 (19.5-24.2); stylet = 15 (14-16) μm; spicules = 17 (16-21) μm; gubernaculum = 4 (3.5-4.5) μm.

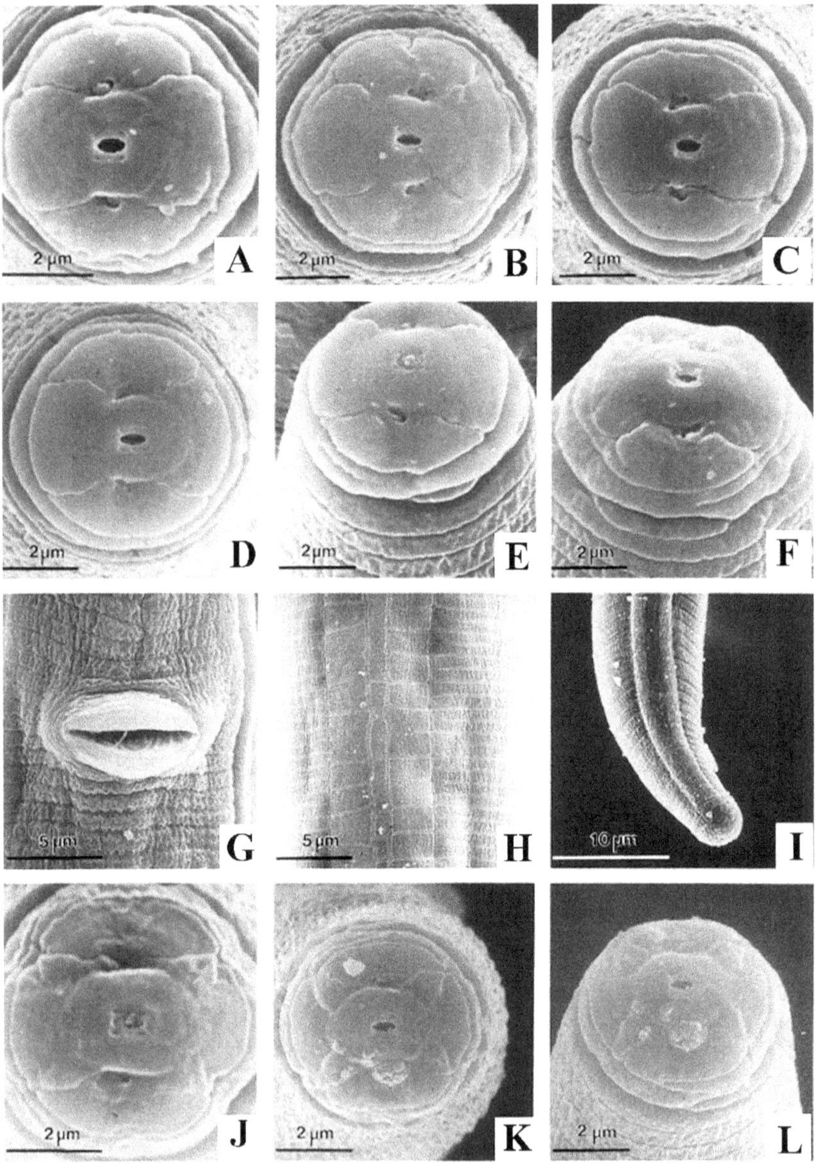

Fig. 77. *SEM micrographs of* Pratylenchus panamaensis *Siddiqi, Dabur & Bajaj, 1991. A-F: Female* en face *views; G: Vulval region; H: Lateral field at mid-body; I: Female tail; J-L: Male* en face *views. After Golden* et al. *(1992).*

DESCRIPTION

Female

Body straight to slightly curved ventrally, at times irregularly, after fixation. Lateral fields occupying one-third to three-sevenths of mid-body diam., with four equidistant lines, not areolated, inner lines usually fusing on tail. Labial region low, flat, with two annuli that are larger than adjoining body annuli. Outer margins of labial framework extending to 1-1.5 body annuli. SEM studies by Golden *et al.* (1992) showed an ovate oral aperture; labial disc and median lips fused, appearing bow-tie shaped; lateral lips large, distinct and partially fused with median lips and labial disc; amphidial apertures small, round to oval, located between labial disc and lateral lips. Stylet with 7-8.5 μm long conus and strong rounded to anteriorly flattened basal knobs, measuring 3 (2.5-3.9) μm across. Opening of dorsal pharyngeal gland 2-3 μm from stylet base. Median pharyngeal bulb well developed, 11-13 $\times$ 9-10.5 μm. Pharyngeal gland lobe overlapping intestine usually for 1.5-2 body diam. posterior to cardia. Nerve ring at middle or posterior region of isthmus. Excretory pore 74 (60-88) μm from anterior extremity, usually opposite pharyngo-intestinal junction. Hemizonid just anterior to excretory pore, 1-2 annuli long. Ovary may extend to pharyngeal glands, oocytes arranged in single row except for zone of multiplication where they are in double rows. Spermatheca longitudinally oval, at 45 (34-59) μm from vulva, filled with sperm; 4-5 sperm occupying its diam. Intra-uterine egg 60 $\times$ 20 μm. Post-vulval uterine sac 16-23 μm long, its tip with 2-3 vestigial cells. Vulva a transverse slit *ca* two-fifths of body diam. in length, located on prominent ventral protuberance. Vulva-anus distance 97 (80-106) μm. Tail subcylindrical to subclavate, 22 (20-24) μm long, with 20 (17-23) annuli on ventral surface and with a broadly rounded, distinctly crenate terminus. Phasmids located at or just anterior to mid-tail 14 (11-16) annuli from tail tip.

Male

As abundant as female. Stylet and labial sclerotisation weaker than that of female. Spicules cephalated, gubernaculum trough-shaped. Tail tip narrow and pointed. Bursa arising slightly anterior to head of spicules and enveloping tail. Phasmids distinct, extending into bursa, at or just anterior to mid-tail.

DIAGNOSIS AND RELATIONSHIPS

Pratylenchus panamaensis is characterised by: labial region low and flat with two annuli, lateral fields with four smooth lines, tail subcylindrical to subclavate, with distinctly crenate terminus.

The matrix code is: A1, B2, C3, D3, E3, F2, G2, H2, I3, J1, K1.

It is close to *P. coffeae*, *P. flakkensis*, *P. loosi* and *P. neglectus* from which it differs by stylet length, tail shape, tail terminus and position of vulva (see the corresponding descriptions).

Pratylenchus gutierrezi Golden, López & Vilchez, 1992 was described from the roots of coffee (*Coffea arabica* L. cv. Caturra), at San Antonio, Naranjo County, Alajuela Province, Costa Rica (Golden *et al.*, 1992). The species was compared and differentiated from *P. flakkensis*, *P. gibbicaudatus* and *P. similis* but was not compared with the almost contemporaneous species *P. panamaensis*. Siddiqi (2000), after comparison of both species, did not find significant differences between them and the former was considered a junior synonym of the latter. We concur with this action and therefore *P. gutierrezi* is maintained as synonym of *P. panamaensis*.

DISTRIBUTION

It has only been recorded from the type locality at Rosa Poás CR, Panama, in the rhizosphere of coffee, *Coffea* sp.

45. *Pratylenchus penetrans* (Cobb, 1917) Filipjev & Schuurmans Stekhoven, 1941

(Figs 78, 79)

MEASUREMENTS

- Female neotype (after Loof, 1960): L = 0.53 mm; a = 26; b = 5.8; c = 16; V = 81.
- 84 females (after Loof, 1960): L = 0.35-0.81 mm; a = 19-32; b = 5.3-7.9; c = 15-24; V = 75-84; stylet = 15-17 μm.
- 34 males (after Loof, 1960): L = 0.31-0.57 mm; a = 23-34; b = 5.4-7.3; c = 16-22; T = 36-58; stylet = 13-16 μm; spicules = 14-17 μm; gubernaculum = 3.9-4.2 μm.
- Females (after Sher & Allen, 1953): L = 0.43-0.65 mm; a = 17-30; b = 5.7-6.5; c = 15-21; V = 78-83; stylet = 17-19 μm.

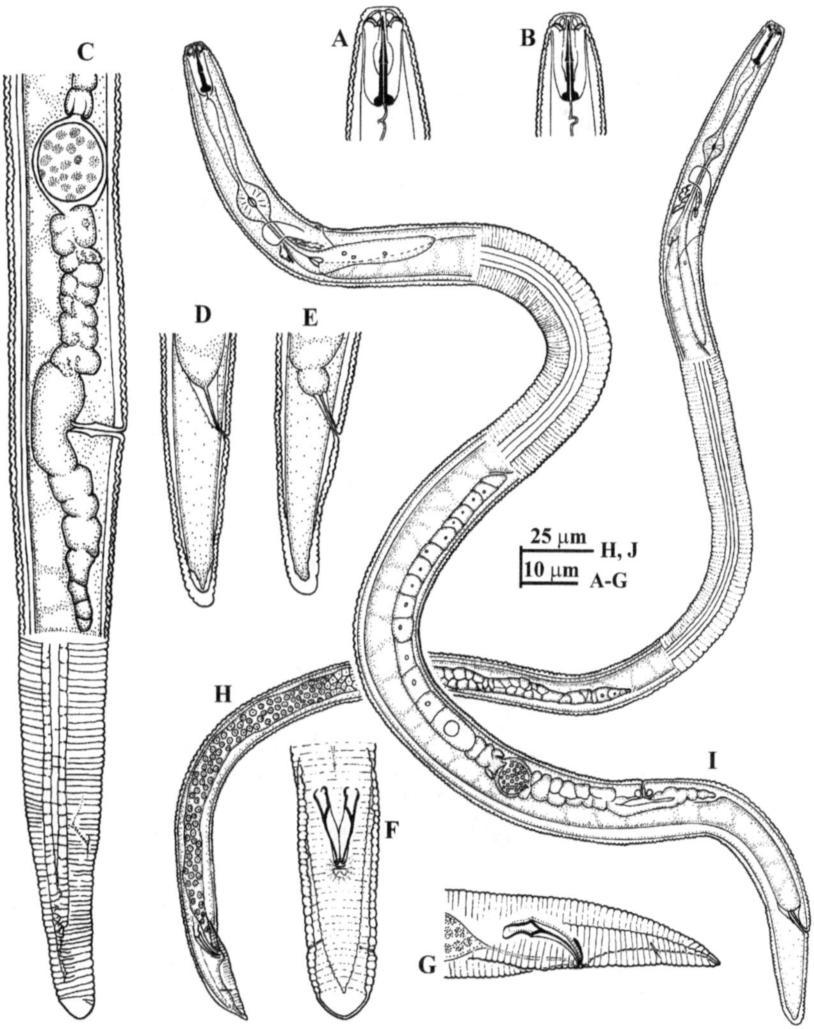

Fig. 78. Pratylenchus penetrans *(Cobb, 1917) Filipjev & Schuurmans Stekhoven, 1941. A: Female labial region; B: Male labial region; C: Female posterior region; D, E: Female tails; F, G: Male tails; H: Entire female; I: Entire male. After Corbett (1973).*

- Males (after Sher & Allen, 1953): L = 0.44-0.56 mm; a = 23-30; b = 5.2-6.0; c = 15-20; T = 43-52; stylet = 16-18 μm.
- 200 females (after Taylor & Jenkins, 1957): L = 0.53 (0.43-0.62) mm; a = 25.4 (17.3-30.6); b = 6.4 (4.6-8.1); c = 21.1 (16.1-27.7); V = 81 (77-85).

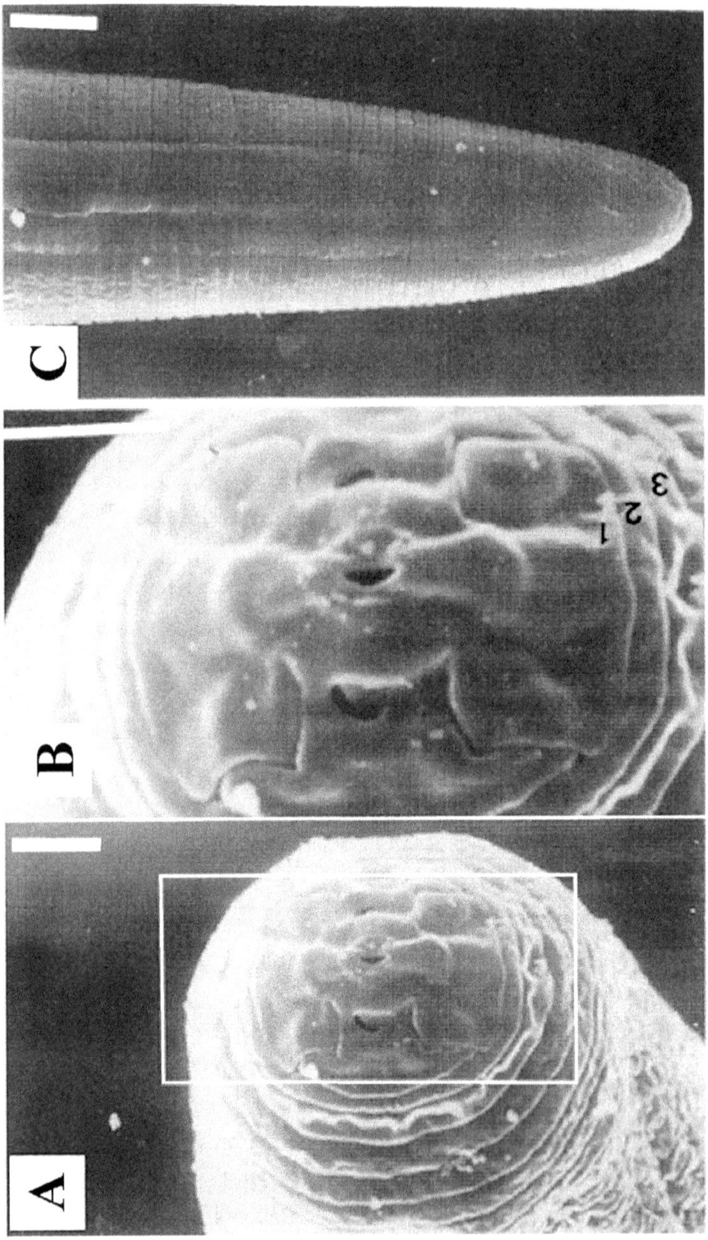

Fig. 79. *SEM micrographs of* Pratylenchus penetrans *(Cobb, 1917) Filipjev & Schuurmans Stekhoven, 1941. A, B: Female en face views; C: Female tail. 1, 2, 3 refers to number of lip annuli. (Scale bars: A, B = 2 μm; C = 5 μm.) After Hernández* et al. *(2000).*

- 10 females (after Ryss, 1988): L = 0.52 (0.41-0.70) mm; a = 24 (19-30); b = 5.6 (5.3-6.7); c = 23 (15-24); c′ = 2.1 (1.5-2.5); V = 80 (77-83); stylet = 15.5 (15-17) μm.
- 16 females (after Vovlas & Troccoli, 1990): L = 0.50-0.67 mm; a = 26-33; b = 5.1-7.6; b′ = 3.6-4.9; c = 20-25; c′ = 1.5-2.2; V = 73-79; stylet = 16-17 μm.
- 10 males (after Vovlas & Troccoli, 1990): L = 0.45-0.47 mm; a = 40-43; b = 6.8-7.3; b′ = 4.1-5.2; c = 19-23; c′ = 2.0-2.4; T = 40-43; stylet = 14-15 μm; spicules = 14-19 μm; gubernaculum = 4-5 μm.

DESCRIPTION

Female

Body moderately slender, almost straight when heat-relaxed. Cuticular annulation fine. Lateral fields normally with four lines, outer bands may be partly areolated, central band sometimes with oblique striae near vulva, becoming areolated posterior to vulva, not extending to tail tip. Labial region slightly offset from body, low, flat anteriorly with rounded outer margins, with three annuli. Labial region with strong, conspicuous skeleton. *En face* view characterised by rectangular subdorsal and subventral lips clearly wider than oral disc and separated from lateral lips by two incisures perpendicular to each other (Hernández *et al.*, 2000). Basal knobs of stylet broadly rounded, sometimes cupped anteriorly. Pharynx overlapping intestine ventrally in a lobe *ca* 1.5 body diam. long. Excretory pore *ca* opposite pharyngo-intestinal junction with hemizonid occupying *ca* two body annuli immediately anterior. Post-vulval uterine sac short, undifferentiated, *ca* 1-1.5 vulval body diam. long. Spermatheca spherical, or nearly so. Tail generally rounded, tip smooth, with 15-27 annuli on ventral surface.

Male

Common. Slightly smaller than female, but similar in form. Lateral fields with four lines ending on bursa, occasionally with oblique lines in central band near mid-body. Spicules slender, with well-marked manubria and ventrally arcuate shafts, 14-17 μm long; gubernaculum simple, 3.9-4.2 μm long. Tail *ca* twice as long as cloacal body diam.; bursa irregularly crenate along margin, enveloping tail tip.

DIAGNOSIS AND RELATIONSHIPS

Pratylenchus penetrans is characterised by: labial region slightly offset, low, flat in front, with rounded outer margins, with three annuli, pharynx overlapping intestine ventrally in a lobe *ca* 1.5 body diam. long, post-vulval uterine sac short, undifferentiated, and tail generally rounded, tip smooth.

The matrix code is: A2, B2, C3, D2, E3, F4, G2, H1, I3, J1, K1.

It can be distinguished from closely related species (*P. fallax*, *P. pseudofallax* and *P. subpenetrans*) by body and stylet length, number of lip annuli, labial framework, position of the vulva, shape of spermatheca and tail terminus (see the corresponding descriptions).

DISTRIBUTION

It has been recorded from the type locality at Rhinebeck, New York, USA, in glasshouse soil. It has been recorded on over 350 hosts mainly in temperate areas in Europe, North America, Central and South America, Africa, Asia and Australia (Corbett, 1973). It is a major pathogen of fruit and conifer nurseries in many areas and causes serious losses in tobacco, apple and cherry orchards and in roses (Corbett, 1973).

It is widespread in Europe: Belgium on potato (Pelsmaeker & Coomans, 1987); Bulgaria on carnation (Katalan-Gateva & Milkova, 1982) and apricot (Ivanova & Choleva, 1999); Cyprus on potato (Philips, 1997a); France on artichoke (Caubel *et al.*, 1975) and peppermint (Esmenjaud *et al.*, 1989); Denmark in rose nurseries (Jacobsen, 1976); Germany on alfalfa (Knot & Schaefer, 1991), barley (Dowe *et al.*, 1990); Greece on *Dahlia* (Koliopanos & Vovlas, 1977) and artichoke (Vovlas & Roca, 1981); Hungary on chrysanthemum (Farkas *et al.*, 1985); Italy on wheat (Ambrogioni *et al.*, 1992), stone and pome fruits (Tacconi & Talamé, 1995), cereals (Inserra *et al.*, 1979), lettuce (Mancini & Moretti, 1974); Moldavia on corn (Nesterov & Lizogubova, 1972); The Netherlands on coastal dune grass, *Ammophila arenaria* (L.) Link (Karssen *et al.*, 2001), *Taxus baccata* L. (Bertrums, 1998), conifers and roses (Schepman, 1994); Norway on roses (Stoeen, 1974); Poland in apple orchards (Szczygiel & Zepp, 1998), on several weed species (Kornobis & Wolny, 1997), rose nurseries (Wojtowicz & Sobio, 1994); Portugal (Abrantes *et al.*, 1987); Russia on gramineaceous fodder crops (Kruchina & Chizhov, 1984) and *Malus domestica* Borkh. (Ryss, 1988); Serbia on tobacco (Grujicic & Jovicic, 1988); Slovakia (Liskova *et al.*,

1988); Slovenia on corn and apple orchards (Urek *et al.*, 2003); Spain in olive nurseries (Nico *et al.*, 2002), horticultural crops (Espárrago & Navas, 1995), cotton and chickpea (Castillo *et al.*, 1991, 1996a) and several hosts and localities (Gómez Barcina *et al.*, 1989; Peña Santiago *et al.*, 2004); Switzerland on small fruits (Vallotton, 1989); United Kingdom on oilseed rape (Evans & Webb, 1989), wind toppled trees (Hooper *et al.*, 1990) and alfalfa (Green, 1979).

It has been recorded widely in North America: California on apple (Jaffee *et al.*, 1982), *P. ponderosa* (Viglierchio, 1979); Connecticut on strawberries (LaMondia, 2002), corn (Miller & Rich, 1980), tomato (Miller, 1977); Michigan on cherry rootstocks (Melakeberhan *et al.*, 1994); Minnesota on apple (Wallace & MacDonald, 1979); New York on cherry and apple (Mai *et al.*, 1994); Ohio on potato (Brown *et al.*, 1980); Oregon on Easter lily, *Lilium longiflorum* Thunb. (Westerdahl *et al.*, 2003), peppermint (Merrifield & Ingham, 1996); Wisconsin on potato (Morgan *et al.*, 2002a); Canada on several crops (Bélair *et al.*, 2001), cereals (Riley & Kelly, 2002), raspberries (Forge *et al.*, 1998), barley (Edwards & Kimpinski, 1997), alfalfa and strawberries (Townshend, 1991; Kimpinski & Martin, 1994), potato (Olthof & Wolynetz, 1991); *Pinus radiata* D. Don (Townshend *et al.*, 1978a; Suatmadji & Marks, 1983).

It is recorded from some countries of Central and South America: Argentina on various crops (Torres & Chavez, 1999); Brazil on chrysanthemum (Ferraz & Monteiro, 1983), on Peruvian carrot, *Arracacia xanthorrhiza* Bancr. (Mendes *et al.*, 2001), artichoke (Rossi & Monteiro, 2001); Costa Rica (Lopez & Salazar, 1990); Trinidad on various crops (Singh, 1973); Venezuela on various crops (Crozzoli, 2002), ornamental plants (Petit & Crozzoli, 1995).

Recorded also in Africa: Algeria and Tunisia on several crops (Troccoli *et al.*, 1992; Troccoli & Di Vito, 2002); Kenya on pyrethrum, *Tanacetum cinerariifolium* (Trev.) Schultz Bip. (Anyango, 1988); Libya on date palm, *P. dactylifera* (Edongali, 1996); Morocco on wheat and barley (Meskine & Abbad Andalousi, 1993); Namibia on corn and pear millet (De Waele *et al.*, 1998); South Africa on wheat (Jordaan *et al.*, 1992), sunflower (Bolton *et al.*, 1989) and sorghum (De Waele & Jordan, 1988b).

It is widely distributed in Asia: China on fruit trees (Wang, 1993), cotton (Yang *et al.*, 1992), various crops (Yin, 1991); India in tea plantations (Jauhari & Lal, 2002), on chrysanthemum (Ramakrishnan

& Vadivelu, 1995), potato (Zaki *et al.*, 1991), almond (Khan & Sharma, 1992b); Japan on sweet potato (Iwahori & Sano, 2003), tobacco (Orui & Mizukubo, 1999a), various crops (Gotoh, 1974), coniferous seedlings in forest nurseries (Mamiya, 1969); Korea on oriental melon, *Cucumis melo* L. (Park *et al.*, 2002), chrysanthemum (Lee *et al.*, 2006), apple orchards (Park *et al.*, 1999), ornamental crops (Kim & Minagawa, 1996) and ginseng (*Panax ginseng* C.A. Meyer) (Chung *et al.*, 2004); Pakistan on potato (Khan & Hussain, 2004), coconut nurseries (Khan *et al.*, 1992), apple (Anwar *et al.*, 1991), pistachio (Qasim & Hashmi, 1988), tobacco (Saeed *et al.*, 1986); Taiwan on strawberries (Wu *et al.*, 2002); Turkey on chickpea, lentil and vetch (Di Vito *et al.*, 1994), various crops (Erdal *et al.*, 2001); Vietnam on potato and cabbage (Ryss & Fam-Tkhan, 1989). It has been also recorded in New Zealand on kaki, *Diospyros kaki* L. f. (Knight, 2001), *Apium graveolens* L. (Boesewinkel, 1977); and in Australia from agricultural soils and native vegetation (Riley & Wouts, 2001), on ornamentals (Suatmadji, 1988) and carrots (Hay & Pethybridge, 2005).

46. *Pratylenchus pinguicaudatus* Corbett, 1969
(Fig. 80)

MEASUREMENTS

- Female holotype (after Corbett, 1969): L = 0.56 mm; a = 28; b = 6.1; c = 16.7; V = 79; stylet = 17 μm.
- 12 females (after Corbett, 1969): L = 0.58 (0.48-0.63) mm; a = 26 (23-29); b = 6.0 (5.6-6.7); c = 19 (17-20); V = 80 (79-81); stylet = 17.5 (17-18) μm.

DESCRIPTION

Female

Body straight or slightly curved, with body posterior to vulva slightly bent ventrally when heat-relaxed. Body tapering slightly from mid-body to labial region, narrowing slightly posterior to vulva, thereafter almost cylindrical to broadly rounded tail tip. Lateral fields originating at *ca* level of stylet knobs as a single line soon becoming two lines, then three *ca* halfway down procorpus and four just posterior to median bulb. Lateral fields with four lines, irregularly areolated along the whole of its

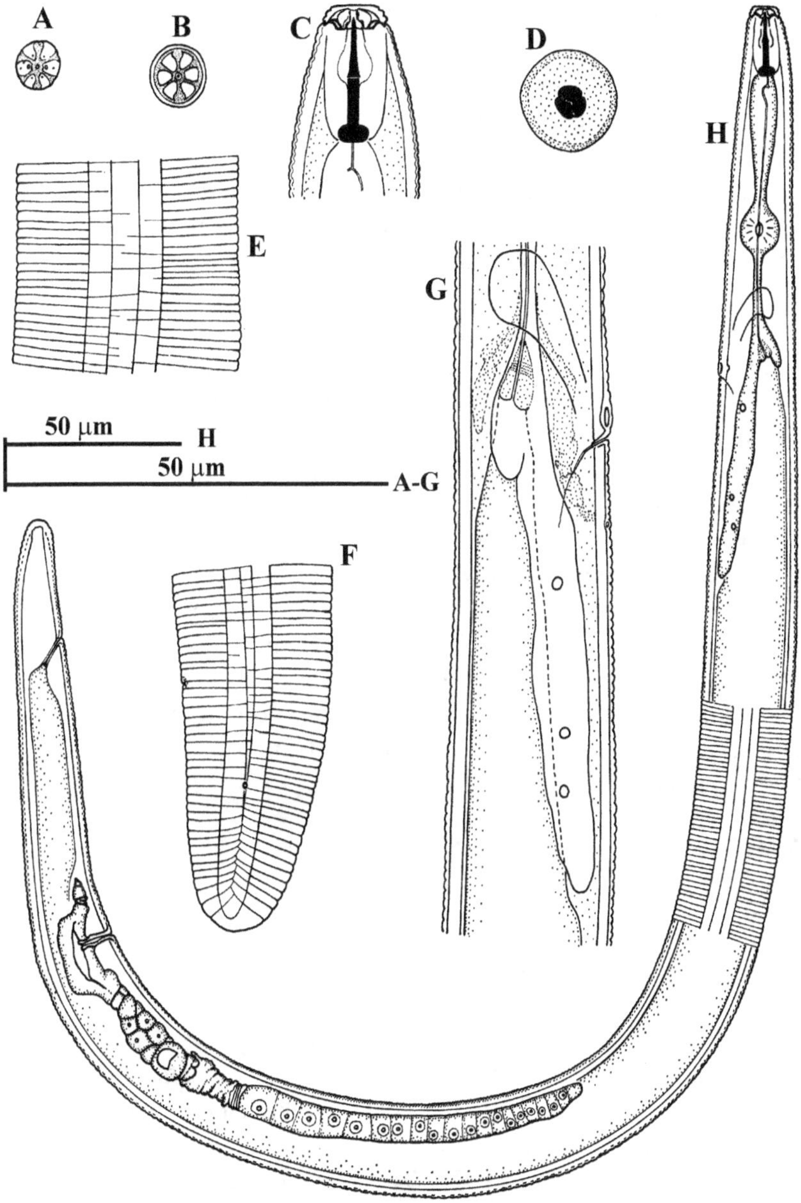

Fig. 80. Pratylenchus pinguicaudatus *Corbett, 1969. A, B: Female* en face *views; C: Female labial region; D: Stylet knobs as seen from the front; E: Lateral field at mid-body; F: Female tail; G: Female pharyngeal region; H: Entire female. After Corbett (1969).*

length, with fewer lines between middle two lines than across outer two bands. Some large specimens with oblique striae instead of areolations for a short length of lateral fields just anterior to vulva. Areolations very fine, not seen in glycerin mounts, only seen in immersion oil mounts. Labial region with three, sometimes indistinct, annuli, offset from body by slight constriction. Labial skeleton strong, amphidial openings and one papilla on lateral lips, one papilla on each subventral and on each subdorsal lip. Stylet in two parts with large rounded basal knobs, 16-20 μm long, front part slightly shorter than rear. Excretory pore *ca* 86-99 μm from anterior end, just posterior to hemizonid. Hemizonid occupying two annuli; hemizonion seen in some specimens *ca* ten annuli posterior to hemizonid. Nerve ring surrounding isthmus just anterior to posterior glandular portion of pharynx. Pharyngo-intestinal junction at *ca* excretory pore level. Pharyngeal glands overlapping intestine ventrally as a long narrow lobe more than two body diam. long. Gonad anterior to vulva, lying in left ventral quadrat of body. Ovary consisting of a single row of oocytes. Conspicuous rounded, non-functional, spermatheca. Vulva located at *ca* 78-81%, post-vulval uterine sac short, usually *ca* one body diam. or less in length. Tail subcylindrical, 19-25 annuli long, with smooth terminus. Phasmid at *ca* mid-tail, with two middle lines of lateral fields coalescing just posterior to it to become one irregular line at junction of areolations. Outer two lines continuing almost to tail tip.

Male

Not found.

DIAGNOSIS AND RELATIONSHIPS

Pratylenchus pinguicaudatus is characterised by: labial region with three annuli, lateral fields with four lines, irregularly areolated along the whole of its length, with fewer lines between the middle two lines than across the outer two bands, a smooth tail tip and absence of males.

The matrix code is: A2, B1, C3, D1, E3, F1, G2, H1, I4, J1, K2.

It is close to *P. andinus*, *P. delattrei*, *P. thornei* and *P. zeae* from which it differs by shape of labial region, length of pharyngeal overlap, position of the vulva and tail shape (see the corresponding descriptions).

DISTRIBUTION

It has been recorded from the type locality at Broadbalk field, Rothamsted Research, in soil around the roots of wheat. It has also been recorded in England and Wales on cereals (Corbett, 1970); several countries of North Africa (Di Vito *et al.*, 2002a; Troccoli & Di Vito, 2002); in Bulgaria (Ryss *et al.*, 1991); in Poland (Brzeski & Szczygiel, 1977); Spain on several herbaceous hosts (Talavera & Tobar, 1997; Talavera & Navas, 2002); and in Australia (Mathur & McLeod, 1977).

47. *Pratylenchus pratensisobrinus* Bernard, 1984
(Fig. 81)

MEASUREMENTS

- Female holotype (after Bernard, 1984): L = 0.48 mm; a = 28.0; b = 5.1; b′ = 3.7; c = 13.4; c′ = 3.0; V = 76; stylet =16 μm.
- 15 females (after Bernard, 1984): L = 0.48 (0.39-0.56) mm; a = 28.4 (25.0-31.5); b = 5.3 (4.4-6.0); b′ = 3.7 (3.5-4.1); c = 13.9 (11.8-15.1); c′ = 3.1 (2.8-3.7); V = 77 (75-80); stylet = 16 (15-17) μm.
- 8 males (after Bernard, 1984): L = 0.47 (0.44-0.50) mm; a = 31.2 (26.5-41.0); b = 5.7 (5.1-5.9); b′ = 3.9 (3.6-4.2); c = 17.0 (15.7-19.7); c′ = 2.5 (2.3-2.9); T = 47 (34-61); stylet = 15 (15-16) μm; spicules = 18 (17-19) μm; gubernaculum = 6 (5-6) μm.

DESCRIPTION

Female

Body slightly curved when heat-relaxed, rather slender. Labial region of medium height, rounded, with three annuli. Lateral fields consisting of four lines with no indication of regular markings between lines; lines usually becoming three on tail, but often remaining four at least to phasmid; lateral fields usually open near tail terminus, rarely closed. Stylet stout, knobs large, cupped anteriorly. One pair of cephalids visible, 3-4 annuli posterior to labial capsule. Median bulb oval, valve conspicuous. Pharyngeal glands elongate, overlapping intestine; dorsal gland nucleus near middle of lobe, subventral nuclei in posterior portion.

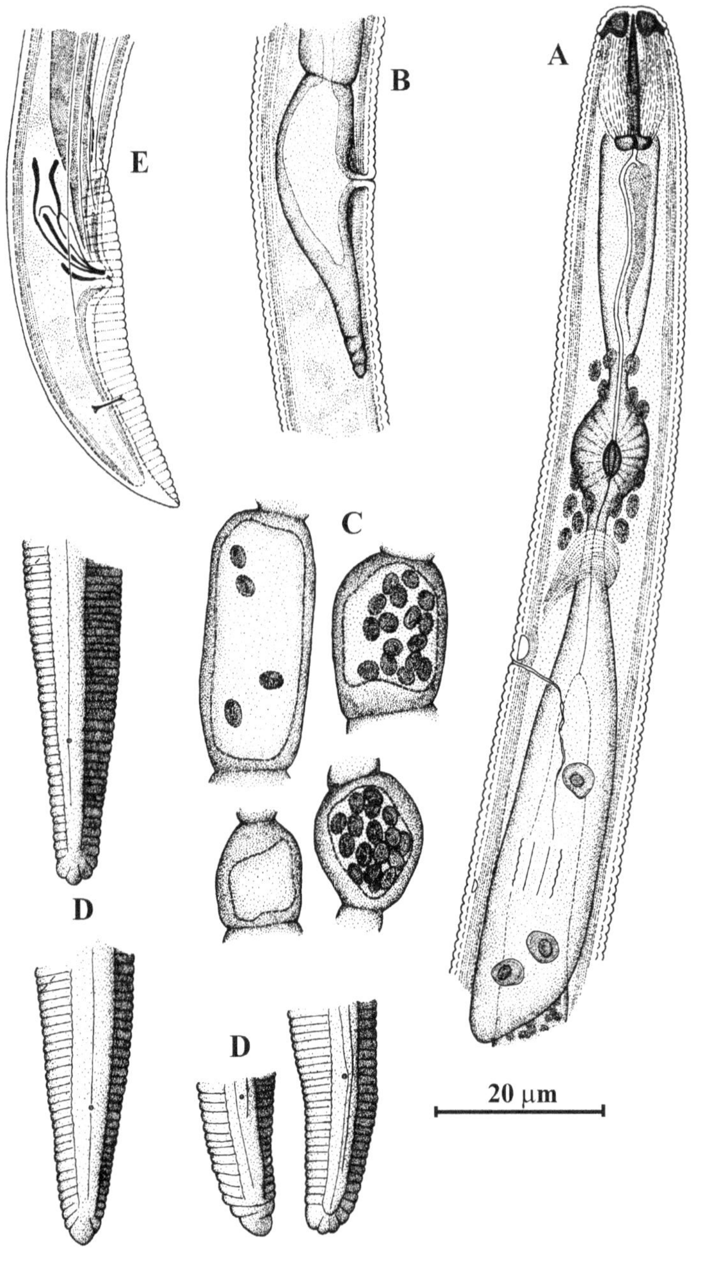

Fig. 81. Pratylenchus pratensisobrinus *Bernard, 1984. A: Female pharyngeal region; B: Vulval region; C: Variability in spermatheca; D: Female tails; E: Male tail. After Bernard (1984).*

Hemizonid in vicinity of pharyngo-intestinal junction, excretory pore immediately posterior to hemizonid; hemizonion distinct, at *ca* 60% of gland lobe. Gonad outstretched; spermatheca square or oval to elongated, usually filled with sperm but occasionally empty; post-vulval uterine sac elongate, 2.0 (1.6-2.7) vulval body diam. long, with a few discrete cells distally. Tail conoid and rather elongated, terminus coarsely annulated or consisting of a large terminal annulus; phasmid centred in lateral fields *ca* midway on tail.

Male

Similar to female, but shorter, more slender and with a slightly more delicate stylet. Gonad outstretched. Spicules relatively robust, evenly curved, partially cephalated; velum prominent, well developed, gubernaculum curved, linear. Caudal alae crenate, enveloping tail. Phasmids arising near mid-tail; lateral fields hardly extending beyond spicules.

DIAGNOSIS AND RELATIONSHIPS

Pratylenchus pratensisobrinus is characterised by: labial region with three annuli, lateral fields consisting of four lines usually becoming three on the tail, but often remaining four at least to the phasmid and tail conoid and rather elongated.

The matrix code is: A2, B2, C2, D3, E2, F3, G3, H2, I1, J1, K1.

It closely resembles *P. pratensis* and could conceivably be considered an extreme variant of that species. However, it differs in the following characters: stylet generally longer (15-17 *vs* 12-16 μm); tail longer ($c = 12\text{-}15$ *vs* 13-24; $c' = 2.8\text{-}3.7$ *vs* 2.4-3.1); tail annuli more numerous (23-37 *vs* 20-28). Although there is some degree of overlap in all of these differentiating characteristics, the differences noted here are deemed sufficient to separate the two species.

This species was considered as a junior synonym of *P. pratensis* by Frederick and Tarjan (1989). However, based on morphology and morphometry, we feel that although both species are closely related, we agree with the differential diagnosis of Bernard (1984) and consider it as a valid species subject to molecular data becoming available.

DISTRIBUTION

It has been recorded from the type locality at Finger Bay, Adak Island, Alaska, USA, from the rhizosphere of *Viola langsdorffii* Fisch. It has also been recorded in citronella grass (*Cymbopogon nardus* (L.) R.) in Thailand (Toida *et al.*, 1996) and from hazelnut orchards in the West Black Sea Region of Turkey (Kepenekci, 2002).

48. *Pratylenchus pseudocoffeae* Mizukubo, 1992
(Figs 82, 83)

MEASUREMENTS

- Female holotype (after Mizukubo, 1992): L = 0.58 mm; a = 30.2; b = 7.1; b′ = 3.3; c = 19.0; c′ = 2.3; V = 79; stylet = 16.5 μm.
- 50 females (after Mizukubo, 1992): L = 0.51 (0.41-0.62) mm; a = 27.5 (22.6-32.1); b = 5.8 (4.7-7.0); b′ = 3.1 (2.6-3.4); c = 19.3 (18.1-20.8); c′ = 2.2 (1.8-2.5); V = 81 (79-82); stylet = 16 (15-17) μm.
- 11 males (after Mizukubo, 1992): L = 0.49 (0.39-0.54) mm; a = 30.6 (25.6-37.0); b = 6.1 (4.6-6.9); b′ = 3.3 (2.9-3.8); c = 20.3 (18.1-23.9); c′ = 2.4 (1.8-2.9); T = 43 (38-48); stylet = 15 (14.0-15.5) μm.

DESCRIPTION

Female

Body curved ventrally when heat-relaxed. Annuli 1.4 (1.0-1.9) μm wide at mid-body. Labial region relatively low, 2.6 (2.1-3.0) μm high, 7.5 (6.9.-8.1) μm diam., flattened anteriorly, weakly offset. Labial framework moderately developed, laterally extending into body for one annulus. Labial region with two annuli on both sides (66% specimens), two and three annuli (28.5%) and three annuli on both sides (5.5%). SEM *en face* view showing division between submedian and lateral segments; lateral edges of submedian segments curving, never angular, assigned into Group 2 of Corbett and Clark (1983). Amphidial apertures narrow, oblique. Cephalids not seen. Stylet 2.2 times (2.0-2.5) as long as labial region diam. Lateral fields 5.4 (4.0-7.3) μm wide, or 28.8% (23.1-34.5) of max. body diam., consisting of three bands or four

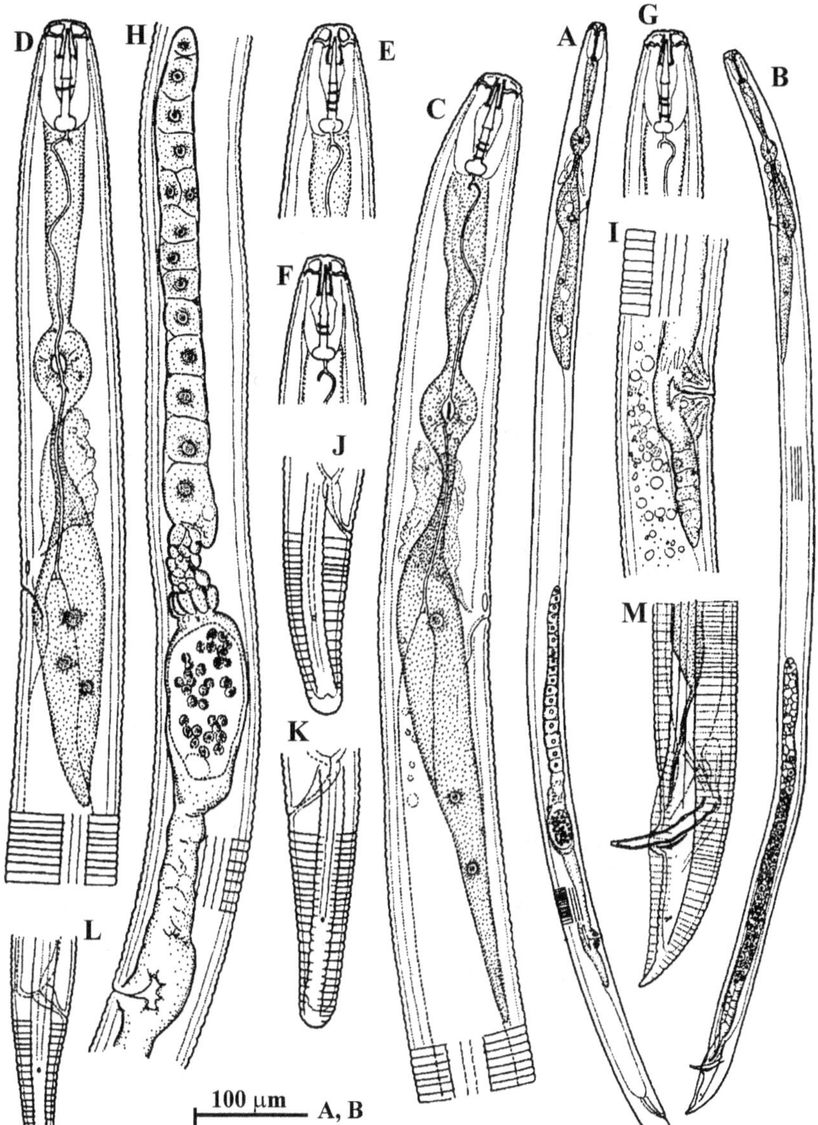

Fig. 82. Pratylenchus pseudocoffeae *Mizukubo, 1992. A: Entire female; B: Entire male; C, D: Female pharyngeal regions; E-G: Female labial regions; H: Female reproductive system; I: Vulval region; J-L: Female tails; M: Male tail. After Mizukubo (1992).*

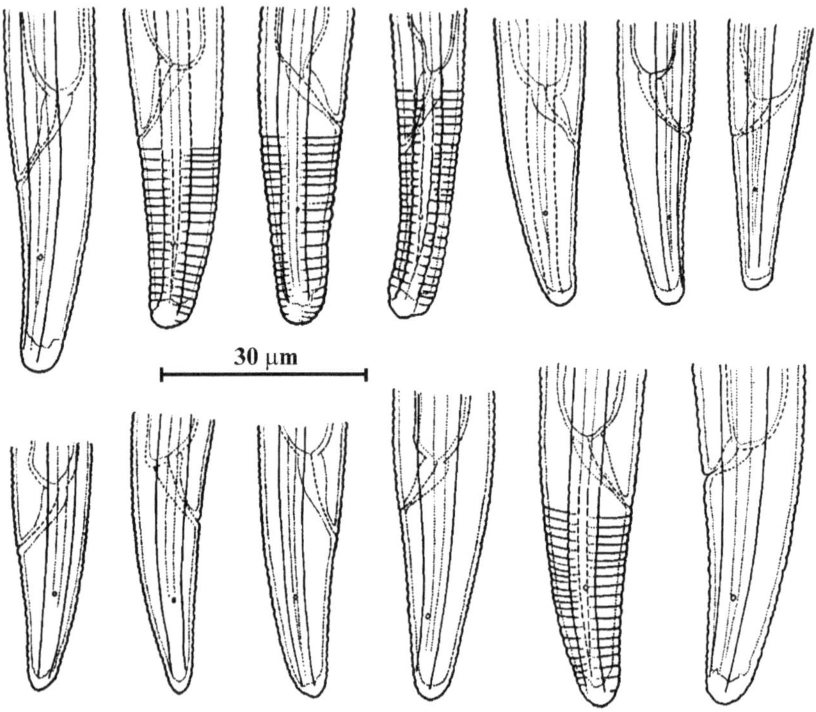

Fig. 83. *Variation of female tails of* Pratylenchus pseudocoffeae *Mizukubo, 1992. After Mizukubo (1992).*

lines; inner band appearing plain in transmitted light but incised by 1-2 faint broken lines; outer bands appearing almost plain in mid-body region and areolated in tail region. Lateral fields extending to tail tip and continuing around tail terminus. Stylet knobs massive, 2.0 (1.7-2.4) μm high, 3.7 (3.3-4.3) μm across; shape variable but mostly broadly rounded or flattened anteriorly. Of the 37 individuals, the following shapes were observed: rounded and sloping posteriorly (2.7%); broadly rounded (40.5%); flattened anteriorly (29.7%); slightly indented (16.2%); indented (10.8%). Metacorpus oval, 13.1 (11.9-15.8) μm long, 9.5 (8.6-10.6) μm across, length : diam. ratio 1.4 (1.3-1.7), occupying *ca* half corresponding body diam.; valve conspicuous, 57 (48-62) μm from anterior body end, or 65.0 (46.0-79.3)% of pharynx length. Pharynx basal lobe extending 182 (150-241) μm from anterior end, ventrally overlapping intestine by 93 (64-155) μm or for 5.1 (3.0-9.0) corresponding body diam.; rarely overlapping dorso-laterally (7.7%) or laterally (15.4%). Pharyngeal nuclei usually in tandem. Excretory pore 100.5 (78-123)% of pharyngeal length or 63 (58-68) body annuli from anterior body end; annuli between anterior body end and excretory pore 1.4 μm (1.2-1.5) in average width. Hemizonid flat, 2-3 annuli long, level with or immediately anterior to excretory pore; hemizonion lenticular, 8 (6-10) annuli posterior to excretory pore. Gonad outstretched, 159 (123-227) μm long; ovary with oocytes generally in a row (rarely partially doubled); spermatheca oblong, 63 (52-77) μm from vulva, 26 (18-38) μm long, 11.7 (9.2-14.5) μm diam., length : diam. ratio of 2.4 (1.6-3.1), packed with sperm. Post-vulval uterine sac usually less than twice vulval body diam. long, with terminal rudimental ovary. Vagina narrow-walled, perpendicular to body axis, 7.1 (5.9-8.3) μm long, or 43.6% (36.4-52.2) of vulval body diam.; vaginal opening wide slit. Annuli between vulva and anus 57 (48-66) in number, 1.4 (1.1-1.7) μm in average width. Tail tip with some variation in shape: using tail tip shape codes of Frederick and Tarjan (1989), hemispherical = 3%, subhemispherical = 52%, bluntly pointed = 45%, but not finely pointed in the 31 individuals with exact lateral orientation; terminus smooth, rarely wrinkled, with no evident annuli. Tail terminal cuticle 2.8 (1.8-3.6) μm thick; phasmids pore-like, centred in lateral fields, 13 (7-19) μm posterior to anus, or 16.0 (10.6-21.1) μm from tail tip.

Male

Body straight to arcuate when heat-relaxed. Annuli finer than in female, 1.2 (0.9-1.5) μm wide at mid-body. Labial region relatively low, 2.6 (2.2-3.0) μm high, 6.7 (5.9-7.3) μm diam., flattened anteriorly, weakly offset or continuous with body. Labial framework moderately developed, laterally extending into body for one annulus. Labial region with two annuli in transmitted light, sometimes three annuli. Cephalids not seen. Stylet shorter than in female, 2.2 times (2.0-2.5) as long as labial region diam. Stylet knobs smaller than female, 3.0 (2.8-3.4) μm across, 2.0 (1.7-2.3) μm high, width : height ratio 1.5 (1.4-1.7); shape varying from rounded to flattened anteriorly, but mostly broadly rounded. Dorsal pharyngeal opening close to stylet base. Metacorpus oval, 11.6 (10.6-12.5) μm long, 8.1 (7.3-9.6) μm diam., length : diam. ratio 1.4 (1.2-1.6), occupying *ca* half corresponding body diam.; valve conspicuous, 53 (47-56) μm from anterior end, or 65.9 (49.3-72.8)% of pharynx length. Pharyngeal basal lobe extending 150 (114-169) μm from anterior body end, ventrally overlapping intestine by 70 (40-90) μm or 4.4 (2.5-6.0) corresponding body diam. Pharyngeal nuclei usually in tandem. Excretory pore at 104 (81-115)% of pharyngeal length; body annuli at metacorpus 1.4 (1.1-1.5) μm wide. Hemizonid flat, same level or immediately anterior to excretory pore. Testis outstretched, with spermatogonia in double or triple rows, 208 (165-244) μm long. Spicules arch-shaped, slightly longer than stylet; gubernaculum crescent-shaped. Bursal alae 39 (32-43) μm long, 3.9 (3.4-4.5) anal body diam. long. Terminal cuticle of tail 6.2 (4.3-7.9) μm thick, ending in finely rounded terminus; phasmids 10.6 (7.3-13.9) μm posterior to anus or 13.3 (9.2-17.2) μm from tail tip. Lateral fields 3.8 (3.0-4.3) μm wide or 24.0 (21.8-27.3)% of max. body diam.

DIAGNOSIS AND RELATIONSHIPS

Pratylenchus pseudocoffeae is characterised by: labial region with two or three annuli, vulva located at 79-82% of body, presence of males, spermatheca oblong, long, differentiated, post-vulval uterine sac, tail subhemispherical or bluntly pointed with smooth terminus.

The matrix code is: A1, B2, C2, D3, E3, F4, G2, H1, I4, J1, K1.

It resembles *P. coffeae* and *P. loosi* from which it differs by the high frequency of three-lip annuli (observed in one-third of individuals examined) and divided pattern in *en face* view, Group 2 of Corbett and Clark (1983), position of the vulva and the shape of the tail tip (see the corresponding descriptions).

DISTRIBUTION

It has been recorded from the type locality at Touganji, Takachiho, Miyazaki Prefecture, Japan, in soil around the roots of *Chrysanthemum morifolium* Ram. and *Artemisia feddei* Lev. et Van. It has also been recorded in Florida, USA, from aster (Inserra *et al.*, 1998).

49. *Pratylenchus pseudofallax* Café-Filho & Huang, 1989
(Fig. 84)

MEASUREMENTS

- Female holotype (after Café-Filho & Huang, 1989): L = 0.46 mm; a = 33; b = 5.5; b′ = 3.6; c = 18.5; c′ = 2.8; V = 81; stylet = 15.5 μm.

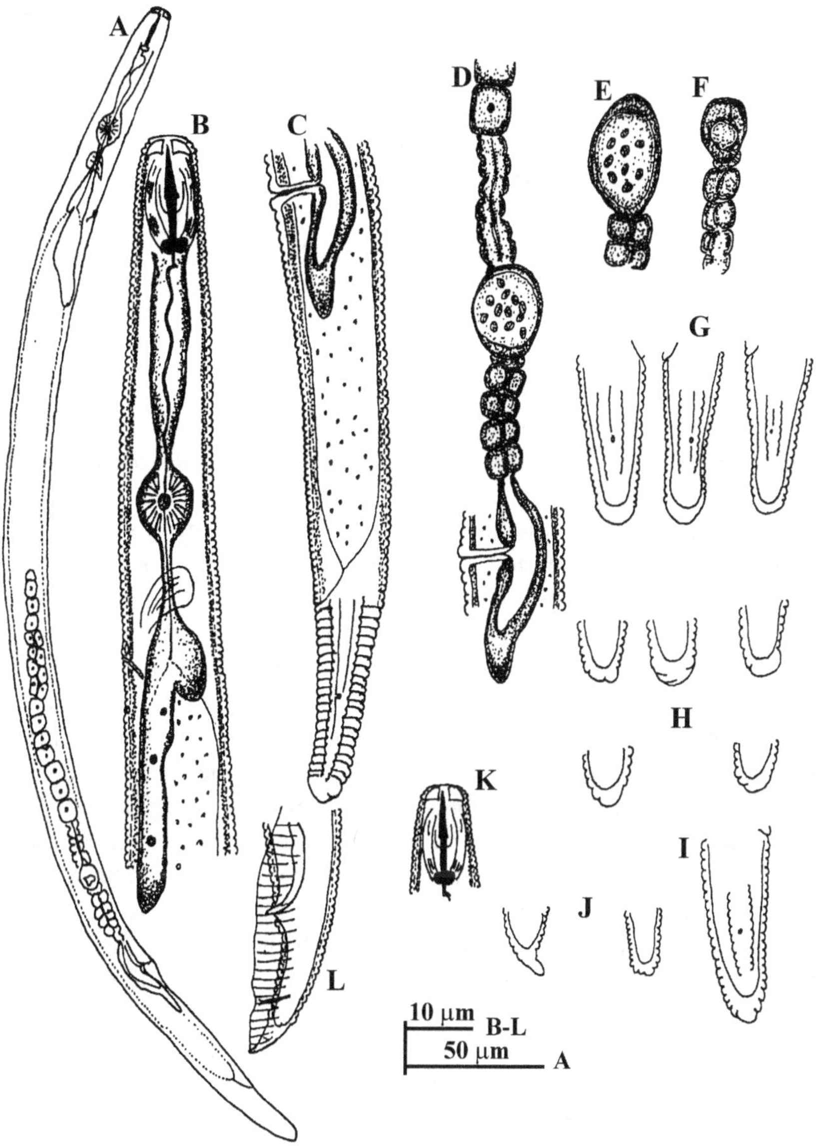

Fig. 84. Pratylenchus pseudofallax *Café Filho & Huang, 1989. A: Entire female; B: Female pharyngeal region; C: Female posterior region; D: Detail of reproductive system showing round spermatheca; E: Spermatheca (frequent); F: Spermatheca (not frequent); G-J: Female tails; K: Male labial region; L: Male tail. After Café Filho and Huang (1989).*

- 22 females (after Café-Filho & Huang, 1989): L = 0.47 (0.42-10.53) mm; a = 29.0 (24.7-36.3); b = 6.3 (5.4-7.6); b′ = 4.1 (3.6-4.5); c = 17 (14.5-19); c′ = 2.5 (1.7-3.4); V = 80 (77-82); stylet = 15 (14-16) μm.
- 8 males (after Café-Filho & Huang, 1989): L = 0.42 (0.37-0.48) mm; a = 28.3 (24.7-31.5); b = 6.3 (5.4-7.3); b′ = 4.1 (3.5-4.6); c = 17.1 (15.7-19.0); c′ = 2.7 (2.0-3.2); T = 28-63; stylet = 14 (13.5-15) μm; spicules = 16 (13-19) μm; gubernaculum = 13-17 μm.

DESCRIPTION

Female

Body relatively slender, cylindrical, tapering in both extremities, slightly curved, very rarely straight. Body annuli strongly marked, mean width 1.1 (0.8-1.3) μm, more deeply demarcated than other species of genus. Lateral fields with four lines, beginning in pharyngeal region. Usually only external lines extending posterior to phasmid, but sometimes one, or even both lines continuing for some micrometres posterior to phasmids. Phasmids 7-14, usually 9-11 annuli from caudal end. External lines conspicuously crenate in caudal region and usually also in rest of body. In some specimens even internal lines were faintly crenate in posterior region of body. Labial region offset with three annuli, sometimes not very distinct. SEM *en face* views revealed one submedian and two lateral wedge-shaped segments fitting in Group 3 of Corbett and Clark (1983). Labial framework strongly developed. Stylet massive; stylet knobs anteriorly flattened, sometimes retrorse or, less often, forward pointed. Dorsal pharyngeal gland orifice 2-3 μm posterior to stylet knobs. Metacorpus oval. Isthmus short, surrounded by nerve ring. Pharyngo-intestinal junction in vicinity of excretory pore. Hemizonid one annulus anterior or immediately anterior to excretory pore, 74-85 μm from anterior end. Hemizonion *ca* eight annuli posterior to hemizonid, sometimes indistinct. Posterior pharyngeal glands overlapping intestine as subventral lobe. Ovary not reaching pharynx. Oocytes in a single row. Spermatheca generally rounded, rarely more ovate, with spermatozoa or without sperm. Tricolumella well developed, with four groups of three cells. Post-vulval uterine sac with non-cellular terminus. Tail conoid-rounded, tail terminus variable, from clearly crenate to plainly smooth. Tail annuli numbering 16-24 (usually 17-20).

Male

The proportion of males to females was 1 : 4 in two populations and not found in a third population. Body slender, almost straight, or slightly curved. Lateral fields with four lines. Labial region almost continuous with body, with three annuli. Labial framework strongly developed. Stylet less massive than in female; stylet knobs rounded, slightly anteriorly flattened or rarely directed anteriorly. Dorsal pharyngeal gland orifice 2-3 μm posterior to stylet knobs. Metacorpus ovate. Isthmus short, surrounded by nerve ring. Pharyngo-intestinal junction in vicinity of excretory pore. Hemizonid immediately anterior or few annuli anterior to excretory pore, 68-88 μm from anterior end. Hemizonion 7-11 annuli posterior to hemizonid. Posterior pharyngeal glands overlapping intestine in a subventral lobe. Testis outstretched, with multiple row of spermatocytes. *Vas deferens* longer than testis. Spicules curved. Bursa strongly crenate. Phasmids posterior to mid-tail.

DIAGNOSIS AND RELATIONSHIPS

Pratylenchus pseudofallax is characterised by: labial region offset with three annuli, lateral fields with four lines, external lines conspicuously crenate in the caudal region and usually also in the rest of the body, presence of males, tail conoid-rounded and tail terminus variable, from clearly crenate to plainly smooth.

The matrix code is: A2, B2, C2, D2, E3, F2, G3, H2, I2, J1, K1.

It is close to *P. fallax*, *P. penetrans* and *P. subpenetrans* from which it differs by body length, the number of labial region, presence of males, stylet length, position of the vulva, the shape of the spermatheca and the shape of tail terminus (see the corresponding descriptions).

DISTRIBUTION

It has only been recorded from the type locality at Lageadinho, Municipality of Veranopolis, RS, Brazil, in soil around the roots of *Malus silvestris* Mill.

50. *Pratylenchus pseudopratensis* Seinhorst, 1968
(Fig. 85)

Measurements

- Female holotype (after Seinhorst, 1968): L = 0.52 mm; a = 29; b = 6.2; c = 18; V = 80; stylet = 17 μm.
- 10 females (after Seinhorst, 1968): L = 0.42-0.56 mm; a = 24-33; b = 5.2-6.7; c = 18-24; V = 77-81; stylet = 16-17 μm.
- 10 males (after Seinhorst, 1968): L = 0.40-0.50 mm; a = 26-33; b = 5.0-6.2; c = 16-25; stylet = 15 μm; spicules = 14-16 μm.
- 30 females (after Fortuner, 1973): L = 0.45 (0.40-0.52) mm; a = 27 (25-31); b = 5.8 (5.0-6.8); b′ = 4.0 (3.2-5.4); c = 21 (19-24); V = 78 (77-81); stylet = 14.5 (13.5-16) μm.
- 5 males (after Fortuner, 1973): L = 0.38 (0.36-0.42) mm; a = 25 (22-26); b = 5.1 (4.9-5.6); b′ = 3.7 (3.2-4.2); c = 18.3 (18.0-19.7); stylet = 13.5 (13.0.5-14.5) μm; spicules = 13.5-16 μm; gubernaculum = 5 μm.
- 10 females (after Ryss, 1988): L = 0.44 (0.35-0.53) mm; a = 24 (20-32); b = 6.2 (5.5-6.8); c = 23 (20-25); c′ = 1.7 (1.2-2.2); V = 78 (76-82); stylet = 14 (12-15) μm.
- 10 males (after Ryss, 1988): L = 0.39 (0.31-0.48) mm; a = 28 (25-30); b = 6.0 (4.7-6.5); c = 23 (17-29); c′ = 2.0 (1.7-2.8); stylet = 13.5 (13.5-14.0) μm; spicules = 15 (14-16) μm; gubernaculum = 3-4 μm.

Description

Female

Labial region with three often rather flat and obscure annuli. Under SEM observations most specimens with three lip annuli, some with three annuli on one side and four on other; rare specimens with two, three or four on one side and five on the other side (Baujard *et al.*, 1990). *En face* view by SEM shows first annulus with prominent labial disc and perioral pad (Baujard *et al.*, 1990). Lateral fields with four parallel lines, usually some additional lines running obliquely between inner two. Only outer two lines proceeding past phasmids. Lateral fields sometimes areolated posterior to phasmid. Labial framework extending *ca* two body annuli into body. Stylet knobs flattened anteriorly. Dorsal pharyngeal gland opening 2-3 μm posterior to stylet knobs. Excretory pore at or

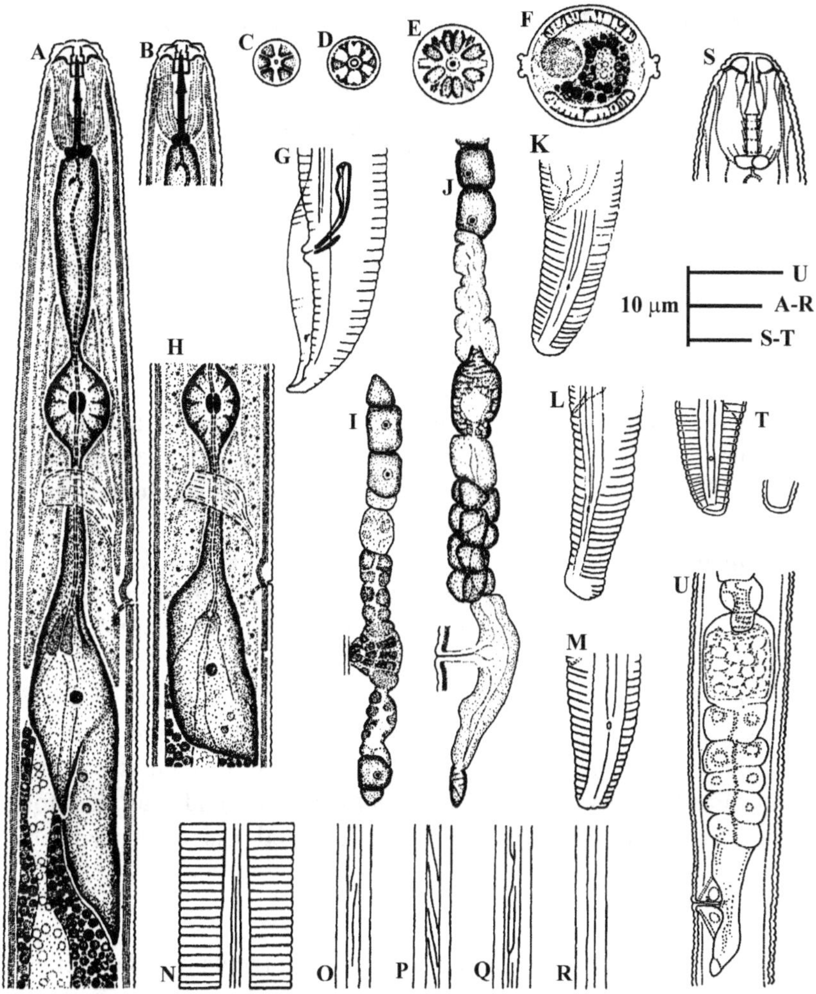

Fig. 85. Pratylenchus pseudopratensis *Seinhorst, 1968. A: Female pharyngeal region; B: Female labial region; C: Female* en face *view; D, E: Transverse sections through stylet; F: Transverse section through mid-body; G: Male tail; H: Female pharyngeal region; I, J: Female reproductive systems; K-M: Female tails; N-R: Lateral field at mid-body; S: Female labial region; T: Female tail and tip; U: Vulval region. After Fortuner (1973); Seinhorst (1968).*

posterior to level of nerve ring. Ovary outstretched, oocytes in single file except for short region near anterior end. Spermatheca rectangular, sometimes empty and then narrower and longer. Distance between vulva and spermatheca 42-71% of that between vulva and anus. Length of post-vulval uterine sac *ca* one-fourth to one-third of distance between vulva and often of two or three rudimentary elements. Tail conical with 16-26 rather narrow annuli, tip rounded or of slightly irregular shape, distinctly crenate to almost smooth. Phasmids between 9-13 annuli from tip.

Male

Testis with spermatocytes in double row. Edge of bursa coarsely crenate. Phasmids at two-fifths tail length from tail tip.

DIAGNOSIS AND RELATIONSHIPS

Pratylenchus pseudopratensis is characterised by: labial region with three annuli, lateral fields with four lines and usually some additional lines running obliquely between the inner two, spermatheca round, males present and tail conical with rounded tip.

The matrix code is: A2, B2, C3, D4, E2, F3, G3, H2, I3, J1, K1.

It is close to *P. fallax*, *P. pratensis* and *P. penetrans* from which it differs by stylet length, shape of the spermatheca, number of tail annuli and tail tip (see the corresponding descriptions).

Frederick and Tarjan (1989) pointed out that *Pratylenchus sefaensis* Fortuner, 1973 was not specifically compared with *P. pseudopratensis* Seinhorst, 1968 when it was described. In fact, *P. sefaensis* cannot be adequately separated from *P. pseudopratensis* and was therefore considered a junior synonym of the latter. This action is maintained here.

DISTRIBUTION

Recorded, apart from the type locality (Doornenburg, The Netherlands, from an apple orchard), in several host and localities: on strawberry in Russia (Belozerova & Metlitskii, 1972); in Poland (Brzeski & Szczygiel, 1977); on fruit crops in Russia (Bezerova & Pokhodenko, 1978; Ryss, 1988); in Slovenia on corn (Urek *et al.*, 2003); on millet (*Pennisetum typhoides* (L.) Leeke) in the Sudano-Sahelian zone of West Africa (Villenave & Cadet, 1998); on camellia in Changweon, China (Kim & Minagawa, 1996); and Brazil (Café-Filho & Huang, 1988); in Pakistan on date palm, *P. dactylifera* (Khan *et al.*, 1987); in Georgia, USA, on soybean (Motsinger & Minton, 1986); and in Illinois, USA, on horseradish (*Armoracia rusticana* Gaertn., Mey., Scherb.) (Walters *et al.*, 2004).

51. *Pratylenchus roseus* Zarina & Maqbool, 1998
(Fig. 86)

MEASUREMENTS

- Female holotype (after Zarina & Maqbool, 1998): L = 0.56 mm; a = 22.8; b = 7.2; b′ = 4.4; c = 20.2; c′ = 1.9; V = 81; stylet = 16 μm.
- 34 females (after Zarina & Maqbool, 1998): L = 0.48 (0.42-0.54) mm; a = 27.5 (23.9-33.4); b = 5.4 (4.2-6.6); b′ = 4.1 (3.6-4.5); c = 18.2 (11.2-24.0); c′ = 2.6 (2.1-3.1); V = 82 (81-83); stylet = 15.5 (15-16) μm.
- 25 males (after Zarina & Maqbool, 1998): L = 0.37 (0.37-0.38) mm; a = 27.9 (26.4-29.3); b = 6.3 (6.0-6.7); b′ = 3.8 (3.6-4.0); c = 18.2 (17.3-19.0); c′ = 2.1 (1.9-2.3); T = 41 (40-41); stylet = 13 (12-14) μm; spicules = 14.5 (13-16) μm; gubernaculum = 3.6 (3.5-4.0) μm.

DESCRIPTION

Female

Body almost straight after heat relaxation, slightly tapering at both extremities. Cuticle finely annulated, transverse striation measuring less than 1 μm. Lateral fields occupying *ca* one-third of body diam., with six lines, outer two lines prominent, smooth, inner lines with oblique striation. Labial margin flattened with elevated oral aperture bearing two annuli, labial framework heavily sclerotised, head 2.4-3.2 μm high and 6.5 μm diam. Stylet small, 15.2-16.0 μm long, basal knobs rounded *ca* 1.6-2.4 μm high, 3.2-4.0 μm wide, dorsal pharyngeal gland opening 1.6-2.4 μm from stylet base. Pharynx 73.6-84.0 μm long, overlapping intestine ventrally by 117.6-129.6 μm, median bulb ovate, well developed, nerve ring encircling isthmus, pharyngo-intestinal junction opposite or slightly posterior to excretory pore which is 72.0-76.0 μm from anterior end, hemizonid two annuli long located just anterior to excretory pore; intestinal fasciculi present. Single anterior reproductive tract outstretched; vulva with lateral flaps or membranes. Spermatheca rounded, with sperm. Tail 24.8-37.6 μm long with 18-22 annuli, subcylindrical; terminus rounded with coarse annulation. Phasmids located at *ca* mid-tail.

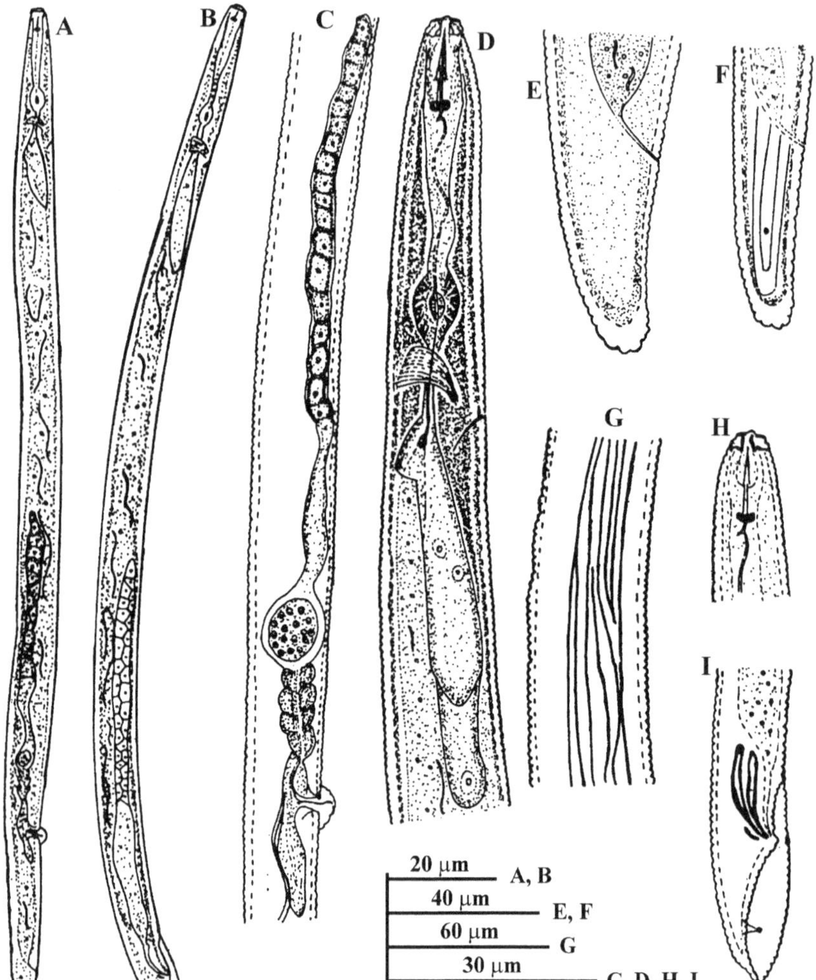

Fig. 86. Pratylenchus roseus *Zarina & Maqbool, 1998. A: Entire female; B: Entire male; C: Female posterior region; D: Female pharyngeal region; E, F: Female tails; G: Lateral fields at mid-body; H: Male labial region; I: Male tail. After Zarina and Maqbool (1998).*

Male

Similar to female in most respects, except smaller in body size and stylet length, stylet and lip sclerotisation weak, testis outstretched, double row of spermatocytes anteriorly. Spicules slender and ventrally curved. Gubernaculum small and curved. Caudal alae crenate, arising anterior to spicules and extending to tail tip. Phasmids near mid-tail.

DIAGNOSIS AND RELATIONSHIPS

Pratylenchus roseus is characterised by: the presence of lateral vulval flaps, labial region with two annuli, lateral fields with six lines, the outer two being smooth and inner lines with oblique striation and tail subcylindrical; terminus rounded with coarse annulation.

The matrix code is: A1, B2, C2, D2, E3, F2, G2, H2, I3, J3, K1.

It is very close to *P. flakkensis*, *P. neglectus* and *P. panamaensis* from which it differs by stylet length, lateral fields, position of vulva, tail terminus and presence of males (see the corresponding descriptions).

DISTRIBUTION

It has only been recorded from the type locality at Umerkot, Nawabshah and Karachi, Pakistan, from soil around the roots of periwinkle (*Catharanthus roseus* (L.) G. Don f.), chilli (*Capsicum annuum* L.), jatropha (*Jatropha podagrica* Hook.) and rose (*Rosa indica* L.).

52. *Pratylenchus scribneri* Steiner *in* Sherbakoff & Stanley, 1943 (Figs 87, 88)

MEASUREMENTS

- Female lectotype (after Loof, 1985): L = 0.53 mm; a = 22; b = 6.2; b′ = 4.7; c = 21; c′ = 2.1; V = 80; stylet = 14 μm.
- 25 females (after Loof, 1985): L = 0.50-0.66 mm; a = 23-33; b = 4.9-7.4; b′ = 3.6-6.0; c = 14-18; c′ = 2.3-3.3; V = 75-79; stylet = 14-15 μm.
- 25 females (after Loof, 1985): L = 0.40-0.58 mm; a = 20-27; b = 4.8-7.6; b′ = 3.7-5.9; c = 13-18; c′ = 2.1-3.0; V = 74-79; stylet = 12-15 μm.
- 7 females (after Loof, 1985): L = 0.42-0.52 mm; a = 22-30; b = 4.6-6.3; b′ = 3.2-4.6; c = 14-22; c′ = 2.1-2.7; V = 72-79; stylet = 15-16 μm.
- 2 males (after Sher & Allen, 1953): L = 0.40-0.47 mm; a = 27-28; b = 5.5-6.6; c = 17-18; T = 42-53; stylet = 12-17 μm.

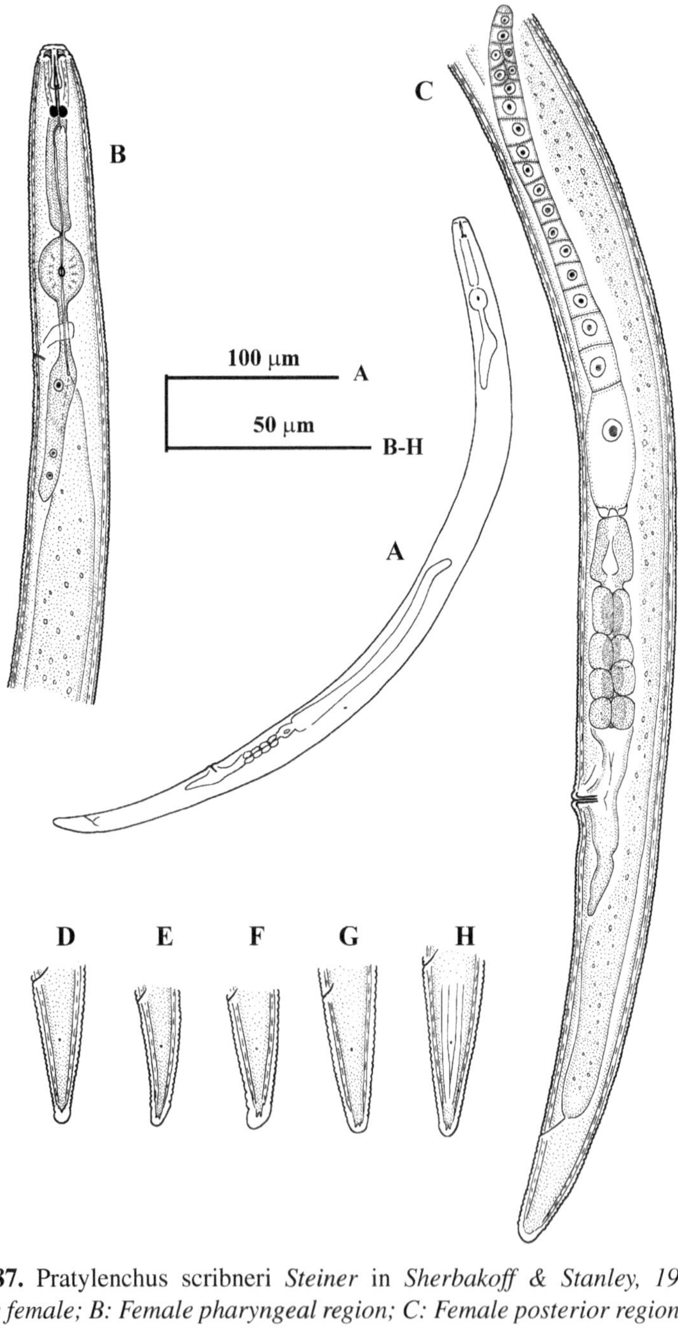

Fig. 87. Pratylenchus scribneri *Steiner* in *Sherbakoff & Stanley, 1943. A: Entire female; B: Female pharyngeal region; C: Female posterior region; D-H: Female tails. After Loof (1985).*

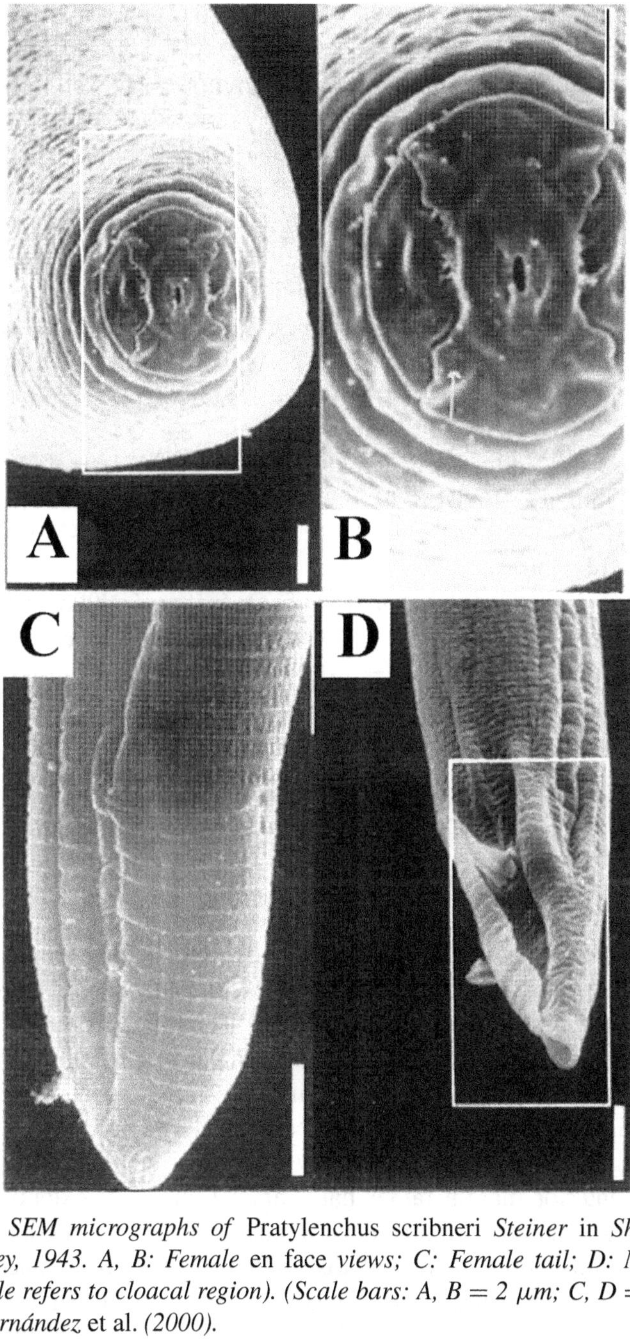

Fig. 88. *SEM micrographs of* Pratylenchus scribneri *Steiner* in *Sherbakoff & Stanley, 1943. A, B: Female* en face *views; C: Female tail; D: Male tail (rectangle refers to cloacal region). (Scale bars: A, B = 2 μm; C, D = 5 μm.) After Hernández* et al. *(2000).*

- 37 females (after Loof, 1960): L = 0.41-0.62 mm; a = 20.0-29.3; b = 5.7-7.7; c = 13.3-18.4; V = 73-79; stylet = 14-17 μm.
- 1 male (after Loof, 1960): L = 0.47 mm; a = 27.6; b = 6.6; c = 17.6; T = 53; stylet = 12 μm.
- Females (after Thorne & Malek, 1968): L = 0.5 mm; a = 24; b = 6.1; c = 18; V = 76; stylet = 16-18 μm.
- 14 females (after Hashim, 1984): L = 0.49 (0.38-0.59) mm; a = 29 (26-32); b = 5.3 (4.5-5.9); c = 18.3 (16.1-25.0); c′ = 2.6 (2.1-3.0); V = 78 (75-79); stylet = 14.6 (14.5-15.0) μm.
- 11 females (after Subramaniyan & Sivakumar, 1991): L = 0.48 (0.41-0.54) mm; a = 23.1 (21.8-25.2); b = 5.7 (4.7-7.3); b′ = 4.6 (4.1-5.6); c = 19.2 (17.2-21.5); c′ = 1.7 (1.5-1.9); V = 76 (74-78); stylet = 16.0 (15.5-17.0) μm.
- 24 females (after Doucet & Cagnolo, 1998): L = 0.45 (0.30-0.50) mm; a = 26 (21.8-29.0); b = 5.7 (4.7-7.3); b′ = 3.9 (3.2-4.6); c = 17.5 (13.2-32.4); c′ = 2.5 (1.8-3.1); V = 74 (69-79); stylet = 15.5 (14-17) μm.
- 21 females (after Van den Berg & Quénéhervé, 2000): L = 0.48 (0.40-0.55) mm; a = 25.7 (21-31); b = 5.3 (4.5-6.5); c = 17.3 (13-19); c′ = 2.5 (2.0-4.0); V = 78 (74-80); stylet = 15.7 (14-17) μm.
- 11 females (after Inserra *et al.*, 2007): L = 0.50 (0.45-0.53) mm; a = 26.6 (22.1-29.8); b = 5.8 (5.5-7.0); c = 19.7 (17.9-22.0); c′ = 2.4 (2.1-2.5); V = 78 (77-79); stylet = 14.6 (14.0-15.5) μm.

DESCRIPTION

Female

Body stout, almost straight when heat-relaxed; young females more slender. Cuticular annulation fine, shallow, indistinct, *ca* 1 μm apart, tending to become more pronounced and further apart on tail. Lateral fields with four longitudinal lines; occasionally a fifth line may be present in pre-vulvar region, especially in younger specimens. Labial region slightly offset from body, composed of two annuli of *ca* same height, anterior one distinctly narrower than second; angles rounded, anterior margin rather flat. Labial framework extending into body for one annulus. *En face* view characterised by subdorsal and subventral lips wider than oral disc and separated from lateral lips by two convex incisures with respect to latter; inner part of lateral lips narrower than outer part (Hernández *et al.*, 2000; Inserra *et al.*, 2007). Stylet stout, with rounded knobs that vary little in shape (Román & Hirschmann, 1969a). Orifice of dorsal pharyngeal gland 2-3 μm posterior to stylet base. Median bulb broadly oval, 11 × 14 μm in lectotype. Nucleus of dorsal gland just posterior to pharyngo-intestinal junction, nuclei of ventrosublateral glands close together near posterior end of gland lobe; in lateral view, all three nuclei lie in tandem. Hemizonid *ca* two annuli long, generally opposite posterior half of isthmus. Excretory pore immediately posterior to hemizonid. Pharyngeal overlap of medium length. Hemizonion and cephalids not seen. Anterior genital branch consisting of: ovary with oocytes in single file except for a short zone near anterior end; short oviduct; oblong, empty spermatheca; uterus a tricolumella (three rows of four cells each). Post-vulval uterine sac generally 1.1-1.4 body diam. long; occupying 25-40% of vulva-anus distance. Dimensions of 14 intra-uterine eggs: 60 (52-68) × 22 (18-26) μm. Vulva-anus distance generally 2.7-3.5 times tail length. Tail tapering slightly, terminus mostly broadly rounded, varying from somewhat narrower to almost truncate; terminus unstriated. Protoplasmic core usually ending in some finger-like processes. Phasmids inconspicuous, invisible in lectotype; lying *ca* mid-tail. SEM view showing inner lateral lines fusing posterior to phasmid.

Male

Spicules of normal shape 17 μm long; gubernaculum slightly curved 7 μm long. Bursa typical. Phasmids 14 μm posterior to cloacal aperture.

DIAGNOSIS AND RELATIONSHIPS

Pratylenchus scribneri is characterised by: labial region with two annuli, slightly offset from body, stylet stout, with rounded knobs, pharyngeal overlap of medium length, spermatheca oblong and tail tapering slightly with smooth terminus.

The matrix code is: A1, B2, C2, D3, E2, F4, G3, H1, I2, J1-3, K1.

It can be distinguished from closely related species (*P. crassi* and *P. hexincisus*) by stylet length, the four lines of the lateral fields, often with oblique striae in the central band at mid-body; distance of the dorsal pharyngeal gland orifice, tail shape, number of tail annuli (see the corresponding descriptions).

Thorne and Malek (1968) described *P. agilis* as being very close to *P. scribneri*. Frederick and Tarjan (1989) pointed out some doubts on the identity of *P. agilis* due to the original description being based on few specimens, with minor differences with respect to *P. scribneri*. However, Handoo and Golden (1989), who examined and described non-type specimens of *P. agilis* from the USDA Nematode Collection (Beltsville, Maryland, USA) in their review, as well as Loof (1991), considered it as a valid species. Comparison of *P. agilis* specimens (from the USDA Collection, collected from corn) with *P. scribneri* material yields no significant differences: the number of labial annuli, the *en face* view, the shape and location of the amphidial openings and female tail are very similar. As reported by Hernández *et al.* (2000), considering that comparison of both species by non-morphological techniques, such as DNA ITS analysis (Powers *et al.*, 1997) and isozyme analysis (Andrés *et al.*, 2000), have recently produced analogous results, *e.g.*, absence of any substantial difference between both species, there is strong evidence that they are one and the same species. Hernández *et al.* (2000) therefore proposed *P. agilis* as a junior synonym of *P. scribneri*, a decision that is maintained here.

Subramaniyan and Sivakumar (1991) described *P. crossandrae* as a new species differentiated from *P. scribneri* based on b and c values and body annulation, all of which are considered minor differences and therefore it is considered herein as a new junior synonym of *P. scribneri*.

Ryss (2002a) reported that *P. jordanensis* was identical to *P. scribneri* in the main diagnostic characters (26 characters). No differences in the body length (410-620 μm in *P. scribneri vs* 380-590 μm in *P. jordanensis*), stylet length (14-18 *vs* 14-15 μm), number of lip annuli (two), values of indices V (73-82 *vs* 75-79) and c (13-21 *vs* 16-25), in cylindrical rounded shape of tail, smooth tail terminus and number of annuli on the ventral surface of the tail (18-30 *vs* 19-24). Thus, the unique diagnostic character of *P. scribneri* is the second lip annulus, which is significantly wider than the first lip annulus and this feature is typical of *P. jordanensis* as well. Consequently, *P. jordanensis* was considered as a new synonym of *P. scribneri*. However, recent studies of LM and SEM microscopy on paratypes of *P. jordanensis* and three populations from Oman by Inserra *et al.* (2005) showed great morphological similarities between *P. jordanensis* and *P. zeae*, suggesting that the former is a junior synonym of the latter and this action is accepted and maintained in this monograph.

DISTRIBUTION

It has been recorded from the type locality at Crossville, Tennessee, USA, from potato tubers. It is widespread in the USA, where it was found on potato in Tennessee (Sherbakoff & Stanley, 1943) and Nebraska (Thorne & Malek, 1968). It has been recorded on several hosts in California: alfalfa (Radewald *et al.*, 1964); Sudan grass (Thomason & O'Melia, 1962), roses (Nesbitt, 1956), *Cymbidium* sp. (Sher, 1959) and onion (Thomason *et al.*, 1964). Siddiqui *et al.* (1973) list 22 host plants in California, including corn, tomato and *Amaryllis*, but not potato or soybean. In Iowa it occurs on corn, soybean and *Setaria* (Norton *et al.*, 1964; Williams, 1982); in Kansas on soybean (Dickerson, 1979); in Maryland, USA, on corn, soybean and tomato (Rebois & Golden, 1985); in South Dakota on corn (Smolik, 1978; Smolik & Evenson, 1987), in Illinois on corn (Willut & Malek, 1979); in Ohio on potato (Brown *et al.*, 1980); in Alabama on white clover, tall fescue, alfalfa, sorghum, Bahia grass, millet (Minton, 1965), pigeon pea (Rodríguez-Kábana & Ingram, 1978) and possibly on several other plants (Minton *et al.*, 1963). Mai *et al.* (1960) list 12 host plants in New Jersey and Pennsylvania. It is very common in Florida on *Amaryllis* (Sher & Allen, 1953; Christie & Birchfield, 1958; Nong & Weber, 1964). In North Carolina it was found on peach (Barker & Clayton, 1973) and in Mississippi and Louisiana on soybean (Rebois & Golden, 1978). In Arkansas it was recorded on cotton (Robbins *et al.*, 1989a, b) and soybean (Robbins *et al.*, 1987). Records from other countries include: Mexico, on grass, poppy and bean (Knobloch & Laughlin, 1973); several hosts and localities from Argentina (Doucet & Cagnolo, 1998; Chaves & Torres, 2001); in Venezuela (Crozzoli, 2002); in Brazil (De Souza *et al.*, 1998); the subtropical districts of Japan, on sugarcane (Gotoh, 1974); in Oman (Mani *et al.*, 1997); Punjab and Himachal Pradesh, India, on barley, *Lathyrus*, *Prunus* and *Berberis* (Sethi & Swarup, 1971) and sugarcane in Tamil Nadu, India (Mehta & Sundararaj, 1990); in Pakistan (Islam *et al.*, 1994, 1996); in Korea (Park *et al.*, 1999); in China (Yin, 1994, 1995; Han *et al.*, 1997); soil around roots of banana, fig, plum and quince, in Israel (Minz, 1957); glasshouses at Izmir, Turkey (Geraert *et al.*, 1975); tobacco, Bulgaria (Katalan-Gateva & Baicheva, 1978); *Hippeastrum*, Sweden (Anderson, 1971); in Slovenia on corn (Urek *et al.*, 2003); in Australia (Riley & Wouts, 2001; Riley & Kelly, 2002); in Italy on corn (Tacconi *et al.*, 1985). It also appears to be widespread in

Africa: Oteifa (1962) recorded it from the Delta region of the Nile Valley, Egypt, on potato, sweet potato, cowpea, strawberry and water melon; it is widespread in corn roots in the Western and Lagos States of Nigeria (Anon., 1975); it was also recorded from South Africa on apple, Cape Province; tobacco, Transvaal (Milne, 1961; Heyns, 1962; Van den Berg, 1971). It has been also recorded from Jordan in soil around the roots of grapevine; Brazil (Café-Filho & Huang, 1988); and from grassland and *Saccharum officinarum* L., in Guadeloupe (Van den Berg & Quénéhervé, 2000).

53. *Pratylenchus sensillatus* Anderson & Townshend, 1985
(Fig. 89)

MEASUREMENTS

- Female holotype (after Anderson & Townshend, 1985): L = 0.68 mm; a = 36; b = 7.8; b′ = 4.9; c = 23; V = 77; stylet = 15.5 μm.
- 20 females (after Anderson & Townshend, 1985): L = 0.62 (0.57-0.69) mm; a = 34 (28-42); b = 7.8 (7.1-8.3); b′ = 4.9 (4.1-5.6); c = 24 (20-31); c′ = 2.3 (1.8-2.5); V = 79 (77-81); stylet = 15.5 (15-17) μm.

DESCRIPTION

Female

Body thin, width 18 (15-22) μm; relaxed posture highly variable, changing from linear to a closed ring, the most prevalent being ventrally arcuate. Body annuli generally indistinct, *ca* 1-1.5 μm wide, subcuticular annulation conspicuous, width 0.5 μm. Lateral fields with four lines, 5.5-7.0 μm wide, *ca* one-third of body diam., margins smooth or weakly crenate. In freshly killed females only, lateral fields irregularly areolated, central zone often with one or two longitudinal lines or oblique striae; with SEM, central zone observed with up to ten longitudinal lines. Labial region low, contour rounded, bearing three annuli, slightly offset from body, appearing continuous in some specimens, apex typically flattened. Labial region diam. 7-8 μm, height 2-3 μm. In SEM micrographs, submedian sectors of oral plate wedge shaped, bearing four cephalic and six labial sensillar pits. Stylet knobs

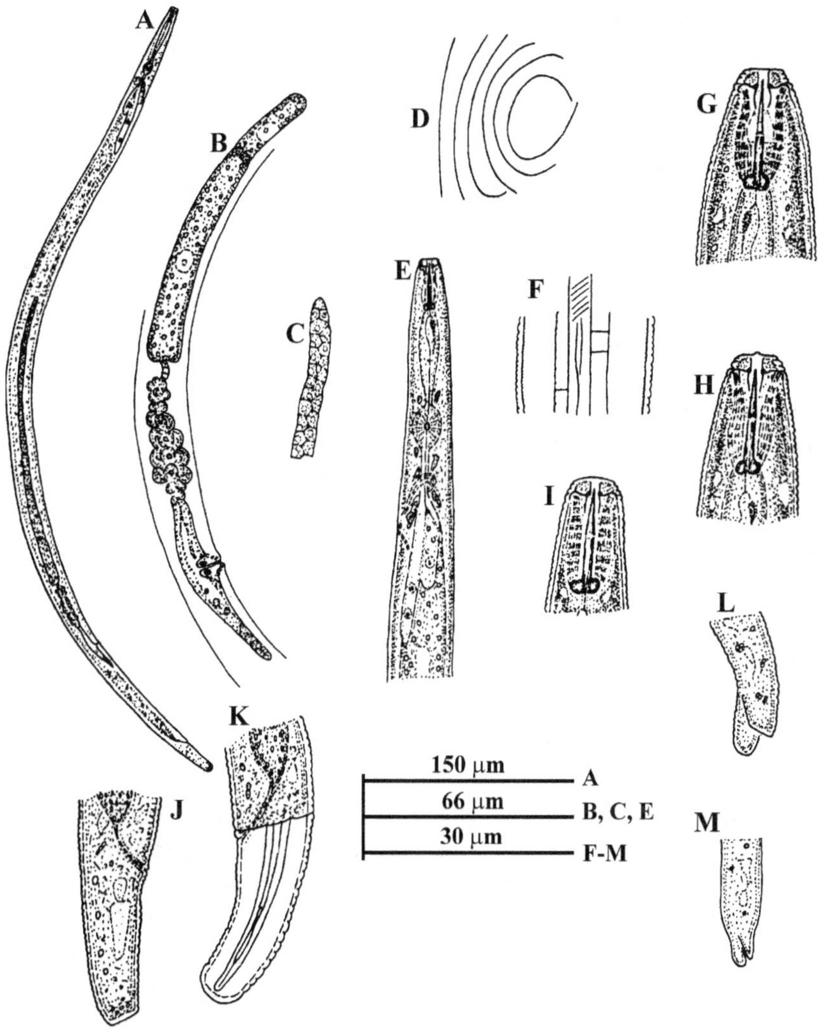

Fig. 89. Pratylenchus sensillatus *Anderson & Townshend, 1985. A, D: Entire females; B: Female reproductive system; C: Detail of pharyngeal region of ovary; E: Female pharyngeal region; F: Lateral field at mid-body; G-I: Female labial regions; J-M: Female tails. After Anderson and Townshend (1985).*

rounded in living specimens, 4.0 (3.5-5.0) μm wide, in glycerin mounts anterior surfaces sometimes sloping or flattened. Stylet conus 7.1 (6.5-8.5) μm long, comprising 45 (42-50)% of total stylet length. Dorsal pharyngeal gland orifice 2.5-3.5 μm posterior to stylet. Pharynx from anterior end to end of basal bulb 127 (106-153) μm, to mid-metacorpal valve 54 (50-60) μm, to pharyngo-intestinal valve 81 (72-87) μm; length of glands 46 (39-66). Excretory pore 87 (81-95) μm from anterior end, 6 (1-14) μm posterior to pharyngo-intestinal valve. Ovary outstretched, extending up to 65 μm from pharynx; oogonia in double row in egg-laying females, single or double in others. Spermatheca indistinct, lacking a discrete structure. Vulval lips usually protuberant. Post-vulval uterine sac 23 (14-32) μm long, distal end cellular. Vulva-anus length 104 (92-121) μm, 79 (72-82)% of vulva-tail length, 3.9 (3.4-4.6) times tail length, vulva-tail length 128 (112-148) μm. Tail with 20 (14-25) annuli, 27 (22-32) μm long, subcylindrical, usually linear, diam. at tail terminus 59 (50-67)% that at anus. Tail terminus broadly rounded in 67% of 63 specimens, truncate in 25% and cleft in 8%. Phasmids in posterior half of tail, at 51 (48-61)% of tail length. Rectum slightly longer than anal body diam., without intestinal overlap. Anal body diam. 12 (10-16) μm.

Male

Not found.

DIAGNOSIS AND RELATIONSHIPS

Pratylenchus sensillatus is characterised by: a low, rounded, offset labial region with three annuli, absence of a functional spermatheca and a subcylindrical tail with a broadly rounded, smooth terminus.

The matrix code is: A2, B1, C2, D1, E2, F5, G2, H1, I3, J1, K2.

It is close to *P. andinus*, *P. bolivianus*, *P. mulchandi*, *P. thornei* and *P. zeae* from which it differs by labial region shape, body length, position of the excretory pore and of the vulva and female tail shape (see the corresponding descriptions).

DISTRIBUTION

It has been recorded, apart from the type locality at Moncrief Farm, Wingham, Ontario, Canada, in soil around the roots of *Phleum pratense* and on oat, also in Canada (Townshend, 1989a).

54. *Pratylenchus silvaticus* Brzeski, 1998
(Fig. 90)

MEASUREMENTS

- Female holotype (after Brzeski, 1998): L = 0.43 mm; a = 25; b = 6.5; b′ = 4.1; c = 18; c′ = 2.4; V = 81; stylet = 15 μm.
- 17 females (after Brzeski, 1998): L = 0.45 (0.42-0.48) mm; a = 25 (21-27); b = 6.3 (5.8-7.0); b′ = 4.2 (3.9-4.7); c = 20 (17-25); c′ = 2.4 (1.9-3.1); V = 81 (80-83); stylet = 15.1 (15-16) μm.
- 5 males (after Brzeski, 1998): L = 0.37 (0.34-0.42) mm; a = 27 (26-29); b = 6.0 (5.7-6.4); c = 18 (17-19.5); stylet = 14.5 (14-15) μm; spicule = 14 (13-15) μm; gubernaculum = 4 (3-5) μm.

DESCRIPTION

Female

Body straight or slightly arcuate ventrad. Cuticular annuli *ca* 1.2-1.3 μm at mid-body. Lateral field with four lines, two additional faint lines often seen in central part of body down to vulva. Labial region continuous with body contour, not offset, with two annuli, although three complete annuli observed in one female and three incomplete in two other females. Labial framework relatively weakly refractive, extending posteriorly for *ca* one annulus. Stylet stout, knobs slightly indented anteriorly. Dorsal pharyngeal gland opening 2.0 (1.5-3.0) μm posterior to base of stylet. Centre of median bulb in females 45 (40-48) μm from anterior end. Pharyngeal gland rather compact, 2.0 (1.7-2.5) body diam. long. Excretory pore at beginning of pharyngeal gland, 71 (66-77) μm from anterior end. Spermatheca 1.4 (0.8-1.9) times as long as diam. when filled with sperm. Post-vulval uterine sac 1.0 (0.7-1.3) body diam. or 28.5 (23-33)% of vulva-anus distance long, cellular elements not usually seen at end of sac; in one female post-vulval uterine sac, 1.8 body diam. or 41% of vulva-anus distance long with two or three terminal cells visible. Female tail tapering considerably, often narrowing near tip and then expanding as bulb-shaped terminal part, terminus irregularly striated, sometimes only indented. Phasmids at *ca* one-third of tail length from anus. Number of tail annuli 20 (16-29), mostly 18-22.

Male

Similar to female but smaller, stylet more delicate, spicules narrow.

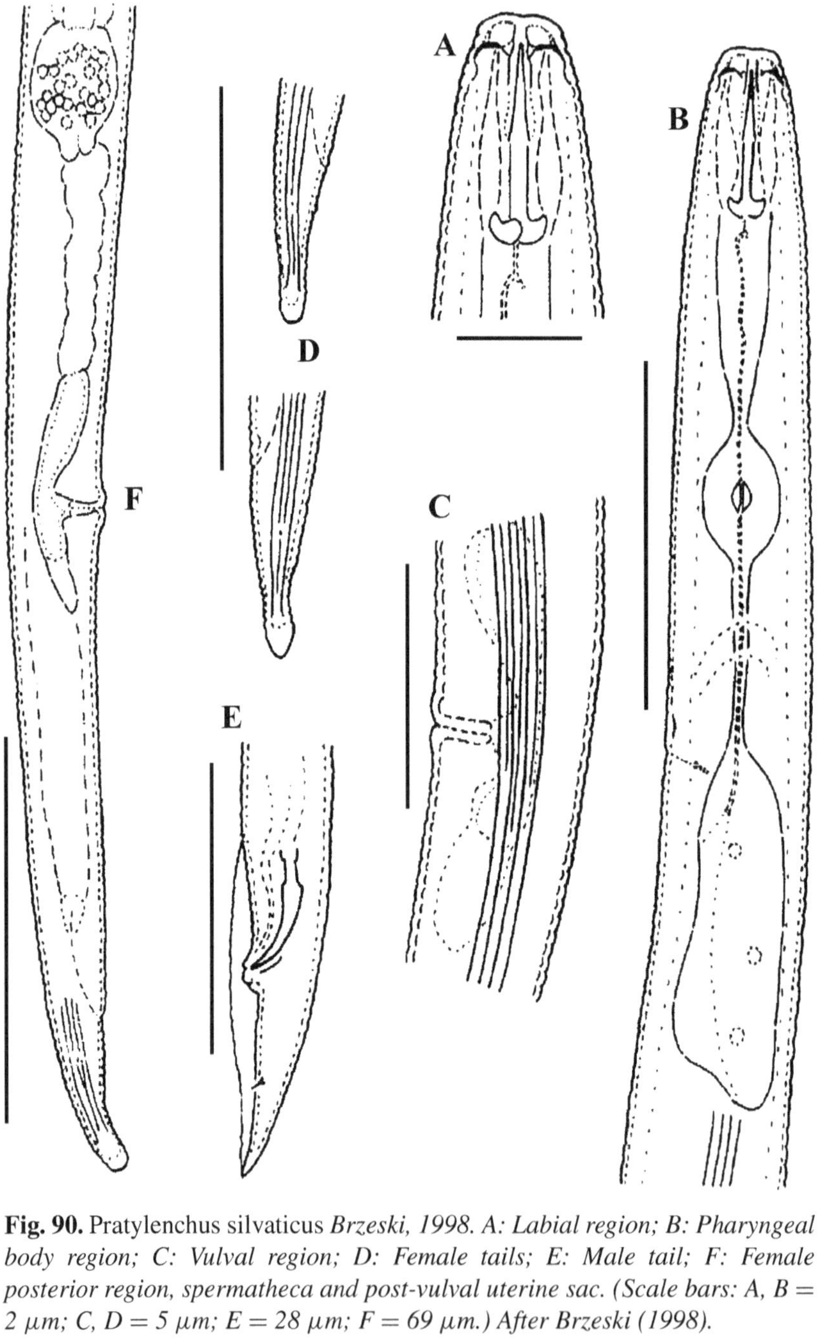

Fig. 90. Pratylenchus silvaticus *Brzeski, 1998. A: Labial region; B: Pharyngeal body region; C: Vulval region; D: Female tails; E: Male tail; F: Female posterior region, spermatheca and post-vulval uterine sac. (Scale bars: A, B = 2 μm; C, D = 5 μm; E = 28 μm; F = 69 μm.) After Brzeski (1998).*

DIAGNOSIS AND RELATIONSHIPS

Pratylenchus silvaticus is characterised by: labial region with two annuli, males common, spermatheca round, female tail tapering with crenate terminus, irregularly striated.

The matrix code is: A1, B2, C2, D2, E3, F2, G3, H1, I2, J1-3, K1.

It can be distinguished from the closely related species *P. flakkensis*, *P. estoniensis* and *P. gibbicaudatus* by the smaller stylet length, posterior position of vulva, presence of males and different tail shape (see the corresponding descriptions).

DISTRIBUTION

It has only been recorded from the type locality, from around *Sphagnum* at Borki, Poland.

55. *Pratylenchus subpenetrans* Taylor & Jenkins, 1957
(Fig. 91)

MEASUREMENTS

- 100 females (after Taylor & Jenkins, 1957): L = 0.40 (0.33-0.48) mm; a = 24.2 (18.4-27.7); b = 6.0 (5.0-7.2); c = 18.2 (16.2-21.4); V = 80 (77-83); stylet = 16 (15-16.5) μm.
- 100 males (after Taylor & Jenkins, 1957): L = 0.39 (0.33-0.47) mm; a = 27.7 (23.7-32.1); b = 6.1 (5.0-7.4); c = 17.9 (14.7-21.0); stylet = 14.5 (14-15.5) μm.
- 12 females (after Choi *et al.*, 2006): L = 0.58 (0.52-0.63) mm; a = 25.0 (23.1-27.8); b = 7.0 (5.7-8.2); b′ = 4.9 (3.7-5.7); c = 18.7 (15.8-24.0); c′ = 2.3 (1.7-3.0); V = 78 (74-81); stylet = 15.2 (13.3-16.8) μm.
- 3 males (after Choi *et al.*, 2006): L = 0.54 (0.51-0.58) mm; a = 25.3 (24.5-25.9); b = 6.1 (5.6-6.4); b′ = 4.2 (4.0-4.3); c = 20.4 (18.8-21.8); stylet = 14.1 (13.3-14.7) μm; spicules = 19.6 (18.2-21.0) μm; gubernaculum = 4.1 (3.8-4.2) μm.

DESCRIPTION

Female

Body somewhat slender, tapering anteriorly and posteriorly. Cuticle marked with transverse striations *ca* 1 μm apart. Lateral fields originat-

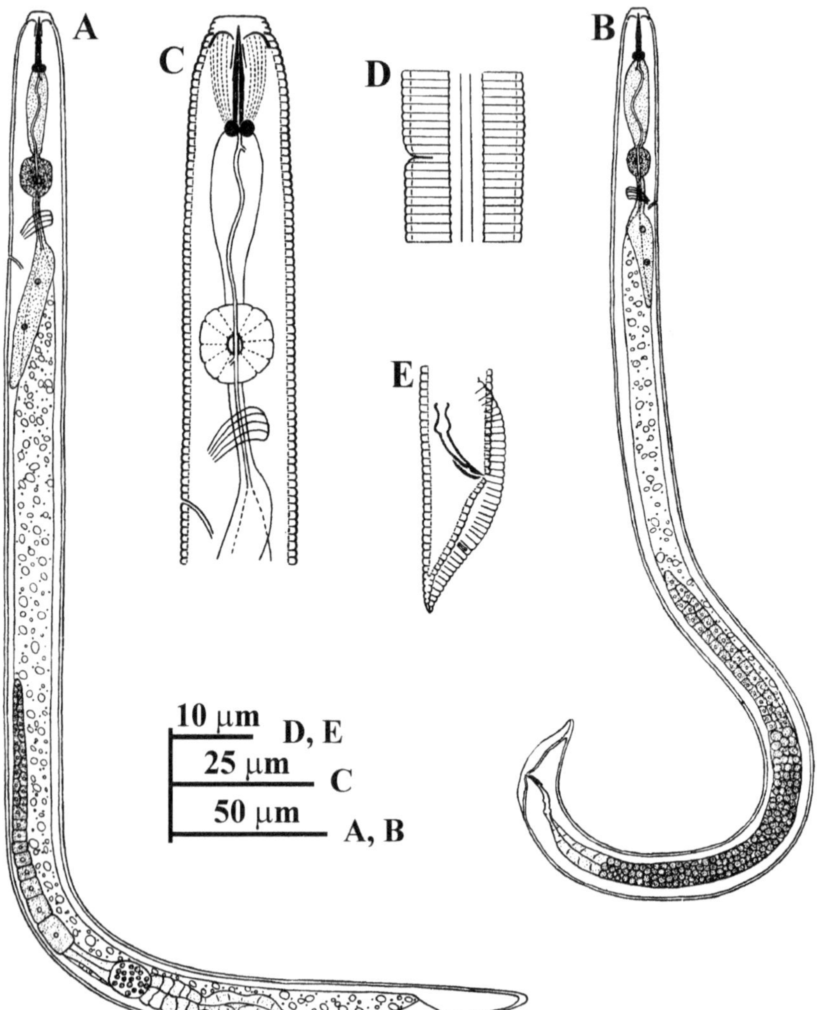

Fig. 91. Pratylenchus subpenetrans *Taylor & Jenkins, 1957. A: Entire female; B: Entire male; C: Female pharyngeal region; D: Lateral field at mid-body; E: Male tail. After Taylor and Jenkins (1957).*

ing as narrow groove at base of stylet guiding piece and occupying *ca* one-third of body diam. in vulval region. Lateral fields lines originating in pharyngeal region so that four are present at junction of pharynx and intestine. These continue longitudinally for entire body length until they disappear on tail. Outer two bands crenate, inner lines straight with no irregularities. Labial region low, flat, slightly offset from body contour and bearing three annuli. Lateral margins of lips rounded. Labial framework extending posteriorly *ca* 2 μm into body and are more heavily sclerotised than rest of structure. Stylet short (14.9-16.5 μm) and stout with spherical basal knobs. Guiding apparatus 3.3-3.8 μm long. Dorsal pharyngeal gland opening 2.0 (1.7-2.2) μm posterior to base of stylet. Anterior portion of distinct pharyngeal corpus broader than posterior portion. Median bulb ovoid, containing a heavily sclerotised valvular apparatus. Subventral pharyngeal gland duct opening into pharyngeal lumen immediately posterior to valvular structure. Nerve ring encircling narrow isthmus *ca* midway between the median bulb and the junction of the intestine. Pharyngeal glands overlap intestine ventrally, long and narrow. Dorsal pharyngeal gland nucleus slightly larger than subventral nuclei. Excretory pore opposite or slightly posterior to junction of pharynx and intestine. Intestinal cells filled with refractive granules *ca* equal in size and evenly distributed. Intestine terminating in oblique rectum opening through a faint anus. Ovary single, outstretched, extending anteriorly *ca* half way to pharynx. Single row of oocytes in ovary except for a double row in anterior portion in region of multiplication. Oviduct opening into a prominent oval spermatheca which in turn is followed by a cellular uterus. Vagina extending transversely *ca* one-half the body diam. and opening by means of a conspicuous vulva, the lips of which do not protrude. Post-vulval uterine sac *ca* 1-1.5 times longer than corresponding body diam. Posterior portion of this structure is cellular. Tail conoid, with smooth terminus. Phasmids at *ca* one-half tail length posterior to anus. Tail tip unstriated, rounded.

Male

Males occurring as frequently as females and although smaller and more slender are morphologically similar, particularly regarding details of alimentary tract. However, stylet is shorter, 14.5 (14.3-15.2) μm and smaller median bulb tending to be flattened somewhat anteriorly. Testis single, outstretched, 45-50% of body length. Testis comprising anterior one-third of system and containing multiple row of spermatocytes. *Vas deferens* and seminal vesicle containing hundreds of small, spherical refractive sperm. Spicules paired, 14 μm long, arcuate dorsally with two swellings, one anteriorly terminal and other subterminal. Gubernaculum 5 μm long. Bursa enclosing tail tip, extending slightly beyond anterior ends of spicules. Phasmids located slightly posterior to mid-tail.

DIAGNOSIS AND RELATIONSHIPS

Pratylenchus subpenetrans is characterised by: labial region with three annuli, males common, spermatheca round, female tail rounded with smooth terminus and spicules arcuate dorsally with two swellings, one anteriorly terminal and the other subterminal.

The matrix code is: A2, B2, C3, D3, E3, F2, G3, H1, I3, J1, K1.

It can be distinguished from the closely related species *P. penetrans* by the smaller body length, longer post-vulval uterine sac with cellular distal part, labial framework extends farther posterior and spicules with two anterior swellings (see the corresponding description).

Recently, Choi *et al.* (2006) described a new species, *P. gongjuensis*, which is closely related to *P. subpenetrans*, and which may be differentiated only by morphometric differences in female and male body length. Detailed comparison of both descriptions as well as drawing and pictures of *P. gongjuensis* are not sufficient to maintain this species as valid and we consider it as a new junior synonym of *P. subpenetrans*.

DISTRIBUTION

It has been recorded, apart from the type locality, Upper Marlboro, Prince Georges Country, Maryland, USA, in soil around the roots of tall fescue (*Festuca elatior* L.) in Canada (Townshend *et al.*, 1978a) and on horticultural crops, chrysanthemum and ginseng in Korea (Kim & Minagawa, 1996; Choi *et al.*, 2006; Kim *et al.*, 2006).

56. *Pratylenchus subranjani* Mizukubo, Toida, Keereewan & Yoshida, 1990
(Fig. 92)

MEASUREMENTS

- Female holotype (after Mizukubo *et al.*, 1990): L = 0.47 mm; a = 27; b = 6.5; b′ = 3.8; c = 14.5; c′ = 3.2; V = 75; stylet = 16.5 μm.

Fig. 92. Pratylenchus subranjani *Mizukubo, Toida, Keereewan & Yoshida, 1990. A: Entire female; B: Female pharyngeal region; C-E: Female labial regions; F, G: Stylets; H: Female posterior region; I: Vulval region; J-Y: Female tails. After Mizukubo* et al. *(1990).*

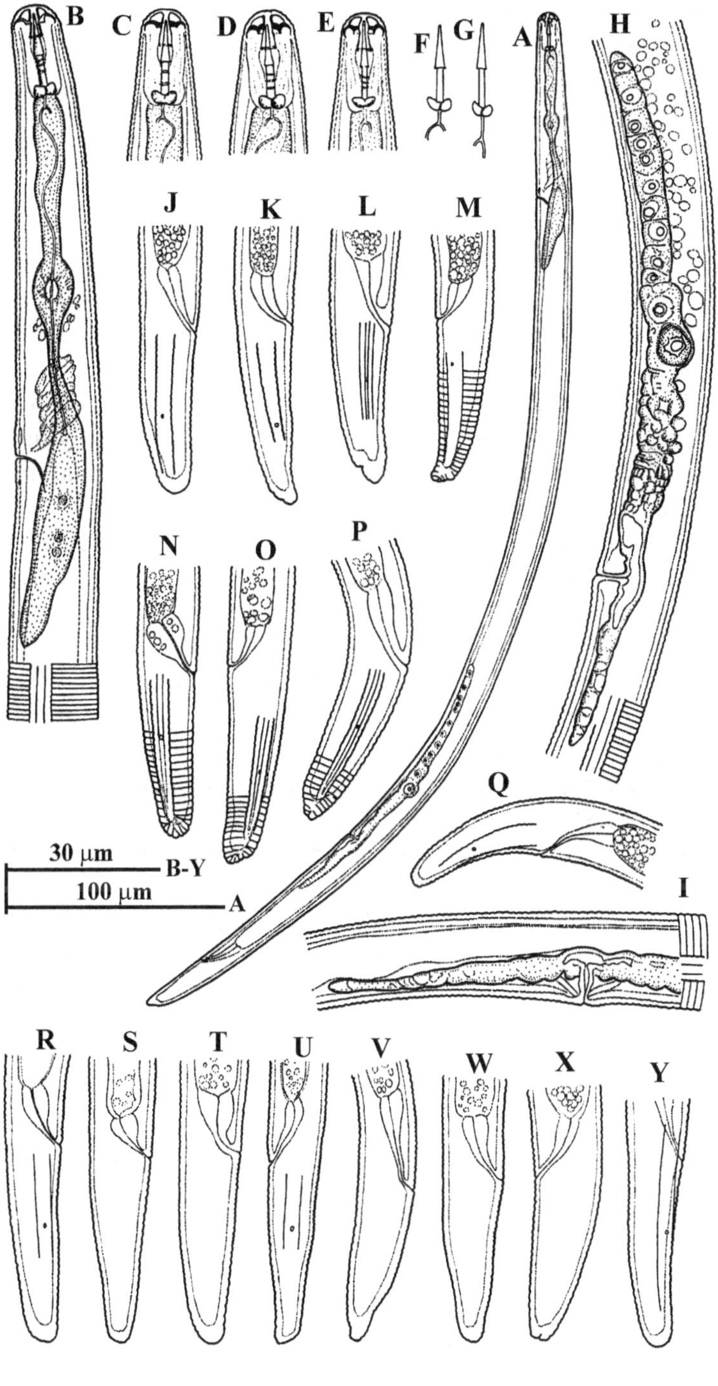
A
B
C
D
E
F
G
H
I
J
K
L
M
N
O
P
Q
R
S
T
U
V
W
X
Y
30 μm
B-Y
100 μm
A

- 40 females (after Mizukubo *et al.*, 1990): L = 0.51 (0.39-0.57) mm; a = 27.2 (23.4-31.5); b = 6.5 (5.4-8.1); b′ = 4.1 (3.3-4.8); c = 15.6 (13.2-18.3); c′ = 2.9 (2.5-3.5); V = 75 (73-77); stylet = 17 (16-18.5) μm.

DESCRIPTION

Female

Body curved ventrally when heat-relaxed. Annuli fine, 1.1 (0.9-1.4) μm wide at mid-body. Labial region medium height, 3.0 (2.3-3.6) μm high, 8.8 (7.6-10.6) μm diam., rounded, barely flattened anteriorly, mostly continuous with body contour or very slightly offset. Labial framework well developed, expanding laterally, highly arched, extending into body for two annuli. Three lip annuli in transmitted light, although it is occasionally difficult to determine whether there are three or four lip annuli due to continuous lip and highly arched labial framework; the fourth annulus from anterior extremity observed by SEM is apparently the first body annulus. SEM *en face* view, though distorted, showed no division between submedian and lateral segments, classifying it into Group 1 of Corbett and Clark (1983). Six inner labial sensilla present. Lateral fields 5 (4-6) μm wide, or 27.9 (23-33)% of widest body diam. In LM, lateral field apparently with weakly crenate margins and three equally spaced plain bands, which gives impression of four lines. However, with SEM photograph, outer bands areolated at intervals of two, three or even more annuli in mid-body region and completely so in tail region. Middle band often sparsely areolated and accompanied by one or two subsidiary lines. In LM, lateral field apparently closing shortly before tail tip, but may extend to tail tip and continue around tail terminus. No cephalids seen. Stylet moderately long, 1.9 times (1.6-2.2) labial region diam. long. Stylet knobs massive, 5.0 (4.0-5.5) μm wide, 1.9 (1.5-2.5) μm high; shape varying from flattened anteriorly to acutely indented. Dorsal pharyngeal gland orifice at 3.3 (2.6-5.3) μm from stylet base. Metacorpus oval, 13 (11-14) μm long, 8.5 (7-109) μm diam., length 1.5 (1.2-1.8) times diam., occupying *ca* half of corresponding body diam.; valve conspicuous, 53 (38-58) μm from anterior body end, or 68.1 (59.1-76.9)% of pharyngeal length. Pharyngeal gland slender, short, extending 123 (97-155) μm from anterior body end, ventrally overlapping intestine for 46 (28-67) μm or 2.5 (1.6-3.4) corresponding body diam.; pharyngeal glands nuclei in tandem. Excretory pore 103.2

(90.8-119.8)% of pharyngeal length or at the 75 (65-85) body annuli from anterior body end; annuli between anterior body end and excretory pore 1.1 (1.0-1.3) μm in average width. Hemizonid flat, 2.5-4 annuli long, 0-3 annuli anterior to excretory pore; hemizonion lenticular, 7.2 (4-12) annuli posterior to excretory pore. Gonad outstretched, 127 (90-214) μm long or 25.1 (17.4-43.5)% of body length; ovary with oocytes in single row; spermatheca empty, rarely observed. Post-vulval uterine sac long, twice as long or longer than body diam., only columnar part of uterus recognisable. Vulva plain; lips slightly protruding; vagina tubular, perpendicular to body axis, rarely slightly directed anteriorly, 8.1 (6.6-9.9) μm long or 45.6 (36.7-56.6)% of vulval body diam. Annuli between vulva and anus 81.6 (67-91) in number, average width 1.1 (0.9-1.3) μm. Tail variable in shape; sorting 40 individuals according to tail tip shape codes of Frederick and Tarjan (1989), their frequencies are: bluntly pointed (BLP) = 53%, subhemispherical (SHM) = 30%, subdigitate (SBD) = 15% and truncate (TRC) = 3% respectively. Form of tail tip annuli, also determined using codes of Frederick and Tarjan (1989), with tails in exact lateral orientation (28 individuals): annulated (ANN) = 71% (including evidently ANN = 50% and barely ANN = 21%), smooth (SHM) = 18% and cleft (CFT) = 11%, respectively. Tail terminal cuticle or hyaline part 3.0 (1.5-4.3) μm thick; phasmids pore-like, centred in lateral fields, 14.2 (7.9-19.1) μm posterior to anus or 18.2 (12.5-27.7) μm from tail tip, or at 56.0 (39.6-77.4)% of tail length.

Male

Not found.

Diagnosis and relationships

Pratylenchus subranjani is characterised by: labial region rounded with three annuli, stylet knobs indented, spermatheca rarely observed, post-vulval uterine sac twice as long as body diam., tail bluntly pointed or subhemispherical with annulated terminus.

The matrix code is: A2, B1, C3, D1, E2, F5, G3, H2, I3, J1, K2.

It is close to *P. delattrei*, *P. goodeyi*, *P. morettoi*, *P. mulchandi* and *P. pratensis* from which it can be distinguished by the number of lip annuli, stylet length, the shape of stylet knobs, position of the vulva, length of post-vulval uterine sac, c ratio, body annuli width, number of tail annuli and the shape of tail terminus (see the corresponding descriptions).

DISTRIBUTION

It has been recorded, apart from the rhizosphere of corn, at type locality at Pukae, Phraphutthabat district, Saraburi province, and other localities of Thailand (Toida *et al.*, 1996).

57. *Pratylenchus sudanensis* Loof & Yassin, 1971

(Fig. 93)

MEASUREMENTS

- Female holotype (after Loof & Yassin, 1971): L = 0.53 mm; a = 29; b = 5.9; c = 18; V = 72; stylet = 16 μm.
- 46 females (after Loof & Yassin, 1971): L = 0.39-0.59 mm; a = 22-31; b = 4.9-7.2; c = 14-23; V = 70-76; stylet = 14-16 μm.
- 30 males (after Loof & Yassin, 1971): L = 0.45-0.56 mm; a = 26-33; b = 4.6-6.3; c = 15-22; T = 32-52; stylet = 13-16 μm.
- 6 females (after Ryss, 1988): L = 0.41 (0.40-0.42) mm; a = 26 (21-31); b = 6.2 (5.0-6.7); c = 18 (15-22); c′ = 2.5 (1.8-3.2); V = 73 (70-76); stylet = 14.5 (13.5-16.0) μm.
- 5 males (after Ryss, 1988): L = 0.48 (0.45-0.52) mm; a = 27 (25-30); b = 5.4 (4.9-5.9); c = 19 (17-21); c′ = 2.4 (1.8-3.0); stylet = 14 (13-16) μm; spicules = 17-18 μm; gubernaculum = 5 μm.
- 7 females (after Zarina & Maqbool, 1998): L = 0.45 (0.37-0.50) mm; a = 21.4 (21.0-21.7); b = 4.0 (3.7-4.3); c = 16.1 (15.5-16.7); c′ = 2.3 (2.1-2.4); V = 74 (71-76); stylet = 14 (13.5-14.5) μm.
- 5 males (after Zarina & Maqbool, 1998): L = 0.35 (0.32-0.39) mm; a = 24.2 (21.1-27.2); b = 3.7 (3.6-3.8); c = 15.4 (15.3-15.6); c′ = 2.4 (2.2-2.7); T = 40 (32-46); stylet = 13 (13.0-13.5) μm; spicules = 16 (16-17) μm; gubernaculum = 4.5 (4.0-5.0) μm.

DESCRIPTION

Female

Body almost straight in death; tapering anteriorly towards labial region. Cuticular annulation fine, inconspicuous. Lateral fields with four longitudinal lines; fifth line visible in some specimens. Labial region rather high, with round edges and three not very distinct annuli. Labial framework extending over 1-2 annuli. Stylet stout, basal knobs flattened

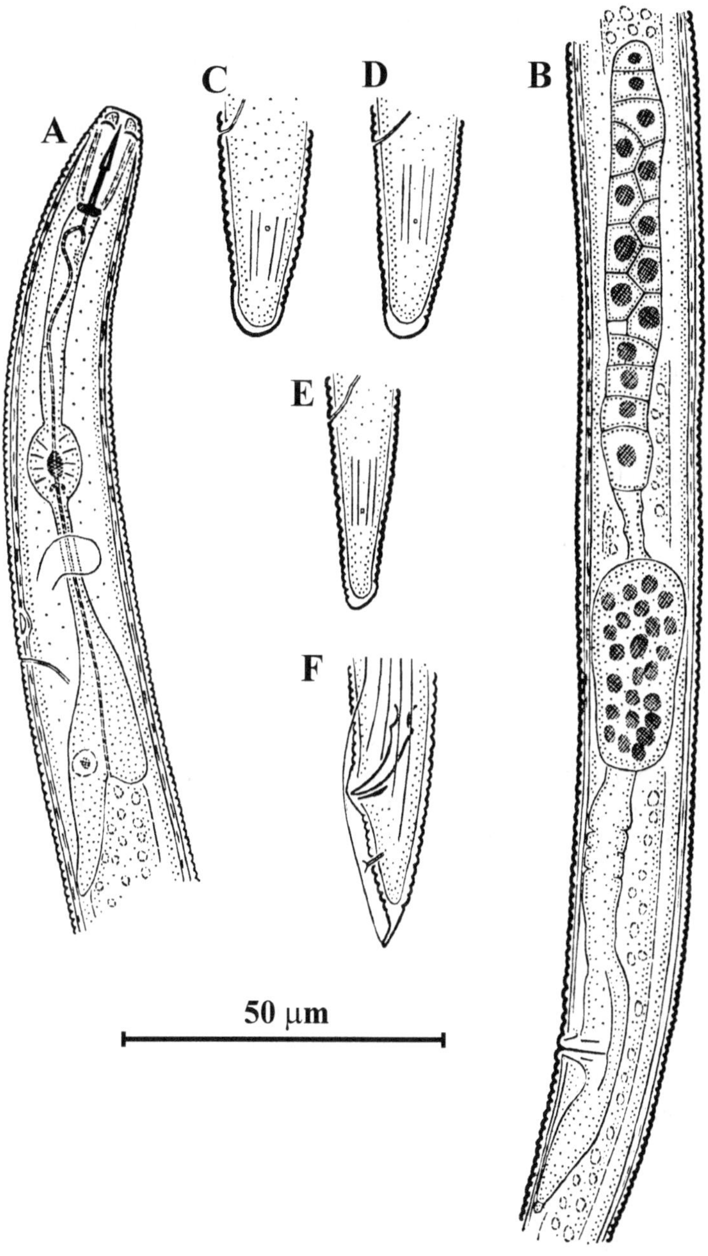

Fig. 93. Pratylenchus sudanensis *Loof & Yassin, 1971. A: Female pharyngeal region; B: Female posterior region; C-E: Female tails; F: Male tail. After Loof and Yassin (1971).*

or slightly concave anteriorly. Orifice of dorsal pharyngeal gland 2.5 μm posterior to stylet base. Median bulb broadly oval, 13 × 10 μm, with central valves 3 μm long. Overlapping gland lobe *ca* two body diam. long. Excretory pore located near pharyngo-intestinal junction, immediately posterior to hemizonid. Gonad single, anterior; oocytes in single row; with large (28 × 16 μm), elongate (in a few specimens almost spherical) spermatheca filled with large sperm. No eggs observed. Post-vulval uterine sac 1-1.5 body diam. in length, scarcely differentiated. Tail stout, subcylindrical, terminus broadly rounded to truncate; mostly smooth, in some specimens with a few striae. Tail 1.7-2.7 anal body diam. long, tail annuli 18-23. Phasmids just anterior to mid-tail.

Male

Similar to female. Testis single, spermatocytes in single row. Spicules tylenchoid, 17-18 μm long; gubernaculum slightly curved, 5 μm long. Bursa enveloping tail, its margins weakly crenate. Tail 2.3-3.2 anal body diam. long.

Diagnosis and relationships

Pratylenchus sudanensis is characterised by: labial region high with three annuli, post-vulval uterine sac 1-1.5 body diam. in length, tail subcylindroid, terminus broadly rounded to truncate; mostly smooth.

The matrix code is: A2, B2, C2, D3, E1, F4, G2, H1, I2, J1, K1.

It can be distinguished from closely related species (*P. pseudopratensis* and *P. vulnus*) by body diam., position of the vulva, the post-vulval uterine sac length, broadly rounded tail terminus and number of tail annuli (see the corresponding descriptions).

Distribution

It has been recorded from the type locality at Gezira, Sudan, from soil around the roots of cotton. It has also been recorded in Sudan on sugarcane (Saadabi *et al.*, 1987; Saadabi, 1988); Thailand on yellow narra (*Pterocarpus indicus* Willd.) (Toida *et al.*, 1996); Pakistan on kantali champa (*Artabotrys hexapetalus* (L.f.) Bhand.); Kenya on potato and sweet potato (Njuguna & Bridge, 1998); Russia on *Tamarix* sp. (Ryss, 1988); and Uganda on yam (Mudiope, 1999; Coyne *et al.*, 2003).

58. *Pratylenchus tenuis* Thorne & Malek, 1968
(Fig. 94)

MEASUREMENTS

- Females (after Thorne & Malek, 1968 and Handoo & Golden, 1989): L = 0.4 mm; a = 25; b = 3.1-3.6; c = 22; V = 79; stylet = 14.5-15 μm.

DESCRIPTION

Female

Body cylindroid, slightly tapering anteriorly. Cuticle finely annulated, coarsest annuli near labial region being only 1.7 μm apart while near mid-body they are less than 1 μm. Lateral fields with four obscure lines, middle two usually disappearing near anal region. Labial region low, flattened, almost four times as wide as high, with two annuli. Stylet 17 μm long with peculiar shaped knobs as illustrated. Isthmus slender, more than twice as long as body diam., with nerve ring slightly anterior to middle. Excretory pore and hemizonid opposite nerve ring. Pharyngeal gland lobe overlapping intestine ventrally, more than twice as long as body diam. Ovary outstretched with oocytes in single file except for short region of multiplication. Post-vulval uterine sac shorter than body diam. Tail conoid-rounded, slightly bent with phasmids near middle.

Male

Not found.

DIAGNOSIS AND RELATIONSHIPS

Pratylenchus tenuis is characterised by: labial region low, flattened with two annuli, lateral fields with four obscure lines, the middle two usually disappearing near anal region and tail conoid-rounded, slightly bent.

The matrix code is: A1, B1, C2, D3, E2, F1, G3, H1, I2, J1, K1.

It can be distinguished from the closely related species *P. scribneri* by stylet knobs shape, pharyngeal gland lobe, post-vulval uterine sac and tail shape (see the corresponding description).

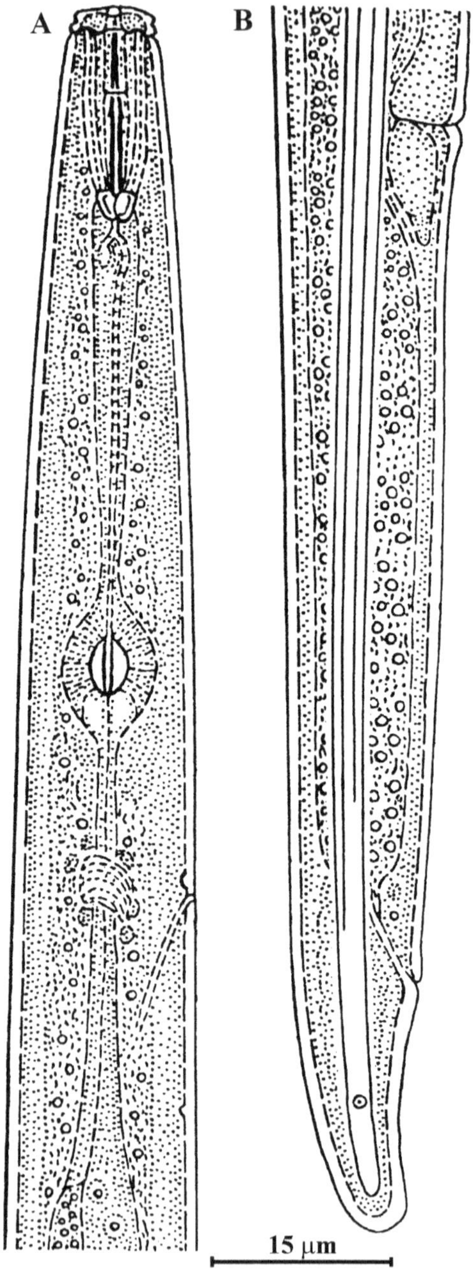

Fig. 94. Pratylenchus tenuis *Thorne & Malek, 1968. A: Female pharyngeal region; B: Female posterior region. After Thorne and Malek (1968).*

Handoo and Golden (1989) checked type specimens (two females) of this species and found that they had a very slender stylet, tulip-shaped stylet knobs which are narrow and high, long isthmus and unusually elongated pharyngeal lobe. Therefore, we consider *P. tenuis* a valid species.

DISTRIBUTION

Recorded, from the type locality (near Avon, South Dakota, USA, in cultivated soil) and from Forest Lake, Minnesota, USA (Crow & MacDonald, 1978).

59. *Pratylenchus teres* Khan & Singh, 1974
(Figs 95, 96)

MEASUREMENTS

- Female holotype (after Khan & Singh, 1974): L = 0.40 mm; a = 21.1; b = 4.2; c = 14.2; c′ = 1.9; V = 70; stylet = 18 μm.
- 17 females (after Khan & Singh, 1974): L = 0.55 (0.42-0.63) mm; a = 30.8 (22.1-39.9); b = 4.6 (3.5-5.6); c = 18.2 (11.5-27.0); c′ = 1.5-2.5; V = 70 (69-78); stylet = 17 (16-18) μm.
- 8 females (after Van den Berg & Quénéhervé, 2000): L = 0.50 (0.47-0.53) mm; a = 30 (29-31); b = 4; c = 14.5 (13-16); c′ = 3; V = 72 (69-74); stylet = 18 (17-18) μm.
- 20 females (after Carta *et al.*, 2002): L = 0.54 (0.47-0.63) mm; a = 30.5 (24.8-34.4); b = 4.1 (3.2-5.5); c = 16.1 (14.7-18.5); c′ = 2.8 (2.5-3.2); V = 73 (69-76); stylet = 15.6 (15.5-16.5) μm.
- 32 females (after Carta *et al.*, 2002): L = 0.56 (0.50-0.64) mm; a = 24 (20.0-29.8); b = 4.4 (3.7-4.9); c = 16.8 (15.4-18.4); c′ = 2.2 (1.9-2.6); V = 75 (71-77); stylet = 15 (15.5-16.5) μm.

DESCRIPTION

Female

Body straight or only slightly arcuate on death, tapering symmetrically and gradually towards vulva. Body striation fine. Lateral fields with six strongly crenate lines with occasional perpendicular striations. First lateral line beginning at stylet level, increasing to two, then to four near

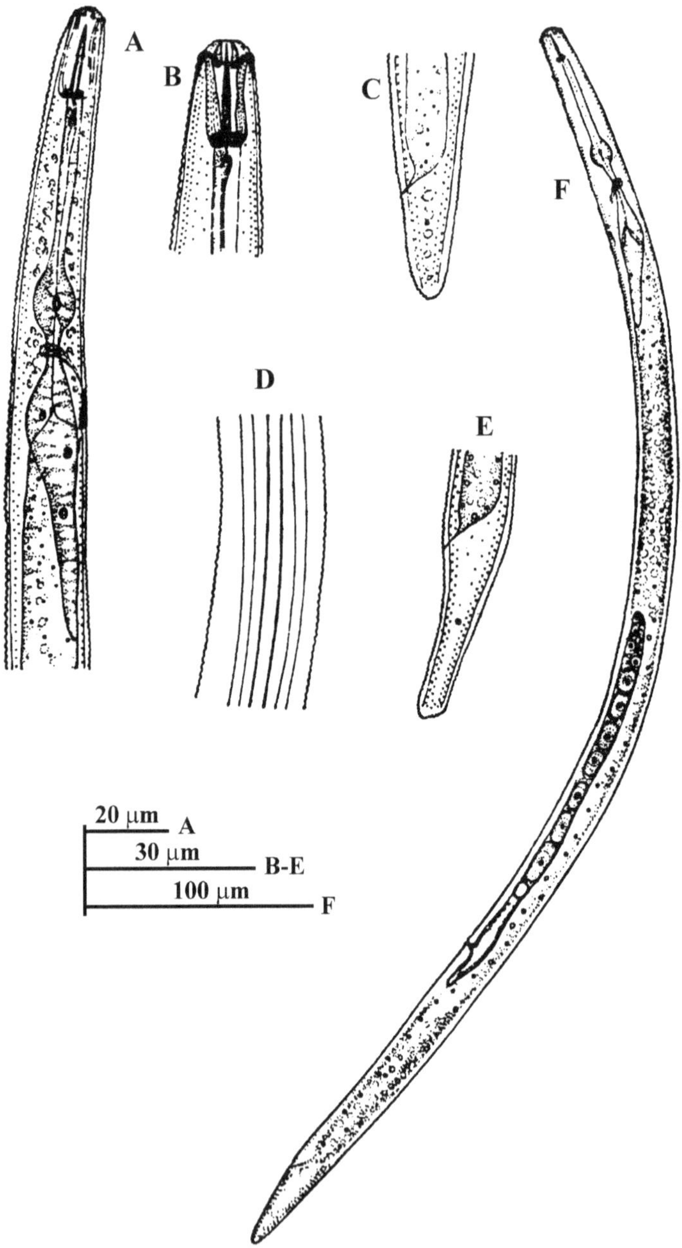

Fig. 95. Pratylenchus teres *Khan & Singh, 1974. A: Female pharyngeal region; B: Female labial region; C, E: Female tails; D: Lateral field at mid-body; F: Entire female. After Khan and Singh (1974).*

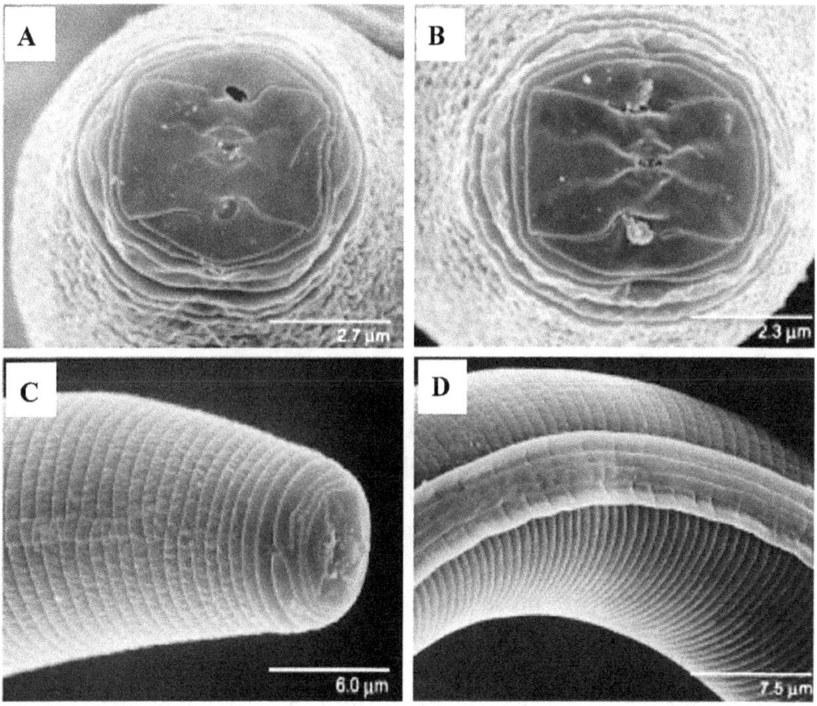

Fig. 96. *SEM micrographs of* Pratylenchus teres *Khan & Singh, 1974. A, B: Female* en face *views; C: Female labial region; D: Lateral field. After Carta* et al. *(2002).*

posterior of metacorpus and then to six in vulval region. Six lines becoming three in area between anus and phasmid after which one or two lines continuing to just before tail tip. Labial region conoid, low, with three annuli. Labial sclerotisation strong, extending up to two body annuli. SEM studies by Carta *et al.* (2002) showed panduriform (broad hourglass shape) *en face* view with fusion of lip sectors. Stylet 18 μm long with slightly anchor shaped knobs. Dorsal pharyngeal gland opening 2.5-3 μm posterior to stylet. Pharynx typical of genus. Isthmus *ca* one corresponding body diam. long, ventral pharyngeal overlap *ca* 3.5 times corresponding body diam. long. One dorsal and two subventral gland nuclei in pharyngeal overlap. Pharyngo-intestinal junction located at 12 (11-13)% of pharyngeal overlap from its anterior end. Pharyngeal gland overlap 35-47 μm. Excretory pore located opposite pharyngo-intestinal junction, hemizonid slightly anterior to excretory pore, extending up to three body annuli. Vulva transverse slit, *ca* five-sixths of vulval body diam. across. Spermatheca inconspicuous, without sperm. Post-vulval uterine sac *ca* 1-1.5 vulval body diam. long. Tail conoid, over two anal body diam. in length (1.5-2.5), bearing 25 (24-30) annuli. Tail tip crenate, phasmids located 18 annuli anterior to tail terminus.

Male

Not found.

DIAGNOSIS AND RELATIONSHIPS

Pratylenchus teres is characterised by: labial region conoid with three annuli, lateral fields with six lines, tail conoid with crenate tip.

The matrix code is: A2, B1, C2, D1, E1, F3, G3, H2, I3, J3, K1.

It is close to *P. crenatus* from which it differs by the fine body annulation, the anterior location of vulva, the straight shape of body when killed and by the presence of six lines in the lateral fields (see the corresponding description).

Carta *et al.* (2002) distinguished a new subspecies named *P. teres vandenbergae* which can be separated from the *P. teres* by the shorter stylet, four lip annules in 53% of the population, a higher, less rounded lip margin with a broader anterior end, only four lateral lines with a central interrupted fifth line near the middle of the body and reduced crenation of the lateral lines, and a less tapered tail region. Nevertheless, molecular analysis on the D3 region of the LSU rDNA gene of *P. teres vandenbergae* and *P. teres teres* demonstrated that both sequences aligned identically (Carta *et al.*, 2002).

DISTRIBUTION

Recorded, apart from type locality (Ludhiana, Punjab, India, soil around the roots of potato), in Nolivier-Subercozeaux, Sainte-Rose, Guadeloupe from sugar cane (Van den Berg & Quénéhervé, 2000), in Barbados (Cadet *et al.*, 1994), in southern Australia (Riley & Wouts, 2001), in China (Liu & Feng, 1995), in Chelyabinsk region, Russia (Savkina, 1989) and South Africa (Fourie *et al.*, 2001; Carta *et al.*, 2002).

60. *Pratylenchus thornei* Sher & Allen, 1953
(Figs 97-99)

MEASUREMENTS

- Female holotype (after Sher & Allen, 1953): L = 0.57 mm; a = 31; b = 5.8; c = 20; V = 77; stylet = 18 μm.
- Females (after Sher & Allen, 1953): L = 0.45-0.77 mm; a = 26-36; b = 5.5-8.0; c = 18-22; V = 73-80; stylet = 17-19 μm.
- 1 male (after Sher & Allen, 1953): L = 0.48 mm; a = 32; b = 5.6; c = 20; T = 30; stylet = 16 μm.
- 37 females (after Loof, 1960): L = 0.41-0.71 mm; a = 25.3-36.4; b = 5.4-8.3; c = 16.8-25.1; V = 74-79; stylet = 15-19 μm.
- 1 male (after Loof, 1960): L = 0.49 mm; a = 29; b = 6.2; c = 20.3; stylet = 16 μm.
- Females (after D'Errico, 1970): L = 0.45-0.61 mm; a = 28-32; b = 4.8-7.8; c = 18.8-27.7; V = 70-79; stylet = 15 μm.
- 1 male (after Fortuner, 1977): L = 0.55 mm; a = 39; b = 6.3; b′ = 4.2; c = 19; stylet = 16 μm.
- 14 females (after Khan & Singh, 1975): L = 0.52 (0.47-0.61) mm; a = 28.3 (24.5-32.1); b = 8.0; b′ = 4.3 (3.5-5.3); c = 16.5 (11.7-21.4); c′ = 2.5; V = 74 (74-80); stylet = 15 (15-17) μm.
- 10 females (after Ryss, 1988): L = 0.65 (0.49-0.75) mm; a = 31 (25-35); b = 6.1 (5.5-8.3); c = 20 (17-25); c′ = 2.8 (2.5-3.0); V = 75 (73-80); stylet = 17 (16-18) μm.
- 20 females (after Yu, 1997): L = 0.61 (0.44-0.79) mm; a = 32.5 (24.1-39.0); b = 6.7 (4.8-8.7); c = 22.3 (16.7-28.1); V = 76 (70-80); stylet = 16 (12-19) μm.
- 48 females (after Pourjam *et al.*, 1999a): L = 0.55 (0.42-0.68) mm; a = 31 (23-37); b = 6.5 (5.2-8.0); b′ = 4.4 (3.3-5.4); c = 20 (15-26); c′ = 2.6 (1.9-3.5); V = 77 (72-82); stylet = 16 (15-17.5) μm.
- 9 males (from carrot disc cultures, a population infecting chickpea in Jerez, Southern Spain, present study): L = 0.65 (0.62-0.69) mm; a = 34.2 (30.0-39.5); b = 7.5 (5.8-8.3); b′ = 5.1 (4.6-5.9); c = 22.3 (18.1-26.0); c′ = 2.1 (1.7-2.8); T = 45.0 (30.0-56.0); stylet = 16.0 (15.5-16.5) μm; spicules = 23.5 (21.5-26.0) μm; gubernaculum = 6.0 (5.0-7.5) μm.

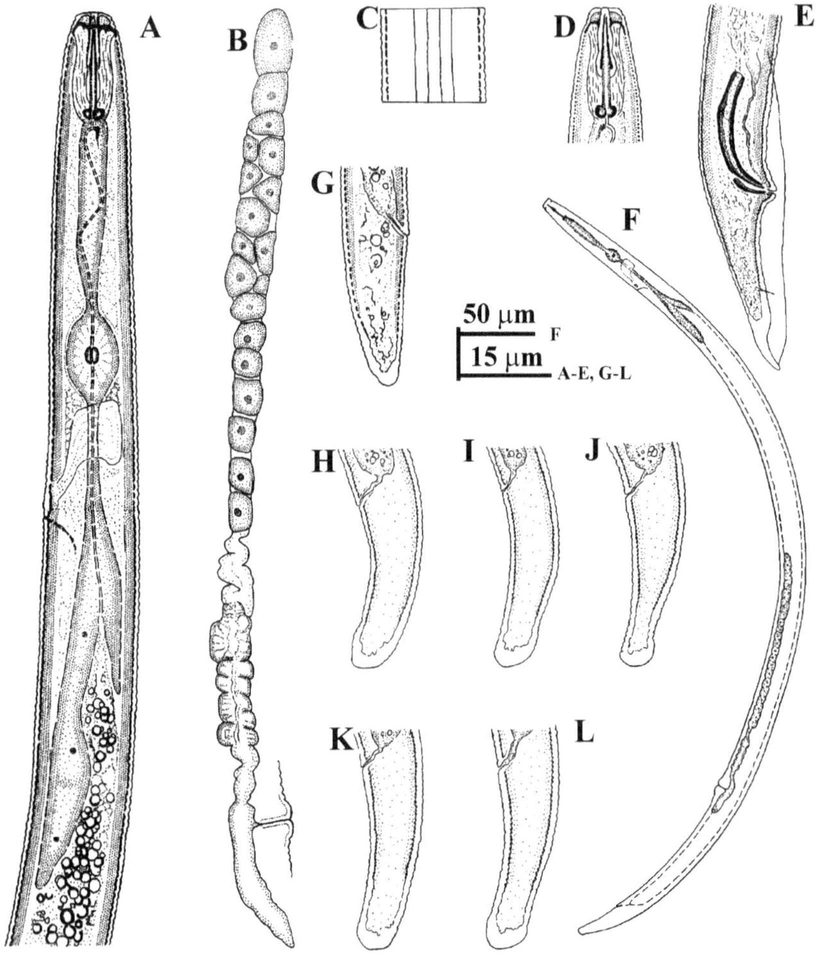

Fig. 97. Pratylenchus thornei *Sher & Allen, 1953. A: Female pharyngeal region; B: Female reproductive system; C: Lateral field at mid-body; D: Male labial region; E: Male tail; F: Entire female; G-L: Female tails. After Fortuner (1977).*

DESCRIPTION

Female

Body large and slender, assuming an open C-shape when killed by gentle heat. Cuticle with transverse striae *ca* 1 μm apart, not conspicuous. Lateral fields with four lines, outer ones straight or weakly crenate. In one specimen, oblique striae were observed by Loof (1960) in central zone. Labial region with three annuli, not offset from body.

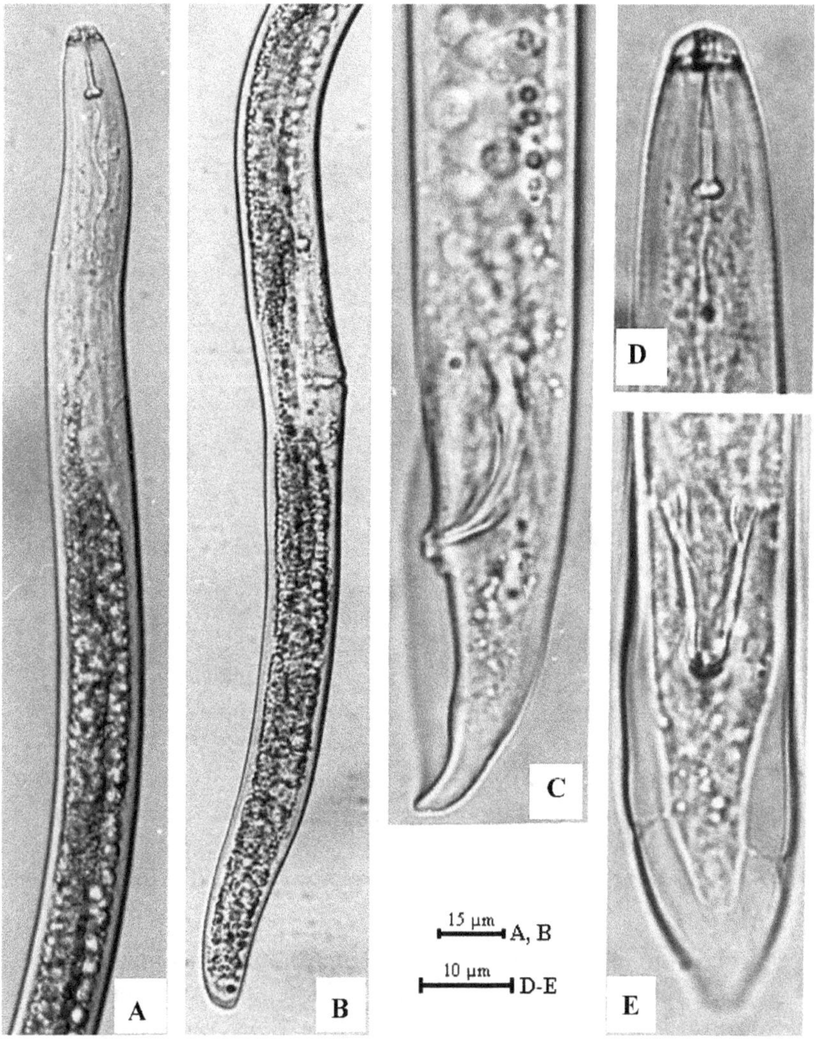

Fig. 98. *Light micrographs of* Pratylenchus thornei *Sher & Allen, 1953. A: Female anterior region; B: Female posterior region; C, E: Male tail; D: Male labial region.*

Outer margin of sclerotised labial framework conspicuously extending *ca* two annuli into body and one annulus into labial region. *En face* view characterised by lateral lips clearly separated from subdorsal and subventral lips by respective incisures which are concave with respect to lateral lips (Hernández *et al.*, 2000). Stylet guiding apparatus extending

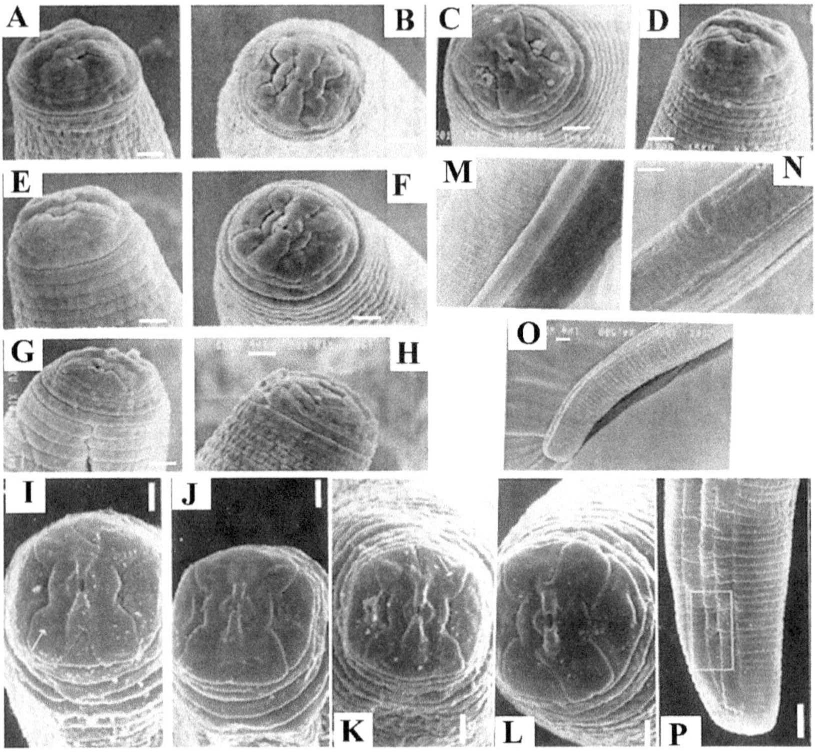

Fig. 99. *SEM micrographs of* Pratylenchus thornei *Sher & Allen, 1953. A-F, I-L: Female* en face *views; G, H: Female labial region; M, N: Lateral field at mid-body; O, P: Female tails (rectangle refers to phasmid region). (Scale bars: A, B, I-L, M-O = 1 μm; P = 3 μm.) After Pourjam* et al. *(1999); Hernández* et al. *(2000).*

posteriorly from basal plate for *ca* four annuli. Stylet medium size (17-19 μm long) with broadly rounded to almost anteriorly flattened basal knobs. Orifice of dorsal pharyngeal gland *ca* 3 μm posterior to stylet base. Nerve ring immediately posterior to pharyngeal bulb, hemizonid *ca* two annuli long, located one annulus anterior to excretory pore. Ovary not extending up to pharynx. Oocytes in single row, except for anterior zone of multiplication; oviduct indistinct, uterus short. Spermatheca difficult to see, not containing spermatozoa (males very rare); post-vulval uterine sac slightly more than 1.5 vulval body diam. long. Phasmids slightly posterior to mid-tail; all four incisures extending posterior to phasmids. Tail dorsally convex-conoid, terminus bluntly rounded to truncate, unstriated.

Male

Very rare (only four specimens previously recorded). Similar to female. Outstretched testis with spermatocytes in single row, followed by a region of multiple rows. Phasmids slightly posterior to mid-tail, not extending into bursa. Spicules very long (21-26 μm), arcuate, hafted, resting upon trough-shaped gubernaculum (5-7.5 μm). Yu (1997) found a male from axenic cultures on excised corn roots and nine additional specimens from axenic cultures on carrot discs from a population infecting chickpea in Jerez, Southern Spain, are reported here completing the morphology and morphometry of this life stage.

DIAGNOSIS AND RELATIONSHIPS

Pratylenchus thornei is characterised by: labial region with three annuli, not offset from body, outer margin of sclerotised labial framework extending conspicuously *ca* two annuli into body and one annulus into labial region, lateral fields with four lines, the outer ones straight or weakly crenate, stylet medium size (17-19 μm long), spermatheca difficult to see, not containing spermatozoa, males very rare.

The matrix code is: A2, B2, C3, D1, E2, F2, G3, H1, I3, J1, K1.

It can be distinguished from the closely related species (*P. mediterraneus*, *P. fallax*, *P. penetrans*, *P. pseudopratensis* and *P. sudanensis*) by labial region shape, stylet length, the low proportion of males, shape of spermatheca and tail shape (see the corresponding descriptions).

The only character differentiating *P. thornei* from *P. ranjani* was the presence of four lip annuli in the latter. However, Loof (1991) and Handoo and Golden (1989) suggested that *P. ranjani* could possibly be conspecific with *P. thornei*. Subsequent SEM studies by Pourjam *et al.* (1999a) on specimens of both species revealed the presence of an incomplete annulus on one side which falsely appeared as a fourth annulus in light microscopy observations. Therefore, both species were synonymised and this action is maintained here.

As previously indicated for *P. brassicae*, *P. allius* (Shahina & Maqbool, 1996) Siddiqi, 2000 was considered as a possible synonym of *P. thornei*. After a precise comparison of both descriptions we concur with that opinion and *P. allius* is considered here as a new junior synonym of *P. thornei*.

DISTRIBUTION

It has been recorded from the type locality at Berkeley, California, USA, in soil around the roots of grass (Sher & Allen, 1953). It has been recorded from several countries in Europe: Belgium on rose (Coolen & Hendricks, 1972a) and chrysanthemum (D'Herde & Brande, 1963); Bulgaria on cereals and tobacco (Katalan-Gateva & Baicheva, 1978; Ryss *et al.*, 1991); Croatia on soybean (Kelic, 1991); Cyprus on carrot and barley (Philis, 1976, 1997); Denmark on barley (Andersen, 1979); England from several hosts and localities (Cotten & Roberts, 1981; Hooper *et al.*, 1990); France on wheat (Esmenjaud *et al.*, 1990); England and Wales on cereals (Corbett, 1970); Germany (Loof, 1960; Decker & Dowe, 1974); The Netherlands on *Iberis* sp., corn, red currant, apple, pear, plum and cherry (Oostenbrink, 1954); on cereals in Italy (D'Errico, 1970; Inserra *et al.*, 1978); Poland on cereals (Jelic, 1992); in Portugal (Abrantes *et al.*, 1987); in Russia on *Ziziphus jujuba* Mill. (Ryss, 1988); Slovakia on cereals (Valocka & Sabova, 1978); Slovenia on corn (Urek *et al.*, 2003); Spain on cereals and legumes (Castillo *et al.*, 1996a; Talavera & Tobar Jiménez, 1997; Nombela *et al.*, 1998), on horticultural crops (Espárrago & Navas, 1995), in the Canary Islands (Guiran & Vilardebo, 1962), in olive nurseries and grass (Nico *et al.*, 2002; Talavera & Navas, 2002), on sugar beet (Arias & Romero, 1975) and several hosts and localities (Gómez Barcina *et al.*, 1989; Peña Santiago *et al.*, 2004); Yugoslavia (Grujicic, 1969, 1974).

It has been recorded from several countries in Africa: Egypt on sugarcane (Oteifa *et al.*, 1963), onion (Oteifa & El-Sharkawi, 1965) and cotton (Anter *et al.*, 1993); Nigeria (Egunjobi & Bolaji, 1979); several North African countries on faba bean (Troccoli *et al.*, 1992; Troccoli & Di Vito, 2002); Morocco on wheat (Meskine & Abbad Andalousi, 1993); South Africa on wheat and soybean (Koen, 1969; Jordaan *et al.*, 1992; Fourie *et al.*, 2001); and in Sudan (Elbadri *et al.*, 2001).

It has been recorded from several countries in North America: in several states of USA, *e.g.*, it is an important pathogen of wheat in Utah (Thorne, 1961); in Ohio on potato (Brown *et al.*, 1980); in dry land field crops in the semiarid Pacific Northwest (Smiley *et al.*, 2004); in Washington State on wheat (Mojtahedi *et al.*, 1988); in Canada on wheat and several hosts (Townshend *et al.*, 1978a; Yu, 1997; Yu *et al.*, 1998); Mexico on cereals, *Agrostis* sp., *Trifolium repens* L. and *Crotalaria juncea* L. (Perez *et al.*, 1970; Van Gundy *et al.*, 1974; Lawn & Sayre, 1992).

It occurs in several countries from South America: Argentina on potato (Doucet, 1988); Chile on strawberry (Aballay *et al.*, 1996); Venezuela on sugarcane (Perichi *et al.*, 2002). It has been recorded in several countries of Asia: Azerbaidzhan (Kasimova & Atakishieva, 1980); China (Yin, 1991); India on several crops including chickpea, safed musli (*Chlorophytum borivillianum* Sant., an important medicinal plant), papaya (Joshi *et al.*, 1970; Sethi & Swarup, 1971; Khan & Singh, 1974; Sabir, 2000; Ali & Askary, 2001; Ali & Sharma, 2003; Pandey, 2003); Iran, on tomato, bean, wheat, alfalfa and tea (Kheiri, 1972; Pourjam *et al.*, 1997; Kheiri *et al.*, 2002a); Israel on wheat (Orion *et al.*, 1984); in Japan (Gotoh & Ohshima, 1963; Gotoh, 1974); Jordan on olive (Hashim, 1983); Korea from several hosts and localities (Choi, 1975; Choo & Choi, 1979; Park *et al.*, 1999); Libya on almond (El-Maleh & Edongali, 1994); Pakistan on wheat (Khan *et al.*, 2001b); Syria on chickpea (Greco *et al.*, 1988, 1992, 1994); Tadzhikistan (Teben'Kova & Ivanova, 1989); from cereals and legumes in several localities from Turkey (Di Vito *et al.*, 1994; Elekcioglu *et al.*, 1994; Elekcioglu & Gozel, 1997; Erdal *et al.*, 2001; Mennan & Handoo, 2006). It has been recorded in Australia on cereals and clover (*Trifolium fragiferum* L.) (Colbran & McCulloch, 1965; Grandison & Wallace, 1974; Riley & Wouts, 2001; Riley & Kelly, 2002); and carrots (Hay & Pethybridge, 2005).

61. *Pratylenchus typicus* Rashid, 1974
(Fig. 100)

Measurements

- Female holotype (after Rashid, 1974): L = 0.59 mm; a = 27.1; b = 5.0; c = 21.3; V = 80; stylet = 17 μm.
- 10 females (after Rashid & Khan, 1976): L = 0.59-0.68 mm; a = 27-37; b = 5.0-6.2; c = 10-26; V = 79-84; stylet = 15-17 μm.

Description

Female

Body almost straight when heat-relaxed. Labial region slightly offset, usually with four labial annuli, in some cases four annuli on one side and five on other side. Body annulation *ca* 1-1.5 μm at mid-body. Labial

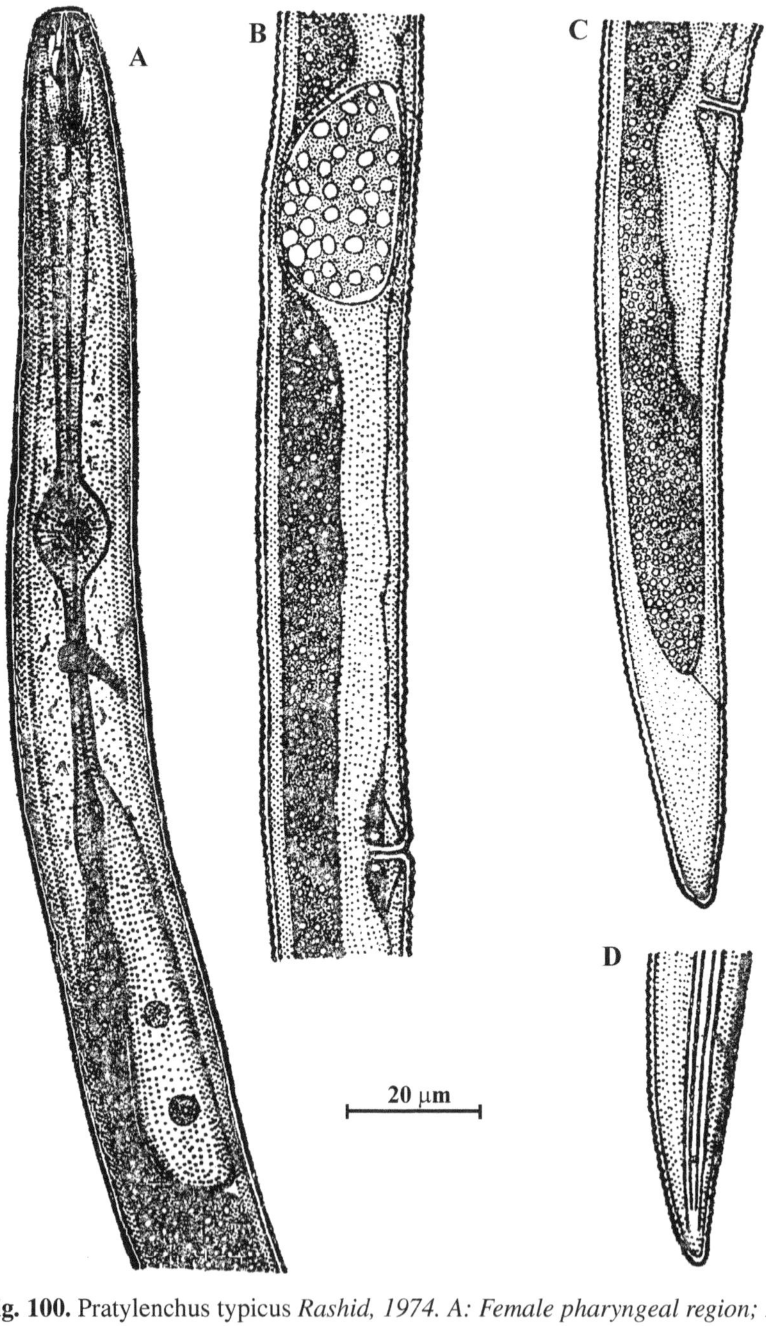

Fig. 100. Pratylenchus typicus *Rashid, 1974. A: Female pharyngeal region; B: Vulval region; C: Female posterior region; D: Female tail. After Rashid (1974).*

framework well developed, strongly sclerotised, margin extending into first body annulus. Lateral fields marked by four lines, extending to tail tip. Stylet robust, basal knob 7.5 μm wide and 2.8 μm high, rounded in shape. Anterior part of stylet slightly shorter than posterior shaft. Orifice of dorsal pharyngeal gland 6 μm posterior to stylet base. Procorpus cylindrical, averaging 48 μm in length and 6.3 μm in diam. Median bulb oval, 15 μm long and 11 μm diam. with well-developed valvular apparatus. Excretory pore at 85 μm from anterior end and hemizonid at level of excretory pore. Pharyngeal gland lobe overlapping intestine ventrally and laterally. Vulva a transverse slit, ovary single, outstretched with single file of oocytes. Spermatheca oblong, measuring 30 μm in length and 18 μm in diam., filled with rounded sperm. Post-vulval uterine sac more than two vulval body diam. long. Distance between middle of spermatheca to vulva = *ca* 95 μm. Tail conoid, terminus rounded and smooth. Tail annuli numbering 24-26. Phasmid 12 annuli posterior to anal latitude. Rectum *ca* 0.5 anal body diam. long.

Male

Not found.

DIAGNOSIS AND RELATIONSHIPS

Pratylenchus typicus is characterised by: rounded stylet knobs, spermatheca elongated and containing sperm, post-vulval uterine sac long and tail terminus smooth.

The matrix code is: A3, B1, C3, D3, E3, F6, G3, H1, I4, J1, K1.

It is close to *P. goodeyi* and *P. pseudopratensis* from which it can be distinguished by the stylet length, shape of stylet knobs, position of vulva, shape of spermatheca, length of the post-vulval uterine sac, shape of tail and tail terminus (see the corresponding descriptions).

DISTRIBUTION

It has only been recorded from the type locality, in the rhizosphere of spinach (*Spinacea oleracea* L.) at Lucknow, India.

62. *Pratylenchus unzenensis* Mizukubo, 1992
(Fig. 101)

MEASUREMENTS

- Female holotype (after Mizukubo, 1992): L = 0.42 mm; a = 25.0; b = 5.3; b′ = 3.3; c = 12.1; c′ = 3.1; V = 75; stylet = 15.0 μm.
- 12 females (after Mizukubo, 1992): L = 0.42 (0.37-0.47) mm; a = 25.8 (21.9-28.5); b = 5.5 (4.9-6.1); b′ = 3.4 (2.9-4.1); c = 13.7 (11.6-16.6); c′ = 2.9 (2.4-3.5); V = 77 (75-79); stylet = 15.0 (14.5-15.5) μm.
- 2 males (after Mizukubo, 1992): L = 0.37-0.40 mm; a = 29.2-30.0; b = 4.9-5.2; b′ = 3.1-3.4; c = 14.2-171; c′ = 2.2-2.8; T = 50; stylet = 14.0-14.5 μm; spicules = 13-13.5 μm; gubernaculum = 4.5-5.0 μm.

DESCRIPTION

Female

Body slightly curved ventrally when heat-relaxed. Annuli 1.2 (1.0-1.3) μm wide at mid-body. Labial region relatively low, 2.3 (2.0-2.6) μm high, 7.7 (7.3-8.3) μm diam., flattened anteriorly, continuous with body contour. Labial framework moderately developed, laterally extending into body for one annulus. Labial region with three annuli on both sides. SEM *en face* micrographs showing division between submedian and lateral segments; lateral edges of submedian segments curved or angular, assigned between Group 2 and Group 3 of Corbett and Clark (1983). Amphidial apertures narrow and oblique. Cephalids not seen. Lateral fields 6.0 (5.3-6.6) μm wide, or 37 (32-40)% of max. body diam., consisting of three bands or four lines; outer bands partially areolated at mid-body region and completely areolated at tail region. Lateral fields extending to tail tip and continuing around tail terminus. Stylet 1.9 times (1.8-2.1) as long as labial region diam. Stylet knobs flattened anteriorly (54%) or broadly rounded (46%), never indented anteriorly, 2.2 (2.0-2.4)

Fig. 101. Pratylenchus unzenensis *Mizukubo, 1992. A: Entire female; B: Entire male; C: Female pharyngeal region; D-F: Female labial region; G: Stylet; H-J: Spermatheca; K, L: Vulval regions; M: Detail of ovary; N: Male pharyngeal region; O: Male tail; P-W: Female tails. After Mizukubo (1992).*

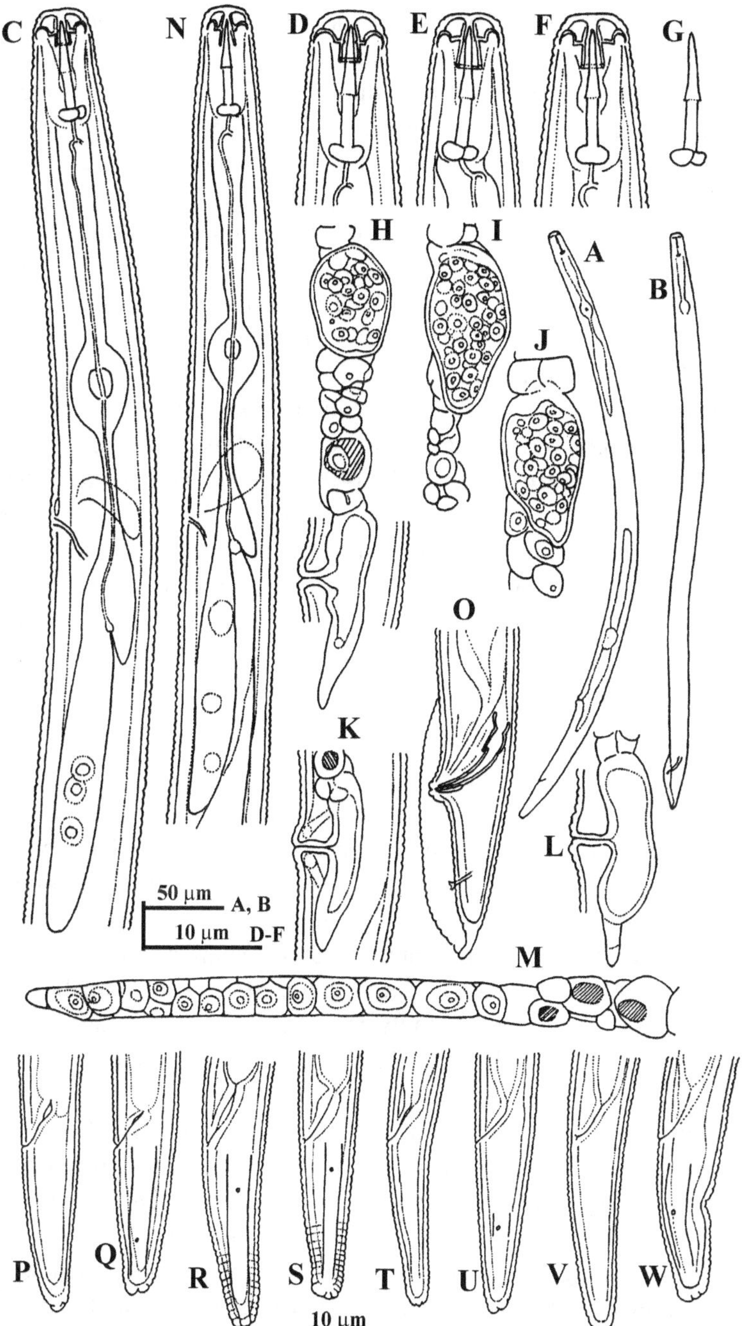
C
N
D
E
F
G
H
I
A
B
J
O
K
L
50 μm A, B
10 μm D-F
M
P
Q
R
S
T
U
V
W
10 μm G-W

μm high, 4.0 (3.5-4.5) μm across. Metacorpus oval, valve conspicuous, 51 (41-58) μm from anterior body end, or 67.2 (55.8-77.6)% of pharynx length. Pharyngeal basal lobe extending 123 (107-135) μm from anterior end, ventrally overlapping intestine 47 (32-65) μm or for 2.9 (2.0-3.9) corresponding body diam. Pharyngeal nuclei in tandem. Excretory pore located at 90-106% of pharyngeal length or at 17.2 (15.0-18.8)% of body length. Hemizonid flat, 1-2 annuli long, immediately anterior to excretory pore. Gonad outstretched, 148 (117-198) μm long; ovary with oocytes in a row or partially double row; spermatheca oval to oblong, 53 (42-84) μm from vulva, 19.6 (13.9-24.4) μm long, 10.7 (8.9-11.2) μm diam., 1.8 (1.5-2.3) times as long as wide, packed with spermatozoa. Post-vulval uterine sac less than two vulval body diam. long, without terminal rudimental ovary. Vagina narrow-walled, perpendicular to body axis, 7.0 (5.9-7.9) μm long, or 44 (40-49)% of vaginal body diam. Annuli between vulva and anus 60 (52-75) μm in number. Tail terminus with some variation in shape: using tail shape codes of Frederick and Tarjan (1989), mostly fall into subhemispherical, rarely truncate, but never bluntly pointed; terminus smooth (38%) or irregularly annulated (62%). Tail terminal cuticle 2.2 (2.0-2.6) μm thick; phasmids pore-like, centred in lateral fields, 16-19 μm posterior to anus.

Male

Body straight to arched when heat-relaxed. Annuli finer than in female, 1.0-1.1 μm apart at mid-body. Labial region relatively low, 2.4-2.6 μm high, 6.6-6.9 μm diam., flattened anteriorly, continuous with body contour. Labial framework moderately developed, laterally extending into body for one annulus. Stylet shorter than in female, 2.0-2.1 times as long as labial region diam. Stylet knobs smaller than female, 3.3-3.6 μm across, broadly rounded. Dorsal pharyngeal gland opening close to stylet base at 2.2-2.8 μm. Metacorpus oval; valve conspicuous, 48-50 μm from anterior end, or 63-66% of pharynx length. Pharyngeal basal lobe extending 116 μm from anterior body end, ventrally overlapping intestine 40-42 μm or for 3.0-3.3 corresponding body diam. Pharyngeal nuclei in tandem. Hemizonid flat, immediately anterior to excretory pore. Testis outstretched, spermagonia in double rows, 198 μm long. Spicules arch-shaped, slightly shorter than stylet; gubernaculum crescent-shaped. Bursa 36 μm long. Terminal cuticle of tail 4.6-5.9 μm thick, ending in finely rounded terminus; phasmids 12.5-

15.8 μm posterior to anus or 9.9-10.6 μm from tail tip. Lateral fields 4.6 μm wide, or 35% of max. body diam.

DIAGNOSIS AND RELATIONSHIPS

Pratylenchus unzenensis is characterised by: labial region with three annuli, short body (L = 0.37-0.47 mm), relatively low V-value (75-79%), moderately long stylet (14.5-15.5 μm), post-vulval uterine sac of moderate length, oval or oblong spermatheca with spermatozoa, tail subhemispherical, rarely truncate, with irregularly annulated or smooth tail terminus and males present.

The matrix code is: A2, B2, C2, D3, E2, F4, G2, H1, I3, J1, K2.

It can be distinguished from closely related species (*P. elamini*, *P. exilis*, *P. kasari*, *P. morettoi*, *P. pratensis*, *P. pratensisobrinus*, *P. teres*, *P. yassini* and *P. zeae*) by the body length, *en face* view, stylet length, position of the vulva, post-vulval uterine sac length, number of tail annuli and shape of the tail terminus (see the corresponding descriptions).

DISTRIBUTION

It has only been recorded from the type locality at Nita-toge, Mt Unzen, Nagasaki Prefecture, Japan, in soil around the roots of *Artemisia* sp.

63. *Pratylenchus ventroprojectus* Bernard, 1984
(Fig. 102)

MEASUREMENTS

- Female holotype (after Bernard, 1984): L = 0.48 mm; a = 28.9; b = 6.0; b′ = 4.3; c = 22.2; c′ = 1.9; V = 80; stylet = 15 μm.
- 10 females (after Bernard, 1984): L = 0.44 (0.39-0.48) mm; a = 30.1 (27.4-34.7); b = 6.0 (5.6-6.4); b′ = 3.9 (3.3-4.4); c = 19.4 (14.3-22.4); c′ = 2.3 (1.9-2.6); V = 79 (78-80); stylet = 15 (14-16) μm.
- 5 males (after Bernard, 1984): L = 0.39 (0.37-0.41) mm; a = 31.1 (30.7-32.7); b = 5.5 (5.2-6.1); b′ = 3.9 (3.8-4.0); c = 20.1 (18.5-22.6); c′ = 2.0 (1.9-2.2); T = 44 (38-49); stylet = 14 (14-15) μm; spicules = 15 (14-17) μm; gubernaculum = 4 (4-5) μm.

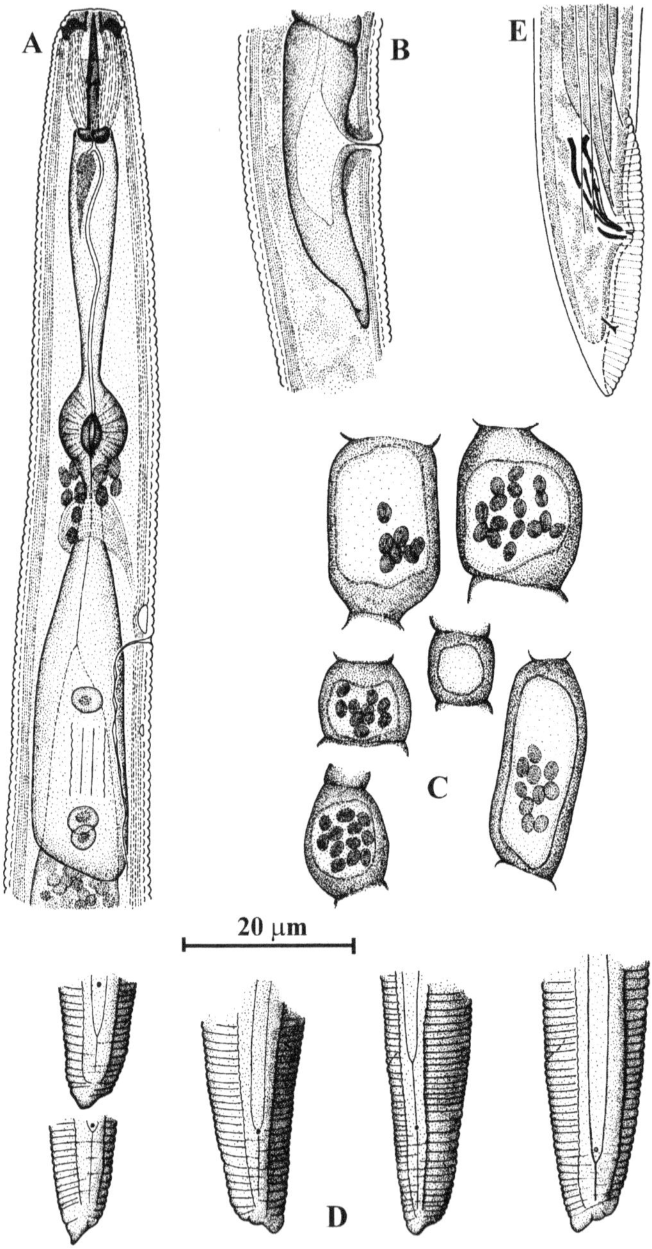

Fig. 102. Pratylenchus ventroprojectus *Bernard, 1984. A: Female pharyngeal region; B: Vulval region; C: Variability in spermatheca; D: Female tails; E: Male tail. After Bernard (1984).*

DESCRIPTION

Female

Body straight or slightly curved when heat-relaxed, small and slender. Labial region low, conoid, flattened, with three annuli. Lateral fields with four lines, without regular markings between them; inner lines meeting near anus or on tail; lateral fields open near tail tip. Stylet stout, knobs variable, usually large and cupped, but sometimes rounded or sloping posteriorly. One pair of cephalids visible 4-5 annuli posterior to labial region. Median bulb broadly oval, valve large. Pharyngeal glands strongly overlapping intestine; dorsal gland nucleus *ca* 50% of distance from median bulb, subventral gland nuclei in posterior region. Hemizonid near pharyngo-intestinal junction; excretory pore immediately posterior to hemizonid; hemizonion *ca* 10-12 annuli posterior to hemizonid. Gonad outstretched; spermatheca oval to rectangular or rarely elongated, usually filled with sperm but occasionally empty; post-vulval uterine sac 1.1 (0.9-1.8) vulval body diam. long, usually with a single demarcated terminal cell. Tail short, broadly conoid, terminus truncate, coarsely or not annulated, with distinct, subventral projection one or two annuli long. Phasmid centred in lateral fields near mid-tail.

Male

Similar to female in most respects. Gonad outstretched or reflexed. Spicules slender, curved, partially cephalated; velum long, narrow, inconspicuous; gubernaculum curved, linear. Caudal alae crenate, arising anterior to spicules and extending to tail tip. Phasmids near centre of tail. Lateral fields usually not extending posterior to cloacal opening.

DIAGNOSIS AND RELATIONSHIPS

Pratylenchus ventroprojectus is characterised by: labial region low, conoid, flattened, with three annuli, lateral fields with four lines, spermatheca oval to rectangular or rarely elongated, tail short, broadly conoid, the terminus truncate, coarsely or not annulated, with a distinct, subventral projection.

The matrix code is: A2, B2, C2, D4, E2, F3, G3, H4, I1, J1, K1.

It can be distinguished from closely related species *P. convallariae*, *P. goodeyi* and *P. pratensis* in lip annulation, labial region shape, body and stylet length, tail shape body and stylet length and position of vulva (see the corresponding descriptions).

DISTRIBUTION

It has only been recorded from the type locality at Finger Bay, Adak Island, Alaska, from the rhizosphere of *Platanthera convallariaefolia* (Fisch.) Lindl.

64. *Pratylenchus vulnus* Allen & Jensen, 1951

(Figs 103, 104)

MEASUREMENTS

- Females (after Allen & Jensen, 1951): L = 0.46-0.91 mm; a = 26.6-39.5; b = 5.3-7.7; c = 14.2-27.7; V = 78-84; stylet = 16-18 μm.
- Males (after Allen & Jensen, 1951): L = 0.46-0.73 mm; a = 28.3-39.2; b = 5.3-7.4; c = 17.5-29.4; T = 36-66; stylet = 15-18 μm; gubernaculum = 4-6 μm; spicules = 14-20 μm.
- 14 females (after Loof, 1960): L = 0.47-0.72 mm; a = 27.8-37.6; b = 5.7-7.7; c = 18.4-24.7; V = 77-82; stylet = 13-16 μm.
- 9 males (after Loof, 1960): L = 0.49-0.61 mm; a = 32.1-38.5; b = 6.3-7.2; c = 20.5-23.6; T = 34-53; stylet = 13-15 μm.
- 25 females (after Román & Hirschmann, 1969a): L = 0.63 (0.53-0.71) mm; a = 31.3 (25.2-35.8); b = 6.7 (5.7-7.7); c = 19.9 (17.4-24.3); V = 79 (77-82); stylet = 15 (14.5-15.5) μm.
- 25 males (after Allen & Jensen, 1951): L = 0.54 (0.48-0.59) mm; a = 32.4 (29.9-36.3); b = 6.1 (5.6-6.7); c = 20.0 (17.7-22.3); T = 45 (36-54); stylet = 14 (13-14.5) μm; spicules = 14-16 μm; gubernaculum = 4.2-6.0 μm.
- 10 females (after Ryss, 1988): L = 0.62 (0.46-0.80) mm; a = 33 (27-37); b = 6.4 (5.3-7.5); c = 21 (14-28); c′ = 2.7 (2.3-3.0); V = 80 (78-83); stylet = 17 (16-18) μm.
- 7 males (after Ryss, 1988): L = 0.57 (0.46-0.71) mm; a = 33 (28-35); b = 6.0 (5.3-7.2); c = 21 (18-26); c′ = 2.7 (2.1-3.1); stylet = 16 (15-18) μm; spicules = 17 (14-20) μm; gubernaculum = 5 (4-6) μm.
- 80 females (after Doucet & Lax, 1997): L = 0.49 (0.43-0.61) mm; a = 30.5 (24.9-37.3); b = 5.5 (4.7-7.0); b′ = 4.1 (3.2-5.2); c =

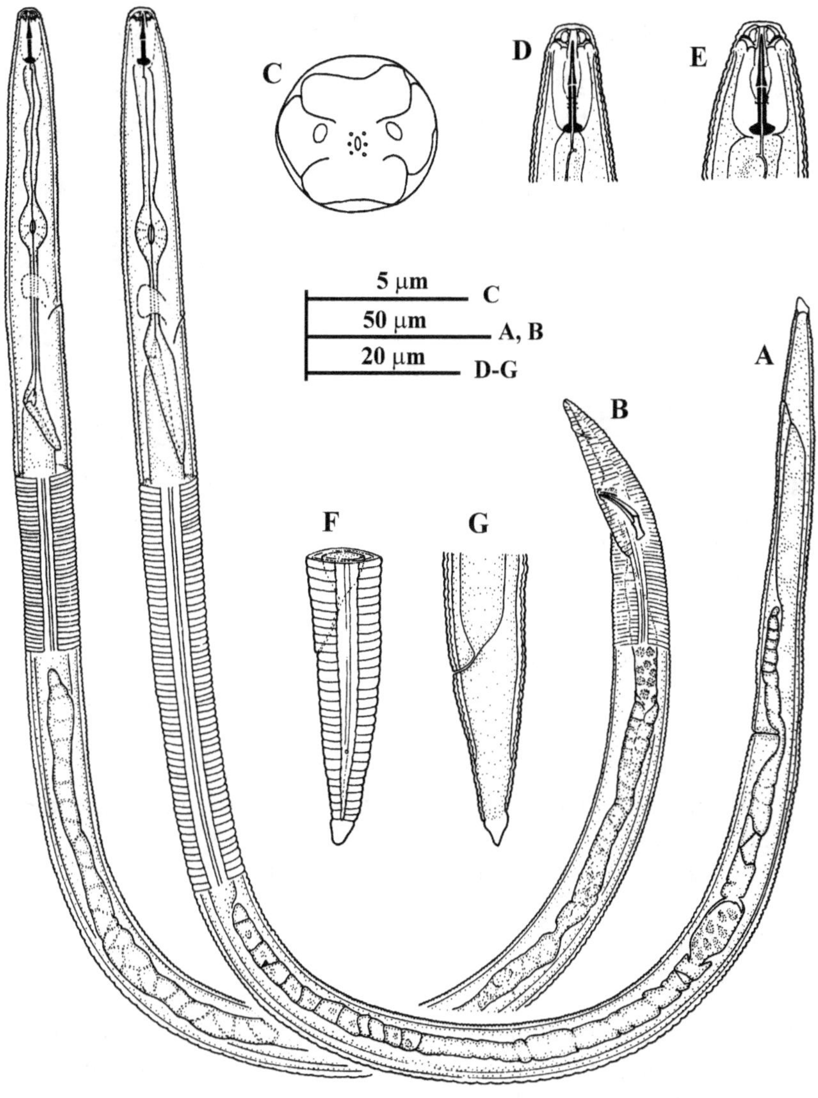

Fig. 103. Pratylenchus vulnus *Allen & Jensen, 1951. A: Entire female; B: Entire male; C: Female* en face *view; D: Male labial region; E: Female labial region; F, G: Female tails. After Corbett (1974).*

20.2 (16.2-25.1); c′ = 2.4 (2.0-3.0); V = 80 (77-84); stylet = 15 (14-16) μm.

- 80 males (after Doucet & Lax, 1997): L = 0.46 (0.41-0.53) mm; a = 31.0 (27.6-37.4); b = 5.3 (4.6-6.0); b′ = 4.1 (3.5-4.6); c = 20.4

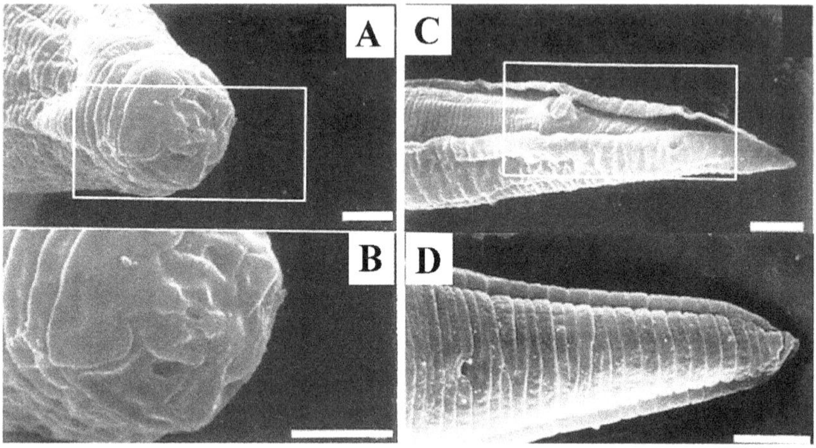

Fig. 104. *SEM micrographs of* Pratylenchus vulnus *Allen & Jensen, 1951. A, B: Female* en face *views; C: Male tail (rectangle refers to cloacal region); D: Female tail. (Scale bars: A, B = 2 μm; C = 5 μm; D = 10 μm.) After Hernández* et al. *(2000).*

(15.9-23.8); c′ = 2.0 (1.9-2.4); stylet = 14.5 (13-16) μm; spicules = 18 (15-21) μm; gubernaculum = 6.5 (5.0-7.0) μm.

- 39 females (after Gao *et al.*, 1999): L = 0.65 (0.50-0.81) mm; a = 26.7 (20.8-30.7); b = 6.8 (4.8-9.0); b′ = 4.7 (3.4-7.1); c = 21.0 (17.5-30.4); V = 80 (73-83); stylet = 16.7 (15-18) μm.
- 20 males (after Gao *et al.*, 1999): L = 0.53 (0.43-0.61) mm; a = 30.7 (21.7-38.4); b = 6.2 (5.2-7.2); b′ = 4.4 (3.7-5.7); c = 19.8 (17.9-25.8); T = 43 (31-55); stylet = 15 (14.5-17) μm.

DESCRIPTION

Female

Body, slender nearly straight when heat-relaxed. Labial region almost continuous with body contour, with three or four annuli, heavily sclerotised skeleton extending into body for *ca* one annulus. *En face* view showing wide subdorsal and subventral lips separated from lateral lips by convex incisures (Hernández *et al.*, 2000). Six small papillae surrounding oral opening, two large amphidial apertures present laterally. Lateral fields with four lines, outer ones smooth or weakly crenate, inner ones closer to each other than to outer ones, occasionally oblique lines present in inner band. Stylet with rounded basal knobs sometimes cupped anteriorly. Median pharyngeal bulb oval, relatively narrow. Excretory pore *ca* opposite pharyngo-intestinal junction. Pharynx overlapping intestine ventrally in a long lobe. Spermatheca functional, oblong when distended with sperm, post-vulval uterine sac *ca* two vulval body diam. long with rudimentary ovary. Tail tapering, with narrowly rounded to subacute smooth tip, occasionally irregular, sometimes with one or two annulations.

Male

Common, generally similar to female in appearance of labial region and pharynx. Lateral fields ending on bursa, which envelop tail. Spicules curved, cephalated; gubernaculum simple.

DIAGNOSIS AND RELATIONSHIPS

Pratylenchus vulnus is characterised by: labial region almost continuous with body contour, with three or four annuli, pharynx overlapping intestine ventrally in a long lobe, spermatheca oblong, post-vulval uterine sac *ca* two vulval body diam. long with rudimentary ovary, tail tapering, with narrowly rounded to subacute smooth tip and males common.

The matrix code is: A2, B2, C2, D3, E2, F6, G3, H3, I2, J1, K1.

It can be distinguished from closely related species (*P. pratensis* and *P. coffeae*) by the number of lip annuli, the absence of striae around the female tail, lateral fields and post-vulval uterine sac (see the corresponding descriptions).

Gao *et al.* (1997) reported that females of *P. vulnus* extracted from roots of peach were markedly longer and plumper than those from the soil around roots, while males extracted from the roots were nearly of the same size as those from soil. Furthermore, females from the roots were significantly longer and plumper than males from the roots, whereas there was no apparent difference between females and males from soil. Similarly, Doucet *et al.* (1996, 1998) reported intraspecific variability on several populations of *P. vulnus* from different geographic origins, although it was not possible to associate particular states with the isolates studies.

DISTRIBUTION

It has been recorded from the type locality at San José, California, USA, in roots of California black walnut, *Juglans hindsii* Jepson. It has been recorded on over 80 species and varieties of plants, many of them woody perennials (Goodey *et al.*, 1965). It has been recorded from several countries in Europe, mostly infecting woody plants. It is generally found in warmer climatic areas (southern Europe) but also occurs in northern Europe as a pathogen of glasshouse crops: in Belgium on rose (Coolen & Hendrickx, 1972a); Bulgaria on peach (Katalan-Gateva, 1980), tobacco (Katalan-Gateva & Baicheva, 1978; Baicheva *et al.*, 1984) and fruit nurseries (Choleva *et al.*, 1984); Denmark on rose (Jakobsen, 1974, 1975); England on rose (Winfield, 1974); Finland (Kurppa, 1988); France in conifer nurseries (Vegh & Bourgeois, 1975), rose (Scotto La Massese, 1971; Berge & Cuany, 1975) and fig tree (Scotto La Massese *et al.*, 1984); Germany (Decker & Dowe, 1974); Norway on rose (Stoeen, 1974); Greece on apricot (Kalyviotis-Gazelas, 1981); Italy on sour orange (Inserra & Vovlas, 1974), olive (Lamberti, 1969; Inserra *et al.*, 1976; Inserra & Vovlas, 1981), rose (Lamberti *et al.*, 1987), stone fruits (Tacconi & Talamé, 1995) and walnut (Ciancio *et al.*, 1995, 1996); Russia on walnut (Tokobaev & Matyashov, 1986) and rose (Ryss, 1988); Spain on plantains (Castillo & Gómez Barcina, 1993), horticultural crops (Espárrago & Navas, 1995), in forest tree nurseries (Talavera *et al.*, 1999), in olive nurseries (Nico *et al.*, 2002); on *Prunus* spp. (Marull & Pinochet, 1991) and from other hosts and localities (Gómez Barcina *et al.*, 1989; Peña Santiago *et al.*, 2004).

It has been recorded on several hosts from different parts of the USA and Canada, where is a major pathogen of deciduous fruit and nut crops: Arkansas on blackberry (Wehunt *et al.*, 1991) and soybean (Robbins *et al.*, 1987); California in prune orchards (Lownsbery *et al.*, 1974), on almond (McKenry & Kretsch, 1987) and *Sequoia* spp. (Maggenti & Viglierchio, 1975); Missouri on cotton (Wrather *et al.*, 1992); North Carolina in peach (Barker & Clayton, 1973) and boxwood nurseries (Barker, 1974); Tennessee (Bernard, 1980); and on roses (Sher, 1957); and from Canada (Townshend *et al.*, 1978a) and Mexico (Teliz & Goheen, 1968). It is also reported from Central and South America: Argentina on walnut and peach (Doucet & Lax, 1997); Brazil on peach (Lordello, 1973), raspberry (Monteiro & Lordello, 1976) and rose (Rossi *et al.*, 2000a); Uruguay on strawberries (Minagawa & Maeso-Tozzi, 1990). It is recorded from Africa in Kenya on pigeon pea (Hillocks & Songa, 1993) and South Africa on peach (Heyns, 1971; Stirling, 1975). It is recorded from several countries of Asia: China on peach (Yin, 1991; Gao *et al.*, 1999), mango (Yin, 1995) and sweet potato (Huan & Xu, 1985; Li, 1985); Iran (Pourjam *et al.*, 1997; Lee *et al.*, 2006); Israel on apple (Nimrod *et al.*, 1999); Japan in crops (Orui & Mizukubo, 1999b) and from uncultivated soils (Gotoh, 1970); Korea on peach (Choi, 1975; Park *et al.*, 1999; Kwon *et al.*, 2004); Pakistan on rose (Anwar *et al.*, 1991) and fruit crops (Anwar *et al.*, 1992a); Sri Lanka on strawberry (Mohotti *et al.*, 1997; Hirata *et al.*, 2000); and Korea (Choi *et al.*, 2006). It has also been recorded in Australia on grapevine (Meagher, 1969; Beaumont, 1975; McLeod, 1978) and strawberry (Colbran, 1974).

65. *Pratylenchus wescolagricus* Corbett, 1983
(Fig. 105)

MEASUREMENTS

- Female holotype (after Corbett, 1983): L = 0.62 mm; a = 29; b = 6.0; b′ = 4.5; c = 25; V = 82; stylet = 18 μm.
- 15 females (after Corbett, 1983): L = 0.60 (0.50-0.68) mm; a = 29 (25-32); b = 5.9 (5.0-6.6); b′ = 4.4 (3.6-5.3); c = 21 (17-25); V = 81 (79-82); stylet = 18 (17-19) μm.

DESCRIPTION

Female

Body straight or slightly curved ventrally on death, conspicuously annulated, tapering from vulva to tail. Lateral fields with four lines; outer two bands irregularly and sparsely areolated; sometimes oblique striae in middle band in mid-body and occasionally an extra line present in middle of outer two bands to give appearance of six lines. Lateral fields widening abruptly at vulva with ventral line dipping towards vulva, narrowing again posteriorly. Lateral fields narrowing still further posterior to phasmid which is small, located in anterior half of tail 12-15 annuli from tip on, or close to ventral-most of two mid-lines of lateral fields posterior to which inner two lines joining to form a single line. Labial region rounded in profile usually with four, sometimes three annuli: of 53 specimens examined, 26 had four annuli; 18 had four on one side and three on the other; and none had three annuli. Labial framework massive, extending two annuli into body. *En face* view by SEM subdorsal and subventral segments of first lip annulus fused to panduriform shape; separated from lateral sectors bearing amphidial apertures on their inner edges. Oral aperture oval,

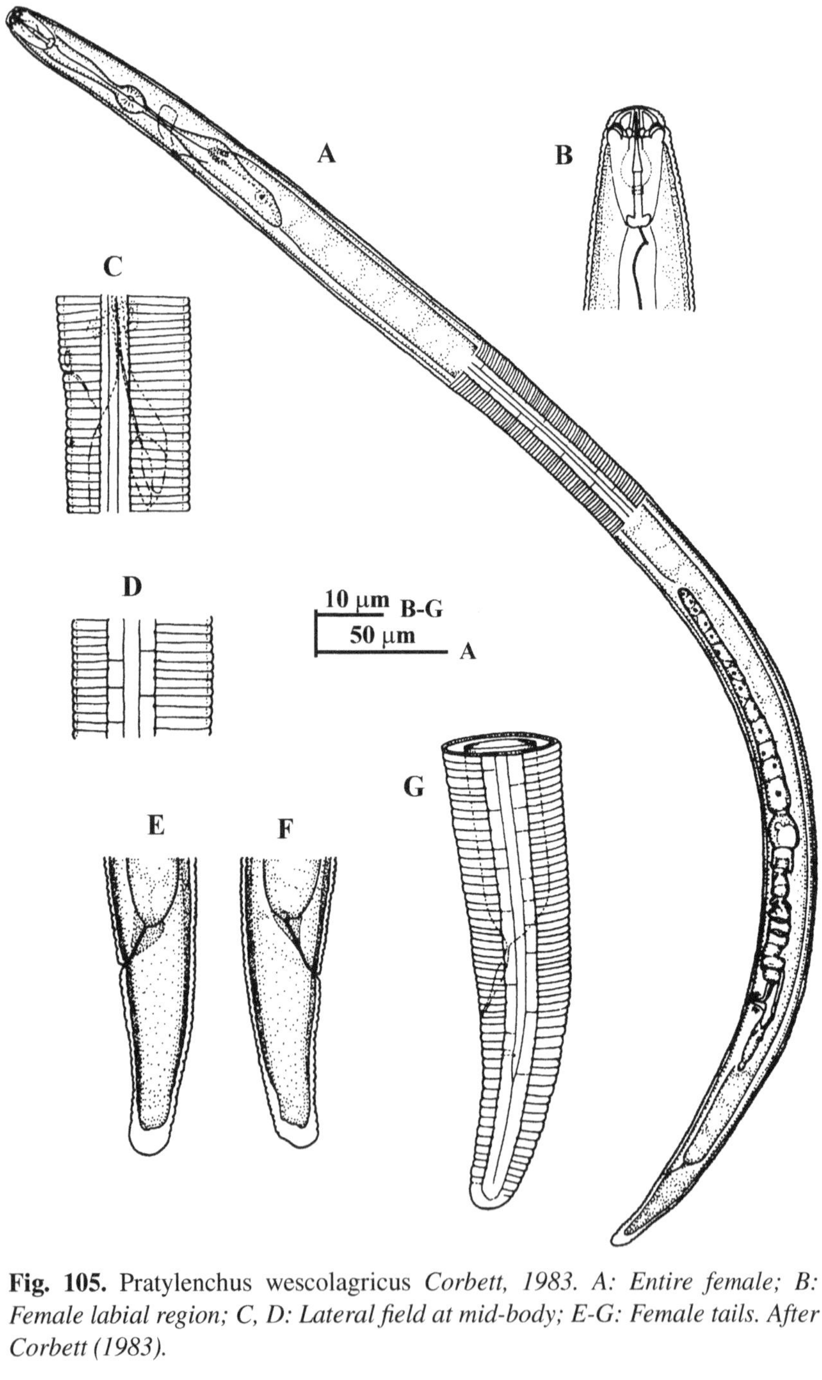

Fig. 105. Pratylenchus wescolagricus *Corbett, 1983. A: Entire female; B: Female labial region; C, D: Lateral field at mid-body; E-G: Female tails. After Corbett (1983).*

surrounded by six small pores in two lines of three. Stylet robust with angular basal knobs, dorsal gland opening 2-3 μm posterior to stylet base. Median bulb ovate, well developed, nerve ring encircling isthmus just anterior to enlarged glandular part of pharynx. Pharyngeal lobe overlapping intestine ventrally and laterally by 21-49 μm. Pharyngo-intestinal junction slightly posterior to excretory pore, which is 82-96 μm from anterior end. Hemizonid immediately in front of excretory pore; in some specimens hemizonion visible 8-16 annuli posterior to hemizonid. Single anterior reproductive tract, spermatheca small, spherical non-functional. Post-vulval uterine sac *ca* 1 body diam. long (17-29 μm), undifferentiated. Cuticle on dorsal body surface opposite anus often with 'double' annuli; annulation on ventral surface of tail irregular, with 16-20 annuli. Tail tip rounded, smooth, occasionally indented, with conspicuous 3-5 μm long hyaline area at tip.

Male

Not found.

Diagnosis and relationships

Pratylenchus wescolagricus is characterised by: labial region rounded in profile usually with four, sometimes three, annuli, lateral fields with four lines, outer two bands irregularly and sparsely areolated, spermatheca small, spherical, non-functional and tail tip rounded, smooth, occasionally indented.

The matrix code is: A3, B1, C4, D2, E3, F3, G2, H1, I2, J1, K2.

It is close to *P. goodeyi*, *P. vulnus* and *P. typicus* from which it differs by stylet length, female tail shape and number of annuli (see the corresponding descriptions).

Distribution

Recorded from the type locality, Ayr Parks Department Glasshouses, Ayr, Scotland, around roots of *Cordyline* sp. and from Germany in glasshouses (Braasch, 1988).

66. *Pratylenchus yamagutii* Minagawa, 1991
(Fig. 106)

MEASUREMENTS

- Female (after Minagawa, 1991): L = 0.41 mm; a = 24.1; b = 4.9; b′ = 3.6; c = 18.2; c′ = 2.2; V = 79; stylet = 16 μm.
- 25 females (after Minagawa, 1991): L = 0.41 (0.33-0.53) mm; a = 23.9 (19.1-27.4); b = 5.2 (4.4-5.9); b′ = 3.5 (2.9-4.2); c = 18.8 (15.7-20.9); c′ = 2.1 (1.5-2.5); V = 79 (77-82); stylet = 16 (15-18) μm.
- 17 males (after Minagawa, 1991): L = 0.36 (0.33-0.41) mm; a = 24.6 (22.7-26.7); b = 4.5 (4.1-5.0); b′ = 3.4 (2.9-4.0); c = 19.7 (18.1-21.0); c′ = 1.9 (1.6-2.3); T = 45 (35-54); stylet = 15 (13.5-16) μm; spicules = 14.5 (12-15) μm; gubernaculum = 4 (3.5-4.0) μm.

DESCRIPTION

Female

Body strongly to slightly curved ventrally when heat-relaxed. Body annuli fine, 1.0 (0.7-1.2) μm wide at mid-body region. Lateral fields 5.1 (4.0-7.0) μm wide, with four lines, outer pair smooth in mid-body and crenate near posterior terminus; inner pair faint, with an additional line in some specimens, fused to one near phasmids. Labial region low, flattened, with two annuli occasionally with three annuli (one out of 25 females), or three annuli on either side (three out of 25 females), slightly setoff from body contour, labial framework generally extending into body by 1-1.5 annuli. *En face* view, subdorsal and subventral segments of first lip annulus fused to oral disc, narrowing at their inner extremity and widening towards rounded outer edge of face. Stylet rigid, knobs rounded to slightly sloping anteriorly or posteriorly in anterior surface, 2.5 (2.0-2.7) μm high and 4.7 (4.3-5.0) μm wide; dorsal orifice of pharyngeal gland at 3.5 (2.7-4.7) μm from stylet base. Median bulb massive, slightly oval, 12.5 (10.3-14.7) μm long, 9.3 (8.7-10.3) μm diam., valve moderately developed. Nerve ring surrounding isthmus. Pharyngeal gland lobes overlapping intestine ventrally and laterally, 37 (26-55) μm long or 1.5-3.4 times corresponding body diam. Hemizonid two or three annuli long. Excretory pore immediately posterior to hemizonid, 75.3 (54.7-95.3) μm from anterior extremity.

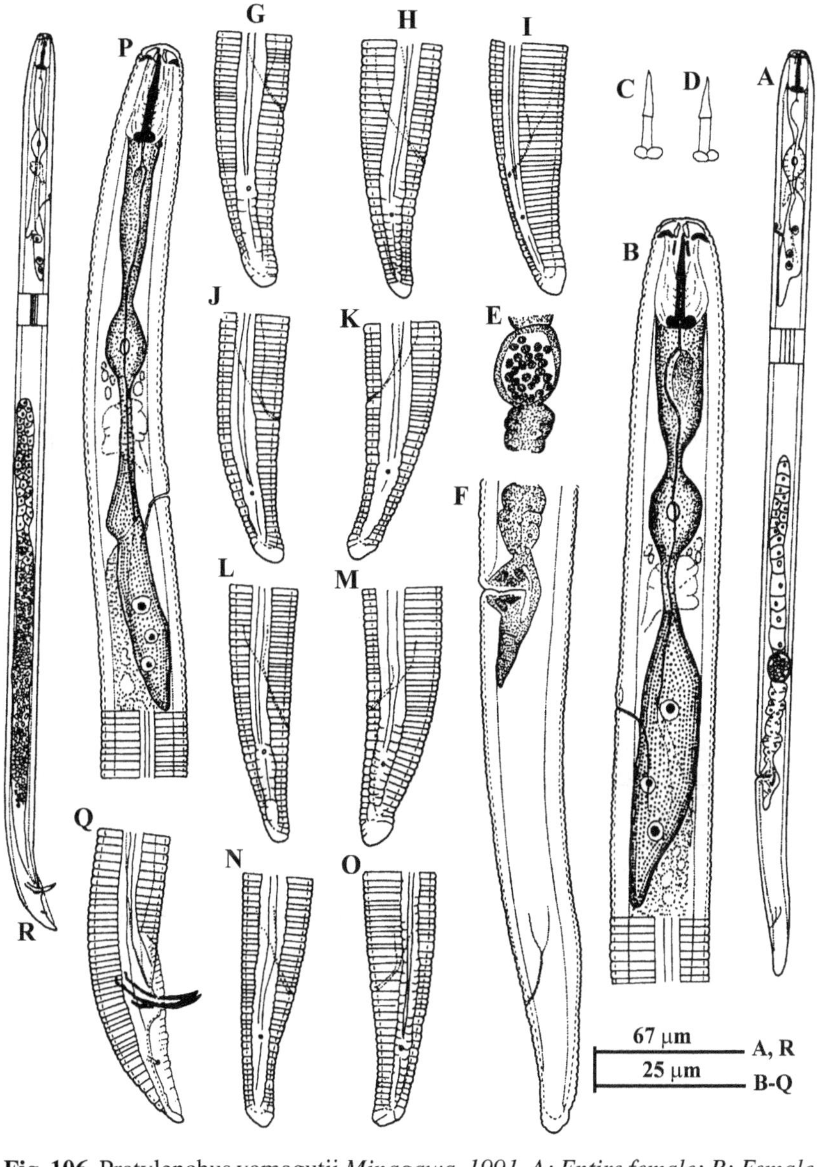

Fig. 106. Pratylenchus yamagutii *Minagawa, 1991. A: Entire female; B: Female pharyngeal region; C, D: Stylets; E: Spermatheca; F: Female posterior region; G-O: Female tails; P: Male pharyngeal region; Q: Male tail; R: Entire male. After Minagawa (1991).*

Reproductive system moderately developed. Spermatheca rounded to oval, 13 (8-17) μm in length, 10 (8-14) μm in diam., filled with rounded spermatozoa. Vulva at *ca* 80%; post-vulval uterine sac 16.7 (13.0-19.3) μm or 1.0 (0.7-1.3) vulval body diam. long. Tail 20.7 (18.0-30.0) μm long with 20.1 (17-25) annuli and terminus rounded without distinct annulation. Phasmids usually located in anterior half of tail, 7.5 (4-12) annuli posterior to anus, 14.3 (12.0-20.0) μm from tail end. *En face* view belonging to Group 2 *sensu* Corbett and Clark (1983).

Male

Body and stylet shorter than those of female, but in other aspects similar to female except for reproductive system. Testis single, outstretched; cloacal prominence small; spicules arch shaped; gubernaculum simple, crescent-shaped. Bursa surrounding tail tip; phasmids rod like, located on bursa, around mid-tail.

DIAGNOSIS AND RELATIONSHIPS

Pratylenchus yamagutii is characterised by: labial region with two annuli, dumbbell-shaped lateral segment of face, rounded and smooth tail terminus, anterior half location of phasmids in tail, rounded to oval spermatheca filled with rounded spermatozoa and common occurrence of males.

The matrix code is: A1, B2, C3, D3, E2, F1, G3, H2, I2, J1, K1.

It is close to *P. alleni*, *P. coffeae* and *P. neglectus* from which it can be distinguished by body and stylet length, b ratio, shape of the lateral face segment, lip annulation, shape of the spermatheca and tail shape (see the corresponding descriptions).

DISTRIBUTION

It has only been recorded from the type locality at Hokkaido, Japan, from the rhizosphere of *Weigela middendorffiana* (Carriere) K. Koch and *Alnus crispa maximowiczii* (Call.) Hult.

67. *Pratylenchus yassini* Zeidan & Geraert, 1991
(Fig. 107)

MEASUREMENTS

- Female holotype (after Zeidan & Geraert, 1991): L = 0.57 mm; a = 30; b = 6.2; b′ = 3.7; c = 15; c′ = 3.1; V = 73; stylet = 16.5 μm.
- 22 females (after Zeidan & Geraert, 1991): L = 0.52 (0.43-0.61) mm; a = 29 (26-34); b = 5.6 (5.0-6.5); b′ = 3.9 (3.2-5.0); c = 16 (13-19); c′ = 3.1 (2.6-3.8); V = 74 (71-76); stylet = 17 (16-18.5) μm.

DESCRIPTION

Female

Moderately long nematodes with rather thin bodies, slightly arcuate to open C-shape when heat-relaxed. Cuticle transversely striated with 1.0-1.5 μm wide annuli. Labial region with three annuli; SEM *en face* view showing an undivided front plate amalgamated with first annulus and followed by two annuli, offset by a fine but deep constriction from rest of body. Oval mouth opening in dorso-ventral plane, *ca* 0.5 μm long. Close to mouth opening are six fine pores of inner labial sensilla and *ca* 1 μm slits of amphidial apertures (usually obscured by exudations). Amphidial apertures are not oblique and not dorsally displaced. Although the *en face* view did not show a line pattern, central area with six papillae is slightly demarcated forming an obscure oral disc; also slightly indicated are a ventral and dorsal segment (terminology of Corbett and Clark, 1983). Lateral fields occupying *ca* one-third of corresponding body diam., four incisures with outer ones areolated; with LM a fifth, diagonally interrupted, line could be observed on seven females; of the six females studied by SEM, all showed this diagonally interrupted line, although four showed only four lines when viewed under LM. Stylet knobs rounded and slightly anteriorly indented. Stylet cone almost as long as or shorter than shaft. Pharynx with elliptical median bulb and elongated glandular lobe overlapping intestine, pharyngo-intestinal junction indistinct. Nuclei in glandular lobe distinct: one at level of pharyngo-intestinal junction; other two others to each other in posterior third of lobe. Hemizonid observed in only one female. Hemizonion and cephalids not seen. Intestine with small central lumen (less than 1 μm), nuclei are 3.5-5.0 μm in size, large, granulated, with one nucleolus;

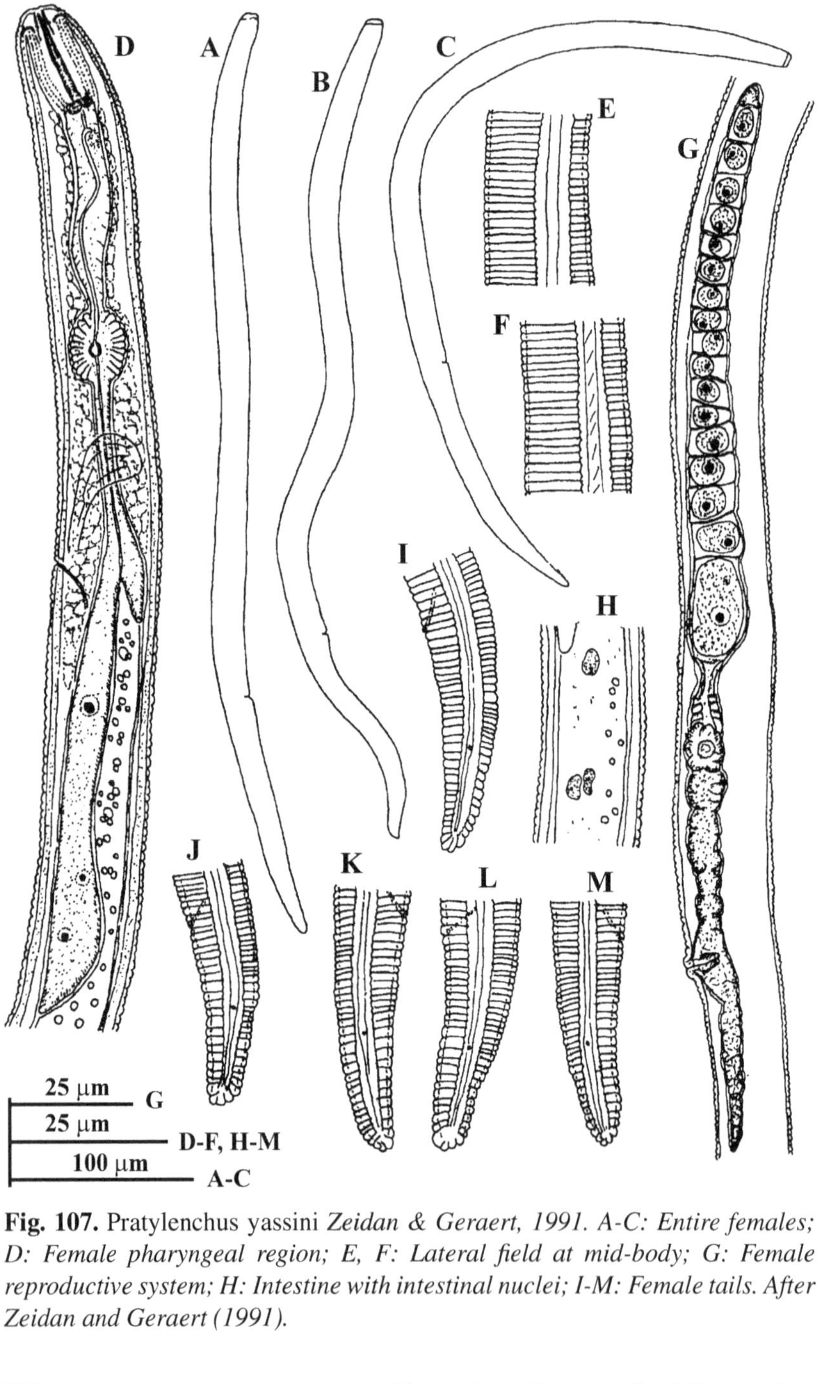

Fig. 107. Pratylenchus yassini *Zeidan & Geraert, 1991. A-C: Entire females; D: Female pharyngeal region; E, F: Lateral field at mid-body; G: Female reproductive system; H: Intestine with intestinal nuclei; I-M: Female tails. After Zeidan and Geraert (1991).*

not all nuclei are distinct but extrapolation from regions where they are suggests there could be some 30 nuclei in all. Female genital tract anteriorly outstretched with long post-vulval uterine sac comprising some cellular material. Oocytes arranged in a single file, spermatheca oval, empty. Vulva slightly raised, transverse slit with thick lips. Tail conical in shape with annulated terminus. Phasmids pore-like.

Male

Not found.

DIAGNOSIS AND RELATIONSHIPS

Pratylenchus yassini is characterised by: the presence of three annuli on the labial region, vulva at 71-75%, lateral fields composed of four lines and occasionally a fifth, diagonally interrupted, one, long post-vulval uterine sac (1.5-2.5 times the corresponding vulval body diam.), a conoid tail with 23-30 annuli and crenate terminus, spermatheca empty and males absent.

The matrix code is: A2, B1, C3, D3, E1, F6, G3, H2, I4, J1, K2.

It can be distinguished from closely related species (*P. crenatus* and *P. teres*) by the number and shape of the lateral field lines, position of the vulva, post-vulval uterine sac, tail shape and length (see the corresponding descriptions).

DISTRIBUTION

It has only been recorded from the type locality at central Sudan from the rhizosphere of a weed (*Philipisara* sp.).

68. *Pratylenchus zeae* Graham, 1951
(Fig. 108)

MEASUREMENTS

- Female neotype (after Fortuner, 1976): L = 0.47 mm; a = 26; b = 5.9; c = 21; V = 70; stylet = 16 μm.
- Females (after Graham, 1951): L = 0.54 (0.36-0.58) mm; a = 27 (25-30); b = 6.5 (5.4-8.0); c = 15.2 (17-21); V = 71 (68-76); stylet = 15.5 (15-17) μm.

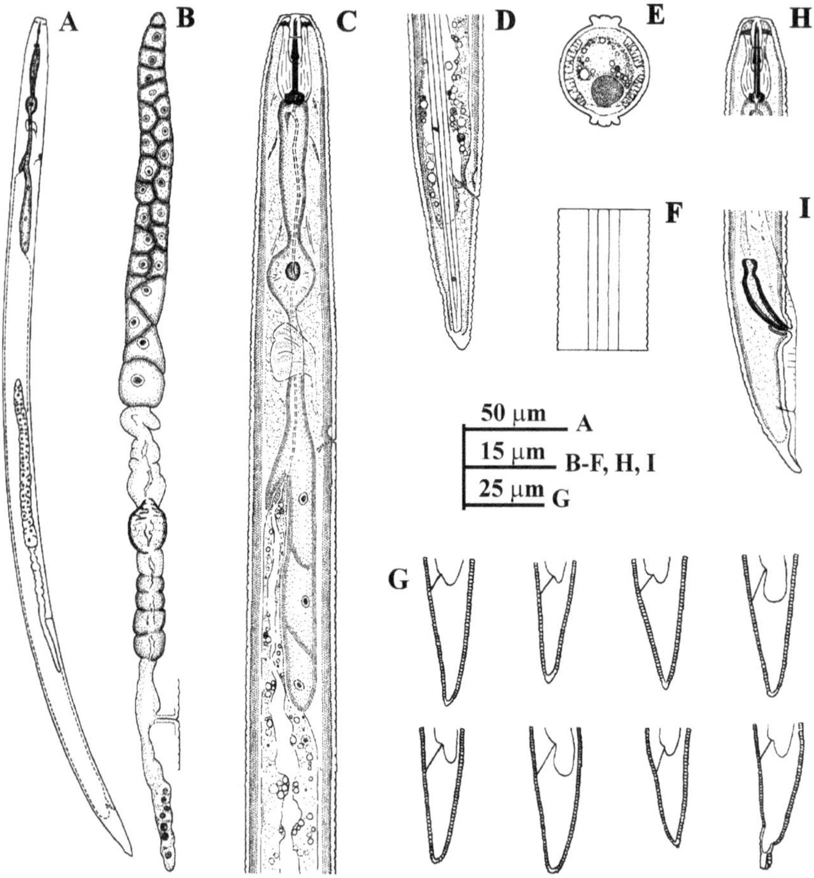

Fig. 108. Pratylenchus zeae *Graham, 1951. A: Entire female; B: Female reproductive system; C: Female pharyngeal region; D, G: Female tails; E: Mid-body transverse section; F: Lateral field at mid-body; H: Male labial region; I: Male tail. After Fortuner (1976).*

- Females (after Sher & Allen, 1953): L = 0.36-0.58 mm; a = 25-30; b = 5.4-8; c = 17-21; V = 68-76; stylet = 15-17 μm.
- 90 females (after Taylor & Jenkins, 1957): L = 0.52 (0.41-0.64) mm; a = 21.1 (17.0-25.4); b = 6.3 (5.0-9.6); c = 17.3 (11.2-24.1); V = 71 (65-75).
- 25 females (after Merny, 1970): L = 0.34-0.55 mm; a = 22-33; b = 3.3-4.9; c = 13-18; V = 69-74; stylet = 15-18 μm.
- 5 males (after Merny, 1970): L = 0.40-0.42 mm; a = 27-32; b = 3.6-5.0; c = 17-21; stylet = 15 μm; T = 30.

- 25 females (after Fortuner, 1976): L = 0.37-0.51 mm; a = 20-30; b = 4.9-6.1; b′ = 3.2-6.1; c = 15-19; V = 69-74; stylet = 15.5-16.5 μm.
- 14 females (after Hashim, 1984): L = 0.49 (0.38-0.59) mm; a = 29 (26-32); b = 5.3 (4.5-5.9); c = 18.3 (16.1-25.0); c′ = 2.6 (2.1-3.0); V = 78 (75-79); stylet = 14.6 (14.5-15.0) μm.
- 10 females (after Ryss, 1988): L = 0.47 (0.35-0.56) mm; a = 28 (25-31); b = 6.5 (5.4-7.5); c = 18 (17-20); c′ = 3.3 (3.0-3.5); V = 71 (68-76); stylet = 16 (15-17) μm.
- 30 females (after Troccoli *et al.*, 1996a): L = 0.53 (0.46-0.60) mm; a = 25 (20-34); b = 6.9 (5.6-7.8); b′ = 4.7 (3.8-5.5); c = 16.2 (13.5-21.6); V = 72 (69-75); stylet = 16 (14-17) μm.
- 15 females (after Doucet & Cagnolo, 1998): L = 0.50 (0.40-0.50) mm; a = 29.2 (25.3-35.0); b = 6.5 (5.5-7.4); b′ = 4.1 (3.5-5.0); c = 17.6 (15.3-19.8); V = 72 (70-79); stylet = 15.0 (14-16) μm.
- 21 females (after Van den Berg & Quénéhervé, 2000): L = 0.48 (0.40-0.55) mm; a = 25.7 (21-31); b = 5.3 (4.5-6.5); c = 17.3 (13-19); c′ = 2.5 (2.0-4.0); V = 78 (74-80); stylet = 15.7 (14-17) μm.

DESCRIPTION

Female

Body slender, almost straight when heat-relaxed, marked by very faint annuli. Labial region not offset from body, bearing three annuli. With SEM, most specimens with three lip annuli; rare specimens with three annuli on one side and four on other (Baujard *et al.*, 1990). *En face* view by SEM showing a first lip annulus without differentiation (Baujard *et al.*, 1990). Outer margins of heavily sclerotised labial framework extending into body *ca* one annulus. Lateral fields with four lines extending along tail beyond phasmids; inner band showing a slight irregularity in mid-body region but no corresponding fifth line seen in transverse section. Stylet 15-17 μm long, with broad, anteriorly flattened basal knobs. Orifice of dorsal pharyngeal gland *ca* 3 μm posterior to stylet base. Hemizonid *ca* two body annuli long, just anterior to excretory pore; hemizonion 9-11 annuli posterior to hemizonid. Ovary not extending to pharynx, oocytes in double row, except for last two or three. Oviduct indistinct; uterus short. Spermatheca round, small, without sperm (even in the only population recorded with males). Vulva at 68-76% of body length. Post-vulval uterine sac short, 1-2 body diam.

long, with rudiments of ovary. Intestine with a short post-rectal extension dorsally. Phasmids slightly posterior to mid-tail. Tail tapering, with 16-25 annuli in a Senegal population (25-27 according to Seinhorst, 1968): terminus variable, generally almost pointed, narrowly rounded to subacute, unstriated.

Male

Extremely rare, found only once in Ivory Coast (Merny, 1970), not essential for reproduction. Similar to female except for sexual dimorphism. Spicules slender, ventrally arcuate, 14-15 μm long; gubernaculum 4-5 μm in length; bursa margins faintly crenate.

DIAGNOSIS AND RELATIONSHIPS

Pratylenchus zeae is characterised by: labial region with three annuli, spermatheca round, small, without sperm, vulva at 68-76% of body length, tail conoid, terminus variable, generally almost pointed, narrowly rounded to subacute, unstriated.

The matrix code is: A2, B2, C3, D1, E1, F5, G3, H3, I2, J1, K1.

It can be distinguished from closely related species (*P. bolivianus* and *P. curvicauda*) by stylet length, position of the vulva, shape of spermatheca and female tail shape (see the corresponding descriptions).

Recent studies using LM and SEM microscopy on paratypes of *P. jordanensis* and three populations from Oman (deposited at the USDA Nematode Collection, Beltsville, MD, USA) and a Florida population of *P. zeae* (from St Augustine grass) by Inserra *et al.* (2005) showed great morphological and morphometrical similarities between *P. jordanensis* and *P. zeae* suggesting that the former is a junior synonym of the latter. This action is accepted herein.

DISTRIBUTION

It has been recorded from the type locality at Florence, South Carolina, USA, from roots of corn (*Zea mays*). It has been recorded from North America in several states of the USA: Arkansas, on soybean (Robbins *et al.*, 1987), blackberry (Wehunt *et al.*, 1991); Florida on potato (Crow *et al.*, 2001), cotton (Robbins *et al.*, 1989a); Georgia on corn (Johnson *et al.*, 1974); Hawaii on coffee (Schenck & Schmitt, 1992); Mississippi on sorghum (Cuarezma-Teran *et al.*, 1984); North Carolina on corn (Barker, 1974) and peach (Barker & Clayton, 1973);

and also in the Pacific Islands on sugarcane (Bridge, 1988) and Canada (Townshend *et al.*, 1978a). It is widespread in several countries of Central and South America: Argentina in corn (Troccoli *et al.*, 1996a), sugarcane and several hosts and localities (Costilla *et al.*, 1976; Doucet & Cagnolo, 1998); Belize on corn and rice (Bridge *et al.*, 1996); Brazil on sugarcane (Moura *et al.*, 1999; Dinardo-Miranda *et al.*, 2003), marandu grass, *Brachiaria brizantha* cv. Marandu (Sharma *et al.*, 2001), bean (Rossi *et al.*, 2000b), vegetable crops and medicinal plants (De Souza *et al.*, 1998), fig (Campos, 1997), *Panicum maximum* and *P. purpurascens* (Lordello & Mello Filho, 1970); Colombia on sugarcane (Agudelo & Volcy, 1998); Costa Rica on sugarcane (Lopez & Salazar, 1990); Cuba on rice (Fernandez Díaz-Silveira & Ortega Herrera, 1998); Dominican Republic on sugarcane (Román & Grullon, 1975); Martinique on sugarcane (Delaville *et al.*, 1996); Panama on sugarcane (Tarté *et al.*, 1977); Puerto Rico on sorghum (Bee-Rodriguez & Ayala, 1977); Venezuela on sugarcane (Siddiqi, 1974; Crozzoli, 2002; Perichi *et al.*, 2002), cocoa (Crozzoli *et al.*, 2001), citrus (Crozzoli *et al.*, 1998). In Europe it has only been recorded in Bulgaria on corn (Stoyanov, 1977); Russia on corn (Ryss, 1988); and in several localities in Spain (Arias & Romero, 1979). It has been widely recorded in Africa: Cameroon on corn (Noupadja, 1997), rice (Samsoen & Geraert, 1975); Egypt on soybean (Salem *et al.*, 1994), corn (Ibrahim *et al.*, 1988); Guinea, Benin and Togo on rice (Coyne *et al.*, 1996, 2000); Ivory Cost on upland rice (Plowright & Hunt, 1994; Coyne & Plowright, 2002), plantain (Adiko, 1988); Kenya on corn (Kimenju *et al.*, 1998; Van den Berg *et al.*, 2001); Madagascar on tobacco (Baudin & Huu-Hai, 1973); Malawi on subsistence crops (Hillocks *et al.*, 1995) and weeds (Jones & Hillocks, 1995); Mozambique on sugarcane, corn and bean (Oever & Mangane, 1992; Oever *et al.*, 1998); Namibia on corn and pearl millet (De Waele *et al.*, 1998); Nigeria on plantains (Speijer *et al.*, 2001) and rice (Babatola, 1984); Papua New Guinea on corn (Bridge & Page, 1984); South Africa on sugarcane (Cadet *et al.*, 2004), soybean (Fourie *et al.*, 2001), peanut (Venter *et al.*, 1992), wheat (Jordaan *et al.*, 1992), corn (Jordaan *et al.*, 1989), sunflower (Bolton *et al.*, 1989), rice (De Waele & Van den Berg, 1988), corn (De Waele & Jordaan, 1988a), sorghum (De Waele & Jordaan, 1988b); and from Zambia on soybean (Lawn *et al.*, 1988a), corn and sunflower (Lawn *et al.*, 1988b).

It is widely recorded in Asia: China (Yin, 1991); India on sugarcane (Mehta & Sundararaj, 1990), corn (Kanwar *et al.*, 1996), pulse crops

(Ali & Askary, 2001), alfalfa (Hasan *et al.*, 2003), coconut and arecanut (Rama & Dasgupta, 2000b), rice (Sharma *et al.*, 1992a; Gaur *et al.*, 1996), several crops (Haidar & Pathak, 2001), *Leucaena leucocephala* (Lam.) and *Albizia procera* L. Benth (Singh, 1999), potato (Waliullah & Bhat, 1990), pigeon pea and peanut (Sharma *et al.*, 1992b), rose (Sundarababu & Vadivelu, 1988); in Japan on tobacco (Orui & Mizukubo, 1999a); Malaysia on tobacco (Sidam, 1989); in Oman on alfalfa (Mani & Al-Hinai, 1997; Mani *et al.*, 1998) and date palm (*P. dactylifera*) (Mani *et al.*, 2005); Pakistan on banana (Khan *et al.*, 2001a), nurseries of floricultural crops (Khan *et al.*, 1997), coconut (Khan *et al.*, 1992); Philippines on corn (Davide, 1988); Thailand on peanut (Toida *et al.*, 1996); Turkey on chickpea (Di Vito *et al.*, 1994), sunflower (Kepenekci, 2001), anise, *Pimpinella anisum* L. (Kepenekci, 2003), tea (Kepenekci & Akgul, 1999); and in Vietnam on peanut (Sharma *et al.*, 1994). It has also been recorded in Australia on sugarcane (Magarey & Bull, 1998; Blair *et al.*, 1999; Pankhurst *et al.*, 2003), cereals (Riley & Kelly, 2002), apple (Stirling *et al.*, 1995; Stirling & Stirling, 2002), rice (Mathur & McLeod, 1977) and white clover (Irwin & Jones, 1977).

Other hosts are sorghum, millet, rye, soybean, tomato, oat, sweet potato, wheat, peanut, barley, strawberry, blue lupin, cowpea, *Amaranthus spinosus* L., *Ambrosia artemisiifolia* L., *Andropogon virginicus* L., *Chenopodium album* L., *C. ambrosioides* L., *Crotalaria mucronata* Desvaux., *C. spectabilis* Roth., *Cynodon dactylon* (L.) Pers., *Dactyloctenium aegyptium* (L.) Beauv., *Digitaria sanguinalis* (L.) Scop., *Diodia teres* Walt., *Echinochloa crusgalli* (L.) P. Beauv., *Eremochloa ophiuroides* (Munro) Hack., *Heterotheca subaxillaris* Lam. Britton & Rusby, *Lespedeza* sp., *Solidago gigantea* Ait., *Tribulus terrestris* L. in the USA (Graham, 1951; Ayoub, 1961), *Capsicum annuum* L. in Trinidad (Singh, 1974), onion and lettuce in Nigeria (Bridge, 1972).

Nematology Monographs & Perspectives, 2007, Vol. 6, 281-293

Chapter 5

Specific diagnostic differentiation

Identification of *Pratylenchus* species is difficult because of the relatively large and significant intraspecific variability in the diagnostic characters. At present the identification of *Pratylenchus* species is based primarily on morphology and, to a lesser extent, on new methodologies (see next chapter). Because of the large number of species within the genus and their variability, identification of these nematodes by morphological characteristics alone is often very difficult. In several revisions of the genus *Pratylenchus*, the morphological characters that have proved to be reliable and useful for species diagnosis and identification include: number of lip annuli (2-4), shape of the labial region (high or low, outer edges rounded or acute, offset or continuous), stylet length, length of overlapping gland lobe, structure of lateral fields, position of vulva, presence and shape of spermatheca, length of the post-vulval uterine sac, shape of female tail and terminus and presence or absence of males (Loof, 1978, 1991; Ryss, 2002a). However, other characters, which have occasionally been used as being diagnostic, are of limited value or even wholly without foundation, *e.g.*, tapering of body anteriorly, position of excretory pore relative to median bulb and pharyngeal gland, posterior extension of basal plate of labial framework (Loof, 1978).

Taxonomy and identification of the root-lesion nematodes belonging to *Pratylenchus* are in a state of constant change. More or less complete identification keys have become obsolete after the description of new species during the last few years or by synonymisation of existing species. Although identification of *Pratylenchus* species has been based on morphological and morphometrical characteristics of adult females, Ryss (1983, 1984) studied the diagnostic characteristics of juvenile stages (J2 to J4) in six species, including *P. convallariae*, *P. crenatus*, *P. flakkensis*, *P. neglectus*, *P. pinguicaudatus* and *P. thornei*, but these have gained little acceptance subsequently. One of the main problems in constructing new identification keys to *Pratylenchus* spp.

is the great number of valid species (currently 68) and the significant intraspecific variability of diagnostic characters of species in the genus. Most available identification keys were constructed according to the dichotomous principle (Corbett, 1969; Loof, 1978, 1991; Frederick & Tarjan, 1989; Handoo & Golden, 1989). However, the application of the polytomous principle for the identification of *Pratylenchus* spp. was used for the first time by Ryss (2002a) who proposed monoentry and a tabular (multientry) keys, the latter being developed on the basis of the stepwise computer diagnostic system BIKEY-PICKEY (Lobanov *et al.*, 1996). For *Pratylenchus* spp. identification, Ryss (2002a) used 26 characters, each subdivided into several classes. Most of these characters are used herein to identify *Pratylenchus* spp. by dichotomous and tabular keys (*e.g.*, number of annuli in the labial region, stylet length, tail shape, post-vulval uterine sac, *etc.*). However, other characters used by Ryss (2002a) were not included in our study because of their limited value for diagnostic purposes, *e.g.*, number of female tail annuli, index a (ratio of body length to max. body diam.), index b (ratio of body length to pharynx length up to the pharyngo-intestinal valve), median bulb, sperm in female spermatheca. With such a large number of species within the genus, we agree with other authors (*e.g.*, Loof, 1991) concerning the complexity involved in creating a workable key to identify such a large number of species but, nevertheless, as in our previous work on *Rotylenchus* (Castillo & Vovlas, 2005), we are confident that, by employing both systems, any user now has the necessary tools to identify the species of *Pratylenchus* with which they are dealing.

In order to improve the usefulness of the tabular and dichotomous keys, the species have been arranged in ascending order within each of the character groups to which the species belongs. Finally, we present in the Appendix several tables where all valid species of the genus are arranged in ascending order for each of the main diagnostic characters used in the tabular and dichotomous keys.

Nevertheless, as pointed out in Chapter 6, with the increasing importance of biochemical and molecular data, it should be possible to construct diagnostic keys based on molecular or biochemical markers, as has been recently proposed for some species of root-knot nematodes of the genus *Meloidogyne* (Adam *et al.*, 2007).

Morphological characters used to distinguish *Pratylenchus* spp. in tabular and dichotomous keys

A) Lip annuli:
   Group 1: two
   Group 2: three
   Group 3: four
B) Male:
   Group 1: absent
   Group 2: present
C) Stylet length:
   Group 1: stylet < 13 μm
   Group 2: stylet 13-15.9 μm
   Group 3: stylet 16-17.9 μm
   Group 4: stylet 18-20 μm
   Group 5: stylet > 20 μm
D) Shape of spermatheca:
   Group 1: absent or reduced
   Group 2: rounded to spherical
   Group 3: oval
   Group 4: rectangular
E) Vulva position, ratio V:
   Group 1: $V < 75\%$
   Group 2: $V = 75\text{-}79.9\%$
   Group 3: $V = 80\text{-}85\%$
   Group 4: $V > 85\%$
F) Post-vulval uterine sac (PUS):
   Group 1: <16 μm
   Group 2: 16-19.9 μm
   Group 3: 20-24.9 μm
   Group 4: 25-29.9 μm
   Group 5: 30-35 μm
   Group 6: >35 μm
G) Female tail shape:
   Group 1: cylindrical
   Group 2: subcylindrical
   Group 3: conoid
H) Female tail tip:
   Group 1: smooth

Group 2: striated
Group 3: pointed
Group 4: with ventral projection

I) Pharyngeal overlapping length:
Group 1: <30 μm
Group 2: 30-39.9 μm
Group 3: 40-50 μm
Group 4: >50 μm

J) Lateral field lines at vulval region:
Group 1: four
Group 2: five
Group 3: six to eight

K) Lateral field structure at vulval region:
Group 1: smooth bands
Group 2: partially or completely areolated bands

Tabular key for the identification of *Pratylenchus* spp.

Table 2. *Tabular key for the identification of* Pratylenchus *spp. arranged alphabetically.*

Species	Character group									
	A	B	C	D	E	F	G	H	I	J	K
acuticaudatus	1	1	3	1	2	3	3	1	3	1	1
alleni	1	2	2	2	3	2	2	1	3	1	1
andinus	2	1	3	2	3	4	1	1	3	1	1
angulatus	1	1	1	1	3	1	1	1	1	1	1
arlingtoni	2	1	3	3	3	2	3	2	1	3	1
artemisiae	1	2	2	3	2	3	2	1	2	1	1
bhattii	2	2	2	2	1	2	2	1	2	1	1
bolivianus	2	1	4	1	3	4	3	1	2	1	2
brachyurus	1	2	4	1	4	3	3	1	4	2, 3	1
brzeskii	1	2	4	4	2	4	3	1	4	1	1
coffeae	1	2	2	3	3	6	2	1	2	1, 2, 3	1
convallariae	2	2	3	2	2	6	2	2	3	1	1
crassi	1	1	3	3	1	3	1	1	3	1	1
crenatus	2	2	3	1	3	3	2	2	1	3	1
cruciferus	2	1	2	1	2	3	2	1	3	1	1
curvicauda	2	1	3	1	1	4	3	2	4	1	1
delattrei	2	1	3	1	2	3	3	1	1	1	1
dunensis	1	2	3	2	2	4	1	1	4	1	2
ekrami	2	2	1	3	3	5	2	3	2	1	1
elamini	2	1	2	3	2	5	3	1	2	1, 2	2
estoniensis	1	1	3	2	3	1	3	2	2	3	1
fallax	2	2	3	2	2	3	3	2	2	1	1
flakkensis	1	2	3	2	2	2	3	2	1	1	1
gibbicaudatus	1	2	2	2	1	5	2	2	3	1	1
goodeyi	3	2	3	4	1	1	3	3	3	1	2
hexincisus	1	2	2	1	2	2	3	1	1	3	1
hippeastri	1	1	2	4	2	5	3	1	3	1	1
jaehni	1	2	2	2	2	4	2	1	2	1	1
japonicus	2	1	4	1	4	2	3	1	3	1	1
kasari	2	2	3	3	2	3	3	2	3	1	1
kralli	2	2	2	2	2	1	3	1	1	1	1
kumaoensis	2	1	2	1	3	3	3	2	1	1	1
loosi	1	2	3	4	3	2	3	3	2	1	1

Table 2. *(Continued).*

Species	Character group										
	A	B	C	D	E	F	G	H	I	J	K
macrostylus	1	2	5	2	4	2	3	1	3	1	1
manaliensis	2	2	2	4	3	3	3	2	4	3	1
mediterraneus	2	2	2	2	2	3	2	1	3	1	1
microstylus	2	1	1	1	2	2	3	1	2	1	1
morettoi	2	2	3	3	2	6	3	3	4	1	1
mulchandi	2	2	3	3	2	5	3	1	2	1	1
neglectus	1	2	3	1	2	1	3	1	1	1	1
neobrachyurus	1	2	2	3	3	1	3	1	2	1	1
okinawensis	1	2	3	3	2	4	3	2	4	1	1
panamaensis	1	2	3	3	3	2	2	2	3	1	1
penetrans	2	2	3	2	3	4	2	1	3	1	1
pinguicaudatus	2	1	3	1	3	1	2	1	4	1	2
pratensis	2	2	2	4	2	3	3	2	1	1	1
pratensisobrinus	2	2	2	3	2	3	3	2	1	1	1
pseudocoffeae	1	2	2	3	3	4	2	1	4	1	1
pseudofallax	2	2	2	2	3	2	3	2	2	1	1
pseudopratensis	2	2	3	4	2	3	3	2	3	1	1
roseus	1	2	2	2	3	2	2	2	3	3	1
scribneri	1	2	2	3	2	4	3	1	2	1, 3	1
sensillatus	2	1	2	1	2	5	2	1	3	1	2
silvaticus	1	2	2	2	3	2	3	1	2	1, 3	1
subpenetrans	2	2	3	3	3	2	3	1	3	1	1
subranjani	2	1	3	1	2	5	3	2	3	1	2
sudanensis	2	2	2	3	1	4	2	1	2	1	1
tenuis	1	1	2	3	2	1	3	1	2	1	1
teres	2	1	2	1	1	3	3	2	3	3	1
thornei	2	2	3	1	2	2	3	1	3	1	1
typicus	3	1	3	3	3	6	3	1	4	1	1
unzenensis	2	2	2	3	2	4	2	1	3	1	2
ventroprojectus	2	2	2	4	2	3	3	4	1	1	1
vulnus	2	2	2	3	2	6	3	3	2	1	1
wescolagricus	3	1	4	2	3	3	2	1	2	1	2
yamagutii	1	2	3	3	2	1	3	2	2	1	1
yassini	2	1	3	3	1	6	3	2	4	1	2
zeae	2	2	3	1	1	5	3	3	2	1	1

Table 3. *Tabular key for the identification of* Pratylenchus *spp. arranged by increasing values of different characters.*

Species	Character group										
	A	B	C	D	E	F	G	H	I	J	K
angulatus	1	1	1	1	3	1	1	1	1	1	1
tenuis	1	1	2	3	2	1	3	1	2	1	1
hippeastri	1	1	2	4	2	5	3	1	3	1	1
acuticaudatus	1	1	3	1	2	3	3	1	3	1	1
estoniensis	1	1	3	2	3	1	3	2	2	3	1
crassi	1	1	3	3	1	3	1	1	3	1	1
hexincisus	1	2	2	1	2	2	3	1	1	3	1
gibbicaudatus	1	2	2	2	1	5	2	2	3	1	1
jaehni	1	2	2	2	2	4	2	1	2	1	1
alleni	1	2	2	2	3	2	2	1	3	1	1
roseus	1	2	2	2	3	2	2	2	3	3	1
silvaticus	1	2	2	2	3	2	3	1	2	1	1
artemisiae	1	2	2	3	2	3	2	1	2	1	1
scribneri	1	2	2	3	2	4	3	1	2	1	1
neobrachyurus	1	2	2	3	3	1	3	1	2	1	1
pseudocoffeae	1	2	2	3	3	4	2	1	4	1	1
coffeae	1	2	2	3	3	6	2	1	2	1	1
neglectus	1	2	3	1	2	1	3	1	1	1	1
flakkensis	1	2	3	2	2	2	3	2	1	1	1
dunensis	1	2	3	2	2	4	1	1	4	1	2
yamagutii	1	2	3	3	2	1	3	2	2	1	1
okinawensis	1	2	3	3	2	4	3	2	4	1	1
panamaensis	1	2	3	3	3	2	2	2	3	1	1
loosi	1	2	3	4	3	2	3	3	2	1	1
brachyurus	1	2	4	1	4	3	3	1	4	2	1
brzeskii	1	2	4	4	2	4	3	1	4	1	1
macrostylus	1	2	5	2	4	2	3	1	3	1	1
microstylus	2	1	1	1	2	2	3	1	2	1	1
teres	2	1	2	1	1	3	3	2	3	3	1
cruciferus	2	1	2	1	2	3	2	1	3	1	1
sensillatus	2	1	2	1	2	5	2	1	3	1	2
kumaoensis	2	1	2	1	3	3	3	2	1	1	1
elamini	2	1	2	3	2	5	3	1	2	1	2
curvicauda	2	1	3	1	1	4	3	2	4	1	1

Table 3. *(Continued).*

Species	Character group										
	A	B	C	D	E	F	G	H	I	J	K
delattrei	2	1	3	1	2	3	3	1	1	1	1
subranjani	2	1	3	1	2	5	3	2	3	1	2
pinguicaudatus	2	1	3	1	3	1	2	1	4	1	2
andinus	2	1	3	2	3	4	1	1	3	1	1
yassini	2	1	3	3	1	6	3	2	4	1	2
arlingtoni	2	1	3	3	3	2	3	2	1	3	1
bolivianus	2	1	4	1	3	4	3	1	2	1	2
japonicus	2	1	4	1	4	2	3	1	3	1	1
ekrami	2	2	1	3	3	5	2	3	2	1	1
bhattii	2	2	2	2	1	2	2	1	2	1	1
kralli	2	2	2	2	2	1	3	1	1	1	1
mediterraneus	2	2	2	2	2	3	2	1	3	1	1
pseudofallax	2	2	2	2	3	2	3	2	2	1	1
sudanensis	2	2	2	3	1	4	2	1	2	1	1
pratensisobrinus	2	2	2	3	2	3	3	2	1	1	1
unzenensis	2	2	2	3	2	4	2	1	3	1	2
vulnus	2	2	2	3	2	6	3	3	2	1	1
pratensis	2	2	2	4	2	3	3	2	1	1	1
ventroprojectus	2	2	2	4	2	3	3	4	1	1	1
manaliensis	2	2	2	4	3	3	3	2	4	3	1
zeae	2	2	3	1	1	5	3	3	2	1	1
thornei	2	2	3	1	2	2	3	1	3	1	1
crenatus	2	2	3	1	3	3	2	2	1	3	1
fallax	2	2	3	2	2	3	3	2	2	1	1
pseudopratensis	2	2	3	4	2	3	3	2	3	1	1
convallariae	2	2	3	2	2	6	2	2	3	1	1
penetrans	2	2	3	2	3	4	2	1	3	1	1
kasari	2	2	3	3	2	3	3	2	3	1	1
mulchandi	2	2	3	3	2	5	3	1	2	1	1
morettoi	2	2	3	3	2	6	3	3	4	1	1
subpenetrans	2	2	3	3	3	2	3	1	3	1	1
typicus	3	1	3	3	3	6	3	1	4	1	1
wescolagricus	3	1	4	2	3	3	2	1	2	1	2
goodeyi	3	2	3	4	1	1	3	3	3	1	2

Dichotomous key for the identification of *Pratylenchus* spp.

Key to species of *Pratylenchus*

1. Labial region with two annuli 2
   Labial region with three annuli 28
   Labial region with four annuli 65
2. Males absent, spermatheca empty 3
   Males generally common, spermatheca filled with sperm 8
3. Stylet length < 13 μm . . . . *P. angulatus*
   Stylet length > 13 μm 4
4. Spermatheca absent or reduced . . . . *P. acuticaudatus*
   Spermatheca well developed 5
5. Vulva position < 75% . . . . *P. crassi*
   Vulva position $\geqslant$ 75% 6
6. Post-vulval uterine sac < 16 μm 7
   Post-vulval uterine sac 30-35 μm long . . . . *P. hippeastri*
7. Female tail tip smooth . . . . *P. tenuis*
   Female tail tip striated . . . . *P. estoniensis*
8. Stylet length > 20 μm . . . . *P. macrostylus*
   Stylet length < 20 μm 9
9. Stylet length 18-20 μm 10
   Stylet length < 18 μm 11
10. Spermatheca reduced, labial region offset from body, with angular anterior margin . . . . *P. brachyurus*
    Spermatheca developed, rectangular . . . . *P. brzeskii*
11. Stylet length 16-17.9 μm 12
    Stylet length < 16 μm 18
12. Spermatheca reduced, second annulus of labial region wider than first . . . . *P. neglectus*
    Spermatheca developed 13
13. Spermatheca rounded to spherical 14
    Spermatheca oval or rectangular 15
14. Post-vulval uterine sac < 20 μm . . . . *P. flakkensis*
    Post-vulval uterine sac 25-30 μm . . . . *P. dunensis*
15. Post-vulval uterine sac < 16 μm . . . . *P. yamagutii*
    Post-vulval uterine sac > 16 μm 16
16. Female tail subcylindrical . . . . *P. panamaensis*
    Female tail conoid 17

17. Pharyngeal overlap < 40 μm, labial region rounded, spermatheca rectangular, tail tapering gradually to smooth conical tip, usually with narrowly rounded terminus, males abundant . . . . *P. loosi*
    Pharyngeal overlap > 50 μm . . . . *P. okinawensis*
18. Spermatheca reduced, lateral fields with six lines on most of body. . . . . *P. hexincisus*
    Spermatheca developed. 19
19. Spermatheca rounded to spherical 20
    Spermatheca oval 24
20. Vulva position < 75% . . . . *P. gibbicaudatus*
    Vulva position > 75% 21
21. Post-vulval uterine sac < 20 μm 22
    Post-vulval uterine sac > 20 μm. . . . . *P. jaehni*
22. Female tail conoid. . . . . *P. silvaticus*
    Female tail subcylindrical 23
23. Female tail tip smooth. . . . . *P. alleni*
    Female tail tip striated . . . . *P. roseus*
24. Vulva position < 75% 25
    Vulva position > 75% 26
25. Female tail subcylindrical . . . . *P. artemisiae*
    Female tail conoid, labial region slightly offset from body, stylet stout, with rounded knobs, spermatheca oblong, tail tapering slightly with smooth terminus . . . . *P. scribneri*
26. Post-vulval uterine sac < 16 μm. . . . . *P. neobrachyurus*
    Post-vulval uterine sac > 16 μm 27
27. Pharyngeal overlap < 40 μm, labial region slightly offset, post-vulval uterine sac 1.0-1.5 body diam. long with a terminal rudimentary ovary, female tail tip truncate or hemispherical . . . . *P. coffeae*
    Pharyngeal overlap > 50 μm . . . . *P. pseudocoffeae*
28. Males absent, spermatheca empty 29
    Males generally common, spermatheca filled with sperm 43
29. Stylet length < 13 μm . . . . *P. microstylus*
    Stylet length > 13 μm 30
30. Stylet length 18-20 μm. 31
    Stylet length < 18 μm 32
31. Post-vulval uterine sac < 20 μm . . . . *P. japonicus*
    Post-vulval uterine sac > 20 μm. . . . . *P. bolivianus*
32. Stylet length < 16 μm 33

Stylet length > 16 μm 37
33. Spermatheca reduced or absent 34
Spermatheca oval .... *P. elamini*
34. Vulva position < 75% .... *P. teres*
Vulva position > 75% 35
35. Post-vulval uterine sac < 25 μm 36
Post-vulval uterine sac > 25 μm .... *P. sensillatus*
36. Female tail tip smooth .... *P. cruciferus*
Female tail tip striated .... *P. kumaoensis*
37. Spermatheca reduced or absent 38
Spermatheca developed 41
38. Vulva position < 75% .... *P. curvicauda*
Vulva position > 75% 39
39. Post-vulval uterine sac < 16 μm .... *P. pinguicaudatus*
Post-vulval uterine sac > 20 μm 40
40. Female tail tip smooth .... *P. delattrei*
Female tail tip striated .... *P. subranjani*
41. Vulva position < 75% .... *P. yassini*
Vulva position > 75% 42
42. Post-vulval uterine sac shorter than 20 μm .... *P. arlingtoni*
Post-vulval uterine sac longer than 20 μm .... *P. andinus*
43. Stylet length < 13 μm .... *P. ekrami*
Stylet length > 13 μm 44
44. Stylet length < 16 μm 45
Stylet length > 16 μm 55
45. Spermatheca rounded to spherical 46
Spermatheca oval or rectangular 49
46. Vulva position < 75% .... *P. bhattii*
Vulva position > 75% 47
47. Post-vulval uterine sac < 16 μm .... *P. kralli*
Post-vulval uterine sac > 16 μm 48
48. Female tail tip smooth .... *P. mediterraneus*
Female tail tip striated .... *P. pseudofallax*
49. Spermatheca oval 50
Spermatheca rectangular 53
50. Vulva position < 75% .... *P. sudanensis*
Vulva position > 75% 51
51. Post-vulval uterine sac > 30 μm, labial region almost continuous with body contour, pharynx overlapping intestine in a long lobe,

spermatheca oblong, post-vulval uterine sac *ca* 2 vulval body diam. long with rudimentary ovary, tail tapering, with narrowly rounded to subacute smooth tip . . . . *P. vulnus*
Post-vulval uterine sac < 30 μm 52

52. Female tail tip smooth . . . . *P. unzenensis*
Female tail tip striated . . . . *P. pratensisobrinus*

53. Vulva position < 80% 54
Vulva position 80-85% . . . . *P. manaliensis*

54. Female tail tip striated, cuticle finely annulated, oval to rectangular spermatheca, post-vulval uterine sac length similar to body diam., tail annulated until terminus . . . . *P. pratensis*
Female tail tip with a ventral projection . . . . *P. ventroprojectus*

55. Spermatheca reduced or absent 56
Spermatheca developed 58

56. Vulva position < 76%, spermatheca round, vulva at 68-76%, tail conoid, terminus generally almost pointed, narrowly rounded to subacute, unstriated . . . . *P. zeae*
Vulva position > 76% 57

57. Female tail tip smooth, labial region not offset from body, outer margin of sclerotised labial framework conspicuously extending *ca* two annuli into body and 1 annulus into labial region, lateral fields with four lines, the outer ones straight or weakly crenate, stylet medium size (17-19 μm long), spermatheca difficult to see, not containing spermatozoa, males very rare . . . . *P. thornei*
Female tail tip striated . . . . *P. crenatus*

58. Spermatheca rounded to spherical 59
Spermatheca oval 62

59. Post-vulval uterine sac longer than 35 μm . . . . *P. convallariae*
Post-vulval uterine sac between 25-30 μm 60

60. Female tail tip smooth, labial region slightly offset, low, flat in front, with rounded outer margins, pharynx overlapping intestine in a lobe *ca* 1.5 body diam. long, post-vulval uterine sac short, undifferentiated and tail generally rounded, tip smooth . . . . *P. penetrans*
Female tail tip striated 61

61. Spermatheca rectangular, female tail tip rounded to truncate, smooth . . . . *P. pseudopratensis*
Spermatheca rounded, female tail tip rounded, smooth or slightly irregular contour . . . . *P. fallax*

62. Post-vulval uterine sac $> 35\ \mu m$ .... *P. morettoi*
    Post-vulval uterine sac $< 35\ \mu m$ 63
63. Female tail tip smooth 64
    Female tail tip striated .... *P. kasari*
64. Post-vulval uterine sac $< 20\ \mu m$ .... *P. subpenetrans*
    Post-vulval uterine sac $> 30\ \mu m$ .... *P. mulchandi*
65. Males absent, spermatheca empty 66
    Males generally common, spermatheca filled with sperm, lateral fields with four inconspicuous lines, the two outer bands partially areolated, tail conoid, ventrally concave .... *P. goodeyi*
66. Post-vulval uterine sac $> 30\ \mu m$ .... *P. typicus*
    Post-vulval uterine sac $< 30\ \mu m$ .... *P. wescolagricus*

Nematology Monographs & Perspectives, 2007, Vol. 6, 295-303

Chapter 6

Biochemical and molecular approaches in *Pratylenchus* diagnosis

The genus *Pratylenchus* is stenomorphic because of the small number of diagnostic characters at the species level and the intraspecific variability of some of these characters. Nevertheless, although morphology continues to be the mainstay in identification in this genus, new technologies based on biochemical and molecular analyses are becoming increasingly important for nematode systematics and practical identifications (Al-Banna *et al.*, 1997; Duncan *et al.*, 1999; Andrés *et al.*, 2000; De Luca *et al.*, 2004). These new approaches are less influenced by environmental parameters and more representative of nematode genotype; however, for maximum usefulness the development of these new approaches should be integrated with morphological and host range data. This chapter summarises all biochemical and molecular technologies applied in the identification of *Pratylenchus* spp.

Biochemical identification

Biochemical techniques for protein analysis have been used to separate *Pratylenchus* spp. Isoelectric focusing electrophoresis (IEF) was shown to be a useful technique for studying protein polymorphism and genetic diversity among *Pratylenchus* populations (Payan & Dickson, 1990). IEF was effective enough to detect enzyme activity of malate dehydrogenase (MDH), phosphoglucomutase (PGM) and phosphoglucose isomerase (PGI) from five populations of *P. brachyurus* and one population of *P. scribneri* (*ca* 250 nematodes per population were used in each run) (Payan & Dickson, 1990). However, three distinct phenotypic groups were found in the MDH and PGM systems for *P. brachyurus* populations although only a single electromorph was detected for PGI. Multiple electromorphs for MDH, PGM and PGI were detected for *P. scribneri* and there was no similarity among these patterns with those from *P. brachyurus*. No phenotypic differences in PGI were observed between females and mixed juveniles of one population of *P. brachyurus*. Payan and Dickson (1990) did not find MDH activity in extracts of *P. brachyurus* and *P. scribneri* separated by IEF and exhaustive attempts failed to detect esterase (EST) in any *P. brachyurus* population or *P. scribneri*.

Protein patterns and isozyme phenotypes of EST, glucose-6-phosphate dehydrogenase (G6-PDH), isocitrate dehydrogenase (IDH), PGI and PGM were analysed to discriminate between six *Pratylenchus* species indigenous to Great Britain, *viz.*, *P. crenatus*, *P. fallax*, *P. neglectus*, *P. penetrans*, *P. pinguicaudatus* and *P. thornei* (Ibrahim *et al.*, 1995), but a great number of specimens was needed for each run (1000 nematodes per species). Although in that study only one population per species was compared for interspecific differentiation, banding patterns of IDH, PGM and G6-PDH revealed six distinct phenotypes that were useful for differentiating the suite of species studied. Analysis of PGI yielded the greatest number of bands although many were very close to each other, thus making it difficult to determine which were diagnostic for a given species.

Jaumot *et al.* (1997) determined the reliability and usefulness of total protein patterns for diagnosis of five *Pratylenchus* species (five populations of *P. vulnus*, four populations of *P. goodeyi* and one each of *P. scribneri*, *P. coffeae* and *P. thornei*) common in tropical and subtropical environments using sodium dodecyl sulphate (SDS) polyacrylamide gel electrophoresis (PAGE). The SDS-PAGE technique yielded protein markers that resulting in at least one to three distinct protein bands for differentiating *P. vulnus*, *P. goodeyi*, *P. scribneri*, *P. coffeae* and *P. thornei*. Nevertheless, a larger number of populations of the last three species would have been desirable to confirm the consistency of SDS-PAGE for species differentiation. In addition, the sample used for analysis can be a limiting factor, since large numbers of nematodes (5000-10 000 nematodes of different life stages) are required to accumulate sufficient total proteins to perform electrophoresis.

Similarly Andrés *et al.* (2000) found species-specific enzyme markers that allowed the identification of nine *Pratylenchus* species, determined the variation between populations of the same species from different geographical origins and, when possible, established species relatedness. The isozyme phenotypes obtained by isoelectrofocusing were successful in specific differentiation of *P. vulnus*, *P. goodeyi*, *P. thornei*, *P.*

scribneri, *P. penetrans* and *P. neglectus* populations from different geographic and host origins. Analysis of PGM was effective for identifying seven *Pratylenchus* species, comprising 33 populations. Although PGM did not distinguish two populations of *P. coffeae* from the single population of *P. neglectus*, this enzyme was particularly useful for differentiating *P. vulnus* (ten populations) and *P. goodeyi* (ten populations). The MDH system yielded a great number of bands, especially for phenotype patterns in *P. vulnus*, *P. thornei* and *P. scribneri*, where many bands were intense and close to each other, making it difficult to analyse intraspecific variability. Uehara (2003) separated six species of *Pratylenchus* indigenous to Japan (*P. penetrans*, *P. coffeae*, *P. vulnus*, *P. loosi*, *P. brachyurus* and *P. zeae*) by isoenzyme patterns. Clear signals were obtained from MDH and PGI. The results demonstrated that isoenzyme phenotypes were useful for supplementing the morphological characterisation of these species.

In conclusion, because of the large number of nematodes required for analyses, methodologies using protein patterns and isozyme phenotypes present a major difficulty for species diagnostics in *Pratylenchus*. In addition, for some techniques it may also be possible that the protein examined may not be present in all life stages of *Pratylenchus* and that expression of a protein may be influenced by unforeseen environmental factors that may cast doubt on the accuracy of the system (Jones *et al.*, 1997).

Molecular identification

DNA-based techniques have been used to differentiate species of *Pratylenchus* since the early 1990s. Random amplified polymorphic DNA (RAPD) PCR analysis showed its potential usefulness in distinguish seven isolates of *P. vulnus* and one isolate of *P. neglectus*, the patterns of amplified DNA bands of the *P. neglectus* isolate being clearly different from those of the *P. vulnus* isolates (Pinochet *et al.*, 1994). However, the study presented a high level of polymorphism in *P. vulnus* populations.

Comparative analysis of regions of ribosomal DNA (rDNA) is a useful tool for *Pratylenchus* species identification. The eukaryotic rDNA repeat consists of three genes (18S, 5.8S and 28S), which are separated by two internal transcribed spacers (ITS1, ITS2); and each repeat is separated by external transcribed spacer (ETS) and by an intergenic spacer (IGS) (Jones *et al.*, 1997). The sequences of the rDNA genes are highly conserved, whereas there is less conservation within the ITS regions and little homology is found in the non-transcribed spacer regions. The more conserved sequences are most useful for classification at higher taxonomic levels (genus to phylum), whereas the ITS sequences are useful at species levels (Hyman & Powers, 1991). Nonetheless, a remarkable ITS size difference exists within the genus *Pratylenchus*; approximately 350 bp separate the smallest and largest amplified ITS region among the species. In addition, intraspecific ITS variation has been observed in *Pratylenchus*, particularly within isolates of *P. coffeae* and *P. vulnus* (Orui, 1996; Uehara *et al.*, 1998b; Waeyenberge *et al.*, 2000; Mizukubo *et al.*, 2003).

Analysis of restriction fragment length polymorphisms (RFLP) in ITS of rDNA amplified by polymerase chain reaction (PCR) was a reliable and precise method for differentiating the main *Pratylenchus* species in Japan: *P. penetrans*, *P. coffeae*, *P. vulnus* and *Pratylenchus* sp. (unidentified species near *P. coffeae*) (Orui, 1996). The DNA extraction method was applied on single specimens of all developmental stages of each species. These four species were easily discriminated from each other by digesting the amplified products with endonucleases, *Alu*I, *Dde*I, *Hha*I, *Hinf*I and *Taq*I. Orui (1996) also reported evidence that different ITS sequences were present within single individuals of *P. vulnus*. Specific primers for *P. penetrans* and *P. scribneri* were developed using universal primers in conjunction with PCR to amplify equivalent fragments of the major sperm protein (*msp*) gene from any nematode (Setterquist *et al.*, 1996). Also, Uehara *et al.* (1998b) indicated that the 5.8s gene sequences were conserved whereas ITS sequences were divergent. Sequence differences in the ITS were used to synthesise specific primer sets (Table 4) *Pc*1 (selected from ITS1 of *P. coffeae*), *Pc*2 (selected from ITS2 of *P. coffeae*), *Pl*1 (selected from ITS1 of *P. loosi*) and *Pl*2 (selected from ITS2 of *P. loosi*). These specific-primers were able to distinguish between *P. coffee* and *P. loosi*, the same amplification products being obtained using DNA extracted from single females, males or juveniles. Uehara *et al.* (1998a) developed species-specific primers (*Pp*1 and *Pp*2) based on the sequences of the internal transcribed spacer regions (ITS1 and ITS2) of the rDNA in order to identify *P. penetrans* (Table 4) *Pp*1 (selected from ITS1 of *P. penetrans*) and *Pp*2 (selected from ITS2 of *P. penetrans*). The purified DNA from

Table 4. *Forward species-specific primers designed from ITS1 and ITS2 rDNA regions of* Pratylenchus *species.*

Pratylenchus spp.	Primer code	Primer sequence	rDNA region	Reference
P. coffeae	*Pc*1	5′-ATGCGCACATTGCATTCAGC-3′	ITS1	Uehara *et al.* (1998b)
P. coffeae	*Pc*2	5′-GAGCGAGAAACACCTCTCAC-3′	ITS2	Uehara *et al.* (1998b)
P. loosi	*Pl*1	5′-CAGTCAGCTAGCTGCTGGAT-3′	ITS1	Uehara *et al.* (1998b)
P. loosi	*Pl*2	5′-ATGAGAGCATAGTCGCTGTG-3′	ITS2	Uehara *et al.* (1998b)
P. penetrans	*Pp*1	5′-ATGATGGAAGTGTCCGCCT-3′	ITS1	Uehara *et al.* (1998a)
P. penetrans	*Pp*2	5′-CCAACGACGGTCAAAAGG-3′	ITS2	Uehara *et al.* (1998a)

seven *Pratylenchus* species was subjected to PCR using these primers, the 462 bp fragment only being generated for *P. penetrans*. In addition, the same fragment was obtained with DNA from the crude lysate of single juvenile and of adult (male and female) *P. penetrans*. Later, Uehara *et al.* (1999) utilised the reverse dot-blot hybridisation, using specific oligonucleotides designed from the sequence of the ITS region, for the simultaneous identification of several *Pratylenchus* spp. The amplified fragments were hybridised with the membrane-immobilised oligonucleotide and the hybridisation was detected non-radioactively. All the *Pratylenchus* species targets in the study (*P. penetrans*, *P. coffeae*, *P. vulnus*, *P. loosi*, *P. brachyurus*, *P. crenatus* and *P. zeae*) were identified in a single hybridisation by the reverse dot-blot assay and cross hybridisation was not observed. One of the main advantages of the reverse dot-blot technique is the potential for the integration of results from several other studies into a single assay.

Subsequently, Waeyenberge *et al.* (2000) demonstrated the usefulness of ITS-RFLP in distinguishing the following *Pratylenchus* species: *P. bolivianus*, *P. brachyurus*, *P. coffeae*, *P. crenatus*, *P. fallax*, *P. goodeyi*, *P. loosi*, *P. mediterraneus*, *P. neglectus*, *P. penetrans*, *P. pratensis*, *P. pseudocoffeae*, *P. scribneri*, *P. subranjani*, *P. thornei*, *P. vulnus* and *P. zeae*; and in revealing intraspecific variation within *P. coffeae*. PCR amplified the ITS regions from all species and populations examined and revealed large differences in length, ranging in size from approximately 900-1250 bp. The rDNA fragments were digested with five restriction enzymes (*Cfo*I, *Dde*I, *Hin*dIII, *Hpa*II and *Pst*I). All tested *Pratylenchus* species could be differentiated from each other by a combination of at least two enzymes; *e.g.*, *Cfo*I differentiated all nematode species with the exception of *P. fallax*, *P. penetrans* and *P. pseudocoffeae*. *Pratylenchus fallax* was further separated by a *Dde*I restriction and *P. pseudocoffeae* by a *Pst*I digestion. Similarly, upon *Cfo*I, *Dde*I, *Hin*dIII, or *Hpa*II digestion, it was possible to separate the three *P. coffeae* populations studied from each other. However, the ITS-RFLP of some populations of *Pratylenchus* may be identical as was demonstrated for *P. dunensis* and *P. penetrans* (Waeyenberge *et al.*, 2000; de la Peña *et al.*, 2006a).

Multiplex PCR reactions use several primers simultaneously in a single PCR reaction to test for the presence of target nematodes, although this technique is limited by the number of primers that can be used in a single reaction. Saeki *et al.* (2003) detected *Meloidogyne incognita* (Kofoid & White, 1919) Chitwood, 1949 and *P. coffeae* using specific primer sets (*Mi*1-*Mi*2 and *Pc*1-*Pc*2) designated based on the sequence of the ITS region of *M. incognita* and *P. coffeae*. Sequences of the specific primers of *P. coffeae* were designed based on the report by Uehara *et al.* (1998b) and the sequence of *M. incognita* obtained by Saeki *et al.* (2003). The multiplex PCR test enabled the detection of two distinctive and specific bands derived from the two plant-parasitic species. Furthermore, these two specific bands were detectable, when the genome of other nematodes was mixed in the template, with at least one nematode of *M. incognita* or *P. coffeae* being present with hundreds of other soil nematodes, *e.g.*, *Mononchus*, *Dorylaimus*, *Globodera*, *Helicotylenchus*, *Longidorus* and *Subanguina* (Saeki *et al.*, 2003).

Molecular approaches to species-specific identification based on PCR amplification are frequently applied in diagnostic, ecological, and phylogenetic studies and make use of universal-primer sequences during the development of the assays or even in routine applications. However, this practice, particularly with field samples, can lead to erroneous conclusions, because samples may be contaminated with a wide range of diverse DNA sequences (Volossiouk *et al.*, 2003). The data presented by Volossiouk *et al.* (2003) pointed out that subsequent sequence analysis of *P. penetrans*, *P. neglectus* and *P. crenatus* revealed several examples of contaminating DNA. Some reactions with universal primers did amplify rDNA sequences from the target nematodes but other reactions preferentially amplified rDNA sequences from the few contaminating organisms associated with the nematode samples. These results illustrate the fact that different samples, obtained from various localities, different laboratories, or at different times, that produce similar test results are no guarantee of accuracy since *Pratylenchus* spp. may be associated with specific microfloral and microfaunal populations. Thus, DNA sequencing provides a valuable check which may eliminate false positives of this type (Volossiouk *et al.*, 2003).

More recently, De Luca *et al.* (2004) carried out a sequence analysis of a specific portion of DNA, the D3 region of the 26S gene, in several *Pratylenchus* species (five populations of *P. thornei*, three populations of *P. neglectus* and one each of *P. mediterraneus*, *P. penetrans*, *P. pinguicaudatus* and *P. vulnus*) and compared these with similar sequences available in databases. The results suggested the hypothesis that *P. penetrans* may represent a species complex, whilst in *P. neglectus* the intra-species heterogeneity observed is due to intra-individual variability. Furthermore, the specific conservation of some nucleotides in different *P. thornei* populations indicates their stability in the rDNA repeats in this species. The presence of these nucleotides, the molecular signature of *P. thornei*, may be useful in determining the nature of nematode infections (De Luca *et al.*, 2004).

Similarly, Al-Banna *et al.* (2004) designed five forward species-specific primers from the internal variable portion of the D3 expansion region of the 26S rDNA in several *Pratylenchus* species, each being used with the single common reverse primer D3B (5′-TCGGAAGGAACCAGCTACTA-3′). In this study, they used two populations of the following species: *P. brachyurus*, *P. neglectus*, *P. penetrans*, *P. scribneri*, *P. thornei* and *P. vulnus*. Every one of the optimised specific primers, each of which was 20 nucleotides in length (Table 5), amplified a unique PCR product from its respective target *Pratylenchus* species and did not produce an amplicon from other *Pratylenchus*. The method was tested by using single females of *P. penetrans*, *P. vulnus* and *P. neglectus*. Results confirmed the overall reliability of the PCR for a given specimen, although this should be validated on a wide collection of field isolates. This method generates a 'yes or no' result, obviates the need for subsequent RFLP or sequence analysis of the PCR product, and can be used as a rapid diagnostic tool in epidemiological and management studies. In fact, the D2D3 LSU sequences have recently been used

Table 5. *Forward species-specific primers designed from the DNA sequence of the D3 expansion region of the 26S rDNA of* Pratylenchus *species (Al-Banna* et al., *2004).*

Pratylenchus spp.	Primer code	Primer sequence	Annealing temperature (°C)	PCR product size (bp)
P. neglectus	*PNEG*	5′-ATGAAAGTGAACATGTCCTC-3′	63	290
P. penetrans	*PPEN*	5′-TAAAGAATCCGCAAGGATAC-3′	62	278
P. scribneri	*PSCR*	5′-AAAGTGAACGTTTCCATTTC-3′	63	286
P. thornei	*PTHO*	5′-GAAAGTGAAGGTATCCCTCG-3′	68	288
P. vulnus	*PVUL*	5′-GAAAGTGAACGCATCCGCAA-3′	68	287

to differentiate between *P. dunensis* and *P. penetrans* (de la Peña *et al.*, 2006a).

Qiu *et al.* (2005) developed a multiplex PCR method with species-specific primers to identify *P. penetrans* and *P. vulnus* that allowed amplification of DNA from target species or separate species in a mixed population. The developed primers also amplified target DNA from soil extracts of field samples at levels as low as an individual nematode amongst other free-living or plant-parasitic nematodes.

More recently, Carrasco Ballesteros *et al.* (2007) provided two different DNA-based methods to identify *P. thornei*: the SCAR-PCR method and identification of a specific satDNA sequence. Both methods have the potential to be applied to routine diagnostic purposes using DNA extracted from different life cycle stages, soil samples or even infected plant material. Additionally, the satDNA sequence can be used as a sensitive and reliable probe to separate *P. thornei* from other closely related, particularly *P. mediterraneus*, populations (De Luca *et al.*, 2004).

An excellent example confirming that new diagnostic techniques are essential in diagnosis of *Pratylenchus* infestations, and which may have consequences for development of decision-making in the integrated management of *Pratylenchus* spp., was the identification of certain populations of *Pratylenchus* spp. that were found on weeds in Florida. For many years, these populations were considered to be *P. coffeae* and consequently the infested areas were not approved for citrus nurseries. However, the low incidence of *P. coffeae* in Florida citrus orchards and its presence on weeds suggested that these nematodes may comprise different races or species. This suspicion was confirmed when molecular and morphological analysis of these nematodes confirmed the identification of *P. loosi* and *P. pseudocoffeae* from the weeds. As neither species parasitises citrus (Inserra *et al.*, 1996, 1998, 2001), regulatory approval for citrus nursery sites was thereby facilitated.

Finally, the high throughput generation of expressed sequence tags (ESTs) from numerous nematode cDNA libraries is providing thousands of new gene sequences, and their availability in public databases will facilitate broad characterisation of their function. Mitreva *et al.* (2004) generated and analysed a total of 1928 5′ ESTs from a mixed life stages library of *P. penetrans* generated by splice-leader 1 (SL1) amplification. The ESTs were grouped into 420 clusters and classified by function using the Gene Ontology (GO) hierarchy and the Kyoto KEGG database. Approximately 80% of all translated clusters show homology to *Caenorhabditis elegans* proteins and 37% of the *C. elegans* gene homologues had confirmed phenotypes as assessed by RNA interference tests. Use of an SL1-PCR approach, while ensuring the cloning of the 5′ ends of mRNA, demonstrated bias toward short transcripts. Putative nematode-specific and *Pratylenchus*-specific genes were identified. Although this approach cannot provide an overview of the entire genome, it does allow a comparison of gene expression profiles among species and is amenable to large-scale analyses. Thus, the information developed with ESTs is also relevant in the development of novel strategies to generate plants that are resistant to *Pratylenchus* spp.

Nematology Monographs & Perspectives, 2007, Vol. 6, 305-324

Chapter 7

Biology and ecology of *Pratylenchus*

Pratylenchus spp. are migratory endoparasites that enter the host root for feeding and reproduction and move freely within the tissue. Consequently, they spend much of their life cycle in roots and are found in soil when the host plants are senescing, stressed or diseased, or when their hosts have been ploughed up after harvest (Stirling, 1991). *Pratylenchus* spp. do not become sedentary in the roots and feeding is restricted almost entirely to the root cortex. Like all plant-parasitic nematodes, *Pratylenchus* have six life stages, *i.e.*, egg, four juvenile stages and the adult. The duration of the *Pratylenchus* nematode life cycle lasts from 3-8 weeks, but this may be influenced by environmental conditions such as adequate temperature and moisture. After embryonic development within the egg to the first-stage juvenile (J1), the nematode moults to the second-stage juvenile (J2) which hatches from the egg. All juvenile and adult life stages of *Pratylenchus* are vermiform and mobile and all life stages (except for the egg and J1) can infect host plants. As reported in Chapter 4, adult males are common in some species yet rare in others and it is believed that *Pratylenchus* usually reproduce by parthenogenesis. Although *Pratylenchus* nematodes can invade plant tubers, rhizomes, pods and infrequently some above-ground plant structures, they are primarily parasites of roots.

Biology

EMBRYOGENESIS AND POST-EMBRYOGENESIS

Before egg laying, the vulva regularly opens and closes and the frequency increases as the onset of egg laying approached (Zunke, 1990a). The egg is rapidly expelled when half the egg has emerged, but it is not clear whether this is caused by muscular contraction of the body wall, internal hydrostatic pressure, or contraction of the vulval musculature (Zunke, 1990a). Embryogenesis in *P. thornei* has been recently studied in our laboratory from gravid females selected from carrot disc cultures (Castillo *et al.*, 1995b) and maintained in Petri dish culture to obtain freshly deposited eggs. These eggs were mounted in transparent water-agar under a cover slip at 24°C and microscopic observations made every 12 h. Eggs were laid at the one-celled stage (without divisions), the first division into two blastomeres was transverse to the longitudinal axis of the egg and completed 12 h after laying (Fig. 109). The second cleavage (three-celled stage) occurs in the anterior blastomere in the next 12 h (Fig. 109). The division of the other blastomere to arrive at the four-celled stage takes place in a period of 36 h after laying. In the next 24 h the 12-celled stage appears and the pregastrulation stage, which is maintained for the 5 days before the gastrula develops (Fig. 109). Gastrula develop to the J1 2 days later, the J1 having no internal differentiation and being coiled three or four times within the egg shell. Finally, some 2 days later, the first moult occurs to produce the J2 with internal differentiation in stylet, pharynx, intestine, *etc.* (Fig. 109). Thus, embryogenic development of *P. thornei* was completed in 10 days and was similar to that of *P. scribneri* (Román & Hirschmann, 1969b). When the J2 is ready to emerge from the egg it ceases movement and the stylet is thrust toward one pole of the egg so as to penetrate the egg shell. Although embryogenesis is favoured by optimal temperatures (around 20°C), Dunn (1972) showed that embryogenic development of *P. penetrans* also occurs at suboptimal temperatures during the winter period where temperatures ranged from 0-4°C.

Postembryogenic studies on *P. brachyurus* and *P. scribneri* by Román and Hirschmann (1969b) showed that prior to the second moult the active movement of J2 ceases, the body adopts a straight pose and this is followed by separation of the cuticle from the anterior end of the nematode. Formation of the new sclerotised parts started with the stylet, labial framework, the pharynx became visible again and the new cuticle of the third-stage juvenile (J3) separated completely from the old cuticle (Román & Hirschmann, 1969b). Third and fourth moults proceed similarly to the previous moult, except for changes in the development of the reproductive system (Román & Hirschmann, 1969b). The sex in *Pratylenchus* could be recognised early in the second moult by the presence of four specialised ventral chord nuclei opposite the genital primordium in female juveniles, such nuclei being absent in male

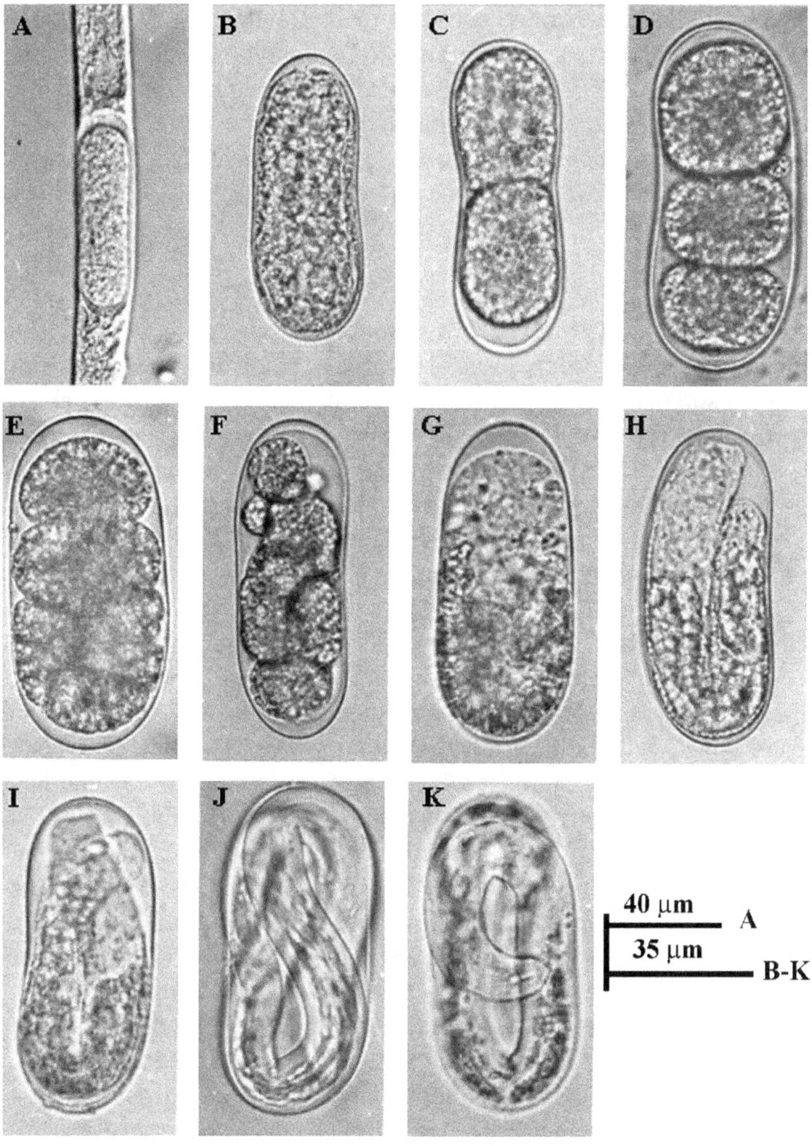

Fig. 109. *Embryogenic development in* Pratylenchus thornei. *A: Gravid female; B: Undivided egg; C: Two-celled stage; D: Three-celled stage; E: Six-celled stage; F: Pregastrula stage; G: Gastrula stage; H, I: First-stage juvenile; J, K: Second-stage juvenile.*

juveniles (Román & Hirschmann, 1969b). Although hatching of eggs in *Pratylenchus* spp. is not influenced by plant root exudates, De Waele *et al.* (1988b) demonstrated under glasshouse conditions that hatching in *P. brachyurus* and *P. zeae* was influenced by maize roots. Subsequently, detailed *in vitro* studies showed that exposure to the amino acid lysine, a substance secreted by maize roots, caused a significant increase in hatching of *P. zeae* over that in water (De Waele *et al.*, 1989b).

LIFE CYCLE

The life cycle of *Pratylenchus* is simple and direct (Fig. 110). The female lays its eggs singly or in small groups in the host root (Fig. 111) or in the soil near the root surface. Little information is available about the true length of *Pratylenchus* life-cycles under natural field conditions. However, on the basis of laboratory observations, life-cycle duration has been estimated in several nematode-host plant combinations. In red clover, *P. penetrans* completed a generation in 54-65 days and produced 16-35 eggs per female at a rate of 1-2 eggs per day (Turner & Chapman, 1972). The complete life cycle of *P. vulnus* was studied by Chitimbar and Raski (1985) on carrot discs under controlled conditions. Second-stage juveniles with four-celled developing gonads were first observed 9 days after inoculation. The moult from second- to third-stage juvenile occurred 11 days after inoculation. The third-stage juvenile was stouter and slightly longer than the second-stage juvenile and had a multicellular, oblong, developing gonad. The third moult began 14 days after inoculation and was completed by day 17. Sexes were first distinguishable by the length of the developing gonads. Fourth-stage juveniles with developing gonads occurred 17 days after inoculation. Fourth-stage males showed the earliest sign of moulting 17.5 days after inoculation. Moulting of fourth-stage juveniles was completed by the 18th day. Twenty-six days after inoculation, mature females had laid one or two eggs; by 28 days, eggs were being produced in abundance. Therefore, the life cycle of *P. vulnus* under controlled conditions took 3-4 weeks. No developmental stage may be called the infective stage because apparently both adults and juvenile stages may enter and leave the roots. Several studies have shown that all mobile life stages of *Pratylenchus* spp. are able to penetrate host roots (Townshend & Wolynetz, 1991; Castillo *et al.*, 1996b). Nevertheless, MacGuidwin (1989) determined that a greater percentage of second- and third-stage juveniles of *P.*

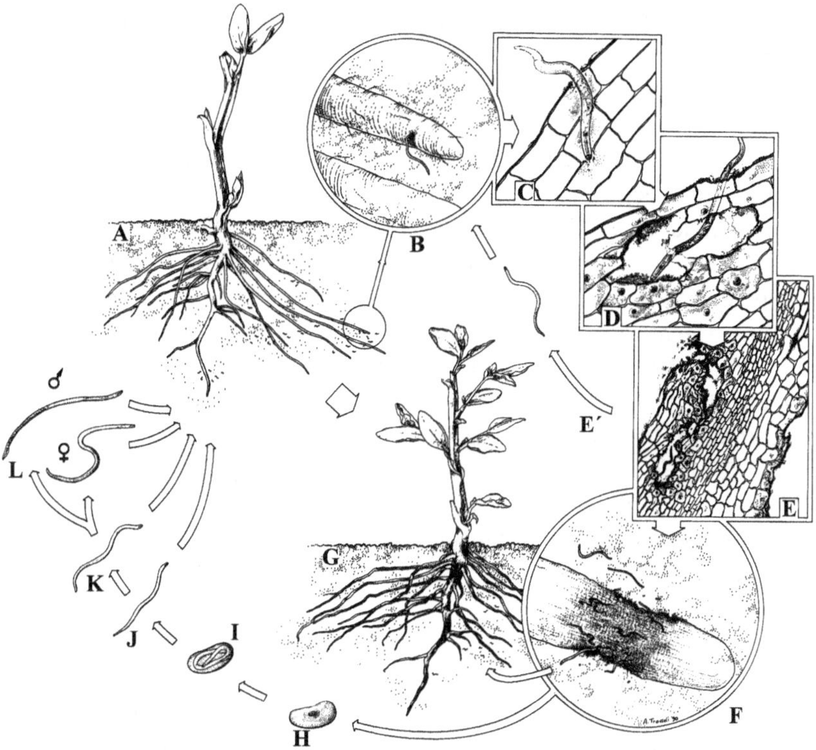

Fig. 110. *Schematic representation of the disease cycle of* Pratylenchus penetrans *on faba bean. A: Root system of healthy plant; B-D: Nematode penetration on roots; E, F: Cortical damage; G: Infected root system necrotic lesions; H, I: Eggs; J, K: Juvenile specimens; L: Adult specimens. After Vovlas and Troccoli (1990).*

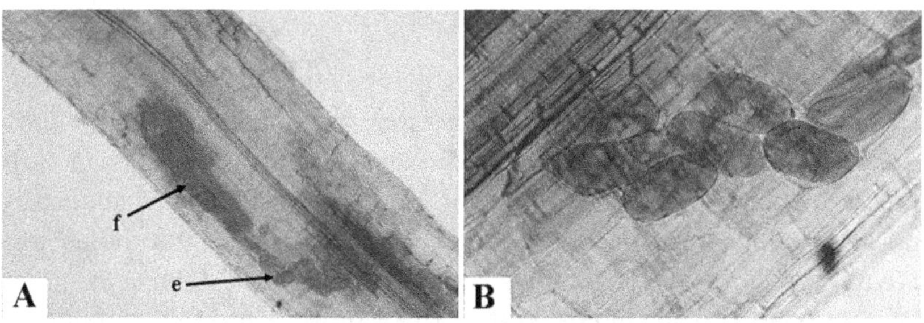

Fig. 111. *A, B: Roots of chickpea infected by* Pratylenchus thornei *showing a group of eggs (e) recently deposited by a female (f) inside the root tissue.*

scribneri were found in roots of potato than fourth-stage juveniles and adults.

The time required to complete the life cycle varies considerably depending on temperature and moisture. The generation time of *P. penetrans* was estimated as 46, 38, 28, 26 and 22 days at 17, 20, 25, 27 and 30°C, respectively (Mizukubo & Adachi, 1997). Juvenile mortality was higher at 17°C (50.4%), 20°C (50.3%) and 30°C (58.4%) than at 25°C (34.6%) and 27°C (37.6%). The developmental zero degrees (°C) and the effective accumulative temperatures (degree-days) required for hatching, female emergence and onset of oviposition (completion of one generation) of *P. penetrans* were estimated to be 2.7 and 200, 4.2 and 548 and 5.1 and 564, respectively (Mizukubo & Adachi, 1997). Similarly, the complete life cycle of *P. coffeae* cultivated at 30°C on carrot callus was 27-28 days, that of *P. penetrans* at 24°C was 34-35 days and that of *P. loosi* at 20°C was 45-46 days (Wu *et al.*, 2002). The life cycle of *P. goodeyi* on a susceptible banana (cv. Nakyetengu) was completed in 24 days at 24°C and in a resistant banana (cv. Sukalindizi) in 30 days; female fecundity was higher (30 eggs per female) in susceptible than in resistant banana (19 eggs per female) (Prasad *et al.*, 1999). *Pratylenchus mediterraneus* completes its life cycle in about 4 weeks on potato root tips (Orion *et al.*, 1995). Similarly, the life cycle of *P. thornei* was completed in about 25-35 days on carrot discs at 20-25°C (Castillo *et al.*, 1995a) and in 25-29 days on corn at 30°C (Siyanand *et al.*, 1982). *Pratylenchus zeae* completed the life cycle in about 5 weeks at 28°C under controlled conditions (Meyer, 1985) and *P. mulchandi* takes 24-36 days to complete its life cycle at 25-30°C (Nandakumar & Khera, 1974).

Epidemiology

Pratylenchus nematodes are mobile animals with the ability to move no more than 1-2 m from the root zone that they infect, although many agricultural operations favour their dispersal. Migration occurs in soil and about plants only when water is available, soil texture adequate and soil temperatures lie between the lower and upper thresholds for activity (Wallace, 1973). Studies on the migration of *P. penetrans* in different types of soils indicated that active movement of the nematode under optimum conditions was only 2 cm in 7 days (Townshend & Webber, 1971). Therefore, in fruit trees *Pratylenchus* nematodes may travel from plant to plant through roots. In most cases, the movement of *Pratylenchus* is defined as 'contagious' as small foci of infested areas gradually enlarge to encompass significant areas of disease. A good example of aggregated distribution has been recently reported with *P. coffeae* infecting coffee trees in South America (Herve *et al.*, 2005). This report indicated that aggregation intensity appeared to be linked to the size of the population considered (aggregation was greater when numbers of nematodes were lower). Areas of disease become more pronounced in adverse environmental conditions such as water and nutrient stress, or if secondary pathogens simultaneously infect the roots. However, this lack of long distance movement does not mean *Pratylenchus* cannot rapidly spread from field to field by other means. Farm equipment and even muddy shoes contaminated with *Pratylenchus*-infested soil can rapidly disperse the nematodes and the movement of water during floods and irrigation can disperse *Pratylenchus* over long distances. Morgan *et al.* (2002b) by means of geostatistics (variography) demonstrated that the spatial distribution of *P. penetrans* was associated with farm machinery entry locations and suggested farm machinery may contribute to *P. penetrans* dispersal and reintroduction in fields.

Likewise the movement of nematode-infected plants can give *Pratylenchus* international tickets to travel the world if plant quarantine officials are not vigilant. Essentially any process that moves soil or plant tissue, especially with perennial plants and annuals grown as transplants, has the ability to disperse *Pratylenchus*, making them difficult plant pathogens to quarantine. The movement of infected plants or propagative parts has been responsible for disseminating *Pratylenchus* nematodes long distances, even intercontinentally. In addition, since many species of *Pratylenchus* are endemic to native vegetation in many locations, new planting sites may already be infested with the nematode prior to cultivation and this may be one of the main reasons why several species are now cosmopolitan.

Flooding is also an effective and potent means of *Pratylenchus* dispersal. Evidence of this was found in a brief test on flooded water at the Guadalquivir River in Córdoba, southern Spain (Fig. 112), which yielded a large quantity and variety of plant-parasitic nematodes, including *P. neglectus*, *P. thornei* and *P. vulnus* (Castillo & Vovlas, unpubl. data). In fact, the soil substrates used in olive nurseries in southern Spain from diluvial sandy soil were found to be infested by

Fig. 112. *A general view of the Guadalquivir River at Córdoba (southern Spain) as an effective means of* Pratylenchus *nematode dispersal.*

several plant-parasitic nematodes, including *P. penetrans* and *P. vulnus* (Nico *et al.*, 2002).

Pratylenchus species develop several generations per growing season (each generation taking about 1 month) and, although large numbers of specimens can be detected from infected roots at early growing season, nematode soil populations are rather low, especially in the absence of a host crop (Loof, 1991). When rains occur early in autumn, the nematodes may complete one or more generations. In the absence of a host crop, *Pratylenchus* spp. survive on weeds or in the soil as eggs, juveniles or adults by mean of several survival strategies. It is well known that weeds play a significant role in the survival of many *Pratylenchus* species. Several weeds have been reported as good hosts of *Pratylenchus* spp. and help in the maintenance, multiplication and spread of the nematode within a field, thereby resulting in greater damage to susceptible crops. Winoto Suatmadji (1988) reported several weed species as good, intermediate and poor hosts for *P. penetrans* in a sandy soil at Keysborough, Victoria. Anwar *et al.* (1992b) found that several weed species (*e.g.*, *Amaranthus blitoides* S. Wats., *Chenopodium album* L., *Portulaca quadrafila* L.) were alternative hosts of *P. brachyurus*, *P. penetrans* and *P. zeae* to their corresponding host crops (peanut, cotton and corn, respectively). Similarly, Mani and Al-Hinai (2003) reported that *Cyperus* sp., *Echinochloa colona* (L.) Link, *P. quadrafila* and *Setaria intermedia* Roem. & Schult. acted as reservoirs of *P. scribneri* populations in alfalfa fields at Oman, suggesting a need for implementing efficient weed management practices to reduce the severity of nematode damage. Similarly, Kornobis and Wolny (1997) reported several weed species belonging to Polygonaceae, Chenopodiaceae, Caryophillaceae, Cruciferae, Compositae, *etc.*, as good hosts of *P. crenatus*, *P. flakkensis*, *P. neglectus*, *P. penetrans* and *P. pratensis*. Similarly, Vanstone and Russ (2001a, b) evaluated several graminaceous and dicotyledonous weeds common in crops, pastures, rotation and fallows in southern Australia and found that *Malva parviflora* L., *Rumex crispus* L., *Emex australis* Steinh., *Tribulis terrestris* L., *Brassica tournefortii* Gouan and *Raphanus raphanistrum* L. were good hosts of *P. neglectus*; *Hordeum glaucum* Steudel, *Hordeum leporinum* Link, *Lolium rigidum* Guadin and *Bromus diandrus* Roth were non-hosts; *Vulpia bromoides* (L.) Gray and *V. fasciculata* (Forssk.) Gray, *L. rigidum* and *A. fatua* were resistant to *P. thornei*; and *Heliotropium europaeum* L. was a non-host of *P. thornei*.

Tixier *et al.* (2006) described a new model (SIMBA-NEM) to simulate the population dynamics of two major plant-parasitic nematode species of banana, namely *R. similis* and *P. coffeae*. SIMBA-NEM is able to predict long-term nematode population size, while taking interspecific competition into account, and may be used to optimise the effect of nematicide applications and for designing sustainable and more environmentally-friendly banana cropping systems.

Ecology

Soil texture

Soil texture is one of the factors most often reported as influencing distribution of *Pratylenchus* species. Some *Pratylenchus* species are associated with lighter sandy soil (with higher availability of oxygen) than with heavier soils, which may be related to nematode requirements (Wallace, 1973). In potato fields, variations in species distribution were associated with differences in soil types; *P. crenatus* was usually found in loams and silt loams, but *P. penetrans* was found more frequently in sandy soils (Florini *et al.*, 1987). Similarly, soil type significantly influenced infection and severity of root necrosis caused by *P. coffeae* in banana (Sundararaju & Jeyabaskaran, 2003). Maximum root-lesion severity and reproduction were observed in silt clay and alluvial soils whereas minimum levels were recorded in red soils with low soil porosity (Sundararaju & Jeyabaskaran, 2003). However, no correlation was detected between both soil types and nematode populations in soil or roots (Sundararaju & Jeyabaskaran, 2003). Similarly, Endo (1959) showed that sandy loam soils favoured infection and reproduction of *P. brachyurus* whilst clay loam soils were least suitable. Studies by Griffin (1996) demonstrated consistent effects of soil texture on the pathogenicity of *P. neglectus* on rangeland grasses. Soil texture favoured root growth but was detrimental to nematode reproduction and thus pathogenicity was lower in fine silt loam than in fine loam soils. Also, the incidence of *P. penetrans* was positively related to a combination of sand percentage and long-term average annual rainfall (Jordaan *et al.*, 1989).

SOIL TEMPERATURE

Pratylenchus nematodes are poikilothermic organisms and, consequently, temperature influences the rates of physiological processes, such as movement, growth and reproduction, sex determination, and relative abundance of food and expression of nematode damage to plants (Freckman & Caswell, 1985). The optimal temperature for population development may differ among *Pratylenchus* spp. Acosta and Malek (1979) showed that optimum temperature of *P. alleni*, *P. brachyurus*, *P. coffeae*, *P. neglectus*, *P. scribneri* and *P. zeae* on soybean was 30°C, while that for *P. penetrans* and *P. vulnus* was 25°C. The latter two were the only species that reproduced at the relatively low temperature of 15°C. Populations of all species increased to some extent over the range of 20-30°C, except those of *P. neglectus* at 20°C. Only *P. brachyurus*, *P. neglectus*, *P. scribneri* and *P. zeae* reproduced at 35°C. The optimal temperature for all but *P. alleni* and *P. penetrans* was well defined by substantially reduced population development at ±5°C of the optimum. Similar studies with *P. scribneri* indicated that the optimal temperature for this species is quite elevated (33-34°C) in several host plants such as sugar beet, Sudan grass, snap bean and tobacco (Thomason & O'Melia, 1962). *Praty-*

lenchus penetrans and *P. vulnus* showed similar requirements for relatively cool temperatures, with some degree of population increase at temperatures as low as 15°C but little development at 30°C and none at 35°C (Acosta & Malek, 1979). The lower threshold temperature for development of a population of *P. neglectus* on barley was 7.75°C, whilst temperatures in excess of 25°C were unfavourable for this population on barley (Umesh & Ferris, 1992). Also, Sher and Bell (1965) found that *P. vulnus* reproduced on rose throughout a range of 18-32°C but that populations increased most rapidly between 24-29°C. Dickerson (1979) reported that even though reproduction of *P. scribneri* and *P. alleni* was best at a soil temperature of 35°C on both tomatoes and soybeans, both species reproduced at relatively low temperatures on tomatoes, so that the temperature limitation depends on the nematode-host interaction and not solely on the nematode itself. *Pratylenchus penetrans* may remain active in roots at low soil temperatures (7-13°C) and continue to cause injury to onion seedlings in late autumn and early spring (Ferris, 1970). Similarly, soil populations of *P. brachyurus* decrease during winter as a result of the low temperatures (5-8°C) and are maintained at higher (20-27°C) temperatures (Koen, 1967). Recently, Van den Bergh *et al.* (2006a) demonstrated that reproduction of *P. coffeae* was higher during the summer than during the cool winter under screenhouse and field conditions.

Nevertheless, low temperatures may be lethal to *Pratylenchus* species and have been used to reduce populations of *P. penetrans* in tubers of potato after storage at low temperatures (Olthof & Yu, 1999). Numbers of *P. penetrans* in potato tubers held at 1-5°C for 8 months decreased to nearly zero, regardless of the suitability of cultivar. However, numbers of *P. penetrans* in potato tubers were constant at 7°C for 19 weeks, but declined rapidly below 5°C (Olthof & Wolynetz, 1991). Also, populations of *P. brachyurus* in potato tubers were also markedly reduced after 15 weeks of storage at 5°C (Koen & Hogewind, 1967). Similarly, under controlled conditions, mortality of *P. penetrans* at −4°C was about 90% after 4 weeks and about 92% at −12°C after 4 h of exposure (Kimpinski & Dunn, 1985). Similar results were found for *P. penetrans* and *P. neglectus* by Townshend (1973) who reported that survival of both species at a subzero temperature (−2°C) was negligible.

SOIL MOISTURE

Soil moisture is necessary for many of the life processes of *Pratylenchus* and thus is doubtless amongst the more important factors governing nematode populations. Studies on the relationships between soil moisture and various nematode activities have usually indicated that 70-80% of soil field capacity provides optimum conditions (Wallace, 1973). Furthermore, it has been suggested that seasonal fluctuations in nematode population densities largely result from variations in soil moisture (Wallace, 1973). Soil moisture, as determined by rainfall, significantly influenced populations of *P. neglectus* in wheat (Kimpinski *et al.*, 1976). The number of nematodes was highest at early spring when soil moisture was high and the lowest populations were in mid-summer when soil moisture was low (Kimpinski *et al.*, 1976). Kable and Mai (1968) also reported that reproduction, invasion into alfalfa roots and survival of *P. penetrans* were greatly suppressed under very low or very high soil moisture conditions. However, the influence of soil moisture on population density of *P. penetrans* differed for different types of soil; *e.g.*, in clay loam soil there was no increase of nematode population with increasing soil moisture (Kable & Mai, 1968). Another positive relationship was found between the incidence of *P. brachyurus* on corn at South Africa and long-term average annual rainfall (Jordaan *et al.*, 1989). Soil moisture may also influence the survival of several stages of the *Pratylenchus* life cycle; *e.g.*, survival of *P. penetrans* in soils with saturated moisture was poor between 12-30°C, but improved in drier soils (Kable & Mai, 1968). These findings may be due to differences in osmotic potential as in wet soils nematodes are immersed in water and must maintain an osmotic balance through energy-utilising metabolism, but in dry soils the osmotic potential may be close to that of the nematode (Wallace, 1973). Norton and Burns (1971) reported that populations of *P. alleni* on soybean roots were higher under dry (field capacity to temporary wilting) than under wet (field capacity or above) conditions. In addition, the proportion of males to females was closer to a 1 : 1 ratio in the dry than in the wet soil, indicating that dry soils favour either male production, inhibition of females or differential survival. Townshend (1973) reported that survival of *P. penetrans* and *P. neglectus* increased as soil moisture declined, probably because under these conditions nematodes have a lower rate of metabolism and deplete their food reserves more slowly. Influence of soil moisture may be differential in the life stages of *Praty-*

lenchus, *e.g.*, Townshend (1973) found that adults and fourth-stage juveniles of *P. penetrans* and *neglectus* showed higher survival percentages than those of third- and second-stage juveniles. Flooding has been shown to decrease populations of *P. coffeae* in banana fields, this effect probably being due to an oxygen deficit (Sarah *et al.*, 1983; Van den Bergh *et al.*, 2006b).

Soil moisture may also influence vertical migration or survival of *Pratylenchus* spp. Vertical migration of *P. scribneri* to deeper soil layers was observed when the top soil layer dried out (Mani, 1999). High numbers of nematodes survived at both 0-15 and 16-30 cm depths from September to November when there was high moisture in the soil (Mani, 1999). Similarly, cultural practices like ploughing may result in a rapid decline of *P. zeae* population, particularly at 0-15 cm depth, which might be correlated to the loss of moisture resulting in desiccation of the nematodes (Mani, 1999).

SOIL PH

Soil acidity is a major abiotic stress factor that may influence nematode development. *Pratylenchus* nematodes are adapted to a wide range of soil conditions and thus present many challenges for crop production at lower pH. In any case, different studies on this matter showed that the optimum pH range for *Pratylenchus* species varies with the species of host plant. An example of a remarkable tolerance to pH was shown by a *P. brachyurus* population, which was not affected by soil pH values 5.0-7.3, although pH 1.0 was lethal and at pH 3.0 only 39% of the specimens survived in comparison with 95% in the control (Koen, 1967). Also, populations of *P. brachyurus* in pineapple decreased as soil pH increased, the lowest populations being at pH 6.0 (Osseni *et al.*, 1997). Similarly, Shukla *et al.* (1998) found that reproduction of *P. thornei* on mint (*Mentha piperita* Huds.) was higher at pH 6.0 than at pH 3.0 or 9.0, with the highest reduction being at pH 9.0. Brzeski (1969) found that on cabbage *P. crenatus* occurred more frequently in acid than in neutral or alkaline soils.

Studies by Morgan and MacLean (1968) and Willis (1972) in vetch and alfalfa, respectively, indicated that *P. penetrans* does best in the range of pH 5.2-6.4, with reproduction dropping as the pH approaches 7. However, Myers (1979), working with tomatoes, recovered more nematodes at pH 6 than at pH 5. Similarly, Kimpinski and Willis (1981) reported that the main response to higher soil pH levels in alfalfa was a sharp drop in the numbers of *P. crenatus* and a significant increase in the numbers of *P. penetrans*; however, in timothy (*Phleum pratense*) the numbers of both nematode species were reduced significantly as soil pH was increased. Also, population densities of *P. penetrans* in roots of sweet cherry declined significantly with decreasing soil pH, with or without significant reduction in plant growth, indicating a direct effect of pH on the nematode (Melakeberhan *et al.*, 2000). Similarly, *P. penetrans* prevailed in soils with a high pH and *P. crenatus* in soils with a low pH (Pelsmaeker & Coomans, 1987), and *P. alleni* was recovered in greater numbers from soybean roots grown in soil pH 6.0 than those growing at pH 4.0 or 8.0 (Burns, 1971). In other cases, *Pratylenchus* species showed a narrow range of pH, *e.g.*, *P. zeae* was also negatively correlated with soil pH between 6.5-7.5 in sorghum fields (Trevathan *et al.*, 1985).

Seasonal fluctuations

Models to predict the population dynamics of a nematode species have been developed, although problems with nematode identification, compounded by the irregular distribution of nematodes in soil, have made it difficult to obtain reliable data on nematode distribution and abundance to refine these models. Population densities of *Pratylenchus* spp. usually decline between growing seasons in regions with harsh winters. In these areas decisions for managing *Pratylenchus* spp. are generally made in the autumn. Consequently, information on the rate and degree to which populations of *Pratylenchus* spp. decline during the winter months can be used to predict numbers of nematodes the following spring. The winter survival of *Pratylenchus* spp. varies between different crops or between cropped and fallow land (Olthof, 1971). The winter decline in population densities of *P. scribneri* associated with corn ranged from 50-63% (MacGuidwin & Forge, 1991); similar reductions (43-68%) were reported for *P. penetrans* associated with red clover (Kimpinski & Dunn, 1985) and for *P. penetrans* associated with rye (40-65%) in Canada (Olthof, 1971). However, the winter survival of *P. scribneri* in potato was more variable, ranging from 15-84% and being lower than that estimated for corn (MacGuidwin & Forge, 1991). Kimpinski and Dunn (1985) showed that the rate of overwinter survival of *P. penetrans* under field conditions at northern USA averaged about 50% when mean minimum and maximum temperatures were about −1.1 and −0.8°C, respectively. Verschoor *et al.* (2001) reported that in the winter season most *P. crenatus* and *P. fallax* nematodes were found inside the roots of permanent grass where they might be less affected by the adverse winter conditions. Similar results were found by Nombela *et al.* (1993) in arable fields, *Pratylenchus* being found in the deeper soil layers during the winter. These authors suggested that *Pratylenchus* migrate to deeper soil layers to escape from low temperatures.

Seasonal fluctuations of *P. penetrans* during an 8-year period in British Columbia (Canada) under a perennial crop such as raspberry (*Rubus idaeus* L.) were studied by Vrain *et al.* (1997). They did not find seasonal patterns, although they observed several reductions of soil populations in late autumn and late spring when soil temperatures were conducive for infection, feeding and development. The lack of seasonal fluctuations suggests that under these conditions *P. penetrans* was in dynamic equilibrium with biotic and biotic factors. Similarly, Szczygiel and Hasior (1972) reported that *P. penetrans* densities under strawberries in Poland did not exhibit seasonal variation. However, LaMondia (2002) reported seasonal fluctuations of *P. penetrans* in strawberry plantations at Connecticut with populations peaking at about the end of May, with a subsequent decline in numbers that was suggested as corresponding to changes in total strawberry root weight and root type distribution. Such changes in nematode habitat result both from loss of roots due to disease and also from a natural change in root type from structural root to suberised perennial root (LaMondia, 2002). Similarly, *P. brachyurus* in millet soils exhibited seasonal population fluctuation with a decline during the growing season probably caused by the nematode leaving the soil and entering the roots. However, during the winter an increase in population occurred, probably as the nematodes were released, or migrated, from rotting roots.

Several investigations reported winter mortality of *Pratylenchus* spp. of approximately 50% under annual (MacGuidwin & Forge, 1991) and perennial (Kimpinski & Dunn, 1985) cropping systems in temperate climates. The highest mortality of *P. scribneri* occurs before the soil freezes, suggesting that a portion of the population may be susceptible to chilling (MacGuidwin & Forge, 1991), as demonstrated by Kimpinski and Dunn (1985) under laboratory conditions.

VERTICAL DISTRIBUTION

The vertical distribution of *Pratylenchus* in soil is highly variable and may be influenced by many biotic and abiotic factors. Root distribution, soil moisture, temperature, soil texture, rainfall and depth of subsoil greatly influence vertical distribution patterns. Wallace (1973) suggested that root distribution is the principal factor in the vertical distribution of plant-parasitic nematodes and that physical factors play an important secondary role. Knowledge of vertical distribution of *Pratylenchus* species in cultivated soils is of great importance as it shows how deeply nematicides, organic amendments, or biological control agents should be applied to obtain maximum effect and to what depth soil samples should be taken during diagnostic and faunistic studies.

The root abundance or root biomass largely determines the vertical distribution of *Pratylenchus* species in soil profile. Several studies have demonstrated that the vertical distribution of *Pratylenchus* species parallels that of host-root growth and distribution. Studies on *P. brachyurus* on corn in South Africa showed that the highest population density during summer was found in the upper 30 cm of soil, whilst during winter the highest population was found at a depth of 20-40 cm (Koen, 1967). This nematode was found at all depths (0-105 cm), but population densities were greatest at 45-75 cm where the soil was 78-79% sand, 6% silt and 15-16% clay (Brodie, 1976). Soil populations of *P. brachyurus* in sandy soils under soybean and corn were significantly higher at 0-30 cm depth and lower at 30-45 cm depth (McSorley & Dickson, 1990a, b). Szczygiel and Hasior (1972) studied the vertical distribution of *P. crenatus* in strawberries and found that as much as 70% of specimens between 0-62 cm deep were distributed in the upper layers of soil (up to 22 cm deep). Also, Ogiga and Estey (1973) found *Pratylenchus* spp. as deep as 80 cm, the highest densities being between 20-40 cm. Similarly, Sohlenius and Sandor (1987) measured population densities of *Pratylenchus* spp. under grass and barley and found the highest densities at depths of 20-30 cm. Vertical distribution (sampled to 60 cm deep) of *P. neglectus* and *P. thornei* in different soil types in South Australia was maximal (94%) in the top 20 cm (Taylor & Evans, 1998). Mani and Al-Hinai (2003) showed that *P. zeae* populations in alfalfa fields were significantly larger in the upper layers of soil (0-15 cm) than those in lower layers (15-20 cm depth), which is coincident with a higher fibrous root distribution at this depth. Populations of *P. zeae* in sugarcane fields in southern India were significantly higher in the upper layers of soil (20 cm) than at deeper levels (40 cm) (Mehta *et al.*, 1992). Also, the greatest densities of *P. goodeyi* were associated with the greater root biomass of banana in the upper layers (Kashaija *et al.*, 2004). Rössner (1972) found that vertical distribution of *P. penetrans* under controlled conditions may extend to 50 cm deep. Recently, the host root system was also clearly revealed as a limiting factor for the vertical distribution of *P. penetrans* in corn, black salsify, carrot and potato in Belgium (Pudasaini *et al.*, 2006a).

The vertical distribution of *Pratylenchus* spp. changes through time as a result of differential reproduction or mortality at particular depths (MacGuidwin & Stanger, 1991), or from directional movement induced by changes in environmental conditions or the presence of host roots. The factors responsible for these changes may be related to vertical migration in order to seek favourable conditions, or to resource allocation or edaphic conditions that differentially favour nematode reproduction at various depths. The vertical distribution of *P. penetrans* under raspberry changes through time but does not follow a seasonal pattern that is similar in successive years. The most notable changes in vertical distribution were reduced population densities near the surface (0-5 cm) and increased population densities at lower depths during winter, when the soil froze at the 0-5 cm depth (Forge *et al.*, 1998). Vertical distribution of *P. scribneri* was also influenced by season and host plant. The numbers of *P. scribneri* early and late in the season decreased with depth under corn and increased with depth to 30 cm under potato (MacGuidwin & Stanger, 1991). Few *P. scribneri* specimens occurred near the soil surface, despite an abundance of corn and potato roots, and few were found below 30 cm, despite temperature and moisture conditions conducive for nematode development and reproduction.

Survival strategies

Soil may appear to be a safe environment, but to a microscopic nematode it is a hostile world filled with danger. In order to survive, a nematode must be able to circumvent several obstacles such as competition with voracious predators, changes in soil temperature and moisture and the death of its host plant. *Pratylenchus* evade these biotic and abiotic obstacles by employing a combination of behavioural and physiological survival strategies. *Pratylenchus* species possess a relatively thick cuticle and, by living inside plant tissue, are provided with some protection from predation, although they are at some risk of death if their host plant succumbs to disease. Nevertheless, *Pratylenchus* survival is affected not only by biotic factors, but also by abiotic ones such as temperature and water availability. The onset of winter or the drying of the soil can be disastrous for *Pratylenchus*.

The egg stage is probably one of the most important stages able to survive adverse conditions for a sufficiently long time in the absence of host plants (Van Gundy, 1965). Eggs of *Pratylenchus* are covered by three layers: *i*) the external vitelline layer, which has a membrane-like structure secreted by the uterine wall; *ii*) a chitinous membrane, or true egg shell secreted by the egg itself; and *iii*) a vitelline layer or lipid membrane, which has a membrane-like structure beneath the chitinous membrane (Endo *et al.*, 1999). The chitinous membrane supplies the structural strength, impermeability to most substances except gases and protection of the lipid membrane beneath. The shell allows the eggs to resist drying for long periods and facilitates indirect dispersion to new environments.

Interestingly, many *Pratylenchus* species are well adapted to abiotic stress and are capable of cryptobiosis, *i.e.*, the ability to enter a state of suspended metabolic activity during unfavourable environmental conditions (drying, heat, cold). The ability of some *Pratylenchus* species to undergo anhydrobiosis is one of the reason these nematodes are very difficult to eradicate from an infested field. Anhydrobiosis is undergone by all life stages, including the egg, as demonstrated in *P. penetrans* by Townshend (1984b). Its ability to survive freezing as an anhydrobiote is likely due to the loss of body water. Anhydrobiosis increases the ability of *P. penetrans* to survive subzero temperatures as loss of moisture reduces ice needle formation, a process that is so destructive to cells (Sayre, 1964). However, anhydrobiosis does not increase the ability of *Pratylenchus* nematodes to survive temperatures above 40°C. Infectivity and reproduction of rehydrated specimens are not usually affected by anhydrobiosis (Townshend, 1984b).

Several studies under controlled conditions have demonstrated the anhydrobiotic state of some *Pratylenchus* species; *e.g.*, *P. brachyurus* survived 21 months in fallow glasshouse soil (Feldmesser *et al.*, 1960), *P. crenatus* also survived for 21 months in polyethylene bags (Harrison & Hooper, 1963) and *P. penetrans* survived 11 or 24 months under glasshouse conditions depending on soil type (Rössner, 1971;

Townshend, 1984b). Townshend (1984b) demonstrated that rates of soil and nematode dehydration are critical factors affecting the development of anhydrobiosis in *P. penetrans* and its ability to survive long periods without moisture. Bodies of anhydrobiotic nematodes are coiled (*e.g.*, *P. thornei*, Fig. 113), with coiling beginning at a much lower water potential in loam sandy soils than in silt loam soils. The number of anhydrobiotes increases greatly as soil moisture is depleted and soil moisture tension increases. The occurrence of anhydrobiotes of *P. penetrans* is greater in the surface layer of soil, where soil moisture tension is greater, than in the lower layers (Townshend, 1987). In the northern Negev region of Israel, *P. thornei* was able to survive the dry summer after a winter wheat harvest when all plants had become desiccated (Glazer & Orion, 1983). Desiccated nematodes could withstand temperatures up to 40°C. Reactivated individuals showed intestines apparently devoid of reserve materials. Glazer and Orion (1983) also demonstrated that infectivity rates of anhydrobiotic *P. thornei* were twice that of the fresh (*i.e.*, non-anhydrobiotic) population. Similarly, anhydrobiosis of *P. thornei* was observed in southern Spain, nematodes surviving the dry fallow season in soil or within the root residues in an anhydrobiotic state until reactivated by the autumn rains (Talavera & Vanstone, 2001). Moreover, survival of anhydrobiotic stages of *P. thornei* was significantly greater when nematodes were previously exposed to a good host crop such as wheat (Tobar *et al.*, 1996; Talavera & Valor, 2000). Similarly, Swanepoel and Loots (1988) demonstrated the anhydrobiotic stage in *P. zeae*. The number of *P. zeae* specimens that coiled and the number of specimens that survived 0% relative humidity for 25 h were fitted to quadratic regressions with optimal temperatures between 25-35 and 25-30°C, respectively (Swanepoel & Loots, 1988). Therefore, anhydrobiosis in some cultivated soils is one of the factors responsible for soil populations of *Pratylenchus* decreasing in summer under low soil moisture and increasing when soil moisture is high (Olthof, 1971; Townshend, 1987).

Under controlled conditions, Wergin *et al.* (2000) demonstrated that juveniles of *P. scribneri* were capable of supercooling to some degree and that they survived subzero temperatures only when no external ice nucleation was induced, *i.e.*, they remained in a supercooled liquid. However, exogenous ice nucleation of the liquid, which was triggered by introducing a commercial ice-nucleating activity agent, was lethal to 90% of the juveniles.

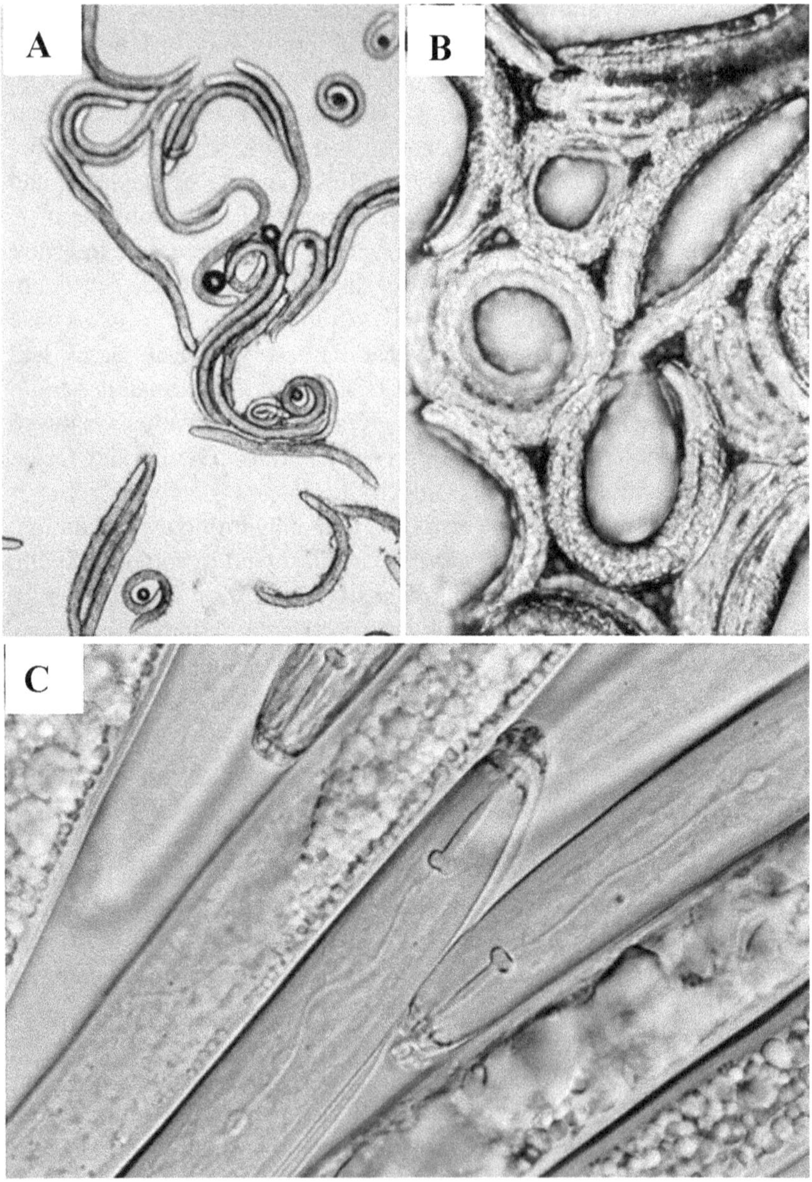

Fig. 113. *Photomicrographs of* Pratylenchus thornei *showing anhydrobiotic stages (A, B)* vs *hydrated and normal specimens (C).*

Nematology Monographs & Perspectives, 2007, Vol. 6, 325-354

Chapter 8

Pathogenicity of *Pratylenchus* species

Pathogenicity is usually used in plant pathology to express the capacity of an organism to induce disease or the amount of physiological damage caused to the host plant by the presence of a pathogen (Shaner *et al.*, 1992). Although controversy persists as to whether plant-parasitic nematodes should be considered as parasites or pathogens, in plant nematology the terminology used in plant pathology for fungi and bacteria is applied (Triantaphyllou, 1987). For that reason, the term 'pathogenicity' in plant nematology has an ambivalence that expresses on the one hand a qualitative concept (characterisation of the capacity of a given nematode species to establish a compatible host-parasite relationship) and, on the other, a quantitative concept (definition of differences in the expression of the damage caused in the plant).

The most obvious above-ground symptoms of root-lesion nematode disease on herbaceous plants are round to oval patches of stunted and chlorotic (yellowish) and woody plants, which give the field a ragged appearance (Fig. 114). Damage is often most severe in the centre of these areas, diminishing toward the edges to normal appearing plants. There is a reduction in leaf size and number of leaves produced on heavily infected plants and yields can be substantially reduced. Symptoms of *Pratylenchus* damage resemble those of other soil-borne diseases, nutrient deficiencies, insect damage, or cultural and/or environmentally induced stress. Nevertheless, damage caused by *Pratylenchus* species is frequently not obvious, so it is necessary to understand their biology, ecology and interaction with other microorganisms in order to determine their impact on crop yield. Thus, a soil analysis and/or diagnostic test is necessary to determine whether a *Pratylenchus* problem exists. In addition, *Pratylenchus* damage to herbaceous plants is greatly enhanced by stress from other factors. Disease symptoms are more pronounced and yield reduction greatest when abnormally high temperatures occur early in the growing season, rainfall is inadequate, soil fertility is low or imbalanced, and/or root-rot organisms attack the plants. However, the

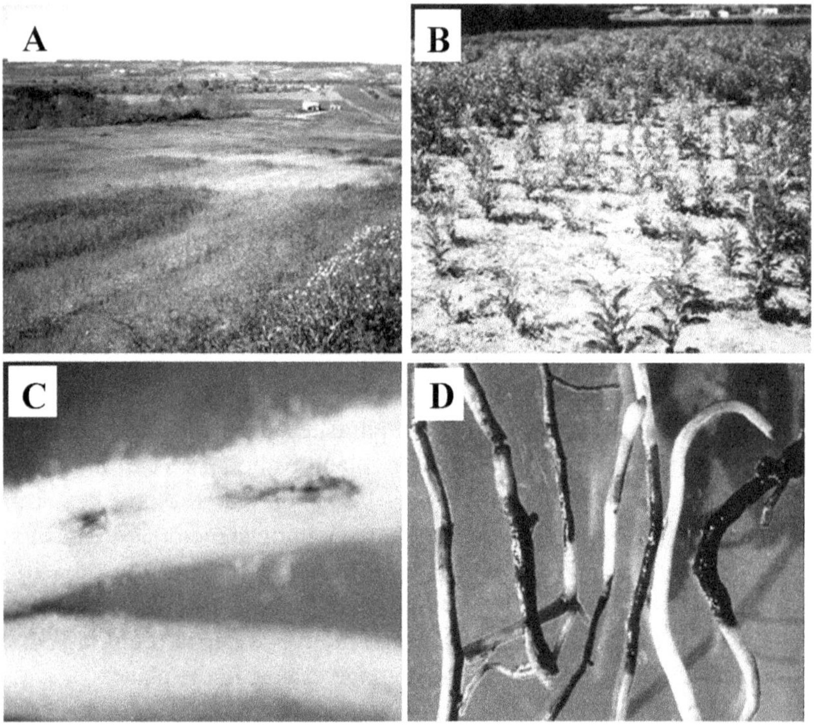

Fig. 114. *Field and root symptoms caused by* Pratylenchus *on herbaceous and fruit tree plants. A: Wheat field with large stunted areas caused by* Pratylenchus neglectus*; B: Nursery showing patch of stunted citrus seedlings infected by* Pratylenchus vulnus*; C: Necrotic root lesions on herbaceous plants; D: Necrotic root lesions on fruit tree.*

most common symptom on susceptible trees or bushes is a slow decline as the nematodes increase to very high numbers over the years (Fig. 114). The foliage of diseased trees may appear lighter green or chlorotic and the seasonal growth of infected plants is less than would be expected in a healthy plant. Flowers or fruit may be reduced substantially in number and quality. As the vigour of the host is reduced, the plant may be predisposed to winter injury or other infectious diseases. Replacing the affected plant without soil treatment frequently results in poor growth, and sometimes in death, of the new plant (Nyczepir & Becker, 1998).

In general, a genotype is considered tolerant if it grows and yields almost as well in the presence of nematode infection as it does in the absence of the nematode (Boerma & Hussey, 1992). Plants differ in their tolerance to *Pratylenchus* nematodes. Some plant species are damaged by a relatively low population, whilst others can support a large population without visible adverse effects. The threshold populations necessary to cause economic damage have been determined for some *Pratylenchus*-host plant combinations. Even so, the pathogenicity of *Pratylenchus* spp. to crops may be influenced by host plant nutrition and/or edaphic factors. Plant nutrition has been demonstrated to influence *Pratylenchus* populations in soil and roots, *e.g.*, numbers of *P. penetrans* in cherry rootstocks were significantly lower under conditions of deficient plant nutrition (Melakeberhan *et al.*, 1997). Increased nutrition generally increased plant growth and nutrient accumulation in plant tissue, thereby influencing nematode populations (Melakeberhan *et al.*, 1997). Nevertheless, although the physiological mechanisms by which increased nutrition influences host response to *Pratylenchus* are unknown, the data suggest that nutrition has a positive effect on host and a negative effect on nematodes (Melakeberhan, 2004). Conversely, parasitism by *Pratylenchus* spp. has been shown to reduce nutrient uptake in plants. Vaast *et al.* (1998) demonstrated that *P. coffeae* reduced NO_3^- uptake rate by 56% and NH_4^+ uptake rate by 24% in susceptible coffee without affecting the proportion of fine roots per root system. Such lower NH_4^+ and NO_3^- uptake rates in the presence of *P. coffeae* were consistent with visual symptoms and foliar analyses indicating N deficiency in long-term field experiments with nematode-infected coffee plants. *Pratylenchus coffeae* seemed directly to affect the root uptake capacity by damaging root cells during nematode penetration, exploration and feeding. Similarly, Mazzafera *et al.* (2004) suggested that parasitism of *P. coffeae* on coffee seedlings caused a rapid detrimental effect on carbon fixation and distribution of photoassimilates in the plant due to direct root damage. Nevertheless, in other plant-*Pratylenchus* interactions, foliar analysis did not detect nutrient deficiencies due to the pathogenic effect of *Pratylenchus* infections, *e.g.*, infections by *P. vulnus* in three rootstocks for peach and plum (Cadaman (*Prunus persica* × *P. davidiana*), Julior (*P. insititia* × *P. domestica*) and Ishtara (*P. belsiana* × *P. cerasifera*)) (Hernández Dorrego *et al.*, 1999).

Several edaphic factors are known to influence the pathogenicity of plant-parasitic nematodes to host plants, *e.g.*, soil temperature, texture, or moisture, all of which may affect population densities of nematodes (Wallace, 1983). Pathogenicity of *P. brachyurus* was greater at lower soil temperatures than at higher soil temperatures (Lindsey & Cairns,

1971) and sandy soils enhanced the damage potential of *P. brachyurus* on soybean (Schmitt & Barker, 1981). Also, soil moisture near field capacity increased pod infection of peanut by *P. brachyurus* (Good & Stansell, 1965). Nevertheless, other studies showed no influence of these factors in several host plant-*Pratylenchus* combinations, *e.g.*, the pathogenicity and reproduction of *P. neglectus* in rangeland grasses and those of *P. brachyurus* on corn and sorghum were not influenced by soil texture (Griffin, 1996) or soil moisture (McDonald & Van den Berg, 1993).

Pratylenchus spp. are obligate, soil-inhabiting parasites and, therefore, documenting their pathogenic capabilities experimentally involves long-term challenges. Acceptable monoxenic culture techniques were not possible until the 1950s work by Mountain (1960) with *P. penetrans* on peach and tobacco. However, several methods for inoculum production have been developed, the monoxenic culture in carrot discs being one of the most widely used techniques by plant nematologists. In this method, carrots are superficially disinfested with 95% alcohol and inoculated with surface-sterilised nematodes (Castillo *et al.*, 1995b). Studies in our laboratory with *P. thornei* indicated that conditions for achieving optimum reproduction were the following: *i*) inoculate 25-100 nematodes/disc; *ii*) incubate at 20-25°C; *iii*) incubation period of least 100 days after inoculation (Castillo *et al.*, 1995b). Under these conditions, a nematode population increase of 1254-3619 times the initial inoculum was obtained (Fig. 115). Also, monoxenic cultures in carrot discs provide standardised conditions for production of nematode inoculum, allowing comparisons of multiplication rates among nematode populations. Additionally, monoxenic carrot discs have been used successfully for culturing *Pratylenchus* spp. such as *P. brachyurus* (Moody *et al.*, 1973) and *P. vulnus* (Townson & Lear, 1982a). Excised corn roots have been used to rear *P. penetrans* (Tiner, 1960) and *P. zeae* (Jordaan & De Waele, 1988). Peng and Moens (1999) demonstrated that surface sterilisation with 0.1% malachite green alone (15 min) or with 0.5% streptomycin sulphate did not significantly reduce *P. penetrans* movement, or attraction to and penetration into *Rosa dumetorum* cv. Laxa seedlings, but cool storage (30 days at 4°C) reduced the survival of *P. penetrans* and its attraction to and penetration of rose seedlings.

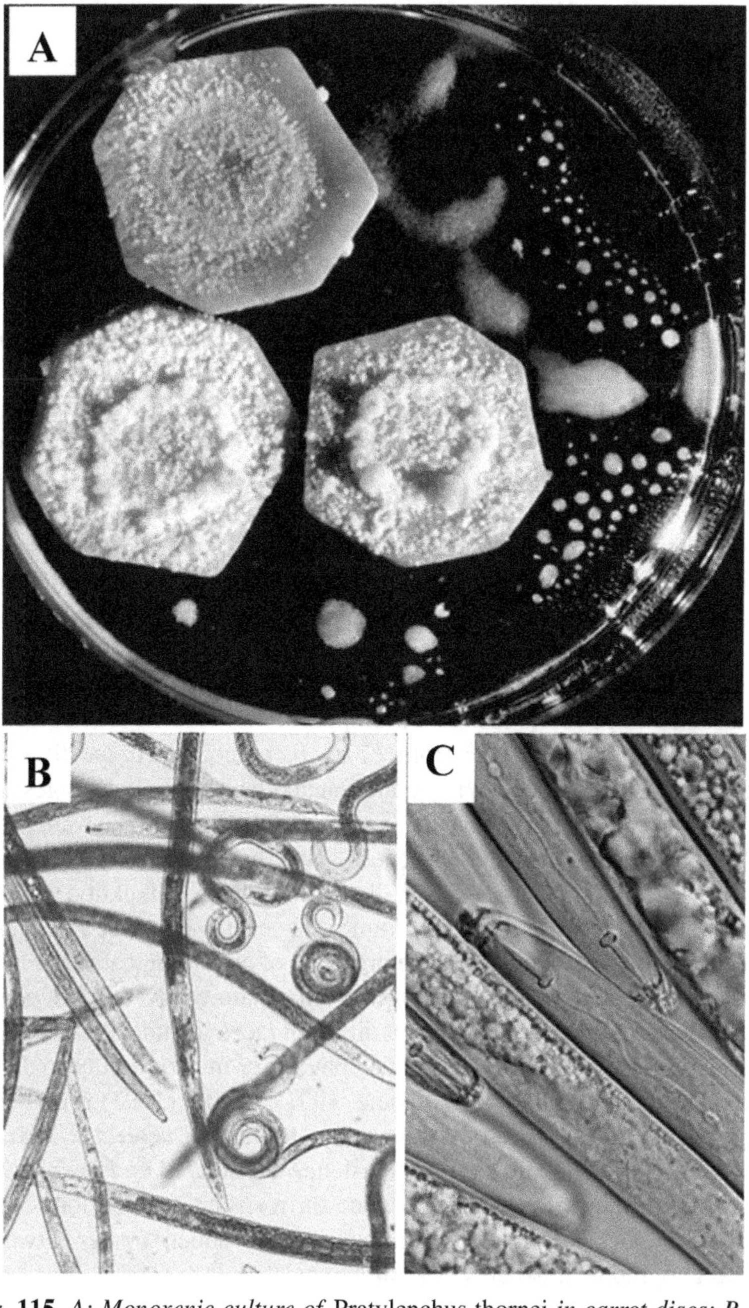

Fig. 115. *A: Monoxenic culture of* Pratylenchus thornei *in carrot discs; B, C: Detail of inoculum.*

Pathogenic variability in *Pratylenchus*

The ability of populations within a given nematode species to parasitise specific hosts or cultivars of crop plants has often been used to classify variants of species into races, pathotypes or biotypes (Triantaphyllou, 1987). The existence of races of *Pratylenchus* spp. was first suggested by Slootweg (1956) who found that certain populations attacked only lilies and not narcissi. Loof (1960) also pointed out the existence of physiological races in *P. neglectus* from The Netherlands. Olthof (1968) later identified two races of *P. penetrans* based on host suitability and pathogenicity on tobacco (*Nicotiana tabacum* L.) and celery (*Apium graveolens* var. *dulce* (P. Mill.) DC.). Morphological traits of the female tail terminus were used to identify two pathotypes of *P. penetrans* (Townshend *et al.*, 1978b). Townshend *et al.* (1978b) proved differences in reproductive fitness and pathogenicity related to the smooth- and crenate-tailed markers, although only in *P. penetrans*, not in *P. crenatus* or *P. neglectus*. Crenate-tailed females of *P. penetrans* were pathogenic to pea, radish and onion and stunted their growth, while in some instances the smooth-tailed females stimulated growth. It is especially noteworthy that only 1-3 crenate-tailed females were required to cause stunting, yet no lesions developed.

Populations of *P. coffeae* from *C. arabica* inoculated on different hosts revealed differences in reproduction and pathogenicity, suggesting a physiological specialisation in this species (Kumar & Viswanathan, 1972). Differences in aggressiveness have been reported in Brazil (Silva *et al.*, 2001; Campos & Villain, 2005) and Mizukubo (1995) found different reproduction levels amongst several Japanese *P. coffeae* populations, thereby indicating the presence of physiological races.

Differences in host suitability of certain potato cultivars to *P. penetrans* also suggested the existence of different races (Dunn, 1973; Kotcon *et al.*, 1987) and races of *P. penetrans* may differ in their ability to damage potato tubers (Olthof & Wolynetz, 1991). Griffin (1991) found different physiological types in virulence on alfalfa in *P. neglectus*, France and Brodie (1995) confirmed the existence of isolates of *P. penetrans* that differ in their ability to reproduce on resistant potato clones and Mudiope *et al.* (2004) reported differences in reproductive rates between several geographical isolates of *P. sudanensis*.

Evaluation of commercial rootstocks conducted in different geographical areas in the last 30 years suggests that differences in both host range and pathogenicity of *P. vulnus* exist, although no races have been described. In Italy, sour orange (*Citrus aurantium* L.) was considered a good host of *P. vulnus* (Inserra & Vovlas, 1974). By contrast, a *P. vulnus* isolate used in a host range study in Spain did not attack six citrus species, including sour orange (Pinochet *et al.*, 1992). These discrepancies were clarified by Pinochet *et al.* (1994) by using seven *P. vulnus* isolates in a host suitability test and demonstrating that some *P. vulnus* isolates readily attacked citrus (sour orange), whereas others were unable to reproduce on this plant and therefore two distinct races could be clearly distinguished within *P. vulnus* populations. Discrepancies in host suitability among American and European populations of *P. vulnus* on *Prunus* have also been reported (Lownsbery & Sher, 1963; Pinochet *et al.*, 1992).

Recently, the pathogenic variability in *Pratylenchus* has also been confirmed by molecular markers. Mizukubo *et al.* (2003) studied the variability of 20 isolates of *P. coffeae* based on PCR-RFLP analyses and reproduction tests on sweet potato (cv. Koganesengan). They found four distinct DNA fragment patterns, which they designated as RFLP phenotypes A, B, C and D, all the Japanese isolates corresponding to RFLP-phenotypes A, B or C, while the Indonesian isolate was phenotype A and the Guatemalan isolate was phenotype D. Phenotype A isolates reproduced in sweet potato whereas phenotypes B and C had lower final populations than initial populations.

Nevertheless, in spite of the previous studies reporting and demonstrating pathogenic variability within some *Pratylenchus* species, there is no general scheme for separating physiological races in this genus, probably because of the great intraspecific variability.

Host range

Host suitability to *Pratylenchus* species can be assessed by measuring their reproduction on plants after artificial inoculation (Lewis, 1987). The reproduction factor (*Rf*) has been widely used in nematological studies to define resistance and susceptibility of plants to plant-parasitic nematodes including *Pratylenchus* spp. (Marull & Pinochet, 1991; Pinochet *et al.*, 1991). When a compatible host-parasite interaction has been established by a nematode, infection of the plant can be followed by a sequence of disruptions in the physiological processes that lead to pathogenesis (Melakeberhan & Webster, 1993). Reductions of normal plant growth parameters may be used as indicators of plant physiology impairment related to pathogenesis.

Although *Pratylenchus* spp. are identified as polyphagous plant-parasitic nematodes with a wide host range, some good examples have shown that the host range of *Pratylenchus* spp. can be a useful tool for the management of these species. For example, host range studies by Endo (1959) showed that corn, crabgrass (*Digitaria sanguinalis* (L.) Scop.), millet, sorghum, rye, soybean and Sudan grass were very good or good hosts of *P. zeae*, while alfalfa, barley, bean, clover, lettuce, oat, tomato, vetch and watermelon were non-hosts. Similarly, Zirakparvar (1980) reported that *P. hexincisus* reproduced well on several crops (cabbage, corn, pea, soybean, tomato, white clover) and the Malvaceae velvetleaf (*Abutilon theoprasii* Medic.) whereas other crops, such as alfalfa, barley, carrot, oat, pepper, red fescue (*Festuca rubra* L.), rye, ryegrass, sorghum, Sudan grass, sunflower, timothy and wheat, were poor hosts and thus may be used as potential alternatives for effective crop rotation.

Host suitability of a wide range of vegetable crops species to *P. brachyurus* has been tested by several authors showing that alfalfa, cabbage, carrot, cauliflower, clover, garlic, lettuce, oat, onion, peanut, pepper, rye, soybean, strawberry, sweet potato and sweet pepper are poor or non-hosts, whereas corn, cotton, cucumber, eggplant, okra (*Abelmoschus esculentus* Moench.), potato, small onion (*Allium fistulosum* L.), Sudan grass, tobacco, tomato and watermelon are susceptible (Endo, 1959; Charchar & Huang, 1981; Machado & Inomoto, 2001). However, Inomoto *et al.* (2001) demonstrated that *P. brachyurus* is a weak pathogen on cotton, although at high population densities (9000 nematodes/plant) and pearl millet (*Pennisetum glaucum* L.) was found to be a poor host by Timper and Hanna (2005).

Pratylenchus coffeae is a major pathogen of coffee in several growing areas and is also pathogenic to many other crops such as banana, citrus, yam and potato in tropical and subtropical countries (Silva & Inomoto, 2002). Das and Das (1986) tested the host range of *P. coffeae* on 63 plant species and reported that banana, black gram (*Vigna mungo* (L.) Welzek., pigeon pea and peanut were good hosts, cabbage, eggplant, mung bean, pearl millet and wheat were poor hosts, and castor bean, chilli, coriander (*Coriandrum sativum* L.), garlic, little millet (*Panicum miliare* Lamk.), litchi, mango, neem, the ornamentals *Eleusine coracana* (L.) Gaertn. and *Polyanthes tuberosa* L., papaya, rice and tamarind were non-hosts.

However, characterisation of the host range of two isolates of *P. coffeae* from Brazil showed that the host preference may differs between isolates (Silva & Inomoto, 2002). In one isolate corn, eggplant, lettuce, melon, millet, okra, rice and sorghum were the best hosts, and banana, bean, cotton, French marigold, onion, peanut, Rangpur lime (*Citrus* limonia Osbeck), sesame (*Sesamum indicum* L.), small onion and sunflower were poor hosts (Silva & Inomoto, 2002), whereas in the other isolate the best hosts were castor oil plant (*Ricinus communis* L.), eggplant, melon, sesame, soybean, sorghum, squash (*Curcubita moschata* Duchesne) and watermelon, and the poorest hosts were banana, bean, coffee, corn, cotton, French marigold, lettuce, millet, okra, onion, peanut, Rangpur lime, small onion, sunflower and sweet pepper (Silva & Inomoto, 2002). *Pratylenchus coffeae* infects weeds that are common in the citrus groves of Florida such as miniature wood-rose (*Merrenia dissecta* (Jacq.) Hallier f.), balsam pear (*Momordica charantia* L.) and Brazilian pepper (*Schinus terebinthifolius* Raddi); and also ornamentals such as parlour palm (*Chamaedorea elegans* Mart.), croton (*Codiaceum variegatum* (L.) Blume) and nephthytis (*Syngonium podophyllum* Schott.). Other weed hosts of the nematode were sandspur (*Cenchrus echinatus* L.), wild horehound (*Eupatorium aromaticum* L.), black nightshade (*Solanum nigrum* L.) and panic grass (*Panicum* spp.) (Inserra *et al.*, 1990).

The host range of *P. goodeyi* was tested in 76 plant species commonly found in banana fields and only five (banana, *Commelina benghalensis* L., *Hyperrhenia rufa* (Nees) Stapf., *Plectranthus barbatus* Andrews and *Tripsacum laxam* Nash) were found to be hosts, indicating that this nematode has a narrow host range (Mbwana *et al.*, 1995). Host range studies of *P. goodeyi* by Namaganda *et al.* (2000) confirmed the narrow host range, with 69 weed species as non-hosts and bean and corn as hosts. Symptoms of *P. goodeyi* infections on banana roots are very similar and sometimes confused with those induced by *Helicotylenchus multicinctus* (Cobb, 1893) Golden, 1956, which also has an endoparasitic behaviour on banana (Vovlas *et al.*, 1994).

Host range studies on *P. penetrans* had shown that nearly 400 plant species are parasitised by this nematode, including stone and pome fruit trees such as apple, cherry, peach, pearl; and other woody crops, such as grapevine, rose and raspberry (Nyczepir & Becker, 1998). Similarly, host suitability studies showed many vegetables, such as alfalfa, bean, cabbage, carrot, celery, chickpea, clover, cowpea, cucumber, faba bean, groundnut, lentil, lettuce, pea, potato, radish, spinach, squash, sorghum, strawberry and tobacco to be good hosts whereas a few, such as pepper and asparagus, were poor or non-hosts, respectively (Mai *et al.*, 1977; Miller, 1978; Johnson, 1998; Di Vito *et al.*, 2002a).

Di Vito *et al.* (2002a) also demonstrated that chickpea, pea, faba bean and durum wheat were good hosts for three populations of *P. thornei* but that sugar beet, pepper, eggplant and sunflower were poor hosts. Similarly, alfalfa, barley, durum wheat, eggplant, faba bean, French bean, lentil, melon, pea, sunflower, sugar beet and tomato were very good hosts for *P. neglectus*, whereas chickpea, pepper and groundnut were poor hosts; *P. pinguicaudatus* reproduced better on faba bean, lentil and pea than on the other plants tested.

In a host range study of *P. vulnus* on a total of 37 commercial fruit tree, grapevine and citrus rootstocks, Pinochet *et al.* (1992) showed that 25 rootstocks were good or moderate hosts of this nematode, including almond (Desmayo Rojo, 1143), apple (EM-9, EM-106), avocado (Hass), cherry (Santa Lucia 64, Camil, M X M 14, Masto de Montanana), grapevine (41-B, 161-49, Fercal, Ritcher 110), hazelnut (Pauetet), loquat (Nadal), peach (GF-305, Montclar, Nemaguard), pear (OHF-333), pistachio (*Pistacia atlantica* Desf., *Pistacia terebinthus* L., *Pistacia vera* L.), plum (San Julian 655-2, Montizo, Pixy, Myrobalan 605) and walnut (Serr). However, all the tested citrus rootstocks (Alemow, rough lemon, Carrizo citrange, sour orange, Troyer citrange, Citrumelo), plus grapevine (SO4, *Vitis rupestris* Scheele, 110-3-P) and the olive rootstock Arbequina were poor or non-hosts. In contrast, Nico *et al.* (2002, 2003) found that an isolate of *P. vulnus* from olive nurseries reproduced well on olive rootstocks Arbequina and Picual, thereby confirming pathogenic variability in this species.

Pratylenchus zeae is a major pathogen of corn and tobacco in several growing areas and is also pathogenic to other crops such as sorghum, soybean and sugarcane (Fortuner, 1976). Host range studies by Hashmi and Hashmi (1989) showed that barley, garlic, onion and potato were poor hosts of *P. zeae*, whereas carrot, chickpea, radish, tomato and wheat were non-hosts.

Damage thresholds of *Pratylenchus* spp.

Reproductive fitness together with virulence are major components of pathogenicity in *Pratylenchus* species (Shaner *et al.*, 1992) and thus important for the assessment and understanding of disease reactions of plants to pathogens. *Pratylenchus* nematodes with higher reproduction fitness may colonise more host tissue and consequently cause more damage to susceptible plants. Genetic diversity and population variability are also known to exist among *Pratylenchus* species (Motalaote *et al.*, 1987; Griffin, 1991; Pinochet *et al.*, 1993b; France & Brodie, 1996).

Assessment of potential crop damage caused by *Pratylenchus* is usually based on population densities in soil at the time of planting or population densities in roots during the growing season. Damage thresholds are effectively used for economic decisions for the control of nematodes (Ferris, 1981). Another example of a damage threshold is the tolerance limit (T) used by Seinhorst (1965) in a biologically descriptive mathematical model relating nematode population density and plant growth or yield, $Y = m + (1 - m)Z^{P-T}$, where Y is the ratio between plant weight at nematode density P at the time of sowing and that in the absence of nematodes, m = the minimum value of Y (Y at a very large initial nematode population density), T is the density below which no reduction of plant growth or yield occurs and Z is a constant <1 reflecting nematode damage, with $Z^{-T} = 1.05$ (Seinhorst, 1965, 1979, 1998; Viaene *et al.*, 1997). Seinhorst's equation has been accepted as a useful basis for describing the relationship between the initial population density and productivity or yield in several host plant-*Pratylenchus* associations, such as coffee-*P. coffeae* (Kubo *et al.*, 2002), coffee-*P. brachyurus* (Oliveira *et al.*, 1999a); faba bean-*P. neglectus* and faba bean-*P. thornei* (Di Vito *et al.*, 2000); chickpea-*P. thornei* (Di Vito *et al.*, 1992); and *Rosa* spp.-*P. penetrans* (Peng & Moens, 2002b). Results from these studies indicated that *Pratylenchus* species are very well adapted to parasitism because minimum value of Y was hardly ever 0, indicating that even with extremely high nematode populations in soils *Pratylenchus* spp. do not kill their host plants (Fig. 116). Nevertheless, these results also indicated that tolerance limits may vary depending on host plant (Fig. 116A) and nematode species (Fig. 116B). In addition, these studies indicated that, although *P. thornei* showed a lower tolerance limit to chickpea than that showed by *P. coffeae* in coffee, at higher inoculum densities the latter was more pathogenic than the former as shown by the minimum value of Y (Di Vito *et al.*, 1992; Kubo *et al.*, 2002) (Fig. 116A). By contrast, on the same host plant (although tolerance limits are similar for *P. neglectus* and *P. thornei* on faba bean) at higher inoculum densities the former species was more

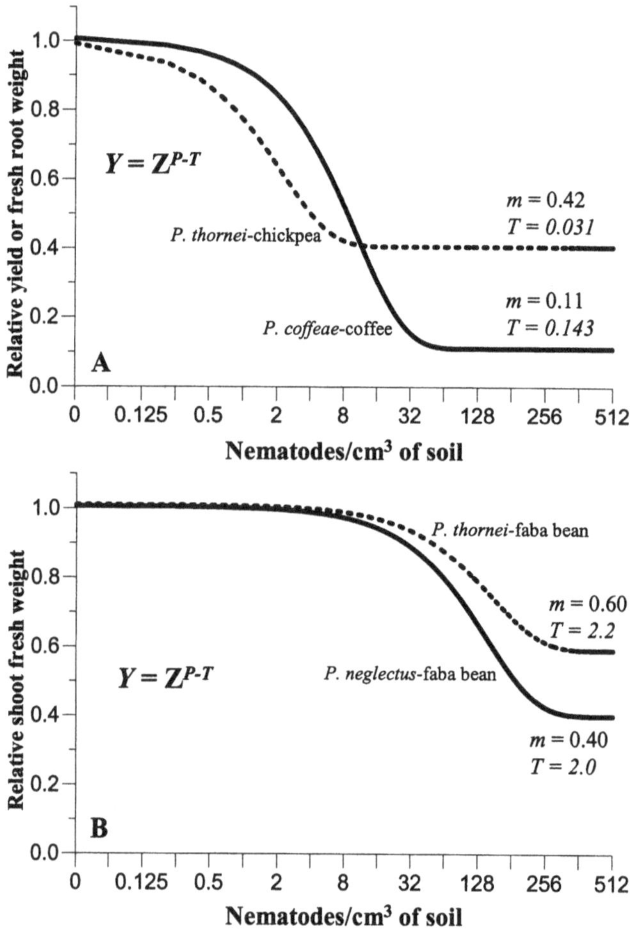

Fig. 116. *A: Relationship between initial population densities (*P*) of* Pratylenchus thornei, P. coffeae *and relative yield or fresh root weight of chickpea cv. Ghab 1 and coffee cv. Mundo Novo, respectively; B: Relationship between initial population densities (*P*) of* P. thornei *and relative shoot fresh weight of faba bean cv. Aguadulce. Lines represent the predicted functions calculated by fitting the Seinhorst model to experimental data. Abbreviations:* m = *minimum relative yield;* P = *initial nematode population on soil;* T = *tolerance limit. After Di Vito* et al. *(1992, 2000); Kubo* et al. *(2002).*

pathogenic than the latter as shown by the minimum value of *Y* (Di Vito *et al.*, 2000) (Fig. 116B). Finally, these results also demonstrated that the tolerance limits of a *Pratylenchus* species are host plant-dependent as occurred with *P. thornei* in chickpea ($T = 0.031$ nematodes/cm^3 soil) and in faba bean ($T = 2.2$ nematodes/cm^3 soil), the latter representing an inoculum density more than 70-fold that of the former (Fig. 116). Similarly, Peng and Moens (2002b) demonstrated that, in the pathosystem *R. corymbifera* cv. Laxa-*P. penetrans*, tolerance limit may also be dependent on the rate of applied fertiliser, *e.g.*, a higher rate of fertiliser application allowed the host to tolerate higher initial nematode densities (Table 6). In addition, Peng and Moens (2002b) demonstrated that, in the pathosystem *Rosa-P. penetrans*, tolerance limits varied from the lowest in *Rosa corymbifera* Borkh. cv. Laxa ($T = 8$ nematodes/cm^3 of soil) to the highest in *Rosa eglanteria* L. ($T = 18$ nematodes/cm^3 of soil) (Table 6).

One of the advantage of yield loss estimation by the Seinhorst curve is that with an initial inoculum density in soil it is possible to estimate the potential damage in a particular host plant-nematode combination. For example, with the previous reported data it is possible to estimate a yield loss of 50% in chickpea when soil population densities reached two nematodes/cm^3 soil (Di Vito *et al.*, 1992) or yield losses of faba bean of about 20 and 40% with soil infestation of 16 nematodes/cm^3 by *P. thornei* and *P. neglectus*, respectively (Di Vito *et al.*, 2000).

There are few documented cases of plants that are intolerant to *Pratylenchus* spp., although in *C. arabica* cvs Mundo Novo and Catuaí, an interesting intolerant reaction to *P. brachyurus* was detected in which very few nematodes (two nematodes/cm^3) caused great damage to coffee seedlings (Inomoto *et al.*, 1998).

Damage to herbaceous and vegetable crops

Most herbaceous and vegetable crops may be attacked by one or more species of *Pratylenchus*. The above-ground symptoms of *Pratylenchus*-infected plants are similar to those previously mentioned in this section. Damage caused by *Pratylenchus* has been documented in a number of herbaceous and vegetable crops, including cereals, carrots, corn, banana, forage and grain legumes, peanut, potato, tobacco, tomato, and wheat (Johnson, 1998). Under favourable growing conditions, especially moisture and fertility, most herbaceous and vegetable crops may grow and produce an acceptable yield whilst supporting a large number of *Pratylenchus*. However, *Pratylenchus* damage to herbaceous and

Table 6. *Damage threshold densities of different host plant-*Pratylenchus *combinations.*

Pratylenchus spp. – host plant	Damage threshold (nematodes/cm^3 soil)	Reference
P. alleni – soybean	14	Acosta (1982)
P. brachyurus – soybean	0.7	Ferraz (1995)
P. brachyurus – coffee	0^T	Oliveira *et al.* (1999a)
P. coffeae – coffee	0.14^T	Kubo *et al.* (2002)
P. coffeae – jute (*Corchorus olitorius* L.)	0.07	Mishra and Sasmal (1988)
P. crenatus – oat	0.33	Barker and Olthof (1976)
P. crenatus – carrot	0.3-1.8	Potter and Olthof (1993)
P. delattrei – crossandra	0.03	Srinivasan and Muthukrishnan (1975)
P. loosi – tea, at 18°C	1	Gnanapragasam and Manuel-pillai (1984)
P. loosi – tea, at 24°C	0.5	Gnanapragasam and Manuel-pillai (1984)
P. neglectus – alfalfa	2	Griffin and Gray (1990)
P. neglectus – barley	1.5	Rivoal and Cook (1993)
P. neglectus – faba bean	2^T	Di Vito *et al.* (2000)
P. neglectus – potato	1.5	Umesh and Ferris (1994)
P. penetrans – alfalfa	1	Griffin (1993)
P. penetrans – apple	0.1	Zepp and Szczygiel (1986)
P. penetrans – bean	0.5	Elliot and Bird (1985)
P. penetrans – Brussels sprouts	0.45	Miller (1978)
P. penetrans – cabbage	6	Olthof and Potter (1973)
P. penetrans – carrot	1	Vrain and Bélair (1981)
P. penetrans – carrot	1.4^T	Seinhorst (1998)
P. penetrans – cauliflower	6	Olthof and Potter (1973)
P. penetrans – celery	0.6	Townshend (1962)
P. penetrans – cherry	0.2	Zepp and Szczygiel (1986)
P. penetrans – clover	1.7	Willis and Thompson (1969)
P. penetrans – corn	0.25	Dickerson *et al.* (1964)
P. penetrans – cranberry	0.75	Bird and Jenkins (1964)
P. penetrans – cucumber	0.45	Miller (1978)
P. penetrans – *Digitalis purpurea*	5.6^T	Seinhorst (1998)
P. penetrans – eggplant	0.45	Miller (1978)
P. penetrans – faba bean	6.2^T	Seinhorst (1998)
P. penetrans – lettuce	6	Olthof and Potter (1973)

Table 6. *(Continued).*

Pratylenchus spp. – host plant	Damage threshold (nematodes/cm^3 soil)	Reference
P. penetrans – *Malus* sp.	1.5^T	Seinhorst (1998)
P. penetrans – onion	0.67	Olthof and Potter (1973)
P. penetrans – peach	0.05	Barker and Olthof (1976)
P. penetrans – pear	0.3	Zepp and Szczygiel (1986)
P. penetrans – plum	3.2	Nyczepir and Halbrendt (1993)
P. penetrans – potato	1-2	Olthof (1987)
P. penetrans – *Rosa corymbifera* (low rate fertiliser)	8^T	Peng and Moens (2002b)
P. penetrans – *Rosa* corymbifera (high rate fertiliser)	12^T	Peng and Moens (2002b)
P. penetrans – *Rosa glauca*	8^T	Peng and Moens (2002b)
P. penetrans – *Rosa eglanteria*	18^T	Peng and Moens (2002b)
P. penetrans – strawberry	0.5	Szczygiel (1982)
P. penetrans – sweet corn	0.67	Olthof and Potter (1973)
P. penetrans – table beet	18	Potter and Olthof (1974)
P. penetrans – tobacco	2	Olthof *et al.* (1973)
P. penetrans – tomato	0.45	Miller (1978)
P. scribneri – bean	0.5	Thomason *et al.* (1976)
P. scribneri – soybean	14	Acosta (1982)
P. thornei – chickpea	0.03^T	Di Vito *et al.* (1992)
P. thornei – faba bean	2.2^T	Di Vito *et al.* (2000)
P. thornei –wheat	0.5-1	Rivoal and Cook (1993)
P. thornei –wheat	0.42	Nicol and Ortiz-Monasterio (2004)
P. thornei –wheat	2.5	Thompson (1993)
P. thornei –wheat	30	Nicol *et al.* (1999)
P. vulnus – grapevine 'Thompson Seedless'	2	Pinochet and Raski (1976)
P. vulnus – grapevine rootstock	8	Chitimbar and Raski (1984)
P. zeae – blue grass (*Cenchrus ciliaris* L.)	0.2	Azmi (1988)
P. zeae – rice	0.5	Sahoo and Sahu (1993)
P. zeae – sorghum	0.5	Onifade and Egunjobi (1996)

^{T} Tolerance limits estimated with the Seinhorst model for plant growth or yield.

vegetable crops is greatly enhanced by stress from other factors, *e.g.*, disease symptoms are more pronounced and yield reduction is greater when abnormally high temperatures occur early in the growing season, rainfall is inadequate, soil fertility is low or imbalanced, or root-rot organisms attack the plants. In some cases, the pathogenic effect of *Pratylenchus* spp. on herbaceous and vegetable crops is reduced to the early stages of the plant, *e.g.*, the main pathogenic effect of *P. crenatus* on cv. Kuroda carrot was at the early stages of seedling emergence, inducing reduced foliage and root growth (Hay & Pethybridge, 2005). This kind of damage may influence the final plant density at harvest as was demonstrated in carrots by the parasitism by *P. crenatus* in Australia (Hay & Pethybridge, 2005) and by *P. mediterraneus* in Israel (Orion *et al.*, 1988).

Corn is severely attacked by several *Pratylenchus* species, including *P. brachyurus*, *P. hexincisus*, *P. penetrans*, *P. scribneri* and *P. zeae* in the major production areas, causing significant losses (Windham, 1998). Although these species have a worldwide distribution, certain species are more conspicuous in some areas than in others. For example, in the midwestern USA the most important species are *P. hexincisus*, *P. penetrans* and *P. scribneri* (Norton *et al.*, 1985; Waudo & Norton, 1986), whereas *P. brachyurus*, *P. pseudopratensis* and *P. zeae* are often associated with warmer climates such as the southeastern USA (Johnson *et al.*, 1975), Nigeria (Afolami & Fawole, 1991; Atu, 1991), Brazil (Lordello *et al.*, 1992) and South Africa (De Waele & Jordan, 1988a). Infected roots of corn show dark brown discrete lesions that may cover the entire root system. More lateral roots may be developed. Nematode infection results in stunted plants with yield losses near 10% (Windham, 1998). Yield losses ranging from 1243 kg ha^{-1} to 2284 kg ha^{-1} were estimated when soil populations at harvest were 100-200 nematodes per 150 cm^3, respectively (Osteen *et al.*, 1988).

Pratylenchus spp. associated with poor growth of potato in most production areas include *P. alleni*, *P. crenatus*, *P. neglectus*, *P. penetrans*, *P. scribneri* and *P. thornei* (Brodie *et al.*, 1993; Ingham *et al.*, 2005; Scurrah *et al.*, 2005). Infected potato plants showed poor growth and occasional yellowing of the foliage, with severe necrosis in roots and tubers. Lesions on tubers are shallow, rarely penetrating the outermost layer of the tuber (Brodie *et al.*, 1993). Tuber surface symptoms vary with species from a scabby appearance with sunken lesions (*P. scribneri*) to wart-like protuberances (*P. penetrans*) (Brodie, 1998).

Also, *P. brachyurus* and *P. coffeae* produced small lesions on the surface of potato tubers and reduced tuber quality in Brazil (Lordello *et al.*, 1954; Kubo *et al.*, 2001). *Pratylenchus penetrans* is one of the most important root-lesion nematodes in western and central Europe, causing a potato disease similar to that caused by the golden cyst nematode, *Globodera rostochiensis* (Wollenweber, 1923) Skarbilovich, 1959 (Loof, 1978). Nevertheless, the greatest economic importance of *Pratylenchus* spp. in potato involves its interaction with Verticillium wilt causing potato early dying (see later in this chapter).

Pratylenchus brachyurus is an economically important pathogen of peanut (*Arachis hypogaea* L.) in most peanut production regions of the USA (Dickson, 1998; Dickson & De Waele, 2005). The nematode attacks the fibrous roots, pegs and pods of peanut causing necrotic lesions (Dickson, 1998). It has also been found to cause up to 30% soybean yield losses in North Carolina (Schmitt & Barker, 1981). Nevertheless, damage to soybean depends on environmental conditions and host plant genotype (Koenning *et al.*, 1985). This nematode has also been reported damaging pineapples in several countries, resulting in decreased plant growth, delayed leaf emergence and reduction and yellowing of leaves (Monteiro & Lordello, 1972; Guerout, 1975; Sipes *et al.*, 2005).

Pratylenchus neglectus was reported to be pathogenic to alfalfa and wheatgrass although some differences in reproductive fitness were detected among populations of different geographical origin (Griffin & Jensen, 1997).

Cereals and grain legumes (chickpea, faba beans and beans) are severely damaged by *P. neglectus*, *P. penetrans*, *P. scribneri* and *P. thornei* resulting in chlorotic and stunted plants (McDonald & Nicol, 2005; Sikora *et al.*, 2005). On susceptible wheat cultivars *P. thornei* has been reported to reduce yield by as much as 32% in Sonora, Mexico (Van Gundy *et al.*, 1974) and by 44-85% in Australia (Doyle *et al.*, 1987; Eastwood *et al.*, 1994; Taylor *et al.*, 1999). Populations of *P. neglectus* increase dramatically when dry land fields are shifted to a higher intensity of cereal cropping (Gair *et al.*, 1969). *Pratylenchus neglectus* caused economic damage to barley in North America (Ferris *et al.*, 1994) and reduced yield of wheat by as much as 20% in Australia (Taylor *et al.*, 1999) and by 8-36% for intolerant wheat cultivars in irrigated, as well as dry land, fields in the Pacific Northwest (Smiley *et al.*, 2005b). Nevertheless, damage may be modified by environmental conditions, *e.g.*, with optimum irrigation *P. thornei* did not affect wheat yield but, with limited irrigation with plants under water stress, yield loss of the CIMMYT cultivars Baviacora 92, Jupateco 73, Nainari 60, Opata 85, Rayon 89, Seri 82 and Tobarito 97 was comparable to that of the susceptible control cv. Warigal (29%) with yield losses of 10-40% occurring at initial *P. thornei* densities of 84-338 per 200 g oven-dried soil (Nicol & Ortiz-Monasterio, 2004). Smiley *et al.* (2005a) reported the first field-derived evidence that *P. thornei* caused economic damage to wheat in the Pacific Northwest. These results showed that in soils with high populations of *P. thornei* there was a greater yield and yield stability for a moderately tolerant variety (cv. Krichauff) than for an intolerant variety (cv. Machete), and responses to nematicide application (aldicarb) showed a late-season reduction in plant stress and improvement in yield. Chickpea growth was also reduced by *P. thornei* infection under growth chamber and field conditions (Castillo *et al.*, 1998a). However, in growth chamber studies, plant growth reduction was only observed when inoculated plants were incubated for a long period (Castillo *et al.*, 1995, 1998b). Similarly, infections by *P. penetrans* and *P. scribneri* damaged beans, resulting in a significant reduction of plant growth and dry bean yield (Thomason *et al.*, 1976; Elliot & Bird, 1985).

Clover, forage legumes and grasses are damaged by several *Pratylenchus* spp. including *P. brachyurus*, *P. crenatus*, *P. neglectus*, *P. penetrans*, *P. pratensis* and *P. scribneri*. Symptoms include root necrosis (Griffin, 1994, 1996) and reduction of plant growth ranging from 7-12% in white clover infected with *P. penetrans* (Thies *et al.*, 1995) to 31-41% in red clover infected with *P. coffeae*, *P. neglectus* and *P. penetrans* (Chapman, 1958).

Growth of tomato plants was reduced by 20-66% after 2 months' infection with *P. penetrans* populations ranging from 8-55 nematodes per 100 g of soil at planting and the severity of root necrosis increased with nematode populations (Miller, 1975). Potter and Olthof (1977) reported that *P. penetrans* caused significant losses to tomatoes at population densities of two nematodes/cm^3 of soil which is comparable to those populations commonly found in coarse-textured soils in southwestern Ontario. Coosemans (1975) also reported that an initial population density of 0.1 *P. penetrans*/cm^3 of soil caused 75% of the carrots to be branched, whilst one nematode/cm^3 of soil killed 40% of the carrots. Vrain and Bélair (1981) found that 2-4 *P. penetrans*/cm^3 of soil induced severe branching in taproots of carrot, superficial brown lesions and severe reduction of plant growth and yield.

Pratylenchus zeae plays a key role in sugarcane yield decline which is characterised by the diminishing ability of cane land to produce sugar per harvested hectare. It reduces the density, length and weight of the roots and induces yellowing of leaves, stunted growth and fewer stalks on the shoot system (Tarté *et al.*, 1977; Cadet & Spaull, 2005). Yield decline in sugarcane has been reported in sugarcane crops grown in continuous monoculture in the USA (Hoy, 1999) and Australia (Pankhurst *et al.*, 2005a). The root damage caused by *P. zeae* slows the growth rate of the primary shoot and the subsequent production and growth of secondary shoots, and ultimately results in a poorer yielding plant (Pankhurst *et al.*, 2005a). This is usually a complex problem associated with long-term sugarcane monoculture with many similarities to replant diseases in horticultural crops (see later) and may involve several factors associated with the site where the sugarcane is established and co-infestations by *P. zeae* and *Meloidogyne* spp. (Magarey & Bull, 1998).

Tobacco is severely damaged by *P. brachyurus*, *P. neglectus*, *P. penetrans* and *P. pratensis* (Johnson, 1998). In Canada, infections of flue-cured tobacco cv. Delgold by *P. penetrans* resulted in considerable economic losses (Bélair *et al.*, 2004). Initial population densities of six and 18 nematodes/g of soil caused losses in flue-cured tobacco of 11 and 27.5%, respectively (Olthof *et al.*, 1973). In Germany, Lange *et al.* (1964) also found that population densities higher than two nematodes/cm^3 of soil damaged the crop and in Nigeria, 1.2 nematodes/cm^3 of *P. brachyurus* significantly reduced plant growth of tobacco NTC 95 (Egunjobi, 1978).

Damage to fruit trees and woody plants

In pome (apple, pear and quince) and stone (almond, peach, cherry and plum) fruit crops, the most economically important *Pratylenchus* species are *P. penetrans* and *P. vulnus*, the former being common in temperate climates whilst the latter is predominant in the Mediterranean region (Nyczepir & Becker, 1998). *Pratylenchus brachyurus* and *P. coffeae* infect citrus orchards and nurseries (Duncan & Moens, 2006), although virulence of *P. coffeae* (population densities can exceed 10 000 nematodes g^{-1} of roots) is higher than that of *P. brachyurus* (*ca* 1000 nematodes g^{-1} of roots). All of these species destroy the cortical parenchyma of the root causing cavities and lesions and predisposing affected tissue to secondary infections caused by fungi and bacteria (Nyczepir & Becker, 1998). The most common symptom on susceptible trees is a slow decline as the nematodes increase to very high numbers over the years. The foliage of diseased trees may appear lighter green or chlorotic. Seasonal growth of infected plants is less than would be expected in a healthy plant. Flowers or fruit may be reduced substantially both in number and quality. The vigour of the host is reduced and the plant may be predisposed to winter injury or other infectious diseases. Replacing the affected plant without soil treatment frequently results in poor growth and sometimes in the death of the new plant. These symptoms are non-specific, can be easily overlooked or readily mistaken for those of other soil-borne pathogens. Damage can seriously affect early stages of plant development in the nursery or when rootstocks are transplanted into the field. In established orchards damage is expressed as a delay in entering into production, lower yields, smaller fruit size, nutrient deficiencies and a reduction in the longevity of the orchard (Nyczepir & Becker, 1998). New infestations on a field level occur with the introduction of infected plant material or through contaminated soil from nearby infested fields.

Orchards require renovation due to loss in tree vigour, reduction in yield, changes in market demand for new fruit varieties and because of the introduction of new horticultural practices, such as the use of rootstocks that limit tree size conferring an invigorating or dwarfing effect (Felipe, 1989). From a phytopathological point of view, it is common to observe poor orchard development in replanted sites, expressed as a failure in tree establishment, suppressed growth and shortened productive life. Replant problems are complex and are the result of several factors associated with the site where the tree is established. However, *Pratylenchus* spp. play a significant role in the short life and replant problems of fruit trees (Colbran, 1979; Mai & Abawi, 1981). Their main role in replant problems seems to be the ability to incite root degeneration by providing extensive infection sites for other pathogenic soil microorganisms (Mai & Abawi, 1981). In fact, *P. penetrans* and *P. vulnus* are considered as primary causal agents in replant problems in many parts of the world (Nyczepir & Becker, 1998). In Australia, *P. coffeae* and *P. scribneri* are considered responsible for replant problems (Colbran, 1953, 1979; Stirling *et al.*, 1995); and *P.*

penetrans and *Thielaviopsis basicola* (Berk. & Broome) Ferris (syn. *Chalara elegans* Nag Raj & Kendrick) are responsible for a cherry replant problem in Germany (Harr *et al.*, 1976). *Pratylenchus vulnus* was found to be pathogenic in several plum rootstocks (Citation, Marianna 2624, Montizo, PSM 101 and Saint Julien) by causing significant root destruction and reduction of plant growth as a result of high and rapid reproduction in the first years after planting (Pinochet *et al.*, 1993c). Similarly, pathogenicity of *P. vulnus* has been reported in three rootstocks for peach and plum released by INRA (Institut National de la Recherche Agronomique), *i.e.*, Cadaman, Julior and Ishtara (Hernández Dorrego *et al.*, 1999). Finally, Pinochet *et al.* (1976) evaluated the pathogenicity of *P. vulnus* to five grapevine rootstocks (Dogridge, Saltcreek, Harmony, Couderc 1613 and Ganzin). Their results showed that the former two grapevine rootstocks supported low populations of *P. vulnus* and growth of both was not affected; root weight of Harmony and Couderc 1613 was only reduced at the highest inoculum level (10 000 nematodes/plant), while in Ganzin 1 growth reduction was proportional to nematode inoculum (Pinochet *et al.*, 1976). Additional factors that contribute to replant problems are poor soil structure, soil-borne fungi and bacteria, low soil fertility and inadequate agronomic practices.

Pratylenchus brachyurus, *P. coffeae* and *P. vulnus* have been demonstrated to parasitise and damage several citrus species, such as grapefruit (*Citrus* paradisi Macf.), lemon, rough lemon and sour orange (O'Bannon *et al.*, 1972; Duncan, 2005). Pathogenicity investigations under glasshouse conditions revealed that citrus seedlings infected by the nematode showed stunting, reduced foliage and wilting after 3 weeks of growth (MacGowan, 1978).

Pratylenchus coffeae is a major limiting factor for coffee production in several growing areas, with estimated coffee losses of *ca* 15% (Campos *et al.*, 1990; Campos & Villain, 2005). It is also pathogenic to many other crops such as banana, citrus, yam and potato in tropical and subtropical countries (Silva & Inomoto, 2002). Coffee infected plants appear stunted and with few, chlorotic, leaves. Roots of coffee infected by *P. coffeae* turn yellow then brown and most lateral roots become rotten (Campos & Villain, 2005). Infection of several banana cultivars (Grande Naine (AAA) and four local banana cultivars, Ngu Tien (AA), Hot (BB), Ben Tre (AAA) and Tay Tia (ABB)) in North Vietnam with *P. coffeae* did not affect the crop cycle duration or the plant height, the pseudostem girth or the number of standing leaves at harvest of any of the cultivars, but did significantly reduce the bunch weight of cv. Ngu Tien (20% reduction) and the bunch weight of cvs Tay Tia and Grande Naine (19%, 13% reduction, respectively). Similarly, infection with *P. coffeae* resulted in a 34% reduction in the number of fingers of cv. Grande Naine (Van den Berg *et al.*, 2006). *Pratylenchus coffeae* also has a wide distribution in tropical and subtropical citrus growing areas of the world causing a 'spreading decline' through severe damage to the fibrous roots of citrus trees leading to a weakened root system incapable of supporting the above ground portion of the tree which exhibits sparse foliage, small fruit, weakened and dead branches and losses in yield (O'Bannon *et al.*, 1976). In Florida (O'Bannon & Tomerlin, 1973), warmer regions of Japan (Yokoo & Ikegemi, 1966) and in India (Siddiqi, 1964) *P. coffeae* was reported causing severe damage to citrus. *Pratylenchus vulnus* is believed to be a threat to citrus orchards in Italy, where it has been associated with decline of sour orange seedlings and damage to young trees similar to that caused by *P. coffeae* was reported (Inserra & Vovlas, 1977).

Pratylenchus loosi causes severe damage to tea plantations which becomes noticeable at the very early stages of infection by patches of unthrifty plants (Gnanapragasam & Mohotti, 2005). Attack debilitates not only the available feeder roots (inducing slow decline), but also damages the storage roots, thus limiting carbohydrate reserves and the consequent recovery from pruning (Gnanapragasam, 2002).

Pratylenchus penetrans has been found closely associated with the decline of stone fruits in Canada (Mountain & Patrick, 1959) and in northeastern USA (Mai & Parker, 1972; Melakeberhan *et al.*, 1997). *Pratylenchus* spp. have also been associated with poor growth in grapevine (Raski & Krusberg, 1984). In California, it has been reported that *ca* 70% of vineyards are infested with *Pratylenchus* spp. (Nicol *et al.*, 1999b).

Interactions of *Pratylenchus* spp. with other plant-parasitic nematodes

Competition between and among nematodes is a function of initial population density, reproductive potential, host status and edaphic factors that influence host-parasite interactions. The distribution and population densities of *Pratylenchus* species in the field may reflect competitive interactions with other plant-parasitic nematodes and several such examples have been reported.

Pratylenchus coffeae and *Tylenchulus semipenetrans* Cobb, 1913 were mutually exclusive within Florida citrus groves. In an experiment conducted under field conditions, indigenous populations of either species did not preclude infection by the other species and inoculation with either *T. semipenetrans* or *P. coffeae* tended to reduce the population size of the other nematode species (Kaplan & Timmer, 1982). Lamberti *et al.* (2001) reported that the reproduction of *P. vulnus* was suppressed by *M. incognita* in olive (cvs Leccino and Pendolino). Similarly, the infection rate of *H. glycines* Ichinohe, 1952 on soybean decreased with the increasing proportion of *P. penetrans*. The rate of *P. penetrans* infection increased with increasing *Heterodera glycines* proportions up to the 50% level, but declined at the 75% level. The relationship between *H. glycines* and *P. penetrans* indicates that the former may be competitive when present at higher proportions than the latter (Melakeberhan & Dey, 2003). The decrease in *P. penetrans* infection at 75% *H. glycines* inoculum suggests some threshold level above which competition may be accelerated. It is possible that the migratory feeding behaviour of *P. penetrans* may confer an advantage over the sedentary nature of *H. glycines* (Melakeberhan & Dey, 2003). Similarly, in tomato the population level of *P. penetrans* on its own was about four times as great as in combination with *M. incognita*, and *M. incognita* alone reproduced twice as fast as it did in the presence of *P. penetrans*. Root invasion by *P. penetrans* was also significantly inhibited by the presence of *M. incognita* (Estores, 1971).

Brinkman *et al.* (2004) demonstrated that *P. penetrans* suppressed the development of the migratory ectoparasite *Tylenchorhynchus ventralis* Loof, 1963 on the natural dune grass *Ammophila arenaria* (L.) Link, but only when *P. penetrans* was added to the plant in relatively high densities that exceeded the field density of these nematodes. Sequential inoculation of *P. penetrans* did not influence the development of *T. ventralis* more than inoculation at the same time. Thus, it was concluded that competition by *P. penetrans* is not a likely mechanism for the regulation of *T. ventralis*. Similarly, *P. penetrans* was a stronger competitor than the sedentary endoparasite *Heterodera arenaria* Cooper, 1955 and it suppressed the abundance of the latter in natural dune grass (Brinkman *et al.*, 2005).

In interspecific relationships between root-knot nematodes and *Pratylenchus* spp. the close nematode-host relationship established by the root-knot species may make the host either more or less suitable for the latter. For example, *M. incognita* suppressed *P. brachyurus* on soybean (Herman *et al.*, 1988) and *P. penetrans* on tomato (Estores & Chen, 1971); in contrast, *Meloidogyne naasi* Franklin, 1965 stimulated *P. penetrans* on bent grass (Sikora *et al.*, 1972). Similarly, Rivoal *et al.* (1995) recorded suppression of *P. neglectus* in the presence of *Heterodera avenae* Wollenweber, 1924 on oats. After *H. avenae* populations were reduced by growing resistant oat cv. Panema, field populations of *P. neglectus* increased. However, in other cases coinfections of both nematode species did not compete, *e.g.*, *P. brachyurus* did not affect reproduction of *M. incognita* in corn (Dickson & McSorley, 1990), nor did *H. avenae* affect reproduction of *P. thornei* in wheat (Nombela & Romero, 1999, 2001). Interspecific competition has also been suggested between *P. coffeae* and *Meloidogyne exigua* Goeldi, 1892 on coffee in Costa Rica (Bertrand *et al.*, 1998) and Guatemala (Herve *et al.*, 2005).

The results of these interactions are dependent on nematode species and host plant. Villenave and Cadet (1998) reported that reproduction of *P. pseudopratensis* on millet (*Pennisetum typhoides* (L.) Rich) and acacia (*Acacia holosericea* A. cunn. Ex G. Don.) was lower in the presence of *Helicotylenchus dihystera* Cobb, 1893 or *Tylenchorhynchus gladiolatus* Fortuner & Amougou, 1973. However, *T. gladiolatus* reproduction was stimulated in the presence of *P. pseudopratensis* on millet, whereas it was inhibited on acacia at the lowest inoculum levels (200 nematodes/plant).

Nevertheless, these interactions may be also influenced by environmental factors (*e.g.*, soil temperature). A good example of this is the competitive interaction between *P. neglectus* and *Meloidogyne chitwoodi* Golden, O'Bannon, Santo & Finley, 1980 on barley and potato (Umesh & Ferris, 1994). The negative effect of *M. chitwoodi* on *P. neglectus* was greatest at 25°C on barley and potato, whereas the negative effect of *P. neglectus* on *M. chitwoodi* was greatest at 15°C in barley and at 25°C in potato (Umesh & Ferris, 1994).

Interactions of *Pratylenchus* spp. with bacteria

Some disease complex and interactions between *Pratylenchus* spp. and pathogenic bacteria have been reported (Pitcher, 1963; Sitaramaiah

& Pathak, 1993). One of the important roles of *Pratylenchus* spp. in disease complexes involving bacteria is the way in which the nematodes act as a predisposing agent by providing wounds for the entry of the bacterial pathogen. Although some examples have been reported which support this mechanism (*e.g.*, *P. penetrans* produced wounds in alfalfa roots which favour infections by *Pseudomonas viridiflava* (Burkholder, 1930) Dowson, 1939, *Pseudomonas corrugata* Roberts & Scarlett, 1981 and *Pseudomonas marginalis* (Brown, 1918) Stevens 1925 (Bookbinder *et al.*, 1982)), the most important interactions between both pathogens are those in which *Pratylenchus* break down resistance to bacteria, *e.g.*, interactions between *P. flakkensis* and *Ralstonia solanacearum* (Smith, 1896) Yabuuchi *et al.*, 1996 race 3 on potato (Mejia-Anaya & Canto Saenz, 1985), or increase the incidence and severity of bacterial diseases, *e.g.*, *P. penetrans* increased the incidence of crown gall induced by *Agrobacterium* tumefaciens (Smith & Townsend, 1907) Conn, 1942 on root systems of red raspberry (Vrain & Copeman, 1987), the predisposing effect of *P. penetrans* being confirmed by sequential inoculations.

On the other hand, examples of antagonism between *Pratylenchus* spp. and bacteria have also been reported, *e.g.*, a strain of *Pseudomonas chlororaphis* (Guignard & Sauvageau, 1894) Bergey *et al.*, 1930 was found significantly to reduce *P. penetrans* numbers in roots of strawberries by about 40% (Hackenberg *et al.*, 2000). Similarly, various *Pratylenchus*-rhizobia interactions have been reported, including inhibition of nodule formation. Most of the reports concerning the suppression of nodulation, *e.g.*, Wardojo *et al.* (1963), found that *P. penetrans* reduced nodulation in white clover. Germani *et al.* (1984) demonstrated that infection of soybean by *P. pseudopratensis* reduced nodulation and N_2 fixation and Bhatt (1995) pointed out that *P. thornei* reduced nodulation in chickpea. However, stimulation of nodule formation has also been observed in other combinations, *e.g.*, *P. penetrans* increased nodules on soybean, although the nitrogen-fixing capacity was also inhibited by the nematode (Hussey & Barker, 1976). Finally, other reports indicate that *Pratylenchus* infection has no significant effect on nodulation and N_2 fixation, *e.g.*, parasitism of *P. penetrans* on red clover (Amosu & Taylor, 1974), or parasitism of *P. penetrans* on peanut (Barker & Hussey, 1976). Although damage to nodular tissues by *Pratylenchus* species has not, to our knowledge, been reported, the mechanisms involved in these interactions might be related to the capacity of *Pratylenchus* species to mainly colonise the cortical parenchyma of the host plant, the suppression of root growth and/or modifications induced in the physiology of the plant.

Recent *in vitro* studies have demonstrated that a strain of *Lysobacter enzymogenes* Christensen & Cook 1978 (bacteria found in soil and water habitats, characterised by having gliding motility and lytic activity against microorganisms), showed antagonistic effects on plant-parasitic nematodes by rapidly immobilising and lysing juveniles and adults of *P. penetrans* after two days of incubation (Chen *et al.*, 2006).

Interactions of *Pratylenchus* spp. with pathogenic fungi

There have been several comprehensive reviews concerning interactions between plant-parasitic nematodes, including *Pratylenchus* spp., and pathogenic fungi (Powell, 1971; Mai & Abawi, 1987; Khan, 1993; Back *et al.*, 2002); thus, only some current summarised and relevant information regarding interactions between *Pratylenchus* spp. and pathogenic fungi will be included in this chapter.

Three type of interactions can occur between plant-parasitic nematodes and pathogenic fungi influencing plant diseases (Back *et al.*, 2002): *i*) *synergism* as an association between nematode and pathogen resulting in plant damage exceeding the sum of individual damage by both pathogens (*i.e.*, $1 + 1 > 2$); *ii*) *antagonism* when an association between nematode and fungus results in plant damage less than that expected from the sum of the individual pathogens (*i.e.*, $1 + 1 < 2$); and *iii*) *neutral* when nematodes and fungi cause plant damage that equates to the sum of individual damage by both pathogens (*i.e.*, $1 + 1 = 2$).

The most frequently reported interactions between *Pratylenchus* spp. and pathogenic fungi comprise the wilt fungi *Fusarium* and *Verticillium* (Riedel *et al.*, 1985; Sumner & Minton, 1987). Interactions between *Pratylenchus* spp. and several *formae speciales* of *Fusarium oxysporum* have been described on several crops (Sumner & Minton, 1987). These studies demonstrated that the infection by *Pratylenchus* spp. increased the incidence or severity of *Fusarium* wilt on susceptible cultivars. However, those previously published results were not confirmed in other compatible plant-nematode fungus combinations (Hutton *et al.*, 1973). Thus, it appears that modification of *Fusarium* wilt incidence or severity may be related to the specific nematode-fungus combination. In addition, some controversial results reported in herbaceous and fruit and woody plants indicate that interactions between soil-borne fungi and root-lesion nematodes are biological and physiological rather than physical in nature (Castillo *et al.*, 1998a).

In wheat, Taheri *et al.* (1994) demonstrated that *P. neglectus* interacts differentially with the most frequent fungi attacking wheat roots in Australia: *Fusarium oxysporum* E.F. Sm. & Swinge, *Fusarium acuminatum* Ellis & Everh., *Fusarium equiseti* (Corda) Sacc., *Microdochium bolleyi* (R. Sprague) de Hoog & Herm.-Nijh., *Gaeumannomyces graminis* (Sacc.) Arx & D.L. Olivier, *Bipolaris sorokiniana* (Sacc.) Shoemaker, *Pythium irregulare* Buisman, *Pyrenochaeta terrestris* (E.M. Hans.) Gorenz, Walker & Larson, *Phoma* sp., *Rhizoctonia solani* J.G. Kühn and *Ulocladium atrum* Preuss. Results indicated that number of nematodes per plant and severity of root lesions were significantly higher in plants co-infected with the nematode and one of each fungi *R. solani*, *M. bolleyi*, *B. sorokiniana*, *P. irregulare*, or the combination of *G. graminis* + *F. equiseti*. However, with the combination of *G. graminis* + *R. solani*, nematode numbers were significantly reduced. Thus, these authors suggested that some fungi may render the wheat roots more suitable for nematode multiplication.

Black root rot of strawberry is a debilitating disease of the root cortex caused by the binucleate fungus *Rhizoctonia fragariae* Hussain & McKeen, the severity of which has been increased by co-infection with *P. penetrans* under field (LaMondia, 1999a) and controlled conditions (LaMondia, 2003; LaMondia & Cowles, 2005). The mechanism by which *P. penetrans* increases black root rot appears to be local to the root rather than systemic in the plant and is at least partially due to direct effects of nematode feeding such as cortical cell damage and death.

Similarly, in chickpea, infections by *P. thornei* increased the severity of root necrosis on both a *Fusarium* wilt-susceptible and wilt-resistant chickpea cultivar, irrespective of incubation period, nematode and fungal inoculum densities and environmental conditions (Castillo *et al.*, 1998a).

In mint, results of interactions between *Verticillium dahliae* Kleb. and *P. penetrans* were also dependent on fungal genotype, *e.g.*, the vegetative compatibility group (VCG) of the fungus infecting the host plant (Johnson & Santo, 2001). VCG 2B isolates were capable of interacting with *P. penetrans* on mint, causing increased disease severity, whereas a VCG 4A isolate did not (Johnson & Santo, 2001).

However, probably one of the best and well documented examples of synergism between soil-borne fungi and *Pratylenchus* spp. has been shown in potato, with the interaction between *V. dahliae*, the primary causal agent of a vascular wilt disease in potato called potato early dying, and *Pratylenchus* species (Rowe & Powelson, 2002). Three species of *Pratylenchus*, *viz.*, *P. crenatus*, *P. penetrans* and *P. scribneri*, interact synergistically and differentially with *V. dahliae* (Riedel *et al.*, 1985; Rowe *et al.*, 1985). In this pathosystem interactions were also dependent of fungal genotype, *e.g.*, isolates from VCG 4A interacted with *P. penetrans* to increase wilt severity, whereas isolates from VCG 4B did not interact with *P. penetrans* or did so only weakly (Botseas & Rowe, 1994). *Pratylenchus penetrans* is the most important root-lesion nematode enhancing the development of wilt symptoms (Riedel *et al.*, 1985), reducing tuber yield and quality (MacGuidwin & Rouse, 1990), or reducing photosynthesis (Saeed *et al.*, 1998) when population levels of the two pathogens are too low to cause disease alone. In the absence of *V. dahliae*, *P. penetrans* did not reduce plant growth and tuber yield of potato below that of the nematode-free control – or did so only at the highest population densities (7.4-32.4 nematodes/cm^3 soil). However, in terms of yield, Martin *et al.* (1982) calculated that 15, 50 and 150 *P. penetrans* per 100 cm^3 soil in combination with *V. dahliae* would result in 36, 60 and 75% reductions in potato tuber weight, respectively. Although the mechanism(s) of these interactions have not been completely clarified, concomitant infections of both pathogens resulted in impaired gas exchange in Russet Burbank potato at soil population levels that were too low to affect photosynthesis in plants inoculated with only one pathogen (Saeed *et al.*, 1997). Similarly, Rotenberg *et al.* (2004) reported that transpiration in plants infected with both pathogens was significantly reduced, although the combined effect of nematode and fungus was synergistic in one case and additive in the other.

In other cases, the impact of interactions between *Pratylenchus* and pathogenic fungi was related to the time of inoculation. For instance, simultaneous inoculation of black shank-susceptible tobacco with *P. brachyurus* and *Phytophthora parasitica* Dastur var. *nicotianae* (Breda de Haan) Tucker increased black shank development and severity, but inoculation with *P. brachyurus* or *P. penetrans* prior to *P. parasitica* var. *nicotianae* reduced black shank symptom severity and disease incidence (Inagaki & Powell, 1969; McIntyre & Miller, 1978).

Pratylenchus species-pathogenic fungi interactions have also been reported in fruit and woody plants. Kheiri *et al.* (2002b) reported interactions between *P. vulnus*, *Fusarium solani* and *F. oxysporum* on maple (*Acer negundo* L.) seedlings in forest nurseries in Iran. Their results demonstrated the antagonistic interaction between these two fungi and *P. vulnus*. In maple seedlings infected with the nematode alone, growth of seedlings was minimal (with 75% seedling mortality) and population density of nematodes in soil and root tissue was maximum. However, in maple seedlings infected with the nematode and fungi together, the population of nematodes decreased and the growth of seedlings increased, in comparison with the nematode alone, reducing seedling mortality to 25% with *F. oxysporum* and 50% with *F. solani*. In contrast, studies by Lamberti *et al.* (2002) on olive did not find an interaction between *V. dahliae* and *P. vulnus*, suggesting that a longer duration of the experiments might have resulted in more obvious interactions among these pathogens.

Interactions between *Pratylenchus* spp. and soil-borne fungi may also influence nematode reproduction, but once again these interactions are complex and related to the plant-nematode-fungus combination. Reproduction of *P. coffeae* on chrysanthemum was decreased in the presence of *Pythium aphanidermatum* (Edson) Fitzp., whereas it was decreased in the presence of *R. solani*, but when both fungi were present the nematode reproduction was scarcely affected (Hasan, 1988). Reproduction of *P. penetrans* on alfalfa was increased by the co-infection of *V. dahliae* (Vrain, 1987). Similarly, reproduction of *P. thornei* on chickpea was influenced by co-infections with *F. oxysporum* f. sp. *ciceris* race 5 depending on the plant response to the fungus, *e.g.*, reproduction was significantly increased in the wilt-susceptible chickpea cultivar but was not affected in the wilt-resistant one (Castillo *et al.*, 1998a). Although no studies have been developed to clarify these interactions, influences on nematode reproduction might be related to the extension and/or colonisation of the root system by the fungus, the defence mechanisms involved in resistant reactions, or both (Castillo *et al.*, 1998a).

Finally, root infections by *Pratylenchus* nematodes may even affect the survival and development of herbivorous insects feeding on the above plant parts. For example, Van Dam *et al.* (2005) demonstrated that root infections of *Brassica nigra* (L.) Koch by *P. penetrans* affected the shoot feeding and performance of larvae of the lepidopteran *Pieris rapae* (L.). Their results indicated that root infections of *B. nigra* by *P. penetrans* induced changes in the plant shoots (*via* systemic effects), which in turn influenced their response to herbivore attack by changing their chemical profile. In fact, larvae of *P. rapae* grew slower and produced fewer pupae on plants that were infected with *P. penetrans*. This effect was attributed to higher levels of glucosinolate and fenolic levels in *P. penetrans*-infected plants after *P. rapae* started feeding.

Nematology Monographs & Perspectives, 2007, Vol. 6, 355-363

Chapter 9

Host-parasite relationships of *Pratylenchus* species on crops

The responses of plants to infection by plant-parasitic nematodes are closely related to the feeding habits of the parasites. The physiology of parasitism and host-parasite relationships involves an active role for nematode secretions and host response. The general response to parasitism by *Pratylenchus* spp. is necrosis or tissue death and browning that typically involves epidermis, cortical tissues and endoderm cells (Fig. 117). Nevertheless, under field conditions root necrosis may be caused by a complex of organisms rather than *Pratylenchus* spp. alone. *Pratylenchus* spp. cause mechanical destruction of root cells during their migration through roots. Intracellular migration kills cortical and adjacent cells, tannin deposition occurs, membrane integrity is lost and cell organelles degenerate (Townshend *et al.*, 1989; Sijmons *et al.*, 1994).

The wide host range of *Pratylenchus* spp. suggests that their parasitism is a less specialised (*i.e.*, more primitive) relationship, possibly representing an evolutionary intermediate between the highly specialised sedentary endoparasites and free-living forms.

Feeding behaviour and mechanism of pathogenesis

Pratylenchus spp. are primarily endoparasites of the root cortex, migrating through and between parenchyma cells and causing necrotic areas that are visible on washed roots as minute lesions. Infections occur along the entire length of the roots of the host plant except for the root tips (Townshend & Stobbs, 1981; Castillo *et al.*, 1998b). Root lesions coalesce and intensity of discoloration increases with time, although the colour of the lesions varies with the host (Townshend & Stobbs, 1981). Invasion of plant tissue by *Pratylenchus* spp. is generally thought to involve both mechanical force from stylet thrusting and pressure

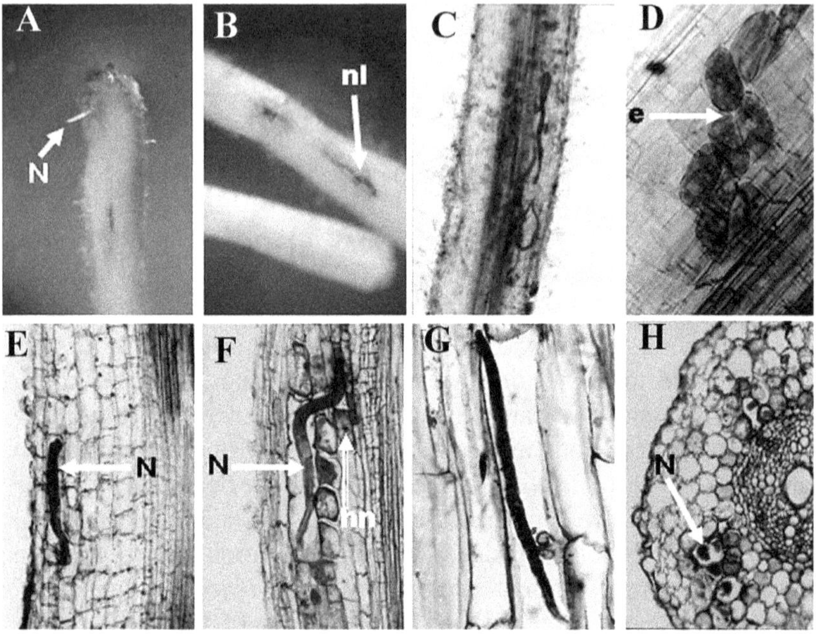

Fig. 117. *Histopathological changes induced by the feeding and migration of* Pratylenchus thornei *in roots of chickpea. A, B: Necrotic root lesions; C: Nematode specimens invading cortical parenchyma; D: Cortical parenchyma showing deposition of eggs; E-G: Longitudinal sections of chickpea roots showing nematode infection in cortical parenchyma, and thickened cell walls, hypertrophied nucleus, and granular cytoplasm of cells adjacent to nematode; H: Transverse section of secondary roots showing extensive damage in cortical tissue. Abbreviations: e = nematode egg; hn = hypertrophied nucleus; nl = necrotic lesion; N = nematode.*

from the labial region and the secretion of cell wall-degrading enzymes through the stylet. *Pratylenchus* nematodes inject secretions synthesised in their pharyngeal glands into plant cells *via* the stylet when feeding and migrating. These secretions may participate in extracorporeal digestion to facilitate ingestion of cytoplasmic components of the infected cell or may degrade the cell wall. In fact, high cellulase and pectinase activities were reported in homogenates of *P. zeae* (Krusberg, 1960) and *P. penetrans* (Morgan & McAllan, 1962). Myers (1965) also reported cellulase activity in *P. penetrans*, as well as a small amount of invertase activity (implicated in the hydrolysis of sucrose into glucose and fructose) and Zuckerman *et al.* (1966) demonstrated phenylalanine deaminase (implicated in the oxidative deamination of amino acids) in homogenates of *P. penetrans* by means of autoradiography with radioactive C^{14}. Currently, β-1,4-endoglucanases are believed to be of prime importance for the process whereby a parasitic nematode commences its interaction with the plant host (Goellner *et al.*, 2001). Endogenous production of cellulases has recently been demonstrated by the isolation of β-1,4-endoglucanase cDNAs from *P. penetrans* (Uehara *et al.*, 2001). These results confirm the presence of β-1,4-endoglucanases in *P. penetrans* and indicate that the invasion and intercellular migration of *P. penetrans* may be facilitated by the secretion of cellulases.

Pratylenchus spp. move through tissues by breaking down successive cell walls, resulting in a loss of cell turgor pressure and a gradual increase in the size of the nucleus, followed by cell death along the nematode's route. Many cortical cells are killed during the endogenous migration with rows of necrotic cells following the path of the nematodes. Neither hypertrophy or hyperplasia occur in adjacent cells, although these show some reactions in the form of increased vacuolation and tannin accumulation (Townshend *et al.*, 1989; Castillo *et al.*, 1998b). However, Townshend and Stobbs (1981) found that hyperplasia occurred in some tissues of trefoil infected by *P. penetrans*, cell multiplication occurring within long, spindle-shaped cells in the cortical parenchyma near the endodermis. Ultrastructural studies of the host-parasite relationship between *P. penetrans* and alfalfa roots revealed that cortical cells penetrated and fed upon by the nematode were generally devoid of cytoplasmic content (Fig. 118) (Townshend *et al.*, 1989). The remaining cytoplasm lacked integrity and recognisable organelles or membrane structure. Cells proximal to the line of cells penetrated by the nematode were typically degenerate and characterised by darkly staining, condensed and vesiculate cytoplasm which was adpressed to the cell walls (Fig. 118). Ultrastructural changes were seen in the cortical parenchyma, endodermis, pericycle and vascular cells, including condensed necrotic cytoplasm, increased tannin depositions in the cytoplasm, on the tonoplast and in the vacuoles, loss of membrane integrity in the plasmalemma, tonoplast and nuclear membrane, and degeneration of the mitochondria, dictyosomes and endoplasmic reticula (Townshend *et al.*, 1989).

Nevertheless, the responses of cortical cells adjacent to nematode feeding included increased rate of cytoplasmic streaming and gradual

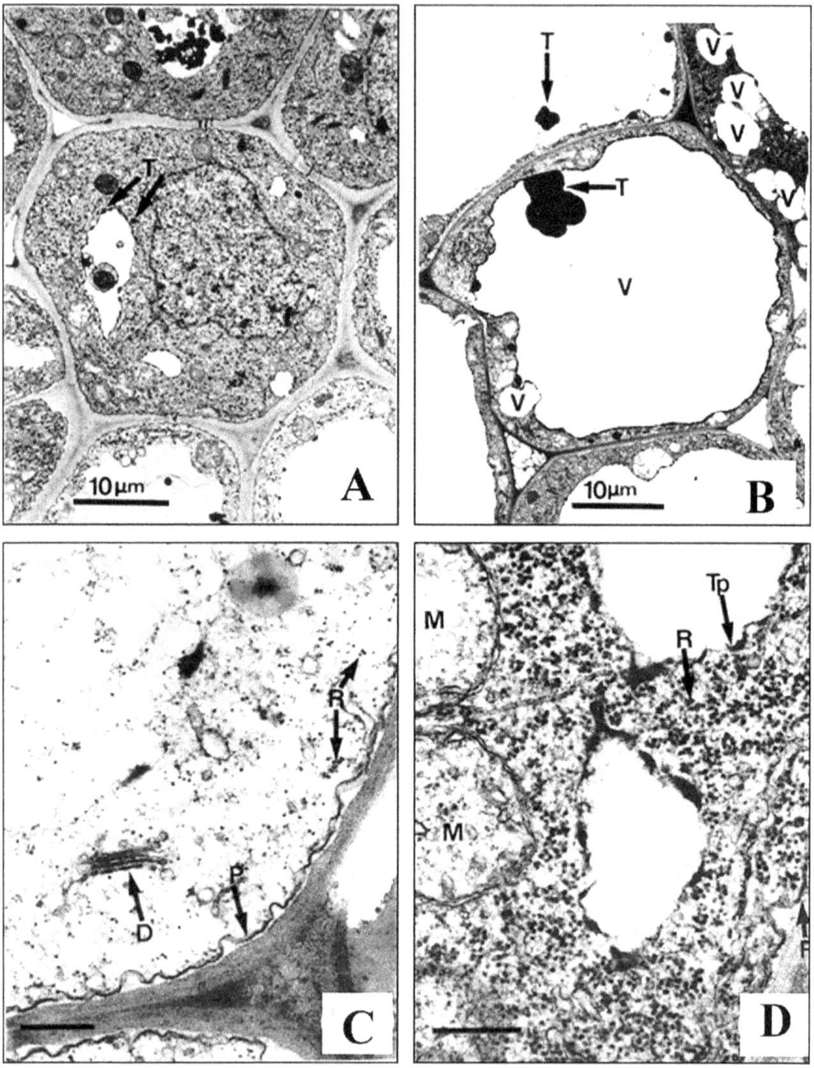

Fig. 118. *Ultrastructure of* Pratylenchus penetrans-*alfalfa root interactions. A: Healthy cortical cell in non-inoculated root; B: Tannin deposits on tonoplast of infected cortical cells; C: Cytoplasmic structure of healthy cortical cell in non-inoculated root; D: Cytoplasmic structure of cell bordering nematode feeding site. Abbreviations: D = dictyosome; M = mitochondria; P = plasmalemma; R = ribosomes; RER = rough endoplasmic reticulum; T = tanin deposition; Tp = tonoplast; V = vacuole. After Townshend* et al. *(1989).*

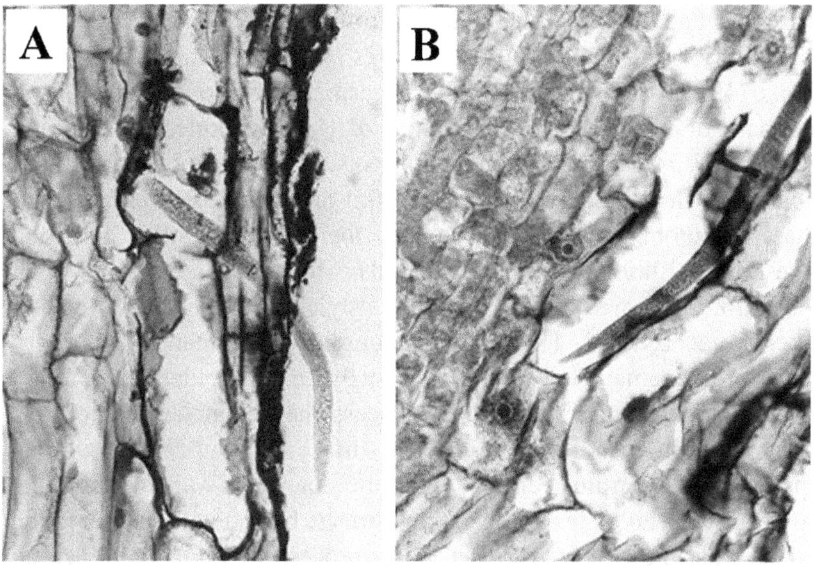

Fig. 119. *A, B: Histopathological changes induced by the feeding of* Pratylenchus thornei *in roots of chickpea showing hypertrophic nucleus and nucleolus in cells adjacent to cavities induced by the nematode.*

hypertrophy of the nucleus (Zunke, 1990a), as seen in chickpea roots infected by *P. thornei* (Fig. 119). Root tissue is also damaged when the nematodes leave the root and numerous dead epidermal cells occur on the root surface (Zunke, 1990a). Thus, in addition to the extensive tissue damage caused by nematode migration and extended feeding, the movement of nematodes into and out of the root provides entry points for other soil-borne pathogens, including bacteria (Kurppa & Vrain, 1985). Zunke (1990a) reported that the amphids of *P. penetrans* nematodes fixed inside the roots contain osmiophilic material, whereas nematodes fixed outside the root tissue have very little osmiophilic material in their amphids. Such material may be cell derived and may assist the nematode in orientation and sensing which cells to use as a food source.

Feeding tubes have been associated with salivation in several plant-parasitic nematodes (Hussey & Davis, 2004) although feeding tubes were not formed at any time during salivation by *P. penetrans* (Zunke, 1990a). The salivation zone formed around the stylet tip of *P. penetrans* is very small compared to other plant-parasitic nematodes and probably reflects the small body size and volume of the dorsal pharyngeal gland cell that can only discharge limited amounts of saliva (Zunke, 1990a).

Rebois and Huettel (1986) demonstrated the ectoparasitic feeding behaviour of *P. scribneri* on root cultures of corn, tomato and soybean. Zunke (1990b) also reported the ectoparasitic feeding of *P. penetrans* on root hairs of corn and rape and Orion *et al.* (1995) reported ectoparasitic feeding of *P. mediterraneus* on potato roots. The feeding behaviour of *P. penetrans* was studied by Zunke (1990a) and could be separated into phases of probing, cell penetration by the stylet, salivation and food ingestion over brief and extended periods. After cell penetration, a small 'salivation zone' was formed around the stylet tip. However, no feeding tubes were observed. These observations indicated that all migratory stages of the nematode fed ectoparasitically on root epidermal cells and root hairs. To initiate the feeding process the nematode positions its labial region perpendicular to the root surface, making contact with its lips and oral aperture. Penetration of the outer cell wall by the stylet tip occurred with as few as ten stylet thrusts, but often the stylet failed to penetrate the cell wall. The feeding process was initiated when an estimated 1-2 μm of the stylet tip had penetrated the cell wall and membrane and contacted the cytoplasm. The nematode body remained immobile throughout feeding. This lack of body movement during external feeding appears to be essential to prevent detachment from the feeding cell and feeding structure. When the stylet tip contacted the cytoplasm, the nematode secreted some material into the cell causing a rapid surge of cytoplasmic components away from the stylet tip. During the next 1-3 min, a small hyaline area or feeding structure appeared to envelop the stylet tip inserted into the cell or root hair. Formation of the feeding structure was followed by a rapid pumping (350 $\pm$ 50 pulses per min) of the metacorpus valve as ingestion began. During feeding the rate of cytoplasmic streaming at the feeding site and in adjacent cells increased slightly. Feeding at a single site lasted from less than 1 min to several hours. Defecation and egg laying occurred periodically while the nematodes were feeding (Zunke, 1990a). The extensive external feeding by migratory stages of *P. scribneri* and the occurrence of all life stages simultaneously and in large numbers without frequent entering or exiting of young roots suggest that this nematode can complete its life as an ectoparasite (Rebois & Huettel, 1986). Similarly, Webb (1996) reported that *P. crenatus*, *P. fallax*, *P. neglectus*, *P. penetrans*, *P. pinguicaudatus* and *P. thornei* fed ectoparasitically on four rape (*Brassica napus* L. var. *oleifera* (Moench) Metzg.) cultivars.

Although damage to endodermal cells occurred in some chickpea cultivars infected by *P. thornei*, it was not found feeding on endodermal cells, even in the most susceptible genotypes (Castillo *et al.*, 1998b). Olowe and Corbett (1976), however, found that *P. brachyurus* and *P. zeae* invaded all parts of corn roots including the stele, where a large amount of necrosis occurred. Thomason *et al.* (1976) hypothesised that a physical barrier or poor food source protected the endodermis of snap and lima beans from parasitism by *P. scribneri*. On the other hand, *P. penetrans* is capable of invading the stele of cabbage and strawberry roots (Townshend, 1963b; Acedo & Rohde, 1971) and heavy attack by *P. fallax* on wheat, barley and sugar beet induced thickening and discoloured endodermis (Corbett, 1972).

The external darkening of root tissues depends largely on the phenol content of the attacked cells and their ability to synthesise phenols after injury (Pitcher *et al.*, 1960; Townshend, 1963a, b). Tannins and other phenol compounds, such as chlorogenic acid, are believed to be important in disease resistance (Krusberg, 1963). β-glucosidase, secreted by *P. penetrans*, has been implicated in the release of phenols bound up in phenolic glycosides (Mountain & Patrick, 1959). Subsequent oxidation of these phenols by cytoplasmic polyphenoloxidases has been identified as the principle factor responsible for the browning and production of necrotic tissues (Acedo & Rhode, 1971). Some investigators have compared the amount of phenolics in plants with colonisation by *Pratylenchus*. MacDonald (1966) found that Sudan grass, a poor host for *P. penetrans*, had 4.7-6.1 times more phenol than winter vetch, a good host of the nematode, and Pitcher *et al.* (1960) found that cortical tissues of apple roots, which were tolerant to *P. penetrans*, were relatively free of phenols. Townshend and Stobbs (1981) found that much of the root discoloration resulted from oxidation of phenols involving the formation of lignin-like substances. In fact, when phenols were maintained in a reduced state by using ascorbic acid, characteristic lesions associated with *P. penetrans* in alfalfa did not form while nematode feeding continued (Townshend & Stobbs, 1981).

Kathiresan and Mehta (2002a) reported peroxidase activity in roots and leaves of a resistant clone (Co 7717) of sugarcane infected with *P. zeae*. In this resistant clone, localised necrosis and cell wall thickening, due to accumulation of peroxidase, were produced around nematode infection sites caused by *P. zeae*. However, in the susceptible sugarcane clone CoC 671, peroxidase activity significantly decreased in *P. zeae*-

infected plants (Kathiresan & Mehta, 2002a). These authors also reported acid phosphatase activity in the resistant clone of sugarcane after infection with *P. zeae* (indicating the stimulation of cytolysis at the infection site), whereas the decreased enzyme activity in the susceptible clone may be related to the increase of ATP, necessary for the protein co-polymerisation that favours a successful host-parasite relationship (Kathiresan & Mehta, 2002a). In addition, the acid phosphatase activity was higher in leaves compared to roots, suggesting that nematode infection induces systemic metabolic changes in the leaves. Kathiresan and Mehta (2003) also reported a decrease of catalase activity in the early and post-infection period of a resistant sugarcane clone. Thus a decline of catalase activity could result in a greater availability of free radicals, increased lipid peroxidation and membrane damage, inducing cell deterioration and resulting in a hypersensitive response. Hence, this response has an essential role in host resistance either by killing the nematode or forcing it to abandon the host. In addition, the nematode-induced catalase may therefore be effective in neutralising H_2O_2, a compound toxic to the parasite and acting as a strong signal which can, in turn, initiate systemic acquired resistance (Levine *et al.*, 1994). Moreover, in a *P. zeae*-infected susceptible sugarcane clone, catalase activity increased, inducing a lower level of superoxides and lipid peroxidation. Hence, the suppression of peroxidase activity could be a consequence of a lessened availability of H_2O_2 in susceptible genotypes because of the substrate-destroying activity of catalase (Zacheo *et al.*, 1990). Recently, Kathiresan and Mehta (2005) demonstrated that the constitutive levels of both phenylalanine ammonia-lyase (PAL) and tyrosine ammonia-lyase (TAL) activities in roots and leaves of sugarcane were greater in clones resistant to *P. zeae* than in susceptible clones. Their results suggest that *P. zeae* resistance in sugarcane clones involves an increase in the level of enzyme activities that produce or utilise H_2O_2. The increased PAL, TAL and esterase (EST) activities are triggered in the infected clones to reinforce the integrity of the cell wall by encoding lignin biosynthesis. Also, superoxide dismutase (SOD) activity decreases and the accumulations of superoxide anion concentration increases. This anion triggers a cascade of events leading to hypersensitivity. From these results they concluded that PAL, TAL, SOD and EST present in sugarcane might play a vital role in resistance to *P. zeae*, either directly or indirectly, through the production of lignin.

Penetration of host roots by *Pratylenchus* spp. is influenced by biotic and environmental factors (Castillo *et al.*, 1996b). In chickpea, root penetration of *P. thornei* increased steadily with increasing time of incubation; and temperature affected penetration rate in chickpea cvs UC 27 and JG 62 but not in line P 2245, being significantly higher at 20-25°C than that at 15°C. Roots of chickpea line P 2245 held at 15°C contained more *P. thornei* than cvs UC 27 and JG 62 and no difference in percentage penetration among host genotypes was observed at 20 or 25°C (Castillo *et al.*, 1996b). Similarly, Khatiresan and Mehta (2002b) reported that penetration of *P. zeae* in six resistant and six susceptible sugarcane clones was not significantly different in resistant and susceptible clones at 12 and 24 h after inoculation. However, after 168 h the number of nematodes that penetrated resistant clones (31 or 39%) was less than in the susceptible clones (45 or 65%). Additionally, several studies *in vitro* and *in planta* conditions have shown that *P. penetrans* may accumulate around a CO_2 source (Edmunds & Mai, 1966, 1967). In fact, these authors suggest that increased attraction of *P. penetrans* towards alfalfa roots was higher in alfalfa plants infected with *Trichoderma viride* Pers. and, particularly, *F. oxysporum*, which showed considerably higher CO_2 measurements than those found in healthy roots. Nevertheless, whether or not CO_2 emissions would produce a similar nematode response within the natural soil environment remains open to debate (Perry, 2005) and remains a largely unknown phenomenon in the field of plant nematology.

Nematology Monographs & Perspectives, 2007, Vol. 6, 365-407

Chapter 10

Management strategies of *Pratylenchus* species

The loss of many nematicides from the market due to environmental concerns and the cost of re-registration have focused attention on the development of alternative methods for managing plant-parasitic nematodes, including *Pratylenchus* spp. The impending complete loss of methyl bromide from the market has increased the urgency to develop such strategies for susceptible fruit and vegetable crops. Crop rotation, host plant resistance, cover crops, irrigation management, fallow periods, soil amendments and other cultural practices can all be used to minimise damage caused by *Pratylenchus* spp. Currently, most nematode management programmes focus on tactics involving the reduction of the initial nematode population density and/or the suppression of their reproduction during the growing season. Most nematode management tactics also have the inherent feature of limiting damage associated with the target crop. The purpose of this chapter is to summarise the management strategies used to reduce soil populations of *Pratylenchus* spp., including strategies for avoiding dispersion of *Pratylenchus* spp. in propagating material, crop rotation and cultural practices, the use of organic amendments and cover crops, the use of nematicidal plants, strategies of physical control (solarisation, thermotherapy), host plant resistance, biological control (beneficial endophytic microorganisms, nematophagous fungi, entomopathogenic and predaceous nematodes, the endospore-forming bacterium *Pasteuria penetrans* (Thorne) Sayer & Starr) and chemical control.

Pratylenchus nematodes are difficult to control once introduced into soil and much effort is spent on prophylactic measures. The choice of management tactic to reduce *Pratylenchus* spp. damage depends upon many factors, but all require accurate diagnosis of the species and population levels of *Pratylenchus* as assessed from soil and root samples taken from any given field or plant material. Both prerequisites of diagnosis and estimation of nematode populations are of great importance as damage thresholds may vary among *Pratylenchus* species and crops depending upon geographic location, crop value and the potential for disease complexes. Unfortunately, breeding for resistance to *Pratylenchus* species is difficult with the result that moderate resistance to *Pratylenchus* nematodes is presently limited to only a few cultivated crops (*e.g.*, forage legumes, potato, fruit rootstocks). Rotations with non-host crops also offer limited opportunities to manage *Pratylenchus* nematode field populations since most *Pratylenchus* species have wide host ranges, including both dicots and monocots. If the species of *Pratylenchus* is accurately diagnosed and a suitable economic non-host can be grown, rotations offer some promise as a management tactic. There are suggestions that microbial antagonists of *Pratylenchus* nematodes, such as soil fungi, can reduce population levels, but this has not yet been proven to be effective in production systems. The two most effective tactics for *Pratylenchus* nematode management remain sanitation and the use of chemical nematicides. However, the best way to manage *Pratylenchus* species is to prevent their introduction into a field or plant material. Choosing an uninfested field site or choosing non-host rotation crops are two ways to circumvent problems with *Pratylenchus* species. Once *Pratylenchus* spp. infest a field or a plant material it is highly unlikely that they can be eradicated. Inoculum levels can be reduced by turning over the soil layer to expose infected roots to the elements. Planting stock should be monitored and certified free of *Pratylenchus* nematode infestation. This is especially important for seedlings of crops that will be grown perennially (*e.g.*, fruits and ornamentals).

Thus, integrating nematode management practices involving exclusion, crop rotation and host resistance (where available), biocontrol practices and selective nematicide application may help to reduce nematicide usage and so delay the onset of loss of efficacy of chemical control agents.

Pratylenchus spp. in propagating plant material

Exclusion is the first control strategy to consider in *Pratylenchus* management as it is easier to deal with *Pratylenchus* nematodes before they become established in soil or plant material than it is subsequently to eradicate or manage them. Fortunately, *Pratylenchus* species have a number of characteristics that limit their long distance dispersal, including their restricted active movement, obligate parasitism, narrow host ranges of certain species, sharp population decline in the absence of hosts and survival depending on environmental conditions and crop management practices. Thus, long distance dispersal of these pathogens is largely passive and by chance. Key means of dissemination include movement of soil on equipment and plant parts, crop transplants, water, animals and contaminated containers. Measures for avoiding dissemination and establishment of new nematode problems should be a component of national or regional nematode-integrated pest management (IPM) programmes. Dispersal and establishment of infestations within soil adhering to seed-tubers, root-stocks or other transplanted material is fairly similar to seed transmission. This type of dispersal induces infestation foci which are small in the first year, extend in subsequent host crops and involve most of the planted area by the next growing season.

There are a number of examples of *Pratylenchus* spp. infecting planting material. *Pratylenchus brachyurus* infects roots, pegs and pods of peanut, overwintering in the shells of peanuts in the soil or in storage (Good *et al.*, 1958). Similarly, peanut shells used as mulch may carry live nematodes and disperse them over long distances (Colbran, 1968). *Pratylenchus coffeae* has been reported infecting planting material of white yam (*Dioscorea alata* L.) (Hutton, 1979) and imported corms of sago (*Cycas revoluta* Bedd.) in China contained *Pratylenchus* spp. (Li *et al.*, 1996). Potato tubers used as seed have been found to be infected by *P. scribneri* in Argentina (Chaves & Torres, 2001), by *P. penetrans* in Pakistan (Khan & Hussain, 2004) and by *P. neglectus*, *P. penetrans* and *P. scribneri* in Canada (Olthof & Wolynetz, 1991). *Pratylenchus brachyurus* heavily parasitises the roots and tubers of potatoes (Koen, 1967). Although tubers become infected as soon as they develop, the visible lesions only appear 4-5 months after planting when the tubers are practically mature. Nevertheless, potatoes lifted early, which may be apparently healthy, may nonetheless be infected and can contribute to nematode dispersion to new areas. Olthof and Yu (1999), however, reported that potato storage at low temperatures reduced populations of *P. penetrans* within the tubers of cvs Superior, Monona, Saginaw, Saginaw Gold, Yukon Gold and Kennebec. The number of live nematodes in potato tubers after 8 months of storage below 5°C was too low to pose a significant threat at planting to already infested fields but would still act as a source of a new infestation.

Pratylenchus damage in fruit crops and vegetable crops may be prevented by using sanitation practices that exclude nematodes from seedlings produced in nurseries and seed beds. Examples of *Pratylenchus* spp. that have been distributed frequently on fruit rootstocks are *P. coffeae* on citrus and coffee (Campos *et al.*, 1990; Duncan & Cohn, 1990); *P. loosi* on tea (Cohn & Duncan, 1990); *P. vulnus* on fig, peach and olive (Campos *et al.*, 1990; Nico *et al.*, 2002) and forest planting stocks of *Acacia* sp., *Cupressus macrocarpa* Hartw., *Juglans regia*, *Ligustrum japonicum* Thunb., *Morus* sp., *Pinus* spp., *Populus* sp., *Salix babylonica* L. and *Ulmus pumila* L. (Talavera *et al.*, 1999); and *P. penetrans* on olive (Nico *et al.*, 2002). Similarly, *P. brachyurus* and *P. vulnus* were introduced into Oman in infected planting material (Mani, 1998). In fact, the damage potential of *Pratylenchus* spp. has been recognised by Spanish, Italian and European Union legislation (BOE, 39/2000; Decreto Legislativo, 151/2000; Directiva Comunitaria, 67/1999). In these, phytosanitary regulations and quarantine measures concern infections of *P. penetrans* on ornamental plants and materials, especially in *Lilium* spp. and *Narcissus* spp., and *Pratylenchus* spp. on *Rosa* sp. Recently, several *Pratylenchus* spp. (*P. coffeae*, *P. crenatus*, *P. penetrans*, *P. vulnus*) have been intercepted in imported rootstocks of apple, pear, peach, plum, apricot and *Rubus* spp. from a number of countries (Arjun-Lal, 2005). Additionally, specific restrictions against several *Pratylenchus* spp. (*e.g.*, *coffeae*, *loosi*, *scribneri* and *zeae*) are imposed by several countries such as Argentina, Brazil, Chile, Egypt, Mexico, Morocco, South Africa, Syria and Uruguay (Hockland *et al.*, 2006).

Thus, the use of *Pratylenchus*-free planting material and uninfested soil during propagation of planting material is essential for minimising the effects of single or concomitant infections by these pathogens during the early years of cultivation and for preventing pathogen spread. The use of certified plant material, *Pratylenchus*-free planting stocks and clean farm equipment should be standard IPM practices and they are particularly important for vegetatively propagated crops such as banana and potato and for forest and fruit crops for which seedlings are transplanted.

Crop rotation and cultural practices

Crop rotation and cultural practices are ancient agronomic practices used by farmers throughout the world in an attempt to circumvent crop failures because of soil-related problems. It has been a management tactic of primary importance for limiting losses due to plant-parasitic nematodes. In general, *Pratylenchus* spp. are not host specific and have a very wide host range, making it difficult to find a suitable non-host crop for rotation. Consequently, the use of crop rotation is a limited option to manage *Pratylenchus* species. Nevertheless, some effective rotations have been developed, *e.g.*, in Sonora (Mexico), populations of *P. thornei* in wheat fields are reduced by rotations which include corn, cotton or soybeans for 2 consecutive years (Van Gundy *et al.*, 1974); in the USA corn is rotated with peanut, cotton and soybean to reduce populations of *P. brachyurus* and *P. zeae* (Johnson *et al.*, 1975); in Germany, sugar and garden beets were found to be suppressive in *Pratylenchus* spp.-infested fields (Decker, 1989); in Queensland (Australia) wheat is rotated with barley cv. Clipper to reduce populations of *P. thornei* (O'Brien, 1983); in potato production 2-year rotations with alfalfa or clover resulted in lower *P. penetrans* populations (Chen *et al.*, 1995); in strawberry, soil populations of *P. penetrans* were reduced by rotation with oat (*Avena strigosa* Schreb.) cv. Saia (LaMondia, 1999b). Similarly, Townshend (1989a) evaluated oat as a rotation crop for tobacco fields infested with several lesion nematode species and found that cv. Saia was more resistant to *P. penetrans* and *P. neglectus* than cv. Woodstock, whereas *P. crenatus* did not develop on either cultivar. Control of *P. zeae* has been reported on rice by crop rotation with non-host crops such as legumes, *e.g.*, *Vigna radiata* (L.) Wilczek, *Vigna mungo* (L.) Hepper, *V. unguiculata* (L.) Walp and *S. indicum* L. (Prasad & Rao, 1978). In fact, control of *P. zeae* infesting rice fields in the Philippines was effective by cropping legumes (cowpea, mung bean), which increased rice yield by 37% compared with cropping with rice, corn and sorghum (Aung & Prot, 1990). Similarly, Pankhurst *et al.* (2005b) demonstrated that populations of *P. zeae* in sugarcane production in Australia were reduced significantly by rotation with sown pasture, alternative crops and bare fallow for different periods of time before being replanted with sugarcane. Furthermore, a step toward quantitative nematological assessment of crop rotations was carried out by Van den Berg and Rossing (2005) using multi-year dynamic models which demonstrated that the six possible rotations comprising lettuce, leeks, carrots and a fallow year reduced populations of *P. penetrans* and that these reductions involved significant economic gains.

Some conifers reduce the population density of *P. penetrans* in nursery stock grown on light sandy soils in The Netherlands and do not suffer damage from the nematode, thus having potential to be used in a crop rotation strategy to control the pathogen. Bertrums (1998) demonstrated that *Taxus baccata* significantly reduced nematode numbers in soil. Although the roots were found to be lightly infested at the end of the growing season, the crop itself did not suffer any nematode damage. Conversely, other crop rotations such as *Picea omorika* (Paèiae) Purkyne, *Ligustrum ovalifolium* Hassk. or *Amelanchier lamarckii* F.G. Schroed. only slightly reduced nematode populations in soil (Bertrums, 1998).

Under controlled conditions Potter and McKeown (2002) reported that populations of *P. penetrans* significantly declined by intercropping with *Rudbeckia hirta* L. and tomato in pot cultivation. Kimpinski and Sanderson (2004) found that populations of *P. penetrans* on carrots (cv. Neptune) in Canada were significantly reduced after rotations with African marigold (*Tagetes erecta* L. cv. Crackerjack). Intermediate reductions were obtained after rotations with barley (*Hordeum vulgare* L. cv. Chapais), pearl millet (cv. Millet 101) and annual ryegrass (*Lolium multiflorum* (Lam.) Parnell cv. Lemtal), but no reduction was detected when carrots were followed with timothy (*Phleum pratense* cv. Common).

Forage and grain pearl millet significantly reduced *P. penetrans* populations in tobacco in Quebec in comparison with rye, the standard rotation crop (Bélair *et al.*, 2004). Moreover, forage pearl millet increased tobacco yields by an average of 103% compared with non-fumigated rye, and was not significantly different from the plots treated with metham sodium following rye. Likewise, Bélair *et al.* (2005) found that a 1-year crop rotation with forage and grain pearl millet (*P. glaucum*) reduced *P. penetrans* populations in three potato (*Solanum tuberosum* cv. Superior) fields in Quebec and provided an increase in potato yields the following year. These crops were compared to oats and (or) barley and showed that forage millet had a suppressive effect on *P. penetrans* populations after a 1-year rotation. Consequently, crop rotation with pearl millet would be an efficient and economically viable alternative to these crops for controlling *P. penetrans* populations and for improving potato yields in this area.

Unfortunately, some crop rotations used to control other plant-parasitic nematodes, such as root-knot nematodes, may result in an increase in *Pratylenchus* populations. A good example is the rotation with barley to reduce the populations of *Meloidogyne chitwoodi* Golden, O'Bannon, Santo & Finley, 1980 in potato production which, however, increased the densities of *P. neglectus* (Ferris *et al.*, 1994). Mixed cropping systems of corn and cowpea significantly reduced populations of *P. pseudopratensis* in Nigeria relative to monocultures of corn (Egunjobi *et al.*, 1986). Similarly, Bolton and De Waele (1989) showed that several commercial sunflower hybrids were poor hosts of *P. zeae* and therefore might be useful as rotation crops in fields of corn heavily infested with *P. zeae*.

The basic principle for the use of crop rotations for diseases caused by *Pratylenchus* spp. is that monoculture of a host plant usually results in increased inoculum density and consequent yield losses. However, some long-term monoculture experiences indicate that monoculture may also reduce *Pratylenchus* populations. Andersen (1975) reported that populations of *P. crenatus* and *P. neglectus* reached highest numbers in the first 3 years of barley monoculture, after which population densities decreased gradually and stabilised at a lower level.

Fertilisation with inorganic sources of N has also been observed to modify populations of *Pratylenchus* spp., *e.g.*, Dmowska and Ilieva (1995) reported that *Pratylenchus* spp. were more abundant in plots of barley fertilised with N over 22 years than in non-fertilised plots; and Sarathchandra *et al.* (2001) found an increase in populations of *Pratylenchus* spp. in N fertilised clover-ryegrass pastures. Melakeberhan (2006) has also reported an increase of *P. penetrans* populations related to higher levels of fertilisation.

Minimum tillage is being adopted rapidly in several agricultural systems, but few reports exist on its effect on population densities of *Pratylenchus*. Thomas (1978) reported that population densities of *P. hexincisus* and *P. scribneri* on corn were significantly higher in no-tillage plots than in any other tillage system including: autumn plough, spring plough, chisel plough, offset disk, till-plant and no-till flat. However, other reports presented contradictory results, *e.g.*, Alby *et al.* (1983) found that populations of *P. scribneri* in conventional tillage soybean were significantly higher than in no-tillage systems and

Bergeson and Ferris (1986) reported that populations of *P. hexincisus* on corn were significantly higher in mouldboard-tilled plough than in no-tillage. Recently, Govaerts *et al.* (2006) reported that in the semi-arid and rain-fed subtropical highlands of central Mexico, *P. thornei* populations were higher in a wheat cropping system that included zero tillage, crop rotations (wheat/corn) and crop residue retention, compared with common farming practices. This lack of agreement in the reports on the influence of tillage on *Pratylenchus* populations might be explained, at least in part, by environmental conditions or soil type.

Planting date also can be used to limit *Pratylenchus* reproduction on crops. *Pratylenchus brachyurus* has a relatively high temperature optimum (28-30°C), which influences its damage potential on soybean. Optimum soil temperatures for *P. brachyurus* activity normally occur relatively late in the growing season, and delayed planting resulted in a linear decrease in the number of *P. brachyurus* at soybean harvest. Effects of planting date on nematode numbers persisted over winter, indicating that survival in the absence of a host is density independent. Compared with winter fallow, winter wheat reduced winter survival of *P. brachyurus* (Koenning *et al.*, 1985).

Fallowing has been suggested as an efficient cultural practice to restore soil fertility as measured by an increase of plant growth and yield. The pathogenicity of certain plant-parasitic nematodes on millet was reduced without the necessity for physical elimination of nematodes but by increasing nematode diversity by fallowing, leading to a less pathogenic nematode community in the environment, *e.g.*, *Scutellonema cavenessi* Sher, 1964 was replaced by *P. pseudopratensis*, which is less pathogenic on millet (Cadet & Floret, 1999). In addition, fallowing, combined or not with other control strategies, has been reported as an effective method to reduce populations of *Pratylenchus* spp., *e.g.*, Ornat *et al.* (1999) reported that fallowing (about 5 weeks) combined with root destruction of French bean was more effective than fallowing alone to reduce populations of *P. neglectus*.

Finally, grafting may be another cultural practice used for the control of *Pratylenchus* spp. Grafting of coffee (*C. arabica*) onto *Coffea canephora* Pierre revealed efficient control of populations of *P. coffeae* in Guatemala and resulted in significantly greater yields compared to non-grafted plants (Villain *et al.*, 2000).

ORGANIC AMENDMENTS

Organic matter in the soil acts as a substrate for microorganisms that are either parasitic on *Pratylenchus* or produce toxins, directly or through breakdown products, that are detrimental to nematodes. The addition of nitrogenous soil amendments results in the formation of ammonia, a compound with nematicidal properties (Rodríguez-Kábana, 1986). Chitin, the most commonly occurring nitrogen-containing polysaccharide in nature (Alexander, 1977), has been used as a soil amendment to control *Pratylenchus* spp. (Bell *et al.*, 2000). Similarly, Sarathchandra *et al.* (1996) demonstrated that soil amendments with crab shell increased the growth of white clover and perennial ryegrass (*Lolium perenne* L.) and significantly reduced soil populations of *Pratylenchus* sp. Although the mode of action of chitin in controlling *Pratylenchus* spp. was not well understood, the suggested mechanisms include increased microbial chitinase activity that may damage chitin-containing egg shells and the nematicidal activity of increased ammonia levels released by chitin hydrolysis (Mian *et al.*, 1982). Remarkable reductions in soil populations of *P. brachyurus* and yield increases of corn in Nigeria have been achieved experimentally with the addition of cocoa pod husks and farmyard manure (Egunjobi & Larinde, 1975).

Oil radish (*R. sativus*) planted in field microplots after harvest of wheat in summer and incorporated as a green manure in autumn significantly reduced soil population densities of *P. neglectus* in the next crop of potato, probably due to a high production of nematicidal compounds, such as breakdown products of glucosinolates (Al-Rehiayani & Hafez, 1998).

A range of organic amendments from farmyard manure, including chicken litter or other animal wastes, have provided considerable protection for plants against plant-parasitic nematodes (Rodríguez-Kábana *et al.*, 1987). Koenning and Barker (2004) showed that the application of poultry litter at 0.0, 6.7, 13.4 and 20.1 t ha^{-1} reduced the population of *P. brachyurus* in two cotton fields in North Carolina. Similarly, Conn and Lazarovits (1999) suppressed *P. penetrans* populations under potato with pre-plant application of poultry manure at 1573 kg total N ha^{-1} or 460 kg NH_4^- N ha^{-1}. Although the mechanisms leading to increased *Pratylenchus* spp. population densities in manured and fertilised soil are not clear, possibilities include increased fecundity of nematodes feeding on nutrient-enriched roots, increased root growth providing more feeding sites and suppression of natural biological control agents (Forge *et al.*, 2005).

COVER CROPS

Cover crops are crops planted between cycles of the main cash crop or intercropped with cash crops to improve soil fertility, soil structure and water infiltration and to control soil erosion, weed and plant-parasitic nematodes (McSorley, 2001). Recently, there has been an increasing interest in using cover crops to develop nematode management strategies in sustainable agricultural systems (Barker & Koenning, 1998). Several examples have demonstrated the usefulness of these cover crops in reducing populations of *Pratylenchus* spp. Cover crops resistant to *P. penetrans*, *e.g.*, redtop (*Agrostis alba* L.), red fescue (*Festuca rubra* L.) and oat cv. Saia, were used in red raspberry plantings (*Rubus idaeus* L.) to minimise soil erosion and to suppress nematode populations (Vrain *et al.*, 1996).

The inclusion of canola (*Brassica napus* L.) in Australian cereal rotations increased some years ago due to improved crop adaptation, increased security of markets and increasing awareness of its effectiveness as a disease break within the rotation. Canola rotations have been associated with reductions in numbers of harmful soil-borne organisms, resulting in earlier crop establishment, increased seedling vigour and increased yields of successive wheat crops (Kirkegaard *et al.*, 1994). Potter *et al.* (1998) found that incorporation of *Brassica* leaf and root tissues into soil containing *P. neglectus* caused a significant reduction in nematode numbers when compared to soil amended with wheat tissues. Also, amendment of soils with equimolar levels of purified 2-phenylethyl isothiocyanate resulted in comparable levels of nematode mortality, suggesting that 2-phenylethyl glucosinolate has a role in the suppressive impact of *Brassica* spp. root tissues (Potter *et al.*, 1998). Winter rapeseed (*Brassica napus* L. ssp. *oleifera* (Metzg.) Sinsk.) cultivar Bienvenue was found to be a non-host of *P. fallax* (Webb, 1990) and cvs Bridger, Gorzanski, H-47, Lindora and Viking were also found to be poor hosts for *P. scribneri* (Bernard & Montgomery-Dee, 1993).

Several leguminous crops have been used as cover crops to reduce *Pratylenchus* spp. Trefoil (*Lotus uliginosus* Schkuhr cv. Maku) significantly decreased the populations of *P. vulnus* (most probably misidentified and could refer to *P. brachyurus*, *P. hexincisus*, *P. penetrans*, *P. scribneri* or *P. zeae*) in corn roots (Boyer *et al.*, 1999). When trefoil was associated with earthworms (*Amynthas corticis* Kinber, 1867), nematode populations inside corn root were significantly lower under trefoil with earthworms compared with corn alone. The presence of earthworms introduced with trefoil strongly affected the relative abundance of nematodes in corn roots and in soil and a high number of nematodes in trefoil roots was observed. Thus, the combined effects of trefoil cover and earthworm inoculation resulted in decreases in nematode populations in corn roots accompanied by increases in corn production and soil carbon mineralisation. Several antagonistic mechanisms attributed to the cover crop and earthworms had been suggested. The high number of nematodes detected in roots of trefoil showed that this cover crop is a good host plant and may be more attractive to this nematode than corn, thereby resulting in fewer nematodes in the corn roots. Earthworms could have two different effects on nematodes. Firstly, the creation of macropores and soil compaction by earthworm activity (Blanchart *et al.*, 1990) could impair the movement of nematodes in soil and so depress populations, and secondly, passive soil ingestion of nematodes by earthworms may alter their parasitic potential during their transit through the gut due to the activity of digestive enzymes (Boyer & Reversat, 1996).

Mazzola and Gu (2000) demonstrated that wheat cover cropping during apple orchard renovation resulted in significant reductions in apple root populations of *P. penetrans*, suggesting that cropping wheat prior to orchard renovation may be useful in managing lesion nematodes on replant sites where *Pratylenchus* spp. are abundant or where this nematode interacts with other soil-borne pathogens to incite apple replant disease.

Canavalia ensiformis (L.) DC. (jack bean), *Lablab purpureus* L. (hyacinth bean) and *Mucuna pruriens* (L.) DC. (velvet bean) significantly reduced *P. zeae* population by about 30% and increased corn yield by about 25% (Arim *et al.*, 2006). *Canavalia ensiformis* was more effective than *M. pruriens* in reducing *P. zeae* damage on maize. The cause of these differences may be related to the fact that *C. ensiformis* takes up more P from the soil than *M. pruriens* (Wortmann *et al.*, 2000), and this is likely to impede nematode reproduction and population increase since low levels of P are associated with a decrease in lesion nematode population (Yeates, 1976).

Unfortunately, there are also indications that some attempts to use cover crops or organic amendments to control *Pratylenchus* spp. have been unsuccessful; *e.g.*, Yamada *et al.* (2003) reported that leguminous green covers, including crimson clover (*Trifolium incarnatum* L.), woolly pod vetch (*Vicia dasycarpa* Ten.) and resistant soybean (cv. Suzuhime), increased the density of *P. penetrans* in Japan. Similarly, sustained use of dairy manure slurry for 4-6 years increased *P. penetrans* population densities in tall fescue relative to the non-treated control (Forge *et al.*, 2005). The mechanisms leading to increase of *P. penetrans* population densities in manured and fertilised soil are not clear, but possibilities include increased fecundity of nematodes feeding on nutrient-enriched roots, increased root growth providing more feeding sites and even suppression of natural biological control agents.

NEMATICIDAL PLANTS

Nematicidal compounds have been isolated from antagonistic plants and have drawn the attention of nematologists in recent years. In fact, many compounds with nematicidal activity against *Pratylenchus* spp., such as dithioacetylenes, glycosides, glucosinolates, have been found in different plant species (Ferraz & De Freitas, 2004). Glucosinolates occurring naturally within the tissues of *Brassica* spp. have been found to suppress *Pratylenchus* populations. Organic amendments of leaf and root tissues of *Brassica* spp. significantly reduced populations of *P. neglectus* and levels of 2-phenylethyl glucosinolate within root amendments were significantly correlated with nematode suppression (Potter *et al.*, 1998, 1999). Similarly, Mazzola *et al.* (2001) demonstrated the impact of canola (*Brassica napus* L.) seed meal with high glucosinolate content on the microbial complex that incites apple replant disease, which significantly enhanced growth of apple and suppressed apple root infection by *P. penetrans*.

The ever popular marigold (*Tagetes* spp.) has been shown to be allelopathic to *Pratylenchus* spp., producing a number of potential bioactive compounds; however, the main active nematicidal compound in marigolds is thiophene α-therthienyl found in the foliage, seed extracts and roots of plants (Ferraz & De Freitas, 2004; Riga *et al.*, 2005) which, either by photoactivation with near ultraviolet light (Bakker *et al.*, 1979) or by the action of peroxidase or other activators (Gommers & Bakker, 1988), generates reactive oxygen species, probably O_3, which eventually kill the nematodes. Gommers and Bakker (1988) suggested that the nematicidal properties of marigolds inside the roots may result from a cascade of chemical reactions triggered by the penetration and movement of the nematodes through the cortex.

French marigold (*T. patula* cv. Creole) and African marigold (*T. erecta* cv. Cracker Jack) have been used in rotation cropping systems in Ontario (Canada) for the biological control of *P. penetrans* infesting fields of flue-cured tobacco (Reynolds *et al.*, 2000). These results indicated that a marigold plant density of about 20 plants m^{-2} reduced *P. penetrans* population densities to levels below the economic threshold for the rotation crop year and the 2 following years (Reynolds *et al.*, 2000). Also, *P. penetrans* soil populations declined about 90% from the preplant level when African marigold was grown in rotation with potato and with tomato (Alexander & Waldenmaier, 2002; Ball-Coehlo *et al.*, 2003). Pudasaini *et al.* (2006b) demonstrated that rotation with French marigold (after a growing period of 105 days) caused an average reduction of *P. penetrans* population of 90% in carrot. In addition, the effect of marigold was persistent as no increase in *P. penetrans* densities over the entire soil profile was noticed after two crop cycles of host plants (Pudasaini *et al.*, 2006b). Likewise, root populations of *P. alleni* from tomato simultaneously cultivated with marigold were significantly fewer than from tomato cultivated alone (Hackney & Dickerson, 1975).

Neem products (*Azadirachtica indica* A. Juss.) have also been shown to be effective against different plant-parasitic nematodes attacking vegetables and legumes (Alam, 1991) and soil amendment with neem seems to be one of the most practical methods for nematode control (Alam, 1993). Musabyimana and Saxena (1999) reported that soil applications of powdered neem seed or neem cake at 100 g/plant at planting and, subsequently, at 3-month intervals, reduced the populations of *P. goodeyi* on balance with a nematicide (carbofuran) applied at 40 g/plant at planting and then at 6-month intervals to banana plants grown under controlled conditions. Banana plants treated with powdered neem cake, seed or kernel or with neem oil had 4-95 times fewer parasitic nematodes than the untreated control and maintained the nematode population below the economic threshold (Musabyimana & Saxena, 1999). Abbasi *et al.* (2005) reported that 1% neem cake (mass/mass soil) caused a 67-90% reduction in the number of *P. penetrans* under glasshouse conditions. The bioactivity of neem materials against pathogens is attributed to the presence of an array of complex compounds, triterpenes or, more specifically, limonoids (Kraus, 1995).

Plant extracts of several plant species have also been found to exhibit nematicidal activity. Green leaves of several species, including *Abutilon indicum* (L.) Sweet, *A. indica*, *Calotropis procera* (Ait.) Ait. f., *Cassia auriculata* L., *Crotalaria juncea* L., *Datura stramonium* L., *Prosopis juliflora* (Sw.) DC., *Tridax procumbens* L., *Vitex negundo* L. and *Xantium indicum* L., were effective in reducing the population of *P. coffeae* on banana in India, significantly increasing plant growth and yield as a result (Sundararaju & Cannayane, 2002; Sundararaju *et al.*, 2003). Some plant species have been reported as having positive and negative results in the control of *Pratylenchus* spp., *e.g.*, *Crotalaria* spp. (Wang *et al.*, 2002). *Crotalaria*, besides being an efficient green manure, is a poor host to many important plant-parasitic nematodes, producing allelopathic compounds toxic to nematodes, and is able to enhance some nematode-antagonistic microorganisms. However, before using *Crotalaria* as a cover crop it is necessary to know the host suitability of the *Crotalaria* sp. regarding the *Pratylenchus* sp. infesting soil. For example, *Crotalaria grantiana* Harvey controlled *P. brachyurus* and *P. zeae* populations on corn (Da Silva *et al.*, 1989), while *Crotalaria paulina* Schrank was ineffective because it was as susceptible as corn to *P. brachyurus* and *P. zeae* (Desaeger & Rao, 2001).

Plant extracts of harmal, *Rhazya stricta* Decne, exhibited a high toxicity against all life stages of *P. zeae* (Mani & Al-Hinai, 1998). Some grasses of the family Poaceae have also shown nematicidal properties, *e.g.*, *Eragrostis curvula* (Schrad.) Nees. significantly reduced the population of *P. loosi* when used in rotation with tea (Gnanapragasam, 1997). The use of dry and fresh leaves of several plants, such as *A. indica*, *Calotropis procera* L., *Datura stramonium* L., *C. juncea* and *Vitex negundo* L., was effective in reducing the populations of *P. coffeae* in banana and significantly increased the yield (Sundararaju *et al.*, 2003). Populations of *P. zeae* were significantly reduced by plant extracts of *A. indica* (Mehta & Sundararaj, 1999) and *C. procera* (Mehta & Sundararaj, 1995) in sugarcane and chickpea seeds coated with latex extracted from *Calotropis gigantea* (L.) W.T. Aiton, *Euphorbia pulcherrima* Willd. *ex* Klotzsch or *Carica papaya* L. significantly reduced soil populations of *P. thornei* in microplots at 45, 90 and 120 days after application (Sebastian & Gupta, 1996). Aqueous rhizome extract of

Acorus calamus L. showed nematicidal activity against *P. thornei* and mortality of the nematode increased with increase in exposure period from 6-48 h (Romabati *et al.*, 1999). Extracts of plant sources of Chinese herbal remedies, *Schizonepeta tenuifolia* Briq., *Artemisia capillaries* Thunb. and *Crataegus pinnatifida* Bge., were effective against *P. vulnus* (Ferris & Zheng, 1999). Oil cake of *A. indica* incorporated into the soil was effective in reducing nematode populations of *P. delattrei* in crossandra (Jothi *et al.*, 2004) whilst Evenhuis *et al.* (2004) reported that French marigold (*T. patula*) culture reduces *P. penetrans* population density in strawberry fields in The Netherlands to almost zero. The yield of strawberry cv. Elsanta was greater from the crop grown 3 years after *T. patula* than that after soil fumigation with 750 l ha^{-1} metam sodium 4 years previously. Possibly French marigold diminishes *P. penetrans* populations at greater depths than soil fumigation because their roots penetrate to a depth of at least 30 cm and therefore are active in that zone, whereas soil fumigation is usually most effective at a depth of 20 cm, activity rapidly declining towards the soil surface and deeper soil layers (Been & Schoemaker, 1999). Aqueous extracts of the tree *Quillaja saponaria* Molina, containing triterpenoid saponins, polyphenols, salts and sugars, have been tested against *P. neglectus* and *P. thornei* in Chile (San Martin & Magunacelaya, 2005), resulting in important nematicidal effects at economically attractive doses (*e.g.*, 100 ppm).

In addition, the use of natural low molecular weight volatile compounds for the control of soil-borne plant pathogens is considered a potential alternative to the use of commercial soil biocides. Calvet *et al.* (2001) evaluated the survival of *P. brachyurus* in saturated atmospheres of several natural chemical compounds: thymol obtained from the essential oil of thyme (*Thymus vulgaris* L.), carvacrol in the oil of oreganum (*Origanum vulgare* L.), borneol found in the oil of rosemary (*Rosmarinus officinalis* L.), p-anisaldehyde, benzaldehyde and cinnamaldehyde found in fennel (*Foeniculum vulgare* L.), in kernels of bitter almond (*Prunus dulcis* L.) and in cinnamon (*Cinnamomum cassia* L.) respectively; salicilaldehyde in *Salix alba* L., geraniol from *Rosa* sp. and citral found in balm (*Melissa officinalis* L.) and in the oil of lemon grass (*Cymbopogon* sp.). Among these compounds, benzaldehyde, salicilaldehyde, borneol, p-anisaldehyde and cinnamaldehyde caused more than 50% mortality of *P. brachyurus*.

Physical control

SOIL SOLARISATION

Soil solarisation, the process of heating soils under transparent plastic tarps to temperatures detrimental to soil-borne pathogens, has successfully controlled a variety of plant diseases (Katan, 1981). Solarisation is an original approach because it targets mesophyllic organisms, which include most plant pathogens and pests, without destroying the beneficial mycorrhizal fungi and growth-promoting rhizobacteria (Stapleton & De Vay, 1984). The efficacy of soil solarisation is dependent on the thermal dose, itself a product of the temperature and exposure time, the thermal sensitivity of the organisms and the chemical, physical and biological characteristics of the soil, including the moisture content. There are many examples indicating the effectiveness of solarisation in controlling *Pratylenchus*. For instance, in Australia, soil solarisation reduced populations of *P. penetrans* in celery (Porter & Merriman, 1985) and in Syria, under field conditions, soil solarisation significantly reduced populations of *P. thornei* and increased yield of chickpea (Di Vito *et al.*, 1991) and faba bean cv. ILB 1814 (Sauerborn *et al.*, 1990). In the San Joaquin Valley of California, population densities of *P. hexincisus* in almond and peach tree roots were significantly reduced by soil solarisation (Duncan *et al.*, 1992) and soil solarisation was useful to disinfest nursery soils infested with *P. vulnus* that attack olive in California's inland valleys (Stapleton *et al.*, 1999). Soil solarisation also reduced, but did not eliminate, *P. penetrans* compared to the non-solarised treatments in the upper 30 cm soil profile of cherry in Western Oregon (Pinkerton *et al.*, 2000). In Croatia, soil populations of *P. crenatus*, *P. fallax* and *P. neglectus* were drastically reduced (by almost 100%) by soil solarisation at a depth of 10 cm, but only 67% reduction was achieved at a depth of 20 cm (Ostrec & Grubisic, 2003). Minagawa *et al.* (2004) reported that *P. penetrans* was greatly reduced in the top 10 cm of the soil layer after solarisation of a carrot field for 4 weeks by 0.02 mm polyethylene film mulch in central Japan. *Pratylenchus penetrans* was not isolated from soil in a polyethylene bag after water bath treatment at 50°C for 1 h, 45°C for 72 h and 40°C for 120 h.

THERMOTHERAPY

Hot water treatment may be also an option to control *Pratylenchus* nematodes in planting material, *e.g.*, infections of banana corms by *P. goodeyi* (Bridge, 1975; Colbran, 1967), or infections of yam (*Dioscorea rotundata* Poir.) tubers by *P. coffeae* (Acosta & Ayala, 1976). Nevertheless, eradication of such infections may be very difficult and dependent on several factors, *e.g.*, treatment of yam tubers with extreme dry rot is less effective and can cause severe physiological damage to plant material (Coates-Bedford & Brathwaite, 1977). Hot water treatments (43°C for 1 h) of Lily-of-the-valley rhizomes was proved to suppress *P. convallariae* (Goffart, 1963). Thermotherapy has proved effective in the control of *Pratylenchus* spp. infecting young fruit trees and woody plants, *e.g.*, hot water treatments (46-47°C for 30 min) of young apple trees controlled infections by *P. penetrans* (Way, 1973) and hot water treatments (48°C for 35 min) of rose plants infected by *P. vulnus* significantly reduced nematode populations (Townson & Lear, 1982b). Elsen *et al.* (2004) reported recolonisation of banana roots in hot water-treated banana planting material sowed in infested soil, the banana plants becoming as severely infected by *P. goodeyi* as untreated planting material 3 years after treatment. In most cases, these treatments must be applied when plants are in a state of absolute dormancy and after hot water immersion the plant material should be immediately submerged in a cold bath in order to circumvent harmful side effects.

Host plant resistance

Due to increasing concern about environmental contamination by pesticides, both plant resistance and tolerance to plant-parasitic nematodes have increased in importance during the past decades. Host plant resistance offers the easiest, safest and cheapest long-term approach to minimise the effects of *Pratylenchus* diseases in crops. In plant nematology, resistance is defined as the ability of a host plant to suppress reproduction of a nematode and is usually governed by a single dominant gene (Cook & Evans, 1987; Trudgill, 1991; Barker, 1993). Tolerance describes the amount of injury caused by the nematode (Cook & Evans, 1987; Wallace, 1987b); and it refers to the ability of the plant to withstand or to recover from the injury caused by the nematode attack (Trudgill, 1991; Roberts, 2002). Resistant cultivars not only suppress nematode reproduction, they also reduce the need for toxic chemicals and shorten rotation periods. To be economically viable, resistant cultivars should yield as much as high-yielding susceptible cultivars treated with nematicides and should have no need for specialised equipment and growing techniques and hence incur no additional costs (Boerma & Hussey, 1992; Young, 1998). The use of resistant cultivars is advantageous in integrated control programmes (Thies *et al.*, 1992). Although resistance is not inherently better than other approaches to manage nematodes, neither are other approaches universally superior to resistance (Starr *et al.*, 2002). As occurs in root-knot nematodes, resistance to *Pratylenchus* spp. may be altered by extreme environmental factors, particularly temperature; *e.g.*, Mountain (1954) showed that a variety of tobacco (Green Briar) was resistant to *P. neglectus* at 21°C, but this resistance broke down when soil temperature was raised to 38°C. Nevertheless, where resistance is lacking, tolerance offers an acceptable alternative. Plant resistance mechanisms include morphological or preformed biochemical barriers and active plant-defence responses (Kaplan & Davis, 1987). However, accurate assessment of nematode infestations and infections is critical for evaluation of plant resistance and tolerance to *Pratylenchus* spp., especially in field situations. Significant advances have been made in detecting resistance to *Pratylenchus* nematodes in various crops using available cultivars, breeding lines and germplasm including related species (De Waele & Elsen, 2002; Peng & Moens, 2003).

FRUIT TREES AND WOODY PLANTS

Studies have showed that *Malus* germplasm varies in resistance or tolerance to apple replant disease caused by *P. penetrans*. Seedling accessions of *Malus sieversii* Roem. and *Malus kirghisorum* Ponom. appear, based on the low relative reproduction index of *P. penetrans*, to have some tolerance to apple replant disease (Isutsa & Merwin, 2000). Likewise, resistance or tolerance to *P. penetrans* was found in seedlings of peach rootstock cultivars Rubira, Pisa, Rutgers, Red Leaf, Tzim Pee Tao and hybrids of Rutgers Red Leaf × Tzim Pee Tao (Potter *et al.*, 1984; Layne, 1987; Ramming & Cociu, 1990).

The search for resistance to *P. vulnus* in *Prunus* spp. has been less successful than for *Meloidogyne* spp.; *e.g.*, Stalin *et al.* (1998) found that of five *Meloidogyne*-resistant and four *Meloidogyne*-susceptible myrobalam clones, all were susceptible to *P. vulnus*. These results illustrate the absence of a resistance relationship between *Meloidogyne* spp. and *P. vulnus* in *Prunus* spp., since the dominant alleles of the *Ma* genes have no major effect on *P. vulnus* reproduction. Nevertheless, potential sources of resistance have been detected in peach Bokhara and Shalil (Okie, 1987), a few wild plum and apricot species and interspecific hybrids (Alcañiz *et al.*, 1996; Pinochet *et al.*, 2000). Five *Prunus* rootstocks showed resistance to one or more *P. vulnus* populations: Torinel (*Prunus domestica* L.), AC-595 (*P. domestica* × *P. insititia* L.), Marianna 4001 (*Prunus cerasifera* Ehr. × *Prunus munsoniana* Wight & Hendrick) and Felinem (*Prunus dulcis* (Mill.) D.A. Webb × *Prunus persica* (L.) Batsch); and Redglow (*Prunus salicina* Lindl Burbank × *P. munsoniana* Jewell) was the only rootstock that showed broad resistance to 11 *P. vulnus* populations from different geographic locations (Pinochet *et al.*, 2000). More recently, Pinochet *et al.* (2002) found that the plum BS N2 was resistant to *P. vulnus*. However, tolerance to *P. vulnus* has been reported in some plum rootstocks, such as Marianna 2624 and Myrobalan 29C, and apricot Royal (Culver *et al.*, 1989).

To date, most of the effort in the search for resistance in *Prunus* spp. has been focused on *Meloidogyne* spp. and some rootstocks have been found to be resistant to one or more species (Marull & Pinochet, 1991; De Waele & Elsen, 2002). However, the plum hybrid Bruce is the only rootstock that exhibits resistance to both *M. incognita* and *P. vulnus* (Pinochet *et al.*, 1996). In fact, Nemaguard, one of the best known peach rootstocks resistant to *M. incognita* and *M. javanica* (Treub, 1885) Chitwood, 1949, is susceptible to *P. vulnus* (McKenry, 1989; Nyczepir & Pinochet, 2001). All three peach and plum rootstocks released by INRA in France in the last decade, Cadaman (*P. persica* × *P. davidiana*), Julior (*P. insititia* × *P. domestica*) and Ishtara (*P. persica* × *P. cerasifera* cv. Belsiana), which are resistant to root-knot nematodes, have been found to be good hosts for *P. vulnus* (Alcañiz *et al.*, 1996).

Resistance to *Pratylenchus* spp. in other fruit trees and woody crops has been developed. For example, Barbados cherry (*Malpighia glabra* L.) was found to be resistant to *P. brachyurus* (Ferraz *et al.*, 1989). Tolerance to *P. penetrans* was detected in several grapevine rootstocks, *e.g.*, Couderc 1202, Couderc 1616, Couderc 3309, Kober 5BB, Teleki 5A, Teleki 5C, Foch, Seyval, Vidal and Vignoles (Ramsdell *et al.*, 1996).

Resistance to *P. coffeae* was also reported in Robusta coffee (*Coffeae canephora* Pierre *ex* Froehner) (Wiryadiputra, 1996) and the presence of high levels of polyphenols in the roots of *C. canephora* was suggested as one of the mechanisms involved in suppressing nematode development (Villain *et al.*, 2002). Anzueto *et al.* (2001) selected two clones for resistance to *P. coffeae* and *Meloidogyne* spp. which were cloned in the laboratory using somatic embryogenesis. Following hybridisation, the rootstock variety Nemaya was developed with resistance to *P. coffeae* and *Meloidogyne* spp. (Anzueto *et al.*, 2001). Oliveira *et al.* (1999b) found that the interspecific hybrids Icatu H 4782-7-514 (artificial hybrid between *C. arabica* and *C. canephora* with several backcrosses with *C. arabica*) and Sarchimor C 1669-33 (natural hybrid between *C. arabica* cv. Vila Sarchi and hybrid of Timor) were poor hosts to *P. brachyurus* among six sources of *Coffea* sp. germplasm. Resistance to *P. vulnus* was also observed in kiwi fruit (*Actinidia deliciosa*) (Simeone *et al.*, 1995) or *Actinidia chinensis* Planch. (Nicotra *et al.*, 2003) and several tea clones have been found to be resistant or tolerant to *P. loosi* (Campos *et al.*, 1990).

Roses

In *Rosa* spp., Coolen and Hendrickx (1972b) identified slight resistance to *P. penetrans* in one (*R. eglanteria*) out of 13 common commercial rose rootstocks in Belgium, showing that *Rosa dumetorum* Thuill., *Rosa canina* L. and *Rosa multiflora* Thunb. *Ex* Murr. were hosts, in descending order of host suitability, for *P. penetrans* under field conditions. Similarly, Santo and Lear (1976) showed that *Rosa noisettiana* Thory cv. Manetti rootstock was a good host of *P. vulnus* whereas *R. multiflora* was less suitable and Ohkawa and Saigusa (1981) found that *Rosa chinensis* Jacq. cv. Major and *R. multiflora* cv. 60-5 proved to be efficient hosts for *P. vulnus* and *P. penetrans*. Subsequently, *R. multiflora* cv. Ludiek was found to be resistant to *P. vulnus* (Schneider *et al.*, 1995). Peng and Moens (2002a) detected partial resistance to *P. penetrans* in *R. virginiana* P. Mill. which supported significantly lower multiplication of the nematode than the control *R. corymbifera* cv. Laxa. Finally, Peng *et al.* (2003) screened 131 *Rosa* accessions and whilst the majority of accessions supported the multiplication of *P. penetrans*, resistance of *R. multiflora* cv. K1 and *R. virginiana* to *P. penetrans* was confirmed and *Rosa laevigata* Michx. *anemoides* supported a significantly low nematode population.

Potato and sweet potato

Breeding for host resistance in potato to control *P. penetrans* has so far not been feasible because of the lack of resistant germplasm to incorporate into commercial cultivars (Brodie, 1998). Nevertheless, some host resistance has been identified; *e.g.*, potato cv. Butte was reported to be highly resistant to *P. neglectus* and to possess some resistance to *P. penetrans* (Davis *et al.*, 1992). Similarly, potato cvs Butte, Hudson and L 118-2, were identified as resistant to one isolate of *P. penetrans* from Cornell and susceptible to one isolate from Long Island (France & Brodie, 1995). Cultivar Hudson was first reported as resistant to *P. penetrans* by Dunn (1973) but then as susceptible by Kotcon *et al.* (1987). France and Brodie (1995) used the same isolate of *P. penetrans* as Dunn (1973) and concluded that after 20 years this isolate still reproduced poorly on cv. Hudson and that Long Island isolate reproduced well on Hudson as established by Kotcon *et al.* (1987). Partial resistance to *P. penetrans* was identified in *Solanum tuberosum* ssp. *andigena* and *Solanum vernei* (Brodie & Plaisted, 1993). The cultivar Russet Burbank was relatively tolerant to infection by *P. penetrans* (Kimpinski & McRae, 1988).

Similarly, several sweet potato cultivars have been developed with resistance to *P. coffeae* in Japan, *e.g.*, Daichinoyume, Tamaotome, Benimasari and Sunny Red (Yamakawa *et al.*, 1998; Ishiguro *et al.*, 2004a, b; Katayama *et al.*, 2004) whilst Anguiz and Canto-Saenz (1991) identified some sweet potato cultivars (*e.g.*, Bakongo or Hert Gold) in Peru as being resistant to *P. flakkensis*.

STRAWBERRIES AND RASPBERRIES

Advances have also been made in breeding for resistance to *P. penetrans* in strawberry (Szczygiel, 1982; Potter & Dale, 1991, 1994; Dale & Potter, 1998; Pinkerton & Finn, 2005). While no strawberry was completely resistant to *P. penetrans*, some resistance has been detected in cvs Annapolis, Guardian, Micmac, Pajaro, Redchief and Senga Sengana (Szczygiel, 1982; Potter & Dale, 1994). Potter and Dale (1994) found sources of resistance or tolerance among several beach strawberry (*Fragaria chiloensis* (L.) Dutch.) and woodland strawberry (*Fragaria virginiana* Dutch.) genotypes, the parents of the cultivated strawberry. Recently, Pinkerton and Finn (2005) showed that some genotypes of wild *Fragaria* L. and commercial cultivars of strawberry *Fragaria* × *ananassa* Duch. were ranked more resistant to *P. penetrans* than the susceptible cv. *F.* × *ananassa* Totem. However, these authors concluded that sources of resistance to *P. penetrans* were uncommon in the evaluated and a better source of resistance can be found in *Fragaria* × *ananassa* cvs that already have commercially important characteristics.

In red raspberry, Bristow *et al.* (1980) and Vrain and Daubeny (1986) identified resistance or tolerance to *P. penetrans* among some genotypes, such as *Rubus idaeus strigosus* Michx., *Rubus crataegifolius* Bge and some raspberry cvs, *e.g.*, Dalhousie Lake, Nootka. Studies on the resistance to *P. penetrans* in red raspberry (Vrain *et al.*, 1994) and alfalfa (Thies *et al.*, 1994) indicated that the inheritance of resistance may be polygenic.

CEREALS AND LEGUMES

The Australian bread wheat line GS50a, selected from wheat variety Gatcher, is partially resistant, reducing considerably the reproduction rates of *P. thornei*, compared to some commercial wheat cultivars (Thompson & Clewett, 1986). Subsequently, Thompson and Haak (1997) found that 39 out of 244 accessions of *Aegilops tauschii* Coss. had resistance to *P. thornei* and that this resistance could be introgressed into cultivated wheat. Similarly, the landrace AUS4930 of wheat from Iraq was identified in Australia as simultaneously resistant against *P. thornei* and *H. avenae* in the field (Nicol *et al.*, 1999a). However, Nombela and Romero (1999, 2001) did not detect resistance to *P. thornei* in the *Cre2* gene present in the wheat/*Aegylops ventricosa* Tausch. introgression line H93-8 and in the *CreAet* gene present in wheat/*Aegylops triuncialis* L. introgression line TR-353, both conferring resistance to *H. avenae* (Delibes *et al.*, 1993). Zwart *et al.* (2004) investigated the inheritance of *P. thornei* resistance in five synthetic hexaploid wheat lines and two bread wheat lines using a half-diallel design of F1 and F2 crosses. Results showed that inheritance of *P. thornei* resistance is polygenic and predominantly conditioned by additive gene action. The synthetic hexaploid wheat lines contain novel sources of *P. thornei* resistance that will provide alternative and more effective sources of resistance to be utilised in wheat breeding programmes. Similarly, Hollaway *et al.* (2000) detected resistance to *P. thornei* in several cereals and legumes in Australia, *e.g.*, barley cvs Tamaroi and Yallaroi and durum wheat cvs Arapiles, Picola and Schooner. Taylor *et al.* (2000) detected resistance to *P. neglectus* in several cereals and legumes in Australia, *e.g.*, durum wheat cv. Yallaroi; faba bean cvs Ascot, Fiord and Icarus;

oat cv. Euro; pea cvs Alma, Bluey, Bonzer, Dundale, Early Dun; rye cv. Bevy; and wheat cvs Excalibur and Krichauff. Also, resistance or tolerance to *P. neglectus* and *P. thornei* was detected in two wheat cvs (Sunvale and Krichauff) by Smiley *et al.* (2003); and the triticale lines Abacus and Muir showed resistance to *P. neglectus* (Farsi *et al.*, 1995). Zwart *et al.* (2005) identified two major quantitative trait loci (QTLs) for resistance to *P. thornei* (*QRlnt.lrc-6D.1*, *QRlnt.lrc-6D.2* and *P. neglectus* (*QRlnn.lrc-6D-1*, *QRlnn.lrc-4D-1*). The QTLs were identified in a doubled-haploid wheat population developed from a cross between the synthetic hexaploid wheat line CPI133872 and the bread wheat Janz and provided strong evidence that these chromosome regions contain true genetic factors of resistance to these nematodes. QTLs showed additive inheritance and thus allow the possibility of marker-assisted selection for nematode resistance.

Hollaway (2002) studied the effect of eight oat cultivars on the population density of *P. thornei.* All of the oat cultivars tested were found to reduce significantly the population densities of *P. thornei* and were similar to, or slightly higher than, the resistant barley cv. Clipper. In Cyprus, Philis (1997) found that barley cv. Athenais was found to be susceptible to *P. thornei* and produced severe damage to plants.

Numerous efforts have been made to identify host resistance in corn to *Pratylenchus* spp. (Windham, 1998). Smolik and Wicks (1987) reported that corn line SD101 (derived by selfing a single plant from the cross of inbred CM65 with a line related to SD37) was resistant to *P. hexincisus* and *P. scribneri* and lines SD102 and SD103 were resistant to *P. hexincisus*. Lordello *et al.* (1985) also identified several corn genotypes resistant to *P. zeae* and *P. brachyurus*, *e.g.*, lines IAC-1 XVIII, IAC Hs 1228, IAC Maya XIX, IAC Phoenyx 1918, Guarani and Palha Roxa. Norton (1989) reported that two wild species of corn, *Zea diploperennis* Ihis, Doebley & Pazy and *Zea perennis* Hitchcock, were resistant to *P. scribneri* and (or) *P. hexincisus*, and significantly reduced nematode populations by 82 and 98%, respectively, when compared to common corn hybrids. The inheritance of resistance in corn to *P. zeae* and *P. brachyurus* was studied by Sawazaki *et al.* (1987) using lines of Col 2(22) (resistant) and Ip 48-5-3 (susceptible). The results indicated that the resistance in line Col 2(22) was due to two dominant genes with an additive effect.

Although a suitable chickpea cultivar with resistance to *P. thornei* is not yet available, numerous investigations have been done on the identification of sources of resistance in cultivated or wild chickpea to this nematode. Tiwari *et al.* (1992) screened 105 chickpea lines against *P. thornei*; three were rated as highly resistant (GNG 543, GF 88428 and PKG-24) and 68 were moderately resistant. Greco *et al.* (1988) screened 24 plant species and 97 chickpea lines against *P. thornei*. All chickpea lines were susceptible to the nematode and it reproduced well on all the leguminous species but not on grass pea (*Lathyrus sativus* L.), lettuce, potato and cauliflower. Very low root infections were observed on carrot, radish, coriander (*Coriandrum sativum* L.), spinach and sugar beet, while turnip, rashad (*Nasturtium fontanum* Asch.), parsley and kumboz (*Cannabis sativa* L.) were free of *P. thornei*. Di Vito *et al.* (1995) screened 249 lines of chickpea belonging to eight wild *Cicer* species for their reaction to *P. thornei*. Six accessions of *Cicer bijugum* Rech. f. (ILWC 32, ILWC 71, ILWC 75, ILWC 76, ILWC 80 and ILWC 227), three of *Cicer cuneatum* Hochst. *Ex* Rich. (ILWC 19, ILWC 37 and ILWC 40), 11 of *Cicer judaicum* Boiss. (ILWC 4, ILWC 31, ILWC 38, ILWC 91, ILWC 93, ILWC 94, ILWC 98, ILWC 99, ILWC 101, ILWC 102 and ILWC 103) and one of *Cicer yamashitae* Kitamura (ILWC 55), were resistant to *P. thornei*. Ali and Ahmad (2000) screened some 600 chickpea lines against *M. javanica*, *P. thornei* and *Rotylenchulus reniformis* Lindford & Oliveira, 1940 under field conditions. Only 17 lines were resistant to *P. thornei* and of these only one entry, IPC 96-69, was resistant to all three nematode species, one entry (ICC 16614) showed dual resistance against *M. javanica* and *P. thornei*, and another (ICC 6928) showed resistance to *P. thornei* and *R. reniformis*.

In faba bean, Di Vito *et al.* (2002b) evaluated the reaction of 99 faba bean lines (40 from Algeria, six from Morocco and 53 from Tunisia) to three populations of *P. thornei* and one each of *P. neglectus*, *P. penetrans* and *P. pinguicaudatus*. Results showed that the lines FRYT98-6 and FRYT98-60 were resistant to *P. neglectus*; FRYT98-35, FRYT98-47 and FRYT98-56 were resistant to *P. thornei* and the lines FRYA98-48 and FRYT98-44 were resistant to *P. penetrans* and *P. pinguicaudatus*, respectively.

In bean, the response to infection by *P. penetrans* varied among cultivars. Gratiot, Kentwood and Saginaw were tolerant whereas Sanilac, Seafarer and Tuscola were highly susceptible (Elliot & Bird, 1985). Similarly, the response of beans to infection by *P. scribneri* was also variable amongst cultivars, *e.g.*, cv. Fordhook 242 was a poor host whereas Kentucky Gonder was highly susceptible (Rich *et al.*, 1977b).

Among other crops carrot, cotton and pepper were resistant, wheat and alfalfa were tolerant, whereas barley, broccoli, cabbage, onion, sorghum, Sudan grass, sweet corn and tomato were susceptible (Rich *et al.*, 1977b). Resistance in bean to *P. scribneri* was related to production of phytoalexins (*e.g.*, coumestrol) in response to nematode infection, coumestrol being reported to inhibit motility of *P. scribneri* (Rich *et al.*, 1977a).

In soybean, Schmitt (1976) screened 25 soybean cultivars against *P. brachyurus* and found some tolerant cvs, *e.g.*, Bragg, Coker 73-375, FFR 555, FFR 6024, Hutton and Lee 68. Zirakparvar (1982) screened 41 soybean cultivars against *P. hexincisus* and found all of them to be susceptible.

Although no resistant cultivars of alfalfa to *P. penetrans* have been developed, two alfalfa germplasms, MNGRN-2 and MNGRN-4, are available showing superior performance in fields infested with large populations of the nematode (Barnes *et al.*, 1990). Similarly, Hafez *et al.* (2000) detected five (ZN 9541, ZN 9651, ZM 9421, ZB 9546, ZC 9751A) out of 37 alfalfa cultivars with tolerance to *P. penetrans* and Thies *et al.* (1994) demonstrated that resistance to *P. penetrans* in alfalfa is conditioned by additive gene action.

Finally, in peanut, Smith *et al.* (1978) reported resistance to *P. brachyurus* in two lines (PI 290606 and PI 295233), Starr (1984) reported resistance in another line (PI 365553) and Timper and Hanna (2004) found that two pearl millet hybrids (HGM-100 and TifGrain 102) were poor hosts for *P. brachyurus*. Motalaote *et al.* (1987) showed that sorghum cv. Pioneer 8222 was a non-host to *P. crenatus* and that *P. brachyurus* was non-pathogenic.

BANANA

Banana (*Musa* AAA) and plantains (*Musa* AAB and ABB) are severely damaged by *P. coffeae* and *P. goodeyi* (Gowen & Quénéhervé, 1990; Gowen *et al.*, 2005). Little progress has been made in the identification of resistant bananas to these nematodes. In most banana- and plantain-growing areas the crop is also severely damaged by *R. similis* and *Fusarium oxysporum* Schlecht. f. sp. *cubense* (E.F. Smith) Snyd. & Hans. (Panama disease), so any attempt in selecting resistant material against *Pratylenchus* spp. should incorporate, to a greater or lesser degree, resistance and/or tolerance to these soil-borne pathogens

(Pinochet & Rowe, 1978; Pinochet, 1998). A triploid (*Musa* AAA) cultivar, Yangambi Km 5 was found to be resistant to *P. goodeyi* and also showed resistance to Panama disease and *R. similis* (Fogain & Gowen, 1998; Pinochet *et al.*, 1998). Devrajan and Rajendran (2001) screened nine banana clones (Matti, Anaikomban, Namarai, Tongat, H 59, H 65, H 84, H 109 and H 110) for resistance to *P. coffeae*, reporting that clones H 109 and H 110 were resistant and clones Anaikomban, Namarai, H 59 and H 65 were moderately resistant. Van den Berg *et al.* (2002) identified possible sources of resistance/tolerance to *P. coffeae* in cultivars Tieu Xanh (AAA), Tieu Mien Nam (AA), Gros Michel (AAA), Com Chua (AAB), Com Lua (AA), Man (AAB), Ngu Thoc (AA) and Grande Naine (AAA) from 26 Vietnamese banana accessions and some wild accessions. Krishnamoorthy and Kumar (2004) screened 18 new synthetic tetraploid banana hybrids and five parental banana clones against *P. coffeae*, recording the lowest nematode population and multiplication rate in H-02-29 followed by H-02-26, H-02-34. However, attempts to detect resistance to *P. coffeae* has been unsuccessful on many occasions, *e.g.*, Collingborn and Gowen (1998) evaluated five banana cultivars, and Stoffelen *et al.* (1999, 2000) evaluated 13 banana genotypes from Malaysia and Vietnam and 25 banana varieties of section *Eumusa* (AA group) and seven of the section *Australimusa* (Fe'i-group) from Papua New Guinea, respectively, all being susceptible to the nematode.

OTHER CROPS

In sunflower, O'Brien (1983) evaluated 19 cultivars which were very poor or non-hosts of *P. thornei*, as well as a commercial cultivar by Tobar *et al.* (1995) whilst Rich and Dunn (1982) found that sunflower was tolerant to parasitism by *P. brachyurus*. Bolton and De Waele (1989) evaluated 21 sunflower hybrids that were non-hosts or poor hosts for *P. zeae* and therefore might be useful as rotation crops in fields heavily infested with this nematode. In addition, several species of *Crotalaria* have been found to be resistant to *P. zeae*, *e.g.*, *Crotalaria grantiana* Harvey, *Crotalaria lanceolata* E. Mey, *Crotalaria mucronata* Desv., *Crotalaria pallida* Ait. and *Crotalaria paulina* Schrank (Da Silva *et al.*, 1989).

Tobacco can be severely damaged by *P. brachyurus*, *P. neglectus*, *P. penetrans* and *P. pratensis* (Johnson, 1998). However, some resistance has been detected in tobacco genotypes, *e.g.*, *P. zeae* was unable to reproduce on tobacco cvs TL 33 and CDL 28 (Van Biljon & Meyer, 2000). Bélair *et al.* (2004) demonstrated that forage pearl millet (*P. glaucum*) hybrid CFPM 101 and grain pearl millet hybrids (CGPM H5 and CGPM H6) in rotation with tobacco cv. Delgold significantly decreased densities of *P. penetrans* compared with rye (the traditional rotation crop) and increased tobacco yields.

Although the breeding of sugarcane resistance to *P. zeae* was a very difficult task, some resistant cultivars have been identified. Dinardo-Miranda and Ferraz (1991) found that sugarcane cvs SP 70-1143 and SP71-1406 were resistant to *P. zeae* and Novaretti (1992) demonstrated that cv. SP 70-1143 was also resistant to *M. javanica*. Mehta *et al.* (1994) screened 46 cultivars of sugarcane from the world germplasm collection at the Sugarcane Breeding Institute in Coimbatore (India) and found three resistant cultivars, *viz.*, Co 88020, Co 89009 and Co 89034. One of the reasons for these responses may be root thickness and hardening as soft roots are more favourable for the penetration and multiplication of *P. zeae*. Dinardo-Miranda *et al.* (1996) evaluated a sugarcane variety and five clones against *P. zeae*, clone IAC77-51 being tolerant. Kathiresan and Mehta (2002b) demonstrated that the resistant reaction to *P. zeae* in six resistant sugarcane clones (Co 62175, Co 7717, CoJ 76, CoT 8201, MS 6847 and Q 69) was a hypersensitive response characterised by necrosis of two or three cell layers immediately around the nematodes. Montasser *et al.* (2002) evaluated four cultivars of sugarcane (GT54-9, GT68-88, G75-368 and NCO-310), all being susceptible to *P. zeae*.

Mechanisms of resistance

The mechanism(s) responsible for host plant resistance to *Pratylenchus* spp. have not been fully researched and understood. Nevertheless, some studies have been undertaken. Slabaugh (1974) reported that in 14 breeding lines of tomato, serine concentrations were 140-280% higher in *P. penetrans*-susceptible than in resistant roots. Serine was thought to have a role in the mechanism of resistance because of its importance in the biosynthesis of tryptophan and N10 formyltetrahydrofolate which are involved in IAA and cytokinin biochemistry, respectively.

Baldridge *et al.* (1998) reported that alfalfa varieties showed antibiosis-based resistance to *P. penetrans* developed by recurrent selection. They found preliminary evidence for involvement of several known plant defence response genes in alfalfa resistance to *P. penetrans*. *Pratylenchus penetrans*-resistant plants have higher constitutive levels of transcripts for key enzymes involved in biosynthesis of isoflavonoid phytoalexins (phenylalanine ammonia-lyase, chalcone synthase, isoflavone reductase), which are known to play a role in fungal resistance and are implicated in resistance to both sedentary and migratory nematodes. Levels of these transcripts fell in *P. penetrans*-infected resistant plants but were induced over 12 h before falling in infected *P. penetrans*-susceptible plants. The validity of phenylpropanoid pathway enzyme transcript suppression in *P. penetrans*-infected resistant alfalfa plants is supported by the simultaneous induction of β-1,3-glucanase transcripts in a resistant *vs* susceptible host pattern typical of many other plant pathogen systems. This work also revealed the role of a phytoalexin, medicarpin, in the resistance to *P. penetrans* since constitutive levels of medicarpin were highest in roots of the two most resistant plants; and additionally, medicarpin inhibited motility of *P. penetrans in vitro* (Baldridge *et al.*, 1998).

Phytoecdysteroids (*i.e.*, plant hormones inducible in response to mechanical damage or parasite herbivores) have been found to inhibit penetration and development of *P. neglectus* in spinach (*Spinacia oleracea* L. cv. Avon) and may confer a mechanism for nematode resistance (Soriano *et al.*, 2004).

Biological control

Biological control is defined as "the reduction of inoculum or disease producing activity of a pathogen accomplished by or through one or more organisms other than man" (Cook & Baker, 1983). The biocontrol agents include not only the parasites and predators of plant-parasitic nematodes, but also the host plant (resistant cultivars) and other plants used either as trap crops or as antagonists. Several antagonistic organisms have been recorded with the capacity to reduce the ability of *Pratylenchus* spp. to survive and reproduce. The most abundant reported antagonists of *Pratylenchus* spp. are populations of soil fungi, entomopathogenic nematodes, bacteria and nematicidal plants. These antagonists have a great diversity and still represent a relatively unexplored field. In this section we deal with reports in the literature concerning the biocontrol of root-lesion nematodes.

BACTERIA AND BENEFICIAL ENDOPHYTIC MICROORGANISMS

Beneficial endophytic rhizobacteria have been reported to have nematicidal effects on *Pratylenchus* species. Some plant growth-promoting rhizobacteria (*Azospirillum* sp. and *Phosphobacterium* sp.) have been reported to reduces populations of *P. zeae* in corn and thereby increase crop yield (Sundarababu *et al.*, 1998). Similarly, in potato soils, Sturz and Kimpinski (2004) have recently demonstrated the nematicidal activity of several bacterial endophytes against *P. penetrans* isolated from African (*T. erecta*) and French (*Tagetes patula* L.) marigold which, in descending order of efficacy, were: *Microbacterium esteraromaticum* (Omelianski, 1923) Takeuchi & Hatano, 1998; *Tsukamurella paurometabola* (Steinhaus, 1941) Collins *et al.*, 1988; *P. chlororaphis*, *Kocuria varians* (Migula, 1900) Stackebrandt *et al.*, 1995; and *K. kristinae* (Kloos *et al.*, 1974) Stackebrandt *et al.*, 1995. Nevertheless, results indicated that no significant differences were found in the total number of *P. penetrans* nematodes within the root tissues of bacterised potato plants compared to the non-bacterised control (Sturz & Kimpinski, 2004). Recently, three metabolites (penipratynolene, 6-methoxy-carbonylpicolinic acid and 2,6-pyridinedicarboxylic acid) isolated from culture filtrates of *Penicillium bilaiae* Chalabuda showed nematicidal activity of 77, 52 and 98%, respectively, against *P. penetrans* (Nakahara *et al.*, 2004). Mazzola *et al.* (2002) reported that the fluorescent pseudomonad, *Pseudomonas putida* Trevisan, 1889, suppressed apple replant disease caused by *P. penetrans* under glasshouse conditions.

Tall fescue (*Festuca arundinacea* Schreb.) and meadow fescue (*Festuca pratensis* Huds. N. Engsvingel) are mutualistically associated with the endophytic fungus *Neotyphodium coenophialum* (Morgan-Jones & W. Gams) Glenn Bacon & Hanlin, which is seed-transmitted and ecologically restricted to growth within the plant (Hinton & Bacon, 1985). This endophyte has been shown to reduce nematode populations of *P. scribneri* in field soils of tall fescue (Kimmons *et al.*, 1990). The mechanism by which the endophyte reduces populations of *Pratylenchus* spp. appears to be related to the production of ergot alkaloids. Timper *et al.* (2005) isolated new strains of *N. coenophialum* from wild tall fescue and these strains produced low to nil levels of ergot alkaloids (referred to as non-ergot strains). Results demonstrated that none of the non-ergot strains of *N. coenophialum* reduced the populations of *P. scribneri* and *P. zeae* as did the endemic strain which produced ergot alkaloids. However, by way of contrast, later studies by Panaccione *et al.* (2006) demonstrated that elimination of certain complex ergot alkaloids (ergovaline and lysergic acid amides) in one gene knockout strain of *Neotyphodium* sp. isolate Lp1 (syn. *Neotyphodium lolii* × *Epichloe typhina*), or complete elimination of ergot alkaloids in another, did not affect the ability of the endophyte to suppress populations of *P. scribneri*.

Walker *et al.* (1966) found that culture filtrates of four isolates of *Streptomyces* spp. were nematicidal to adults and juveniles of *P. penetrans*. When inoculated into soil, an isolate of *Streptomyces costaricanus* Esnard *et al.* from a nematode-suppressive soil in Costa Rica had broad activity against plant-pathogenic fungi and several nematodes including *P. penetrans* in strawberries (Dicklow *et al.*, 1993). Similarly, Samac and Kinkel (2001) found a suppressive effect of a strain of *Streptomyces* in alfalfa fields infested by *P. penetrans*. Results suggested that the *Streptomyces* amendment suppresses nematode reproduction to a similar extent as the protection provided by genetic resistance (Samac & Kinkel, 2001). Moreover, mutant strains that do not produce, or produce greatly reduced quantities of antibiotics, were as effective as the wild type strains in suppressing *P. penetrans* in roots of alfalfa plants (Samac & Kinkel, 2001). This suggests that the mechanism causing a reduction in nematode population may be related to compounds that directly affect the nematode, or that actually induce resistance to the nematode in the host plant. The basidiomycete *Omphalotus olearius* (de Candolle *ex* Fr) Singer produced a metabolite, omphalotin A, which demonstrated *in vitro* nematicidal activity on *P. penetrans* (Mayer *et al.*, 1999).

Several mechanisms have been proposed to explain the process by which beneficial endophytes may mediate nematode attack. For example, certain rhizobacteria are known to promote plant growth so that plants not only grow faster, but are sturdier, with a better developed root mass and are more capable of withstanding low level disease pressure than non-bacterised plantlets (Sharma & Nowak, 1998). However, it has been shown that secondary metabolites produced directly by bacterial isolates can have nematicidal properties, as in the case of *Bacillus thuringiensis* Berliner (Leyns *et al.*, 1995). The effect appears to be more in keeping with the production of a nematistatic compound(s) that interrupts, attenuates, or inhibits processes in the nematode (*P. penetrans*) reproductive life-cycle. This appears to be the case with *Pseudomonas aureofaciens* Kluyver which is known to inhibit hatch of *Criconemoides* (Westcott & Kluepfel, 1993).

NEMATOPHAGOUS FUNGI

Pratylenchus spp. interact with fungi either directly in the soil or indirectly, mediated through the host plant. Nematophagous fungi may be parasitic, as in the case of the endoparasitic nematophagous fungus *Drechmeria coniospora* (Drechsler) Gams & Jansson which was found parasitising the labial region and tail of *P. penetrans* and *P. coffeae* (Jansson *et al.*, 1987). *Hirsutella rhossiliensis* Minter & Brady reduced the number of *P. penetrans* entering roots of young potato plants by 25%. The effect of the fungus on root penetration was dependent on the distance the nematode had to move through *Hirsutella*-infested soil to reach a root and on the density of infective conidia (Timper & Brodie, 1994). In agar culture plates Timper and Brodie (1993) found that *Verticillium balanoides* (Drechsler) Dowsett, Reid. & Hopkin, *D. coniospora and Nematoctonus* sp. were weak or non-pathogens, but that the trapping fungi, *Arthrobotrys dactyloides* Drechsler, *A. oligospora* Fresenius, *Monacrosporium ellipsosporum* (Preuss) Cooke & Dickinson and *M. cionopagum* (Drechsler) Subramanian, killed most of the *P. penetrans* adults and juveniles added to the fungus cultures. Similarly, *Arthrobotrys arthrobotryoides* (Berlese) Lindau, *A. dactyloides*, *Dactylaria thaumasia* Dreschler and *Monacrosporium doedycoides* Dreschler greatly reduced penetration of alfalfa roots by *P. penetrans* (Rao, 1973; Mai *et al.*, 1977). Castro *et al.* (2000) reported that *Arthrobotrys musiformis* (Dreschler) parasitised *P. brachyurus* in about 30% of the specimens under controlled conditions. Also, *Paecilomyces lilacinus* (Thom.) Samson reduced nematode populations of *Pratylenchus* in corn and *P. coffeae* in chickpea and the percent reduction in nematode population increased with increasing application rate of the fungus (Gapasin, 1995; Tiyagi & Shamim, 2004). In addition, *Monacrosporium lysipagum* (Drechsler) Subram has been reported capturing and killing 81% of mobile stages of *P. neglectus in vitro* using adhesive knobs within 20 h of exposure to the fungus (Khan *et al.*, 2006).

Although only moderate mortality have been found in soil (and most *Pratylenchus* specimens are protected within the roots) the verified capability of nematophagous fungi to colonise endophytically monocotyledon and dicotyledon plant roots (Bordallo *et al.*, 2002) may have profound implications in the performance of these microorganisms as biocontrol agents of *Pratylenchus* spp. Future studies involving practical use of nematophagous fungi should therefore address their root colonisation capabilities and also possible induction of plant defence mechanisms.

ENTOMOPATHOGENIC NEMATODES

Although the nature of interaction between entomopathogenic and plant-parasitic nematodes is unknown, application of *Steinernema* spp. reduced nematode populations of certain plant-parasitic nematodes (Bird & Bird, 1986). Tsai and Yeh (1995) reported that the effect of *Steinernema carpocapsae* (Weiser, 1955) Wouts, Mráček, Gerdin & Bedding, 1982 on the infectivity of *P. coffeae* varied with host plant, *e.g.*, on adzuki bean, *Vigna angularis* (Willd.) Ohwi & Ohashi, and tomato seedlings, *S. carpocapsae* reduced the infectivity of *P. coffeae* by 85% and 88%, respectively. However, on radish it increased the infectivity of *P. coffeae* by 160% and there was no effect on edible rape (*Brassica chinensis* Jusl. var. *oleifera* Tsen & Lee). Moreover, no effects of *S. carpocapsae* and *Steinernema feltiae* (Filipjev, 1934) Wouts, Mráček, Gerdin & Bedding, 1982 infective juveniles were detected on *P. penetrans* in strawberry field microplots in Connecticut, USA (LaMondia & Cowles, 2002).

PREDACEOUS NEMATODES

Several nematodes have been reported as predating on *Pratylenchus* spp., *e.g.*, *Seinura tenuicaudata* (de Man, 1865) Goodey, 1960 was found feeding on *P. pratensis* under natural conditions (Linford & Oliveira, 1937); *Iotonchus monhystera* (Cobb, 1917) Jairajpuri, 1970 caused a 50% predation on *P. zeae* under controlled conditions in agar plates and soil tests (Azmi, 1997); *Odontopharynx longicaudata* de Man, 1912 consumed 70-78% of *P. vulnus* in agar plates (Chitambar & Noffsinger, 1989); and *Mylonchulus minor* (Cobb, 1893) Andrássy, 1958 predated on *P. delattrei* (Choudhury & Sivakumar, 2000). Nevertheless, the importance of predation by dorylaimid, diplogastrid and mononchid nematodes on *Pratylenchus* spp. under natural conditions has not been evaluated and their importance as a control measure has not been quantified.

PASTEURIA PENETRANS

Five species of *Pasteuria* that parasitise plant-parasitic nematodes have been described: *Pasteuria penetrans* (Thorne) Sayre & Starr, on

Meloidogyne species, *Pasteuria thornei* Starr & Sayre, on *Pratylenchus* spp., *Pasteuria nishizawae* Sayre, Wergin, Schmidt & Starr on *Heterodera* spp. and *Globodera* spp. (Sayre & Starr, 1989) and *Pasteuria usgae* Giblin-Davis, Willams, Bekal, Dickson, Brito, Becker & Preston, an obligate endoparasite of *Belonolaimus longicaudatus* Rau, 1958 (Giblin-Davis *et al.*, 2003). Nevertheless, several cases report that isolates of *Pasteuria* from *Meloidogyne* spp. are capable of parasitising *Pratylenchus* spp.; *e.g.*, an isolate of *Pasteuria* recovered from *M. javanica* was found parasitising *P. coffeae* (Sharma & Davies, 1996), or a *Pasteuria* isolate parasitic on *Heterodera goettingiana* Liebscher, 1892 at Munster, Germany, parasitise females and juveniles of *P. brachyurus* (Winkelheide & Sturhan, 1996). In other cases, unidentified isolates of *Pasteuria* parasitised migratory and sedentary ectoparasitic nematodes such as *T. semipenetrans*, *Trophonema arenarium* (Raski) Siddiqi, 1999 and *Tylenchorhynchus cylindricus* Cobb, 1913 (Inserra *et al.*, 1992; Kaplan, 1994; Galeano *et al.*, 2003).

Pasteuria thornei was found parasitising *P. pratensis* (Thorne, 1940) and *P. brachyurus* but not *P. scribneri* (Starr & Sayre, 1988). Similarly, parasitism of *P. neglectus* by *Pasteuria thornei* has been found to occur in Germany (Sturhan, 1985), Italy and Croatia (Ciancio *et al.*, 1994) and Spain (Ornat *et al.*, 1999) and a *Pasteuria* sp. was found parasitising juveniles of *P. thornei* and *P. penetrans* in Turkey (Elekcioglu, 1995). Usually, the frequency of association between *Pasteuria* spp. and *Pratylenchus* spp. is lower than that found in *Meloidogyne* spp.; *e.g.*, in pineapple on Hawaii, Ko *et al.* (1995) found frequencies of association between *Pasteuria* spp. and *P. brachyurus* and *M. javanica* of 24% and 52%, respectively. Recently, *P. gibbicaudatus* and *P. microstylus* have been reported with infections of *Pasteuria* spp. in India (Bajaj & Walia, 2005).

Although it has been shown that *Pasteuria penetrans* has potential as a biological control agent for *Meloidogyne* spp. (Stirling, 1991), few data support the usefulness of *Pasteuria thornei* as a biocontrol agent of *Pratylenchus* spp.

ARBUSCULAR MYCORRHIZAL FUNGI (AMF)

Arbuscular mycorrhizal fungi are soil-borne fungi that establish an obligate mutualistic symbiosis with many plant species (Harley & Smith, 1983). They are characterised by transient dichotomouslybranched arbuscules that form inside cortical cells of plant roots from intraradical branch hyphae and are known to improve the growth of plants by enhancing nutrient uptake, growth rate and hormonal activity (Smith & Read, 1997). Several studies have reported that AMF of the genus *Glomus* increase host tolerance in some host plant-*Pratylenchus* pathosystems, such as *P. penetrans* on bean (Elliot *et al.*, 1984), *P. vulnus* on rootstocks of plum (Camprubí *et al.*, 1993), pear (López *et al.*, 1997) and peach (Pinochet *et al.*, 1996), *P. goodeyi* on banana (Jaizme-Vega & Pinochet, 1997), *P. zeae* on corn (Jothi & Sundarababu, 1997), *P. coffeae* on coffee (Vaast *et al.*, 1998) and banana (Elsen *et al.*, 2003a), and *P. penetrans* on carrot (Talavera *et al.*, 2001). Similarly, Ottawa 3 apple rootstocks mycorrhised with *G. intraradices* or *G. mosseae* increased plant growth and leaf concentrations of P, Cu, Mg and Zn, in addition to reducing populations of *P. penetrans* in roots and soil from the root zone (Forge *et al.*, 2001). In fact, a commercial formulation of *G. mosseae* has been demonstrated to be effective in minimising root lesions, reducing soil population of *P. delattrei* and enhancing the flower yield of the ornamental *Crossandra undulaefolia* Salisb. (Sundarababu *et al.*, 2004). Conversely, *Pratylenchus* nematodes are reported to have diverse interactions with AMF in perennial crops. A few studies report the absence of an interaction in which the development and colonisation of pathogen and symbiont are unaffected by the presence of either organism. This has been reported in the association between *P. vulnus* and *G. mosseae* on Saint Julien 655-2 plum rootstock (Camprubí *et al.*, 1993), EMLA-26 apple (Pinochet *et al.*, 1993a) and on Nemared peach (Pinochet *et al.*, 1995a). However, the nematode can be detrimental to mycorrhizal root colonisation as *P. vulnus* reduced the percentage of mycorrhizal infection by *G. mosseae* in Marianna 2624 and Myrobalan 605 plums (Camprubí *et al.*, 1993) and *P. vulnus* significantly reduced mycorrhizal infection and the number of vesicles formed by *G. intraradices* within the roots of BA-29 quince (Calvet *et al.*, 1995). The presence of the nematode also significantly suppressed the formation of internal spores in Santa Lucia 64 cherry rootstock inoculated with *G. intraradices*, although the arbuscular and vesicular phase were not affected in comparison to nematode-free treatments (Pinochet *et al.*, 1995b). The main causes affecting the arbuscular phase and the production of storage structures (vesicles and internal spores) appears to be a reduction of spaces for fungal colonisation (Smith, 1988) and a decrease in supply of carbohydrates in the roots

(Wallace, 1987a) as a result of the nematode feeding and migration and the formation of dark necrotic lesions in the cortical parenchyma. Mechanisms that enabled the beneficial effects of AMF against root-lesion nematode attack remain unknown, but effective protection against *Pratylenchus* is probably a consequence of several mechanisms – improved nutrient status, competition for nutrients and penetration sites, anatomical changes in the roots, microbial changes in the rhizosphere and activation of plant defence mechanisms have all been proposed as possible anti-nematode mechanisms (Azcón-Aguilar *et al.*, 2002).

In coffee, early inoculation with *Glomus clarum* Nicol. & Schenck improved tolerance to *P. coffeae* but this was not coupled with a decrease in root nematode density as observed in most studies on the interactions between *Pratylenchus* spp. and AMF (Vaast *et al.*, 1998). However, population densities of *P. coffeae* were significantly reduced in mycorrhizal plants of *Musa* sp. with *G. mosseae* compared to non-mycorrhizal plants and significantly less root necrosis was observed in mycorrhizal plants (Elsen *et al.*, 2003a). In carrot, *Glomus* sp. reduced soil populations of *P. penetrans* and increased tolerance to infections by the nematode (Talavera *et al.*, 2001) whilst on transformed carrot roots *G. intraradices* suppressed the reproduction of *P. coffeae* (Elsen *et al.*, 2003b).

Although *Glomus* species are the most diverse of the arbuscular mycorrhizae and are found in many soils all over the world, other AMF species have been reported interacting with *Pratylenchus* species and enhancing host tolerance. For instance, in cotton the AMF *Gigaspora margarita* Becker & Hall increased tolerance to *P. brachyurus*; the nematode did not influence reproduction and sporulation of the mycorrhizal fungus and nematode reproduction was similar on mycorrhizal and non-mycorrhizal plants (Hussey & Roncadori, 1978). Similarly, in coffee the AMF *Acaulospora mellea* Spain & Schenck also increased tolerance to *P. coffeae* (Vaast *et al.*, 1998).

Apart from the previous reported interactions between AMF and *Pratylenchus* spp. in different crops, another study confirms that these interactions also occur in natural environments. De la Peña *et al.* (2006b) confirmed that in roots of the coastal dune grass (*Ammophila arenaria*) infection and multiplication of *P. penetrans* were significantly reduced by the native inoculum of AMF (*Scutellospora castanea* C. Walker and several *Glomus* spp.). Plant pre-inoculation with AMF further decreased nematode colonisation and reproduction. Their results suggest that AMF are crucial for the control of *P. penetrans* in natural systems and illustrate that locally operating mechanisms are involved in this process.

INDUCED RESISTANCE

Plants defend against pathogens by constitutive and inducible barriers. Induced resistance may be expressed locally at the site of infection as well as systemically. Hence, induced resistance is a promising new technology for crop protection that offers a long lasting defence against a broad spectrum of diseases in various plant species. Induced resistance can now be brought about by synthetic compounds like BION®. Sonnemann *et al.* (2002) found that BION® treatment significantly reduced the growth of barley roots and increased root infection by *Pratylenchus* sp., but did not cause measurable changes in plant productivity, in the composition of the free-living soil biota or in root infection by mycorrhizal fungi. The increased infection of barley roots by *Pratylenchus* in the BION®-treated plots shows that induced resistance may influence root-inhabiting soil organisms that can adversely affect plant health. Potential mechanisms include changes in root exudation that increase attraction for nematodes, changes in root morphology that facilitate invasion, and a signalling conflict or trade-off between two different defence pathways (Heil, 2001). However, the fact that barley productivity was not affected suggests that the *Pratylenchus* infection was still below the damage threshold.

Chemical control*

Soil disinfestation by chemical compounds to control *Pratylenchus* spp. is mainly used when cultural, physical and biological methods fail to reduce nematode populations sufficiently. However, because of environmental and human health concerns there are increasing

*Mention of a trademark, proprietary product, or vendor does not constitute a guarantee or warranty of the product by the authors, and does not imply its approval to the exclusion of other products and vendors that might also be suitable. Similarly the authors do not accept liability for any error or omission in the content, or for any loss, damage or other accident arising from the use of products mentioned in this chapter.

restrictions on their use. Chemical control of plant-parasitic nematodes, particularly *Pratylenchus* spp., is typical in disinfestation of soils in high-value, mainly horticultural and tree, crops. Nevertheless, the use of any chemical nematicide is dependent upon whether the nematicide is labelled for management of root-lesion nematodes in that crop and if the predicted economic return on investment warrants the use of a nematicide. Moreover, the nematode cuticle is impermeable to many organic molecules. Consequently, most nematicides have tended to be rather toxic or volatile, with poor target specificity and less than perfect human or environmental safety, potential hazards including groundwater contamination or atmospheric ozone depletion (Chitwood, 2002).

Two groups of chemical nematicides have been classically distinguished; fumigants and non-fumigants (Bakker, 1993; Whitehead, 1998). Fumigants are solid or liquid chemicals that vaporise into gases when applied to soil, diffusing through the available pore spaces. The most common fumigant nematicides applied to control *Pratylenchus* spp. are chloropicrin (considered a good fungicide but a poor nematicide), dazomet, 1,3-D, EDB, metham sodium and methyl bromide (which has been categorised as a Class I ozone-depleting substance by the Montreal and Kyoto Protocols). This compound is being phased out from the international market for ecological and toxicological reasons (in industrialised nations by 2005, developing nations by 2015). Non-fumigant nematicides are either organophosphates or carbamates and include aldicarb, carbofuran, disulphoton, ethoprophos, fenamiphos, fensulfothion and oxamyl. There have been several comprehensive reviews of *Pratylenchus* management by chemical compounds (Brown & Kerry, 1987; Whitehead, 1998; Rich *et al.*, 2004) and therefore only some current summarised and relevant information concerning the chemical control of *Pratylenchus* spp. in herbaceous and fruit trees and woody plants will be included in this section.

HERBACEOUS AND VEGETABLE CROPS

The most likely alternatives to methyl bromide for nematode control in vegetables are 1,3-D and to a lesser extent metham sodium (Roberts *et al.*, 1988; Locascio *et al.*, 1997). However, disinfestation of soils in low-value crops such as cereals and legumes cannot be economically justified, and only a few nematicides have been applied experimentally to control *Pratylenchus* spp. in those crops. Nonetheless, the nematicide seed-dressing application method has been shown to have environmental and economic advantages and has been successfully applied to several herbaceous and vegetable crops, including cereals and legumes. One of the advantages is that only a minimal amount of nematicide per unit area is applied and this is degraded during the growing season, as indicated by residue levels below detection levels in wheat grains and straw (Orion & Shlevin, 1989). From an economic point of view seed dressing is a relatively inexpensive nematicide application method and thus economically attractive to low-value crops grown in marginal production areas.

Soaking cut potato seed pieces or whole tubers in various concentrations of oxamyl (1-32 mg ml^{-1}) reduced populations of *P. penetrans* in soil and roots and increased plant growth under glasshouse conditions (Townshend & Olthof, 1988). However, soaking cut potato seed pieces with oxamyl at several concentrations (32, 64 and 96 mg ml^{-1}) under field conditions revealed that only the concentration of 64 mg ml^{-1} reduced populations of *P. penetrans*, whilst 32 mg ml^{-1} was not effective and the highest concentration (96 mg ml^{-1}) was highly phytotoxic (Olthof & Townshend, 1991). Consequently, although potato tuber soak treatments reduced populations of *P. penetrans* in many cases, the reductions under field conditions were relatively small compared to those obtained using soil drenches, foliar sprays, or both (Olthof *et al.*, 1985; Olthof, 1989).

In chickpea, seed coating with paste containing nematicides has resulted in contradictory results. The application of aldicarb, carbofuran and fensulfothion at three concentrations (1-2% active substance (a.s.) seed weight) each to chickpea through seed treatment significantly reduced final populations of *P. thornei* under controlled conditions, the highest concentrations giving the greatest nematode reduction (Walia & Seshadri, 1985). Similarly, chickpea seed soaked in 0.1% carbosulfan significantly reduced soil and root populations of *P. thornei* and increased grain yield by 40% under controlled conditions (Shukla *et al.*, 2001). However, under field conditions chickpea seed coated with aldicarb was ineffective (Greco *et al.*, 1988). Satisfactory results were also obtained in wheat (cv. Bet Lehem) seeds coated with either a seed-dressing formulation of furathiocarb (10 g a.s. kg^{-1} seed), carbofuran (10 g a.s. kg^{-1} seed), or oxamyl (3.6 g a.s. kg^{-1} seed), reducing populations of *P. mediterraneus* and increasing grain yield (Orion & Shlevin, 1989). In some cases, dividing nematicide soil application into treatments before sowing and after emergence, especially for legumes which are sown in autumn, prevents *Pratylenchus* invasion of roots for longer periods and thereby increases yield. In Syria, grain yield of chickpea was doubled when aldicarb was applied before sowing and 1 month after plant emergence (5 kg a.s. ha^{-1} each time) to control *P. thornei* (Greco *et al.*, 1988; Di Vito *et al.*, 1991). In forage legumes such as red clover, alfalfa and bird's foot trefoil, treatments with fensulfothion (11.2-44.8 kg a.s. ha^{-1}) or fenamiphos (5.6-22.4 kg a.s. ha^{-1}) reduced populations of *P. penetrans* in the soil as well as in the roots of the three legumes (Thompson & Willis, 1975). In wheat, applications of aldicarb at a low rate (1.5 kg a.s. ha^{-1}) gave inconsistent nematode control and did not significantly increase yield of susceptible/intolerant wheat varieties (cvs Machete and Janz, respectively), whereas high rate applications (2.5 kg a.s. ha^{-1}) reduced *P. neglectus* and *P. thornei* densities by 70-90%, resulting in up to 23% greater yield for cultivar Machete (Taylor *et al.*, 1999). Application of aldicarb (4 kg a.s. ha^{-1}) in soil infested by *P. thornei* in Italy almost eliminated the nematode from the top 10 cm of the soil but had no effect below 15 cm (Tacconi *et al.*, 1988a).

In corn, good control of *P. hexincisus* was achieved by combined applications of 1,3-D and carbofuran in Iowa (Norton & Hinz, 1976) and applications of EDB, 1,2-dibromo-3-chloropropane (DBCP), 1,3-D-1,2-D, fenamiphos and aldicarb significantly reduced populations of *P. zeae* 3 months after planting and increased growth and yield of sweet corn Seneca Chief in Georgia (Johnson & Chalfant, 1973). However, use of DBCP is declining and considerable safety precautions are in place in countries where some application is still allowed. Pre-plant treatments with 1,2-D-1,3-D or carbofuran significantly reduced populations of *P. brachyurus* and *P. zeae* and increased crop yield under field conditions at Hamilton County, Florida (Rich *et al.*, 1985) and applications with terbufos and carbofuran at planting and post-planting application of carbofuran significantly reduced *P. scribneri* populations in soil and roots (Todd & Oakley, 1995).

The control of *P. zeae* in sugarcane through soil application of carbofuran and terbufos under field conditions has been demonstrated as technically and economically viable. The efficacy of the chemical control, expressed in terms of reduction of nematode population level and the consequent increase of cane production, was dependent on nematicide dose (8.5-10 litres c.p. ha^{-1} and 50-60 kg c.p. ha^{-1}, respectively) (Novaretti *et al.*, 1998).

In peanut, treatments with fenamiphos, aldicarb and chlorpyrifos significantly reduced *P. brachyurus* populations (Minton *et al.*, 1990). Similarly in pineapple, treatments with fenamiphos alone and fenamiphos plus carbofuran significantly reduced *P. brachyurus* populations (Sarah, 1987).

In potato, oxamyl has been shown to provide protection from *P. penetrans* damage and increased marketable yield of tubers by as much as 44.4% (Olthof *et al.*, 1985). Applications of aldicarb to the widely grown cv. Superior in North America have increased tuber yields by up to 40% when *P. penetrans* populations were large (Bernard & Laughlin, 1976; Olthof, 1986; Kimpinski & McRae, 1988). This cultivar is less tolerant to *P. penetrans* damage than cv. Yukon Gold (Olthof, 1986) and responded more positively to nematicide treatments than have cvs Russet Burbank and Kennebec (Kimpinski, 1982; Kimpinski & McRae, 1988). Likewise, treatment of early- and late-maturing potato cultivars with 1,3-D (0.6 kg a.s. ha^{-1}) significantly reduced populations of *P. penetrans* (Kimpinski *et al.*, 2001). Both nematicide applications increased total potato tuber yield by 12% for early-maturing cultivars (AC Chaleur, AC Belmont, Superior and Eramosa), but by less than 2% for the late-maturing cultivars (AC Novachip, Norchip, Russet Burbank, Kennebec and Russet Norkotah) under field conditions at Harrington, Prince Edward Island, Canada (Kimpinski *et al.*, 2001).

In tobacco, the major fumigant nematicides used for *Pratylenchus* control in tobacco growing areas included chloropicrin, 1,3-D, EDB, 1,3-D-methyl isothiocyanate mixtures, metham-sodium and methyl bromide (Rich *et al.*, 1989). Applications of triazophos at 8.4 kg a. i. ha^{-1} or Telone C (1,3-D, related C3 hydrocarbons 85% and chloropicrin 15%) at 90 l ha^{-1} significantly reduced *P. penetrans* on tobacco cv. Delhi 34 (Tu *et al.*, 1996) and treatments with Telone C17 (1,3-D 76.3%, chloropicrin 17.1% and inert ingredients 6.6%), Vorlex (methyl isothiocyanate 20%, mixture of 1,3-D and related C3 hydrocarbons 80%), Vorlex CP (methyl isothiocyanate 23.5%, mixture of chloropicrin, 1,2-dichloropropane and 1,3-D 76.5%), dazomet and fenamiphos, all, with the exception of fenamiphos, increased the yield of tobacco and significantly reduced nematode populations in soils (Tu *et al.*, 1995). Populations of *P. penetrans* in soil and roots of tobacco in Canada were also reduced by applications of 1,3-D and triazophos at 90 l ha^{-1} or 8.5 kg ha^{-1}, respectively (Tu *et al.*, 1996).

In tomato, Reynolds *et al.* (1992) reported that applications of 1,3-D at 68 l ha^{-1} or mixed applications of 20% methyl isothiocyanate + 40% 1,3-D controlled field populations of *P. penetrans* whilst in yam, seed treatment by immersion in oxamyl at 2400 mg a.s. l^{-1} for 15 min at 2 days and 1 month after harvest proved to be a highly effective treatment for control of *P. coffeae* in stored tubers (Oramas, 2002).

Fruit trees and woody plants

Pre-plant fumigation is the method most widely used to reduce population densities of *Pratylenchus* spp. before fruit trees and woody crops are planted in nursery stocks or infested fields. Application of fenamiphos has been reported to reduce *P. penetrans* population densities in established plantings of small-fruit (Vrain & Keng, 1986; Lolas *et al.*, 1992) and tree-fruit (Santo & Wilson, 1990) crops. Preplant chemical treatment with 1,3-D has been recommended to provide effective control of *P. vulnus* attacking peach (Nyczepir, 1991) and aldicarb applied at 12.2 kg ha^{-1} to rose beds controlled *P. penetrans* and increased flower production under glasshouse conditions (Johnson & McClanaham, 1974). In apple replant disease, applications of fenamiphos (Mazzola, 1998) or chloropicrin (Koch *et al.*, 1980) effectively controlled *P. penetrans* from orchard soils. However, improvement in plant growth was observed in response to applications of chloropicrin, but not with applications of fenamiphos. Similarly, significant control of *P. vulnus* on grapevine was achieved with four non-fumigants (carbofuran, oxamyl, fenamiphos and sulfocarb) and one fumigant nematicide (DBCP) (Viglierchio *et al.*, 1977). These nematicide applications reduced nematode population density sufficiently to prevent major root damage during the first years following tree establishment, thus allowing the tree to have a healthy start (Nyczepir *et al.*, 2000). Nevertheless, in other cases a dual nematicide application, usually consisting of a preplant fumigant followed by a non-fumigant applied at planting, has been used because of the severe pathogen pressure caused by both infested soil and infected planting stock (Westerdahl *et al.*, 2003).

In coffee, nematicide treatments with terbufos (1 or 2 g per plant) only suppressed nematode populations of *P. coffeae* in the first year of sampling (before the coffee plants began to yield). Thereafter, nematicide treatment had no effect on nematode populations density, even at the highest dose (Villain *et al.*, 2000). Unfortunately, in spite of the significant reduction of nematode populations, nematicide treatments did not greatly influence coffee yield (Villain *et al.*, 2000).

The long-term effects of nematicides on soil microbial populations has become a matter of concern, as soil fertility may be adversely affected by the continuous application of crop protection chemicals. Sturz and Kimpinski (1999) studied the interactions between the nematicides aldicarb and fosthiazate and populations of *P. penetrans*, plant growth promoting and plant growth inhibiting bacteria in the root zone soils of potatoes (cv. Superior). Results indicated that aldicarb appeared to suppress the populations of plant growth promoting bacteria that contribute to enhanced growth in the crop. However, fosthiazate, but not aldicarb, significantly increased tuber yields and reduced soil populations of nematodes.

Novel nematicide compounds

In the last decade or so novel chemical compounds have been evaluated to control *Pratylenchus* spp. A good example is the exposure of *Pratylenchus* sp. to the vapours of butyric acid for a period of 7 days or their incubation in sand treated with high concentrations of butyric acid (0.1 and 1 M) which resulted in a significant reduction in nematode population (89-100%) as compared to the untreated controls (Browning *et al.*, 2004). These results were later confirmed by drenching strawberry plants infected with *P. penetrans* with butyric acid (0.1 and 1 M) resulting in reductions of nematode populations by 98-100% (Browning *et al.*, 2006). These results suggest that butyric acid may be an alternative to synthetic soil fumigants for control of *Pratylenchus* spp.

Some natural nematicidal substances have been reported. Thiarubrine C, a polyacetylenic 1,2-dithiin isolated from the roots of the Asteraceae *R. hirta*, exhibited strong nematicidal activity against *P. penetrans in vitro* and growth chamber assays, with an LC_{50} of 23.5 ppm (Sanchez de Viala *et al.*, 1998). Also, new nematicidal alkaloids, peniprequinolone and quinolinone, isolated from *Penicillium* cf. *simplicissimum* (Oudemans) Thom. showed nematicidal activity to *P. penetrans* on lettuce and rice (Kusano *et al.*, 2000). Several classes of proteins (*e.g.*, lectins or lectin-related proteins) have been suggested as candidate nematicides or nematostatics. In fact, Carlens *et al.* (2005) reported that transgenic *Ara-*

bidopsis thaliana (L.) Heynh. lines expressing the lectins Banlec and IRIP significantly reduced the reproduction of *P. coffeae*, although this result was not reproducible. Also, seaweed concentrates from the brown kelp *Ecklonia maxima* (Osbeck) Papenfuss gave significant reductions of *P. zeae* populations *in vitro* but failed to reproduce this effect under glasshouse conditions (De Waele *et al.*, 1988a). Synthetic acetylenes have shown activity against *P. coffeae*, several compounds with only one acetylenic bond being active at less than 1.0 μg ml^{-1} (Mori *et al.*, 1982).

HERBICIDES

The commercial application of herbicides (*e.g.*, imazethapyr) may have an influence upon endoparasitic *Pratylenchus* spp. In soil treated with imazethapyr at rates of 0.08 and 0.16 g a.s. ha^{-1} the nematode population decreased by 13.9-42.8% and 27.1-50.7%, respectively. The observed trend in the dynamics of the nematode population persisted for 1 year only and changed with the crop rotation (Trifonova & Peneva, 2003). Nevertheless, Kornobis (2000) found that the influence of herbicides may be variable depending on the specific herbicide-host plant-*Pratylenchus* spp. In fact, *P. neglectus* numbers increased in microplots treated with benazolin (0.45 kg ha^{-1}) in a winter rape field whereas the nematode population decreased in the untreated control. On the other hand, no differences between herbicide treated and control microplots were observed when linuron + bentazone (1.0 kg ha^{-1} + 1.5 kg ha^{-1}, respectively) were applied to pea; and methabenzthiazuron (2.8 kg ha^{-1}) was applied to winter wheat (Kornobis, 2000).

In summary, the current agriculture and agri-food sector is expected to move toward a more sustainable system of production. Greater productivity and competitiveness are anticipated to come from increased efficiency through the acquisition and application of new crop production strategies. With the rising costs associated with agrochemical applications, producers need to become increasingly efficient in their use of agrochemical resources. However, in general, nematicides do not provide long-term suppression of nematodes and environmental and human health concerns are leading to increased restrictions on their use. Some safe procedures for *Pratylenchus* control have been developed based on biological control agents and organic amendments but there is still a need to develop alternative, environmentally friendly, measures or compounds for effective nematode control.

Nematology Monographs & Perspectives, 2007, Vol. 6, 409-414

Appendix

Table 7. Pratylenchus *spp. arranged by number of lip annuli.*

Species	Lip annuli	Species	Lip annuli
acuticaudatus	2	*curvicauda*	3
alleni	2	*delattrei*	3
angulatus	2	*ekrami*	3
artemisiae	2	*elamini*	3
brachyurus	2	*fallax*	3
brzeskii	2	*japonicus*	3
coffeae	2	*kasari*	3
crassi	2	*kralli*	3
dunensis	2	*kumaoensis*	3
estoniensis	2	*manaliensis*	3
flakkensis	2	*mediterraneus*	3
gibbicaudatus	2	*microstylus*	3
hexincisus	2	*morettoi*	3
hippeastri	2	*mulchandi*	3
jaehni	2	*penetrans*	3
loosi	2	*pinguicaudatus*	3
macrostylus	2	*pratensis*	3
neglectus	2	*pratensisobrinus*	3
neobrachyurus	2	*pseudofallax*	3
okinawensis	2	*pseudopratensis*	3
panamaensis	2	*sensillatus*	3
pseudocoffeae	2	*subpenetrans*	3
roseus	2	*subranjani*	3
scribneri	2	*sudanensis*	3
silvaticus	2	*teres*	3
tenuis	2	*thornei*	3
yamagutii	2	*unzenensis*	3
andinus	3	*ventroprojectus*	3
arlingtoni	3	*vulnus*	3
bhattii	3	*yassini*	3
bolivianus	3	*zeae*	3
convallariae	3	*goodeyi*	4
crenatus	3	*typicus*	4
cruciferus	3	*wescolagricus*	4

Table 8. Pratylenchus *spp. arranged by stylet length (μm).*

Species	Stylet length	Species	Stylet length
microstylus	11.5	*neglectus*	16
ekrami	12	*panamaensis*	16
angulatus	12.5	*penetrans*	16
artemisiae	13.5	*pratensisobrinus*	16
bhattii	13.5	*pseudocoffeae*	16
alleni	14	*subpenetrans*	16
elamini	14	*typicus*	16
jaehni	14.5	*yamagutii*	16
kralli	14.5	*zeae*	16
kumaoensis	14.5	*convallariae*	16.5
pratensis	14.5	*dunensis*	16.5
scribneri	14.5	*fallax*	16.5
vulnus	14.5	*morettoi*	16.5
tenuis	14.8	*pseudopratensis*	16.5
hexincisus	15	*kasari*	16.8
manaliensis	15	*arlingtoni*	17
mediterraneus	15	*flakkensis*	17
pseudofallax	15	*goodeyi*	17
sudanensis	15	*okinawensis*	17
unzenensis	15	*subranjani*	17
ventroprojectus	15	*teres*	17
silvaticus	15.1	*thornei*	17
coffeae	15.5	*yassini*	17
cruciferus	15.5	*andinus*	17.5
gibbicaudatus	15.5	*crassi*	17.5
hippeastri	15.5	*delattrei*	17.5
neobrachyurus	15.5	*mulchandi*	17.5
roseus	15.5	*pinguicaudatus*	17.5
sensillatus	15.5	*wescolagricus*	18
acuticaudatus	16	*bolivianus*	19
crenatus	16	*brzeskii*	19
curvicauda	16	*brachyurus*	19.5
estoniensis	16	*japonicus*	19.8
loosi	16	*macrostylus*	23

Table 9. Pratylenchus *spp. arranged by length of pharyngeal gland overlap (μm).*

Species	Overlap	Species	Overlap
hexincisus	14	*crassi*	40
delattrei	19	*gibbicaudatus*	40
arlingtoni	20	*kasari*	40
kralli	20	*mediterraneus*	40
flakkensis	22	*panamaensis*	40
angulatus	23	*pratensisobrinus*	40
crenatus	25	*subpenetrans*	40
pratensis	25	*thornei*	40
neglectus	27	*andinus*	41
pseudopratensis	27	*convallariae*	41
ventroprojectus	27	*cruciferus*	41
kumaoensis	28	*teres*	41
artemisiae	30	*acuticaudatus*	42
ekrami	30	*alleni*	42
estoniensis	30	*hippeastri*	43
fallax	31	*roseus*	45
zeae	31	*japonicus*	46
neobrachyurus	32	*sensillatus*	46
jaehni	33	*subranjani*	46
sudanensis	33	*unzenensis*	47
tenuis	34	*goodeyi*	49
bhattii	35	*macrostylus*	50
bolivianus	35	*penetrans*	50
loosi	35	*brachyurus*	51
scribneri	35	*brzeskii*	54
wescolagricus	35	*dunensis*	54
coffeae	36	*typicus*	54
silvaticus	36	*manaliensis*	55
elamini	37	*okinawensis*	55
microstylus	37	*yassini*	58
mulchandi	37	*pinguicaudatus*	61
pseudofallax	37	*curvicauda*	62
yamagutii	37	*pseudocoffeae*	70
vulnus	38	*morettoi*	76

Table 10. Pratylenchus *spp. arranged by post-vulval uterine sac length (μm).*

Species	Post-vulval uterine sac	Species	Post-vulval uterine sac
angulatus	9	*mediterraneus*	21.5
neobrachyurus	11	*pratensisobrinus*	22
pinguicaudatus	12	*teres*	22
yamagutii	13	*ventroprojectus*	22
goodeyi	14	*acuticaudatus*	23
tenuis	14	*wescolagricus*	23
estoniensis	15	*cruciferus*	24
kralli	15	*kasari*	24
neglectus	15	*curvicauda*	25
alleni	16	*dunensis*	25
bhattii	16	*okinawensis*	25
loosi	16	*andinus*	26
roseus	16	*scribneri*	26
flakkensis	17	*unzenensis*	26
japonicus	17	*bolivianus*	26.5
macrostylus	17	*brzeskii*	27
microstylus	17	*jaehni*	27
pseudofallax	18	*pseudocoffeae*	27
silvaticus	18	*penetrans*	28
subpenetrans	18	*sudanensis*	28
thornei	18	*hippeastri*	30
arlingtoni	19	*zeae*	30
hexincisus	19	*mulchandi*	32
panamaensis	19.5	*sensillatus*	33
brachyurus	20	*subranjani*	33
crassi	20	*ekrami*	33.5
crenatus	20	*gibbicaudatus*	34
delattrei	20	*elamini*	35
fallax	20	*vulnus*	36
kumaoensis	20	*yassini*	37
pseudopratensis	20	*typicus*	40
manaliensis	20.5	*convallariae*	42
artemisiae	21	*coffeae*	45
pratensis	21	*morettoi*	59

Table 11. Pratylenchus *spp. arranged by position of vulva (V).*

Species	V	Species	V
teres	70	*okinawensis*	79
zeae	71	*pseudopratensis*	79
bhattii	73	*sensillatus*	79
curvicauda	73	*tenuis*	79
gibbicaudatus	73	*ventroprojectus*	79
sudanensis	73	*yamagutii*	79
goodeyi	74	*convallariae*	79.5
yassini	74	*vulnus*	79.5
crassi	74.5	*alleni*	80
elamini	75	*coffeae*	80
flakkensis	75	*ekrami*	80
mulchandi	75	*panamaensis*	80
subranjani	75	*penetrans*	80
delattrei	76	*pinguicaudatus*	80
microstylus	76	*pseudofallax*	80
morettoi	76	*subpenetrans*	80
thornei	76.5	*bolivianus*	81
brzeskii	77	*manaliensis*	81
cruciferus	77	*neobrachyurus*	81
hippeastri	77	*pseudocoffeae*	81
kralli	77	*silvaticus*	81
pratensisobrinus	77	*wescolagricus*	81
scribneri	77	*typicus*	81.5
unzenensis	77	*angulatus*	82
acuticaudatus	78	*arlingtoni*	82
artemisiae	78	*crenatus*	82
dunensis	78	*kumaoensis*	82
jaehni	78	*roseus*	82
kasari	78	*estoniensis*	82.5
pratensis	78	*andinus*	83
fallax	79	*loosi*	83
hexincisus	79	*brachyurus*	85.5
mediterraneus	79	*japonicus*	86
neglectus	79	*macrostylus*	87

Table 12. Pratylenchus *spp. arranged by increasing body length (mm).*

Species	Body length (mm)	Species	Body length (mm)
angulatus	0.36	*dunensis*	0.51
neobrachyurus	0.36	*mediterraneus*	0.51
alleni	0.38	*neglectus*	0.51
microstylus	0.39	*pseudocoffeae*	0.51
subpenetrans	0.39	*subranjani*	0.51
tenuis	0.40	*japonicus*	0.52
elamini	0.41	*yassini*	0.52
yamagutii	0.41	*zeae*	0.52
unzenensis	0.42	*coffeae*	0.53
crassi	0.43	*ekrami*	0.53
estoniensis	0.43	*mulchandi*	0.54
hexincisus	0.43	*andinus*	0.55
kumaoensis	0.44	*teres*	0.55
ventroprojectus	0.44	*artemisiae*	0.56
kralli	0.45	*brachyurus*	0.57
silvaticus	0.45	*manaliensis*	0.57
arlingtoni	0.46	*loosi*	0.58
crenatus	0.46	*penetrans*	0.58
delattrei	0.46	*pinguicaudatus*	0.58
okinawensis	0.46	*scribneri*	0.58
pseudofallax	0.47	*bolivianus*	0.59
bhattii	0.48	*hippeastri*	0.59
gibbicaudatus	0.48	*convallariae*	0.60
panamaensis	0.48	*macrostylus*	0.60
pratensisobrinus	0.48	*vulnus*	0.60
roseus	0.48	*wescolagricus*	0.60
fallax	0.49	*thornei*	0.61
jaehni	0.49	*sensillatus*	0.62
pratensis	0.49	*typicus*	0.64
pseudopratensis	0.49	*goodeyi*	0.66
sudanensis	0.49	*kasari*	0.67
curvicauda	0.50	*brzeskii*	0.69
flakkensis	0.50	*cruciferus*	0.73
acuticaudatus	0.51	*morettoi*	0.74

List of abbreviations

- 1,2-D = 1,2 dichloropropane
- 1,3-D = 1,3 dichloropropene
- AMF = Arbuscular mycorrhizal fungi
- DBCP = 1,2-dibromo-3-chloropropane
- EDB = ethylene dibromide
- EST = esterase
- ESTs = expressed sequence tags
- ETS = external transcribed spacer
- GO = Gene Ontology
- G6-PDH = glucose-6-phosphate dehydrogenase
- IDH = isocitrate dehydrogenase
- IEF = isoelectrofocusing
- IGS = intergenic spacer
- INRA = Institut National de la Recherche Agronomique
- IPM = integrated pest management
- ITS = internal transcribed spacer
- LSU = large subunit 26S rRNA gene
- MDH = malate dehydrogenase
- PAGE = poly-acrylamide gel electrophoresis
- PAL = phenylalanine ammonia-lyase
- PCR = polymerase chain reaction
- PGM = phosphoglucomutase
- PGI = phosphoglucose isomerase
- RAPD = random amplified polymorphic DNA
- rDNA = ribosomal DNA
- rRNA = ribosomal RNA
- RFLP = restriction fragment length polymorphisms
- SDS = sodium dodecyl sulphate
- SL1 = splice-leader 1 library
- SOD = superoxide dismutase
- TAL = tyrosine ammonia-lyase

References

ABALLAY, E., JARDEL, A. & MONTEALEGRE, J. (1996). Efectos de la solarizacion y bromuro de metilo sobre *Verticillium dahliae* Kleb. y nematodos fitoparasitos asociados a monocultivos de frutilla. *Fitopatologia* 31, 230-240.

ABBASI, P.A., RIGA, E., CONN, K.L. & LAZAROVITS, G. (2005). Effect of neem cake soil amendment on reduction of damping-off severity and population densities of plant-parasitic nematodes and soilborne plant pathogens. *Canadian Journal of Plant Pathology* 27, 38-45.

ABRANTES, I.M. DE O., FARIA, C.A.T. & SANTOS, M.S.N.A. (1987). Root-lesion nematodes (*Pratylenchus* spp.) in Portugal. *Nematologia Mediterranea* 15, 375-378.

ACEDO, J.R. & ROHDE, R.A. (1971). Histochemical root pathology of *Brassica oleracea capitata* L. infected by *Pratylenchus penetrans* (Cobb) Filipjev and Schuurmans Stekhoven (Nematoda: Tylenchidae). *Journal of Nematology* 3, 62-67.

ACOSTA, A. & AYALA, A. (1976). Hot water and chemical dips for nematode control in tubers of *Dioscorea rotundata*. *Journal of Agriculture of the University of Puerto Rico* 60, 395-402.

ACOSTA, N. & MALEK, R.B. (1979). Influence of temperature on population development of eight species of *Pratylenchus* on soybean. *Journal of Nematology* 11, 229-232.

ADAM, M.A.M., PHILLIPS, M.S. & BLOK, V.C. (2007). Molecular diagnostic key for identification of single juveniles of seven common and economically important species of root-knot nematode (*Meloidogyne* spp.). *Plant Pathology* 56, 190-197.

ADEMOLA, I.A., AKEM, C.N. & DASHIELL, K.E. (1996). Plant parasitic nematodes of soybeans in Nigeria. *Afro-Asian Journal of Nematology* 6, 191-194.

ADIKO, A. (1988). Plant-parasitic nematodes associated with plantain, *Musa paradisiaca* (AAB), in the Ivory Coast. *Revue de Nématologie* 11, 109-113.

ADIKO, A. & N'GUESSAN, A.B. (2001). Evolution of the nematofauna of plantain, *Musa* AAB, in Côte d'Ivoire. *Infomusa* 10, 26-27.

AFOLAMI, S.O. & FAWOLE, B. (1991). Effect of *Pratylenchus sefaensis* Fortuner 1973 on growth and yield of *Zea mays* L. cv. FARZ-7 under continuous cropping. *Plant and Soil* 138, 133-138.

AGU, C.M. (2004). Effect of *Meloidogyne incognita* and *Pratylenchus brachyurus* on leaf growth of sweet potato. *Tropical Science* 44, 48-50.

AGUDELO, M.P. & VOLCY, C. (1998). Población de nematodos fitoparasitos en dos municipios cañeros de Antioquia. *Fitopatologia Colombiana* 21, 89-96.

AILAROVA, Z.S. (1986). [Nematodes of some vegetable crops in Karakalpakiya.] *Uzbekskii Biologicheskii Zhurnal* 3, 64-65.

AKGUL, H.C. & OKTEN, M.E. (2001). A list of Tylenchida (Nematoda) associated with poppy crop (*Papaver somniferum* L.) in Afyon region, Turkey. *Pakistan Journal of Nematology* 19, 63-66.

ALAM, M.M. (1991). Control of plant parasitic nematodes with oilseed cakes on some vegetables in field. *Pakistan Journal of Nematology* 9, 21-30.

ALAM, M.M. (1993). Bioactivity against phytonematodes. In: Randhawa, N.S. & Parmar, B.S. (Eds). *Neem research and development*. New Delhi, India, Society of Pesticide Science, pp. 123-143.

AL-BANNA, L., WILLIAMSON, V. & GARDNER, S.L. (1997). Phylogenetic analysis of nematodes of the genus *Pratylenchus* using nuclear 26S rDNA. *Molecular Phylogenetics and Evolution* 7, 94-102.

AL-BANNA, L., PLOEG, A.T., WILLIAMSOM, V.M. & KALOSHIAN, I. (2004). Discrimination of six *Pratylenchus* species using PCR and species-specific primers. *Journal of Nematology* 36, 142-146.

ALBY, T., FERRIS, J.M. & FERRIS, V.R. (1983). Dispersion and distribution of *Pratylenchus scribneri* and *Hoplolaimus galeatus* in soybean fields. *Journal of Nematology* 15, 418-426.

ALCAÑIZ, E., PINOCHET, J., FERNÁNDEZ, C., ESMENJAUD, D. & FELIPE, A. (1996). Evaluation of *Prunus* rootstocks for root-lesion nematode resistance. *HortScience* 31, 1013-1016.

ALEXANDER, M. (1977). *Introduction to soil microbiology*. New York, USA, John Wiley & Sons, Inc., 467 pp.

ALEXANDER, S.A. & WALDENMAIER, C.M. (2002). Suppression of *Pratylenchus penetrans* populations in potato and tomato using african marigolds. *Journal of Nematology* 34, 130-134.

ALI, S.S. & AHMAD, R. (2000). Screening of chickpea germplasm against nematode. *International Chickpea and Pigeonpea Newsletter* 7, 8-9.

ALI, S.S. & ASKARY, T.H. (2001). Taxonomic status of phytonematodes associated with pulse crops. *Current Nematology* 12, 75-84.

ALI, S.S. & SHARMA, S.B. (2003). Nematode survey of chickpea production areas in Rajasthan, India. *Nematologia Mediterranea* 31, 147-149.

ALLEN, M.W. (1960). The genera *Pratylenchus*, *Radopholus*, *Pratylenchoides*, *Rotylenchulus* and *Nacobbus*; *Tylenchulus*, *Trophotylenchulus*, *Trophonema* and *Sphaeronema*. In: Sasser, J.N. & Jenkins, W.R. (Eds). *Nematology*. Chapel Hill, NC, USA, University of Carolina Press, pp. 181-184.

ALLEN, M.W. & JENSEN, H.J. (1951). *Pratylenchus vulnus*, new species (Nematoda: Pratylenchidae), a parasite of trees and vines in California. *Proceedings of the Helminthological Society of Washington* 18, 47-50.

AL-REHIAYANI, S. & HAFEZ, S. (1998). Host status and green manure effect of selected crops on *Meloidogyne chitwoodi* race 2 and *Pratylenchus neglectus*. *Nematropica* 28, 213-230.

ALVAREZ, M. & FERNANDEZ, E. (1982). Aspectos de la importancia de los nematodos del genero *Meloidogyne* en el cultivo del kenaf (*Hibiscus cannabinus*). *Ciencia y Tecnica en la Agricultura, Proteccion de Plantas* 5, 7-18.

AMBROGIONI, L. & RAPETTI, S. (1992). Un grave caso di deperimento in colture di *Viola odorata* L. associato con *Zygotylenchus guevarai* (Tobar Jimenez) Braun e Loof. *Redia* 75, 415-427.

AMBROGIONI, L., CAROPPO, S. & PALMISONO, A.M. (1992). Prime osservazioni sulla dinamica delle popolazioni di nematodi fitoparassiti in rotazioni colturali comprendeti frumento duro. *Redia* 75, 67-100.

AMOSU, J.O. & TAYLOR, D.P. (1974). Interaction of *Meloidogyne hapla*, *Pratylenchus penetrans* and *Tylenchorhynchus agri* on Kenland red clover, *Trifolium pratense*. *Indian Journal of Nematology* 4, 124-131.

AMSING, J.J. (1996). Population dynamics and damage potential of the root-lesion nematode, *Pratylenchus bolivianus*, on *Alstroemeria*. *Nematologica* 42, 71-79.

ANDERSEN, H.J. (1972). Migratory nematodes on leguminous plants in Denmark. *Tidsskrift for Planteavl* 76, 559-569.

ANDERSEN, H.J. (1975). *Pratylenchus* population densities in barley monocultures. *Nordisk Jordbrugsforskning* 57, 1129-1130.

ANDERSEN, H.J. (1979). Migratory nematodes in Danish barley fields. I. The qualitative and quantitative composition of the fauna. *Tidsskrift for Planteavl* 83, 1-8.

ANDERSON, R.V. & TOWNSHEND, J.L. (1985). A new species of root-lesion nematode (Pratylenchidae: Nematoda) in Canada with a scanning electron microscope study of its head morphology. *Canadian Journal of Zoology* 63, 2378-2382.

ANDERSON, S. (1971). *Pratylenchus scribneri*, a root-lesion nematode in *Hippeastrum*. *Växtskyddsnotiser* 35, 43-47.

ANDRÁSSY, I. (1960). Beiträge zur Kenntnis der freilebenden Nematoden Chinas. *Annals of the Natural History Museum of Hungary* 52, 201-216.

ANDRÁSSY, I. (1976). *Evolution as a basis for the systematization of nematodes.* London, UK, Pitman Publishing, 288 pp.

ANDRÉS, M.F., PINOCHET, J., HERNÁNDEZ BORREGO, A. & DELIBES, A. (2000). Detection and analysis of inter- and intraspecific diversity of *Pratylenchus* spp. using isozyme markers. *Plant Pathology* 49, 640-649.

ANGUIZ, R. & CANTO-SAENZ, M. (1991). Reaction of sweet potato cultivars (*Ipomoea batatas* (L.) Lam.) to *Pratylenchus flakkensis* Seinhorst. *Nematropica* 21, 197-201.

ANON. (1975). Annual Report 1973-74. *Nigeria Federal Department of Agricultural Research*, 178 pp.

ANTER, E.A., ABD-ELGAWAD, M.M.M. & HAMMAD, M. (1993). A population and distribution survey of cotton nematodes in Egypt. *Bulletin of Faculty of Agriculture, University of Cairo* 44, 907-917.

ANWAR, S.A., GORSI, S., ANWAR UL HAQ, M., REHMAN, T. & YOUSUF, P. (1991). Plant parasitic nematodes of some field, vegetable, fruit and ornamental crops. *Journal of Agricultural Research (Lahore)* 29, 233-249.

ANWAR, S.A., KHAN, M.S.A. & TAHIR, A. (1992a). Plant parasitic nematode problems of some field, vegetable, fruit and ornamental crops in the Punjab, Pakistan - II. *Proceedings of Parasitology* 14, 84-98.

ANWAR, S.A., RAUF, C.A. & GORSI, S.D. (1992b). Weeds as alternate hosts of phytonematodes. *Afro-Asian Journal of Nematology* 2, 41-47.

ANWAR, S.A., GORSI, S.D. & SHAKOOR, A. (1993). Effect of *Longidorus elongatus*, *Meloidogyne incognita* and *Pratylenchus brachyurus* on peanut growth. *Pakistan Journal of Nematology* 11, 115-124.

ANYANGO, J.J. (1988). The effect of root-knot (*Meloidogyne hapla*) and lesion (*Pratylenchus penetrans*) nematodes of pyrethrum seedlings in Kenya. *Acta Horticulturae* 218, 355-358.

ANZUETO, F., MOLINA, A., FIGUEROA, P. ETIENNE, H. & VILLAIN, L. (2001). Desarrollo de la variedad porta-injertos 'nemaya' (Robusta) resistente a nematodos. *Boletín PROMECAFE* 91/92, 20-22.

ARIAS, M. & ROMERO, M.D. (1975). Nematodos asociados al cultivo de remolacha (*Beta vulgaris* L.) en la region I: Galicia y Cantabrico. *Boletin Real Sociedad Española de Historia Natural, Vol. Extraordinario del I Centenario (1871-1971), Tomo* 2, 45-54.

ARIAS, M. & ROMERO, M.D. (1979). Nematodos posibles fitoparásitos asociados a los cultivos de cereales en España. *Anales Instituto Nacional de Investigaciones Agrarias, Servicio de Protección Vegetal* 11, 109-130.

ARIM, O.J., WACEKE, J.W., WAUDO, S.W. & KIMENJU, J.W. (2006). Effects of *Canavalia ensiformis* and *Mucuna pruriens* intercrops on *Pratylenchus zeae* damage and yield of maize in subsistence agriculture. *Plant and Soil* 284, 243-251.

ARJUN-LAL, R. (2005). Nematodes intercepted in introduced germplasm of horticultural crops. *Indian Journal of Plant Protection* 33, 282-285.

ATU, U.G. (1991). Nematode problems in cropped reclaimed gully site at Owerri, Nigeria. *Afro-Asian Journal of Nematology* 1, 15-18.

AUNG, T. & PROT, J.C. (1990). Effects of crop rotations on *Pratylenchus zeae* and on yield of rice cultivar UPL Ri-5. *Revue de Nématologie* 13, 445-447.

AYOUB, S.M. (1961). *Pratylenchus zeae* found on corn, milo, and three suspected new hosts in California. *Plant Disease Reporter* 45, 940.

AZCÓN-AGUILAR, C., JAIZME-VEGA, M.C. & CALVET, C. (2002). The contribution of arbuscular mycorrhizal fungi to the control of soilborne pathogens. In: Gianinazzi, S., Schüepp, H., Barea, J.M. & Hasselwandter, K. (Eds). *Mycorrhizal technology in agriculture*. Basel, Switzerland, Birkhäuser Verlag, pp. 187-197.

AZIZOVA, E.P. (1972). [The pathogenic nematodes of grapevines.] *Gel'minty pishchevykh produktov Tezisy dokladov mezhrespublikanskoi nauchnoi konferentsii, 22-25 Dekabrya 1972*, 63-64.

AZMI, M.I. (1988). Effect of *Pratylenchus zeae* on growth of buffel grass (*Cenchrus ciliaris*). *Indian Journal of Nematology* 18, 334-335.

AZMI, M.I. (1993). Plant parasitic nematodes associated with su babool. *Indian Journal of Forestry* 16, 384.

AZMI, M.I. (1997). Studies on biotic potential and predation efficiency of *Iotonchus monhystera* (Mononchida: Nematoda). *Indian Journal of Nematology* 27, 222-232.

BABATOLA, J.O. (1984). Rice nematode problems in Nigeria: their occurrence, distribution and pathogenesis. *Tropical Pest Management* 30, 256-265.

BACK, M.A., HAYDOCK, P.P.J. & JENKINSON, P. (2002). Nematodes and soilborne pathogens disease complexes involving plant parasitic nematodes and soilborne pathogens. *Plant Pathology* 51, 683-697.

BAFOKUZARA, N.D. (1982). Nematodes associated with pineapples in Uganda. *Nematropica* 12, 45-49.

BAICHEVA, O., DASKALOVA, I. & ZHELYAZKOVA, R. (1984). [Distribution of *Pratylenchus* species.] *Rastitelna Zashchita* 32, 40-41.

BAJAJ, H.K. & BHATTI, D.S. (1984). New and known species of *Pratylenchus* Filipjev, 1936 (Nematoda: Pratylenchidae) from Haryana, India, with remarks on intraspecific variations. *Journal of Nematology* 16, 360-367.

BAJAJ, H.K. & WALIA, K.K. (2005). Studies on a *Pasteuria* isolate from an entomopathogenic nematode, *Steinernema pakistanense* (Nematoda: Steinernematidae). *Nematology* 7, 637-640.

BAKKER, J. (1993). Current state of nematicides. In: Zadoks, J.C. (Ed.). *Modern crop protection: developments and perspective.* Wageningen, The Netherlands, Wageningen Agricultural University Press, pp. 21-26.

BAKKER, J., GOMMERS, F.J., NIEUWENHUIS, I. & WYNBERG, H. (1979). Photoactivation of the nematicidal compound α-terthienyl from roots of marigolds (*Tagetes* spp.). *Journal of Biological Chemistry* 254, 1841-1844.

BALA, G. & HOSEIN, F. (1996). Plant-parasitic nematodes associated with anthuriums and other tropical ornamentals. *Nematropica* 26, 9-14.

BALDRIDGE, G.D., O'NEILL, N.R. & SAMAC, D.A. (1998). Alfalfa (*Medicago sativa* L.) resistance to the root-lesion nematode, *Pratylenchus penetrans*: defense-response gene mRNA and isoflavonoid phytoalexin levels in roots. *Plant Molecular Biology* 38, 999-1010.

BALDWIN, J.G. & PERRY, R.N. (2004). Nematode morphology, sensory structure and function. In: Chen, Z.X., Chen, S.Y. & Dickson, D.W. (Eds). *Nematology advances and perspectives, volume I. Nematode morphology, physiology and ecology.* Wallingford, UK, CABI Publishing, pp. 175-257.

BALL-COELHO, B., BRUIN, A.J., ROY, R.C. & RIGA, E. (2003). Forage pearl millet and marigold as rotation crops for biological control of root-lesion nematodes in potato. *Agronomy Journal* 95, 282-292.

BARANOVSKAYA, I.A. & HAQUE, M.M. (1974). [Description of *Pratylenchus clavicaudatus* n. sp. (Nematoda, Pratylenchinae Thorne, 1949).] *Zoologichesky Zhurnal* 47, 759-761.

BARKER, K.R. (1974). Influence of geographic area and previous crop on occurrence and densities of plant-parasitic nematodes in North Carolina. *Plant Disease Reporter* 58, 991-995.

BARKER, K.R. (1993). Resistance/tolerance and related concepts/terminology in plant parasite nematology. *Plant Disease* 77, 111-113.

BARKER, K.R. (1998). Introduction and synopsis of advances in Nematology. In: Barker, K.R., Pederson, G.A. & Windham, G.L. (Eds). *Plant and nematode interactions.* Madison, WI, USA, American Society of Agronomy, Inc., Crop Science Society of America, Inc., Soil Science Society of America, Inc., pp. 1-20.

BARKER, K.R. & CLAYTON, C.N. (1973). Nematodes attacking cultivars of peach in North Carolina. *Journal of Nematology* 5, 265-271.

BARKER, K.R. & HUSSEY, R.S. (1976). Histopathology of nodular tissues of legumes infected with certain nematodes. *Phytopathology* 66, 851-855.

BARKER, K.R. & KOENNING, S.R. (1998). Developing sustainable systems for nematode management. *Annual Review of Phytopathology* 36, 165-205.

BARKER, K.R. & OLTHOF, A.H.A. (1976). Relationships between nematode population densities and crop responses. *Annual Review of Phytopathology* 14, 327-353.

BARNES, D.K., THIES, J.A., RABAS, D.L., NELSON, D.L. & SMITH, D.M. (1990). Registration of two alfalfa germplasms with field resistance to the root lesion nematode. *Crop Science* 30, 751-752.

BASTIAN, H.C. (1865). Monograph on the Anguillulidae, or free nematoids, marine, land, and freshwater; with descriptions of 100 new species. *Transactions of the Linnean Society of London* 25, 73-184.

BAUDIN, P. & HUU-HAI, V. (1973). Les maladies du tabac à Madagascar. *Agronomie Tropicale, II* 28, 189-207.

BAUJARD, P., MOUNPORT, D. & MARTINY, B. (1990). Étude au microscope électronique à balayage de quatre espèces du genre *Pratylenchus* Filip'ev, 1936 (Nemata: Pratylenchidae). *Revue de Nématologie* 13, 203-210.

BEAUMONT, P.B. (1975). Nematodes. *Australian Plants* 8, 183-188.

BEEN, T.H. & SCHOMAKER, C.H. (1999). Fumigation of marine clay soils infested with *Globodera pallida* and *G. rostochiensis* using 1,3-dichloropropene and additional top soil treatments. *Nematology* 1, 3-14.

BEE-RODRIGUEZ, D. & AYALA, A. (1977). Nematodes associated with sorghum in Puerto Rico. *Nematropica* 7, 16-20.

BÉLAIR, G., DAUPHINAIS, N., FOURNIER, Y. & MAULEON, H. (2001). Survey of plant parasitic and entomopathogenic nematodes in vineyards of Quebec. *Phytoprotection* 82, 49-55.

BÉLAIR, G., DAUPHINAIS, N., FOURNIER, Y. & DANGI, O.P. (2004). Pearl millet for the management of *Pratylenchus penetrans* in flue-cured tobacco in Quebec. *Plant Disease* 88, 989-992.

BÉLAIR, G., DAUPHINAIS, N., FOURNIER, Y., DANGI, O.P. & CLÉMENT, M.F. (2005). Effect of forage and grain pearl millet on *Pratylenchus penetrans* and potato yields in Quebec. *Journal of Nematology* 37, 78-82.

BELL, N.I., WATSON, R.N. & SARATHCHANDRA, S.U. (2000). Suppression of plant parasitic nematodes in pastoral soils amended with chitin. *New Zealand Plant Protection* 53, 44-47.

BELOZEROVA, G.S. & METLITSKII, O.Z. (1972). [The species and population dynamics of stylet-bearing nematodes in roots and in the rhizosphere of strawberries in the non-Chernozem zone of the RSFSR (USSR).] *Kul'tura zemlyaniki v SSSR Doklady simpoziuma, 28 iyunya 1 iyulya 1971*, 451-457.

BERGE, J.B. & CUANY, A. (1975). Possibilité de desinfection de plants de rosiers contaminés par deux nématodes endoparasites grâce au trempage dans des solutions nématicides. *Pépinièristes Horticulteurs Maraichers* 153, 15-17.

BERGESON, G.B. & FERRIS, J.M. (1986). Influence of tillage methods on *Pratylenchus* spp. in two soil types. *Plant Disease* 70, 326-328.

BERNARD, E.C. (1980). Identification, distribution, and plant associations of plant-parasitic nematodes in Tennessee. *Bulletin, Agricultural Experiment Station, University of Tennessee* 594, 18 pp.

BERNARD, E.C. (1984). Hoplolaimoidea (Nematoda: Tylenchida) from the Aleutian Islands with descriptions of four new species. *Journal of Nematology* 16, 194-203.

BERNARD, E.C. & LAUGHLIN, C.W. (1976). Relative susceptibility of selected cultivars of potato to *Pratylenchus penetrans*. *Journal of Nematology* 8, 239-242.

BERNARD, E.C. & MONTGOMERY-DEE, M.E. (1993). Reproduction of plant-parasitic nematodes on winter rapeseed (*Brassica napus* ssp. *oleifera*). *Journal of Nematology* 25, 863-868.

BERT, W., VAN GANSBEKE, R., CLAEYS, M., GERAERT, E. & BORGONIE, G. (2003). Comparative morpho-anatomical studies of the female gonoduct within the Pratylenchidae (Nematoda: Tylenchina). *Nematology* 5, 293-306.

BERTRAND, B., CILAS, C., HERVÉ, G., ANTHONY, F., ETIENNE, H. & VILLAIN, L. (1998). Relations entre les populations des nématodes *Meloidogyne exigua* et *Pratylenchus* sp. dans les racines de *Coffea arabica* au Costa Rica. *Plantations Recherche Développement* 5, 279-286.

BERTRUMS, E.J. (1998). Population decline of root lesion nematode (*Pratylenchus penetrans* Cobb) after a *Taxus baccata* crop. *Mededelingen Faculteit Landbouwkundige en Toegepaste Biologische Wetenschappen, Universiteit Gent* 63, 649-653.

BEZEROVA, G.S. & POKHODENKO, A.P. (1978). [*Pratylenchus* on soft fruit crops.] *Sbornik Nauchnykh Rabot, Nauchno Issledovatel'ski Zonal'nyi Institut Sadovodstva Nechernozemnoi Polosy* 12, 22-28.

BHATT, J. (1995). Pathogenic effect of *Pratylenchus thornei* on protein content and yield of gram (*Cicer arietinum* L.). *Advances in Plant Sciences* 8, 201-204.

BIRD, A.F. & BIRD, J. (1986). Observations on the use of insect parasitic nematodes as a means of biological control of root-knot nematodes. *International Journal for Parasitology* 16, 511-516.

BIRD, A.F. & BIRD, J. (1998). Introduction to functional organization. In: Perry, R.N. & Wright, D.J. (Eds). *The physiology and biochemistry of free-living and plant-parasitic nematodes*. Wallingford, UK, CABI Publishing, pp. 1-24.

BIRD, D.M. & KOLTAI, H. (2000). Plant parasitic nematodes: habitats, hormones, and horizontally-acquired genes. *Journal of Plant Growth Regulation* 19, 183-194.

BIRD, G.W. & JENKINS, W.R. (1964). Occurrence, parasitism, and pathogenicity of nematodes associated with cranberry. *Phytopathology* 54, 677-680.

BLAIR, B.L., STIRLING, G.R. & WHITTLE, P.J.L. (1999). Distribution of pest nematodes on sugarcane in South Queensland and relationship to soil texture, cultivar, crop age and region. *Australian Journal of Experimental Agriculture* 39, 43-49.

BLANCHART, E., LAVELLE, P. & SPAIN, A.V. (1990). Effects of species of tropical earthworms (Oligochaeta: Eudrilidae) on the size distribution of aggregates in an African soil. *Revue d'Ecologie et de Biologie du Sol* 26, 417-425.

BLAXTER, M.L., DE LEY, P., GAREY, J., LIU, L.X., SCHELDEMAN, P., VIERSTRAETE, A., VANFLETEREN, J., MACKEY, L.Y., DORRIS, M., FRISSE, L.M., VIDA, J.T. & THOMAS, W.K. (1998). A molecular evolutionary framework for the phylum Nematoda. *Nature* 392, 71-75.

BOE (Boletin Oficial del Estado) (2000). Reglamento técnico de control de la producción y comercialización de los materiales de reproducción de las plantas ornamentales. N° 39, 6849-6856.

BOERMA, H.R. & HUSSEY, R. (1992). Breeding plants for resistance to nematodes. *Journal of Nematology* 24, 242-252.

BOESEWINKEL, H.J. (1977). New plant disease records in New Zealand: records in the period 1969-76. *New Zealand Journal of Agricultural Research* 20, 583-589.

BOLTON, C. & DE WAELE, D. (1989). Host suitability of commercial sunflower hybrids to *Pratylenchus zeae*. *Journal of Nematology* 21, 682-685.

BOLTON, C., DE WAELE, D. & LOOTS, G.C. (1989). Plant-parasitic nematodes on field crops in South Africa. 3. Sunflower. *Revue de Nématologie* 12, 69-76.

BOOKBINDER, M.G., BLOOM, J.R. & LUKEZIC, F.L. (1982). Interactions among selected endoparasitic nematodes and three pseudomonads on alfalfa. *Journal of Nematology* 14, 105-109.

BORDALLO, J.J., LOPEZ-LLORCA, L.V., JANSSON, H.B., SALINAS, J., PERSMARK, L. & ASENSIO, L. (2002). Colonization of plant roots by egg-parasitic and nematode-trapping fungi. *New Phytologist* 154, 491-499.

BOSWELL, T.E. & GRICHAR, W.J. (1983). Nematicide trial for control of lesion nematodes, 1980. *Fungicide and Nematicide Tests* 38, 3.

BOTSEAS, D.D. & ROWE, R.C. (1994). Development of potato early dying in response to infection by two pathotypes of *Verticillium dahliae* and co-infection by *Pratylenchus penetrans*. *Phytopathology* 84, 275-282.

BOYER, J. & REVERSAT, G. (1996). Effects of the gut of *Pontoscolex corethrurus* on second stage juveniles of *Heterodera sacchari*. In: *Proceedings of the Third International Nematology Congress, Gosier, Guadeloupe, 7-12 July 1996*, p. 169.

BOYER, J., MICHELLON, R., CHABANNE, A., REVERSAT, G. & TIBERE, R. (1999). Effects of trefoil cover crop and earthworm inoculation on maize crop and soil organisms in Reunion Island. *Biology and Fertility of Soils* 28, 364-370.

BRAASCH, H. (1988). *Pratylenchus wescolagricus* Corbett, 1983 – ein neuer pflanzenparasitischer Nematode in europaischen Gewachshausern. *Nachrichtenblatt fur den Pflanzenschutz in der DDR* 42, 178-179.

BRAASCH, H. & DECKER, H. (1989). *Pratylenchus acuticaudatus* sp. n. (Nematoda: Pratylenchidae) aus einem gewächshaus in der DDR. *Nematologica* 34, 57-61.

BRIDGE, J. (1972). Plant parasitic nematodes of irrigated crops in the northern states of Nigeria. *Samaru Miscellaneous Paper 42, I.A.R., Ahmadu Bello University, Nigeria*, 17 pp.

BRIDGE, J. (1975). Hot water treatment to control plant parasitic nematodes of tropical crops. *Mededelingen Faculteit Landbouwwetenschappen Rijksuniversiteit Gent* 40, 249-259.

BRIDGE, J. (1976). Plant-parasitic nematodes from the lowlands and highlands of Ecuador. *Nematropica* 6, 18-23.

BRIDGE, J. (1988). Plant-parasitic nematode problems in the Pacific Islands. *Journal of Nematology* 20, 173-183.

BRIDGE, J. & PAGE, S.L.M. (1984). Plant nematode pests of crops in Papua New Guinea. *Journal of Plant Protection in the Tropics* 1, 99-109.

BRIDGE, J., PRICE, N.S. & KOFI, P. (1995). Plant-parasitic nematodes of plantain and other crops in Cameroon, West Africa. *Fundamental and Applied Nematology* 18, 251-260.

BRIDGE, J., HUNT, D.J. & HUNT, P. (1996). Plant-parasitic nematodes of crops in Belize. *Nematropica* 26, 111-119.

BRINKMAN, E.P., VAN VEEN, J.A. & VAN DER PUTTEN, W.H. (2004). Endoparasitic nematodes reduce multiplication of ectoparasitic nematodes, but do not prevent growth reduction of *Ammophila arenaria* (L.) Link (marram grass). *Applied Soil Ecology* 27, 65-75.

BRINKMAN, E.P., DUYTS, H. & VAN DER PUTTEN, W.H. (2005). Competition between endoparasitic nematodes and effect on biomass of *Ammophila arenaria* (marram grass) as affected by timing of inoculation and plant age. *Nematology* 7, 169-178.

BRISTOW, P.R., BARRITT, B.H. & MCELROY, F.D. (1980). Reaction of red raspberry clones to the root lesion nematode. *Acta Horticulturae* 112, 39-46.

BROADLEY, R.A. (1981). Distribution and control of root-knot and root lesion nematodes on peanuts in north Queensland. *Australian Journal of Experimental Agriculture and Animal Husbandry* 21, 223-226.

BRODIE, B.B. (1976). Vertical distribution of three nematode species in relation to certain soil properties. *Journal of Nematology* 8, 243-247.

BRODIE, B.B. (1998). Potato. In: Barker, K.R., Pederson, G.A. & Windham, G.L. (Eds). *Plant and nematode interactions.* Madison, WI, USA, American Society of Agronomy, Inc., Crop Science Society of America, Inc., Soil Science Society of America, Inc., pp. 567-594.

BRODIE, B.B. & MURPHY, W.S. (1975). Population dynamics of plant nematodes as affected by combinations of fallow and cropping sequence. *Journal of Nematology* 7, 91-92.

BRODIE, B.B. & PLAISTED, R.L. (1993). Resistance in potato to *Pratylenchus penetrans*. *Journal of Nematology* 25, 466-471.

BRODIE, B.B., GOOD, J.M. & JAWORSKI, C.A. (1970). Population dynamics of plant nematodes in cultivated soil: effect of summer cover crops in newly cleared land. *Journal of Nematology* 2, 217-222.

BRODIE, B.B., EVANS, K. & FRANCO, J. (1993). Nematode parasites of potatoes. In: Evans, K., Trudgill, D.L. & Webster, J.M. (Eds). *Plant parasitic nematodes in temperate agriculture.* Wallingford, UK, CABI Publishing, pp. 87-132.

BROOKS, F.E. (2004). Plant-parasitic nematodes of banana in American Samoa. *Nematropica* 34, 65-72.

BROOKS, T.L. & PERRY, V.G. (1967). Pathogenicity of *Pratylenchus brachyurus* to citrus. *Plant Disease Reporter* 51, 569-573.

BROWN, D.J.F. (1981). The effect of operator and measuring system error on the morphometrical variability of one nematode specimen. *Revue de Nématologie* 4, 169-171.

BROWN, M.J., RIEDEL, R.M. & ROWE, R.C. (1980). Species of *Pratylenchus* associated with *Solanum tuberosum* cv. Superior in Ohio. *Journal of Nematology* 12, 189-192.

BROWN, R.H. & KERRY, B.R. (1987). *Principles and practice of nematode control in crops*. Sydney, Australia, Academic Press, 447 pp.

BROWNING, M., DAWSON, C., ALM, S.R., GORRES, J.H. & AMADOR, J.A. (2004). Differential effects of butyric acid on nematodes from four trophic groups. *Applied Soil Ecology* 27, 47-54.

BROWNING, M., WALLACE, D.B., DAWSON, C., ALM, S.R. & AMADOR, J.A. (2006). Potential of butyric acid for control of soil-borne fungal pathogens and nematodes affecting strawberries. *Soil Biology and Biochemistry* 38, 401-404.

BRZESKI, M.W. (1969). Nematodes associated with cabbage in Poland. II. The effect of soil factors on the frequency of nematode occurrence. *Ekologia Polska – Seria A* 17, 205-225.

BRZESKI, M.W. (1981). The genera of Anguinidae (Nematoda, Tylenchida). *Revue de Nématologie* 4, 23-24.

BRZESKI, M.W. (1998). *Nematodes of Tylenchina in Poland and temperate Europe*. Warsaw, Poland, Museum and Institute of Zoology, Polish Academy of Sciences, 398 pp.

BRZESKI, M.W. & SZCZYGIEL, A. (1977). Materialy do poznania krajowych nicieni (Nematoda) pasozytow roslin. 1. Rodzaj *Pratylenchus* Fil. (Tylenchida: Pratylenchidae). *Fragmenta Faunistica Warsaw* 23, 1-11.

BURNS, N.C. (1971). Soil pH effects on nematode populations associated with soybeans. *Journal of Nematology* 3, 238-245.

CADET, P. & FLORET, P. (1999). Effect of plant parasitic nematodes on the sustainability of a natural fallow cultural system in the Sudano-Sahelian area in Senegal. *European Journal of Soil Biology* 35, 91-97.

CADET, P. & SPAULL, V.W. (2005). Nematode parasites of sugarcane. In: Luc, M., Sikora, R.A. & Bridge, J. (Eds). *Plant parasitic nematodes in subtropical and tropical agriculture*. Wallingford, UK, CABI Publishing, pp. 645-674.

CADET, P. & VAN DEN BERG, E. (1995). Les nématodes parasites des plantes cultivées en Guyane. *Phytoma* 474, 41-44.

CADET, P., VAN DEN BERG, E., DELATTE, A. & FIARD, J.P. (1994). Comparison de quelques peuplements nématologiques des Petites Antilles. *Biogeographica* 70, 125-138.

CADET, P., GUICHAOUA, L. & SPAULL, V.W. (2004). Nematodes, bacterial activity, soil characteristics and plant growth associated with termitaria in a sugarcane field in South Africa. *Applied Soil Ecology* 25, 193-206.

CAFÉ-FILHO, A.C. & HUANG, C.S. (1988). Nematoides do genero *Pratylenchus* no Brasil. *Fitopatologia Brasileira* 13, 232-235.

CAFÉ-FILHO, A.C. & HUANG, C.S. (1989). Description of *Pratylenchus pseudofallax* n. sp. with a key to species of the genus *Pratylenchus* Filipjev, 1936 (Nematoda: Pratylenchidae). *Revue de Nématologie* 12, 7-15.

CALVET, C., PINOCHET, J., CAMPRUBÍ, A. & FERNANDEZ, C. (1995). Increased tolerance to the root-lesion nematode *Pratylenchus vulnus* in mycorrhizal propagated BA-29 quince rootstocks. *Mycorrhiza* 5, 253-258.

CALVET, C., PINOCHET, J., CAMPRUBÍ, A., ESTAÚN, V. & RODRÍGUEZ-KÁBANA, R. (2001). Evaluation of natural chemical compounds against root-lesion and root-knot nematodes and side-effects on the infectivity of arbuscular mycorrhizal fungi. *European Journal of Plant Pathology* 107, 601-605.

CAMPBELL, C.L. & MADDEN, L.V. (1990). *Introduction to plant disease epidemiology*. New York, NY, USA, John Wiley & Sons, 532 pp.

CAMPOS, V.P. (1997). Nematoides na cultura da figueira. *Informe Agropecuario Belo Horizonte* 18, 33-38.

CAMPOS, V.P. & VILLAIN, L. (2005). Nematode parasites of coffee and cocoa. In: Luc, M., Sikora, R.A. & Bridge, J. (Eds). *Plant parasitic nematodes in subtropical and tropical agriculture*. Wallingford, UK, CABI Publishing, pp. 529-579.

CAMPOS, V.P., SIVAPALAN, P. & GNANAPRAGASAM, N.C. (1990). Nematode parasites of coffee, cocoa and tea. In: Luc, M., Sikora, R.A. & Bridge, J. (Eds). *Plant parasitic nematodes in subtropical and tropical agriculture.* Wallingford, UK, CABI Publishing, pp. 387-430.

CAMPRUBÍ, A., PINOCHET, J., CALVET, C. & ESTAÚN, V. (1993). Effects of the root-lesion nematode *Pratylenchus vulnus* and the vesicular arbuscular mycorrhizal fungus *Glomus mosseae* on the growth of three plum rootstocks. *Plant and Soil* 153, 223-229.

CARLENS, K., SÁGI, L., VAN DAMME, E.J.M., DE WAELE, D. & SWENNEN, R. (2005). The use of transgenic *Arabidopsis thaliana* plants to evaluate the nematicidal effect of lectins and lectin-related proteins upon migratory nematodes. *International Journal of Nematology* 15, 1-8.

CARRASCO BALLESTEROS, S., CASTILLO, P., ADAMS, B.J. & PÉREZ ARTÉS, E. (2007). Identification of *Pratylenchus thornei*, the cereal and legume root-lesion nematode, based on SCAR-PCR and satellite DNA. *European Journal of Plant Pathology* 118, 115-125.

CARTA, L.K., SKANTAR, A.M. & HANDOO, Z.A. (2001). Molecular, morphological and thermal characters of 19 *Pratylenchus* spp. and relatives using the D3 segment of the nuclear LSU rRNA gene. *Nematropica* 31, 193-207.

CARTA, L.K., HANDOO, Z.A., SKANTAR, A.M., VAN BILJON, J. & BOTHA, M. (2002). Redescription of *Pratylenchus teres* Khan & Singh, 1974 (Nemata: Pratylenchidae), with the description of a new subspecies from South Africa, and a phylogenetic analysis of related species. *African Plant Protection* 8, 13-24.

CASTILLO, P. & GOMEZ-BARCINA, A. (1993). Plant-parasitic nematodes associated with tropical and subtropical crops in southern Spain. *Nematologia Mediterranea* 21, 45-47.

CASTILLO, P. & VOVLAS, N. (2005). Bionomics and identification of the genus *Rotylenchus* (Nematoda: Hoplolaimidae). *Nematology Monographs and Perspectives, volume 3*. (Series editors: Hunt, D.J. & Perry, R.N.) Leiden, The Netherlands, Brill Academic Publishers, 316 pp.

CASTILLO, P., GÓMEZ BARCINA, A., VOVLAS, N. & NAVAS, A. (1991). Some plant-parasitic nematodes associated with cotton and chickpea in southern Spain with description of *Amplimerlinius magnistylus* sp. n. *Afro-Asian Journal of Nematology* 1, 195-203.

CASTILLO, P., JIMÉNEZ DÍAZ, R.M., GOMEZ-BARCINA, A. & VOVLAS, N. (1995a). Parasitism of the root-lesion nematode *Pratylenchus thornei* on chickpea. *Plant Pathology* 44, 728-733.

CASTILLO, P., TRAPERO-CASAS, J.L. & JIMÉNEZ-DÍAZ, R.M. (1995b). Effect of time, temperature, and inoculum density on reproduction of *Pratylenchus thornei* in carrot disk cultures. *Journal of Nematology* 27, 120-124.

CASTILLO, P., GÓMEZ-BARCINA, A. & JIMÉNEZ-DÍAZ, R.M. (1996a). Plant parasitic nematodes associated with chickpea in southern Spain and effect of soil temperature on reproduction of *Pratylenchus thornei*. *Nematologica* 42, 211-219.

CASTILLO, P., TRAPERO-CASAS, J.L. & JIMÉNEZ-DÍAZ, R.M. (1996b). The effect of temperature on hatching and penetration of chickpea roots by *Pratylenchus thornei*. *Plant Pathology* 45, 310-315.

CASTILLO, P., MORA-RODRIGUEZ, M.P., NAVAS-CORTÉS, J.A. & JIMÉNEZ-DÍAZ, R.M. (1998a). Interactions between *Pratylenchus thornei* and *Fusarium oxysporum* f.sp. *ciceris* on chickpea. *Phytopathology* 88, 836-844.

CASTILLO, P., VOVLAS, N. & JIMÉNEZ-DÍAZ, R.M. (1998b). Pathogenicity and histopathology of *Pratylenchus thornei* populations on selected chickpea genotypes. *Plant Pathology* 47, 370-376.

CASTRO, J.M.C., LIMA, R.D., FERRAZ, S. & NEVES, J.C.L. (2000). Capacidade de predacao de *Arthrobotrys musiformis* a fitonematoides. *Summa Phytopathologica* 26, 58-62.

CAUBEL, G., CURVALE, J.P. & BOHEC, J.L. (1975). The root-lesion nematode, *Pratylenchus penetrans*, a pest of globe artichoke in Brittany. *Comptes Rendus des Séances de l'Academie d'Agriculture de France* 61, 89-97.

CHANG, S.S. (1991). The investigation of nematodes on banana in Fujian. *Fujian Fruit Tree* 2, 30-31.

CHAPMAN, R.A. (1958). The effect of root-lesion nematodes on the growth of red clover and alfalfa under greenhouse conditions. *Phytopathology* 48, 525-530.

CHARCHAR, J.M. (1997). Nematoides associados a cultura da batata (*Solanum tuberosum* L.) nas principais regioes de producao do Brasil. *Nematologia Brasileira* 21, 49-60.

CHARCHAR, J.M. & HUANG, C.S. (1981). Host range of *Pratylenchus brachyurus*. II. Vegetables. *Fitopatologia Brasileira* 6, 67-72.

CHAU, N.N., THANH, N.V., DE WAELE, D. & GERAERT, E. (1997). Plant-parasitic nematodes associated with banana in Vietnam. *International Journal of Nematology* 7, 122-126.

CHAVES, E. & TORRES, M.S. (2001). Nematodos parasitos de la papa en regiones productoras de papa semilla en la Argentina. *Revista de la Facultad de Agronomia Universidad de Buenos Aires* 21, 245-259.

CHAWLA, M.L. & PRASAD, S.K. (1973). Multiple cropping and nematodes. I. Effect of fertility management and intensive rotations on nematode populations and crop yield. *Indian Journal of Nematology* 3, 34-49.

CHEN, J., BIRD, G.W. & MATHER, R.L. (1995). Impact of multi-year cropping regimes on *Solanum tuberosum* tuber yields in the presence of *Pratylenchus penetrans* and *Verticillium dahliae*. *Journal of Nematology* 27, 654-660.

CHEN, J., MOORE, W.H., YUEN, G.Y., KOBAYASHI, D. & CASWELL-CHEN, E.P. (2006). Influence of *Lysobacter enzymogenes* strain C3 on nematodes. *Journal of Nematology* 38, 233-239.

CHEN, T.A. & WEN, G.Y. (1972). Ultrastructure of the feeding apparatus of *Pratylenchus penetrans*. *Journal of Nematology* 4, 155-161.

CHITAMBAR, J.J. & NOFFSINGER, E.M. (1989). Predaceous behavior and life history of *Odontopharynx longicaudata* (Diplogasterida). *Journal of Nematology* 21, 284-291.

CHITIMBAR, J.J. & RASKI, D.J. (1984). Reactions of grape rootstocks to *Pratylenchus vulnus* and *Meloidogyne* spp. *Journal of Nematology* 16, 166-170.

CHITIMBAR, J.J. & RASKI, D.J. (1985). Life history of *Pratylenchus vulnus* on carrot discs. *Journal of Nematology* 17, 235-236.

CHITWOOD, B.G. (1950). An outline classification of the Nematoda. In: Chitwood, B.G. & Chitwood, M.B. (Eds). *An introduction to nematology I. Anatomy*. Baltimore, MD, USA, Monumental Printing Co., pp. 12-27.

CHITWOOD, D.J. (2002). Phytochemical based strategies for nematode control. *Annual Review of Phytopathology* 40, 221-249.

CHIZHOV, V.N. (1978). [The dynamics of ecto-parasitic nematodes in the roots of fodder grasses.] *Byulleten' Vsesoyuznogo Instituta Gel'mintologii im KI Skryabina, Fitogel'mintologiya* 23, 67-69.

CHOI, D.R., LEE, J.K., PARK, B.Y., HAN, H.R. & CHOI, Y.E. (2006). A new and one unrecorded species of *Pratylenchus* from Korea (Nematoda: Pratylenchidae). *Journal of Asia-Pacific Entomology* 9, 5-9.

CHOI, Y.E. (1975). A taxonomical and morphological study of plant parasitic nematodes (Tylenchida) in Korea. *Korean Journal of Plant Protection* 14, 1-19.

CHOLEVA, B., NEDELCHEV, S. & K'NEV, I. (1984). [Parasitic nematodes in fruit tree nurseries in the Plovdiv region (Bulgaria).] *Rastitelna Zashchita* 32, 21-24.

CHOO, H.Y. & CHOI, Y.E. (1979). A study on the plant-parasitic nematodes (Tylenchida) in Jeju Province. *Korean Journal of Plant Protection* 18, 169-176.

CHOUDHURY, B.N. & SIVAKUMAR, C.V. (2000). Biocontrol potential of *Mylonchulus minor* against some plant parasitic nematodes. *Annals of Plant Protection Sciences* 8, 53-57.

CHRISTIE, J.R. & BIRCHFIELD, W. (1958). Scribner's lesion nematode, a destructive parasite of amaryllis. *Plant Disease Reporter* 42, 873-875.

CHUNG, K.C., PARK, S.D., KHAN, Z. & KIM, B.J. (2004). A survey of plant parasitic nematodes associated with ginseng (*Panax ginseng* C.A. Meyer). *Korean Journal of Medicinal Crop Science* 12, 355-359.

CIANCIO, A., BONSIGNORE, R., VOVLAS, N. & LAMBERTI, F. (1994). Host records and spore morphometrics of *Pasteuria penetrans* group parasites of nematodes. *Journal of Invertebrate Pathology* 63, 260-267.

CIANCIO, A., LANDRISCINA, S. & LAMBERTI, F. (1995). Temporal and spatial dynamics of *Pratylenchus vulnus* parasitising walnut. *Nematologia Mediterranea* 23, 367-371.

CIANCIO, A., FAZIO, P. & LAMBERTI, F. (1996). Nematodes occurring in the rhizosphere of walnut in Italy. *Nematologia Mediterranea* 24, 129-133.

CLAYTON, E.E. & MCMURTREY, J.E. (1950). Tobacco diseases and their control. *Farmer's Bulletin No. 2023, USDA*, 70 pp.

COATES-BECKFORD, P.L. & BRATHWAITE, C.W.D. (1977). Comparison of various treatments for the control of *Pratylenchus coffeae* in yam. *Nematropica* 7, 20-26.

COBB, N.A. (1917). A new parasitic nema found infesting cotton and potatoes. *Journal of Agricultural Research* 11, 27-33.

COBB, N.A. (1919). A new nema, *Tylenchus musicola*, n. sp., said to cause a serious affection of the bluggoe banana in Grenada, British West Indies. *West Indian Bulletin* 17, 179-182.

COBB, N.A. (1920). One hundred new nemas (type species of 100 new genera). *Contributions to Science of Nematology* 9, 217-343.

COHN, E. & DUNCAN, L. (1990). Nematode parasites of tropical and subtropical fruits. In: Luc, M., Sikora, R.A. & Bridge, J. (Eds). *Plant parasitic nematodes in subtropical and tropical agriculture.* Wallingford, UK, CABI Publishing, pp. 347-362.

COLBRAN, R.C. (1953). Problems in tree replacement. 1. The root-lesion nematode *Pratylenchus coffeae* Zimmerman as a factor in the growth of replant trees in apple orchards. *Australian Journal of Agricultural Research* 4, 384-389.

COLBRAN, R.C. (1967). *Hot water tank for treatment of banana planting material.* Division of Plant Industry, Department of Primary Industries, Queensland, Australia, Advisory Leaflet No. 924, 3 pp.

COLBRAN, R.C. (1968). *Nematodes – important pests of peanuts.* Division of Plant Industry, Department of Primary Industries, Queensland, Australia, Advisory Leaflet No. 955, 5 pp.

COLBRAN, R.C. (1974). Nematodes in strawberries. *Queensland Agricultural Journal* 100, 522-525.

COLBRAN, R.C. (1979). Problems in tree replacement. III. The effects of cultural practices and soil fumigation on root-lesion nematodes and on the growth of apple replants. *Australian Journal of Agricultural Research* 30, 113-123.

COLBRAN, R.C. & MCCULLOCH, J.S. (1965). Nematodes associated with wheat in Queensland. *Queensland Journal of Agriculture and Animal Science* 22, 353-356.

COLLINGBORN, F.M.B. & GOWEN, S.R. (1998). Screening Indian cultivars of *Musa* for resistance or tolerance to *Radopholus similis* and *Pratylenchus coffeae*. *Acta Horticulturae* 490, 369-372.

CONN, K.L. & LAZAROVITS, G. (1999). Impact of animal manures on Verticillium wilt, potato scab, and soil microbial populations. *Canadian Journal of Plant Pathology* 21, 81-92.

COOK, R. & EVANS, K. (1987). Resistance and tolerance. In: Brown, R.H. & Kerry, B.R. (Eds). *Principles and practice of nematode control in crops*. Orlando, FL, USA, Academic Press, pp. 179-232.

COOK, R.J. & BAKER, K.F. (1983). *The nature and practice of biological control of plant pathogens*. St Paul, MN, USA, 539 pp.

COOLEN, W.A. & HENDRICKX, G.J. (1972a). Monograph on the nematological situation in Belgian rose culture. *Publikatie nr W 10, Rijksstation voor Nematologie en Entomologie, Merelbeke* 29.

COOLEN, W.A. & HENDRICKX, G.J. (1972b). Investigations on the resistance of rose root-stocks to *Meloidogyne hapla* and *Pratylenchus penetrans*. *Nematologica* 18, 155-158.

COOMANS, A. (1962). Morphological observations on *Rotylenchus goodeyi* Loof & Oostenbrink, 1958. I. Redescription and variability. *Nematologica* 7, 203-215.

COOSEMANS, J. (1975). The influence of *Pratylenchus penetrans* on growth of *Impatiens balsamina* L., *Daucus carota* L., *Linum usitatissimum* L., and *Chrysanthemum indicum* L. *Mededelingen van de Faculteit Landbouwwetenschappen Rijksuniversiteit Gent* 40, 465-471.

CORBETT, D.C.M. (1969). *Pratylenchus pinguicaudatus* n. sp. (Pratylenchinae: Nematoda) with a key to the genus *Pratylenchus*. *Nematologica* 15, 550-556.

CORBETT, D.C.M. (1970). Cereal *Pratylenchus* spp. in England and Wales and their recognition. *Plant Pathology* 19, 6-10.

CORBETT, D.C.M. (1972). The effect of *Pratylenchus fallax* on wheat, barley and sugar beet roots. *Nematologica* 18, 303-308.

CORBETT, D.C.M. (1973). *Pratylenchus penetrans*. *CIH Descriptions of plant-parasitic nematodes*, Set 2, No. 25. Farnham Royal, UK, Commonwealth Agricultural Bureaux, 4 pp.

CORBETT, D.C.M. (1974). *Pratylenchus vulnus*. *CIH Descriptions of plant-parasitic nematodes*, Set 3, No. 37. Farnham Royal, UK, Commonwealth Agricultural Bureaux, 3 pp.

CORBETT, D.C.M. (1976). *Pratylenchus brachyurus*. *CIH Descriptions of plant-parasitic nematodes*, Set 6, No. 89. Farnham Royal, UK, Commonwealth Agricultural Bureaux, 4 pp.

CORBETT, D.C.M. (1983). Three new species of *Pratylenchus* with a redescription of *P. andinus* Lordello, Zamith & Boock, 1961 (Nematoda: Pratylenchidae). *Nematologica* 29, 390-403.

CORBETT, D.C.M. & CLARK, S.A. (1983). Surface features in the taxonomy of *Pratylenchus* species. *Revue de Nématologie* 6, 85-98.

COSTILLA, M.A., GOMEZ, T.A.H. & OJEDA, S.G. (1976). Nematodes identificados en cultivos de cana de azucar en las provincias del Noroeste Argentino. *Revista Industrial y Agricola de Tucuman* 53, 55-59.

COTTEN, J. & ROBERTS, H. (1981). Root-lesion nematodes (*Pratylenchus* spp.) in raspberry-cane beds entered for certification in England. *Plant Pathology* 30, 101-104.

COTTEN, J., BARTLETT, P.W. & WEBB, R.M. (1990). A first record of the root lesion nematode, *Pratylenchus bolivianus* Corbett in England and Wales. *Indian Phytopathology* 43, 311-312.

COURTRIGHT, E.M., WALL, D.H., VIRGINIA, R.A., FRISSE, R.A., VIDA, J.T. & THOMAS, W.K. (2000). Nuclear and mitochondrial DNA sequence diversity in the Antarctic nematode *Scottnema lindsayae*. *Journal of Nematology* 32, 143-153.

COYNE, D.L. & PLOWRIGHT, R.A. (2002). Assessment of the importance of individual plant-parasitic nematode species in a community dominated by *Heterodera sacchari* on upland rice in Côte d'Ivoire. *Nematology* 4, 661-669.

COYNE, D.L., PLOWRIGHT, R.A. & FOFANA, I. (1996). Preliminary investigations of nematodes associated with rice in Guinea, Benin and Togo. *Afro-Asian Journal of Nematology* 6, 70-73.

COYNE, D.L., PLOWRIGHT, R.A., TWUMASI, J. & HUNT, D.J. (1999). Prevalence of plant-parasitic nematodes associated with rice in Ghana with a discussion of their importance. *Nematology* 1, 399-405.

COYNE, D.L., HUNT, D.J., PLOWRIGHT, R.A. & DARBOE, M.L.K. (2000). Further observations on nematodes associated with rice in Côte d'Ivoire, The Gambia, Guinea and Togo. *International Journal of Nematology* 10, 123-130.

COYNE, D.L., TALWANA, H.A.L. & MASLEN, N.R. (2003). Plant-parasitic nematodes associated with root and tuber crops in Uganda. *African Plant Protection* 9, 87-98.

CROW, R.V. & MACDONALD, D.H. (1978). Phytoparasitic nematodes adjacent to established strawberry plantations. *Journal of Nematology* 10, 204-207.

CROW, W.T., WEINGARTNER, D.P., DICKSON, D.W. & MCSORLEY, R. (2001). Effect of sorghum-sudan grass and velvetbean cover crops on plant-parasitic nematodes associated with potato production in Florida. *Journal of Nematology* 33, 285-288.

CROZZOLI, P.R., CASASSA, P.M., RIVAS, G.D. & MATHEUS, C.J. (1991). Nematodos fitoparasitos asociados al cultivo del guayabo en el Estado Zulia, Venezuela. *Fitopatologia Venezolana* 4, 2-6.

CROZZOLI, R. (2002). Especies de nematodos fitoparasiticos en Venezuela. *Interciencia* 27, 354-364.

CROZZOLI, R., LAMBERTI, F., GRECO, N. & RIVAS, D. (1998). Nematodos fitoparasiticos asociados con los citricos en Venezuela. *Nematologia Mediterranea* 26, 31-58.

CROZZOLI, R., LAMBERTI, F., GRECO, N. & RIVAS, D. (2001). Nematodos fitoparasiticos asociados al cultivo del cacao en Choroni, Cumboto y Cuyagua, estado Aragua. *Fitopatologia Venezolana* 14, 5-12.

CUAREZMA-TERAN, J.A., TREVATHAN, L.E. & BOST, S.C. (1984). Nematodes associated with sorghum in Mississippi. *Plant Disease* 68, 1083-1085.

CULVER, D.J., RAMMING, D.W. & MCKENRY, M.V. (1989). Procedures for field and greenhouse screening of *Prunus* genotypes for resistance and tolerance to root lesion nematode. *Journal of the American Society for Horticultural Science* 114, 30-35.

D'ERRICO, F.P. (1970). Some plant parasitic nematodes found in Italy. *Bolletino del Laboratorio di Entomologia Agraria Filippo Silvestri, Portici* 28, 183-189.

D'HERDE, J. & VAN DEN BRANDE, J. (1963). Wortelrot bij chrysanten. *Mededelingen van de Landbouwhogeschool en de Opzoekingsstations van de Staat te Gent* 28, 630-637.

DA SILVA, G.S., FERRAZ, S. & DOS SANTOS, J.M. (1989). Resistance of *Crotalaria* species to *Pratylenchus brachyurus* and *P. zeae*. *Nematologia Brasileira* 13, 81-86.

DABAJ, K.H. & JENSER, G. (1991). Preliminary observation on apricot decline. *Acta Horticulturae* 293, 511-513.

DALE, A. & POTTER, J.W. (1998). Strawberry cultivars vary in their resistance to northern lesion nematode. *Journal of Nematology* 30, 577-580.

DANEEL, M., DILLEN, N., HUSSELMAN, J., JAGER, K. & DE WAELE, D. (2003). Results of a survey on nematodes of *Musa* in household gardens in South Africa and Swaziland. *Infomusa* 12, 8-11.

DAS, S. & DAS, S.N. (1986). Host-range of *Pratylenchus coffeae*. *Indian Journal of Nematology* 16, 180-184.

DAS, V.M. (1960). Studies on the nematode parasites of plants in Hyderabad (Andhra Pradesh, India). *Zeitschrift für Parasitenkunde* 19, 553-605.

DAS, V.M. & SULTANA, S. (1979). Five new species of the genus *Pratylenchus* from vegetable crops of Hyderabad (Andhra Pradesh). *Indian Journal of Nematology* 9, 5-14.

DAVIDE, R.G. (1988). Nematode problems affecting agriculture in the Philippines. *Journal of Nematology* 20, 214-218.

DAVIS, J.R., HAFEZ, S.L. & SORESSEN, L.H. (1992). Lesion nematode suppression with the Butte potato and relationships to Verticillium wilt. *American Potato Journal* 69, 371-383.

DE GOEDE, R.G.M. & BONGERS, T. (1998). *Nematode communities of northern temperate grassland ecosystems*. Giessen, Germany, Focus Verlag, 338 pp.

DE GUIRAN, G. (1965). Nematodes associated with cassava from the south of Togo. *Cultures Tropicals Marseille Mars* 1965, 678-680.

De Guiran, G. & Vilardebo, A. (1962). Le bananier aux îles Canaries. IV. Les nématodes parasites du bananier. *Fruits* 17, 263-277.

De Luca, F., Fanelli, E., Di Vito, M., Reyes, A. & De Giorgi, C. (2004). Comparison of the sequences of the D3 expansion of the 26S ribosomal genes reveals different degrees of heterogeneity in different populations and species of *Pratylenchus* from the Mediterranean region. *European Journal of Plant Pathology* 110, 949-957.

De Souza, J.T., Campos, V.P. & Maximiniano, C. (1998). Ocorrencia e distribuicao de nematoides associados a hortalicas e plantas medicinais. *Summa Phytopathologica* 24, 283-291.

De Souza, J.T., Maximiniano, C. & Campos, V.P. (1999). Nematoides associados a plant as frutiferas em alguns estados Brasileiros. *Ciencia e Agrotecnologia* 23, 353-357.

De Waele, D. & Elsen, A. (2002). Migratory endoparasites: *Pratylenchus* and *Radopholus* species. In: Starr, J.L., Cook, R. & Bridge, J. (Eds). *Plant resistance to parasitic nematodes*. Wallingford, UK, CABI Publishing, pp. 175-206.

De Waele, D. & Jordaan, E.M. (1988a). Plant-parasitic nematodes on field crops in South Africa. 1. Corn. *Revue de Nématologie* 11, 65-74.

De Waele, D. & Jordaan, E.M. (1988b). Plant-parasitic nematodes on field crops in South Africa. 2. Sorghum. *Revue de Nématologie* 11, 203-211.

De Waele, D. & Van den Berg, E. (1988). Nematodes associated with upland rice in South Africa, with a description of *Hemicycliophora oryzae* sp. n. (Nemata: Criconematoidea). *Revue de Nématologie* 11, 45-51.

De Waele, D., McDonald, A.H. & De Waele, E. (1988a). Influence of seaweed concentrate on the reproduction of *Pratylenchus zeae* (Nematoda) on maize. *Nematologica* 34, 71-77.

De Waele, E., Loots, G.C. & Heyns, J. (1988b). Observations on the effect of maize roots on the hatching of *Pratylenchus zeae* and *P. brachyurus*. *Phytophylactica* 20, 135-137.

De Waele, E., Loots, G.C. & Wolmarans, C.T. (1989). Chemical factors of maize root exudates and their influence on the hatching of *Pratylenchus zeae* eggs. *Phytophylactica* 21, 201-203.

De Waele, D., McDonald, A.H., Jordaan, E.M., Orion, D., Van den Berg, E. & Loots, G.C. (1998). Plant-parasitic nematodes associated with corn and pearl millet in Namibia. *African Plant Protection* 4, 113-117.

Decker, H. (1989). *Plant nematodes and their control (phytonematology)*. Leiden, The Netherlands, E.J. Brill, 540 pp.

Decker, H. & Dowe, A. (1974). Über das auftreten von arten der gattungen *Pratylenchus*, *Pratylenchoides* und *Hirschmanniella* (Nematoda: Pratylenchidae) in der DDR. *Helminthologia Bratislava* 15, 829-834.

DECKER, H. & EL-AMIN, E.M. (1980). Some observations on the influence of ecological factors on the occurrence and distribution of plant-parasitic nematodes in the Sudan. *Beitrage zur Tropischen Landwirtschaft und Veterinarmedzin* 18, 181-187.

DECRETO LEGISLATIVO (1999). Norme tecniche sulla commercializzazione dei materiali di moltiplicazione delle piante ornamentali. *Decreto Legislativo* 151, 13-36.

DELAVILLE, L., ROSSI, J.P. & QUÉNÉHERVÉ, P. (1996). Plant row and soil factors influencing the microspatial patterns of plant-parasitic nematodes on sugarcane in Martinique. *Fundamental and Applied Nematology* 19, 321-328.

DELIBES A., ROMERO, D., AGUADED, S., DUCE, A., MENA, M., LÓPEZ-BRAÑA, I., ANDRÉS, M.F., MARTÍN-SÁNCHEZ, J.A. & GARCÍA-OLMEDO, F. (1993). Resistance to the cereal cyst nematode (*Heterodera avenae* Woll.) transferred from the wild grass *Aegilops ventricosa* to hexaploid wheat by a 'stepping-stone' procedure. *Theoretical and Applied Genetics* 87, 402-408.

DEMENT'EVA, S.P. (1971). [Parasitic species of nematodes on solanaceous crops in Moldavia.] *Parazity Zhivotnykh i Rastenii* 7, 139-142.

DESAEGER, J. & RAO, M.R. (2001). The potential of mixed covers of *Sesbania, Tephrosia* and *Crotalaria* to minimise nematode problems on subsequent crops. *Field Crop Research* 70, 111-125.

DEVRAJAN, K. & RAJENDRAN, G. (2001). Reaction of some banana clones to the burrowing and lesion nematodes. *Indian Journal of Nematology* 31, 176.

DI VITO, M., GRECO, N. & SAXENA, M.C. (1991). Effectiveness of soil solarization for control of *Heterodera ciceri* and *Pratylenchus thornei* on chickpea in Syria. *Nematologia Mediterranea* 19, 109-111.

DI VITO, M., GRECO, N. & SAXENA, M.C. (1992). Pathogenicity of *Pratylenchus thornei* on chickpea in Syria. *Nematologia Mediterranea* 20, 71-73.

DI VITO, M., GRECO, N., ORESTE, G., SAXENA, M.C., SINGH, K.B. & KUSMENOGLU, I. (1994). Plant parastic nematodes of legumes in Turkey. *Nematologia Mediterranea* 22, 245-251.

DI VITO, M., ZACCHEO, G. & CATALANO, F. (1995). Response of chickpea lines to *Meloidogyne artiellia* and *Pratylenchus thornei*. *Nematologia Mediterranea* 23, 81-83.

DI VITO, M., ZACCHEO, G. & CATALANO, F. (2000). Effect of *Pratylenchus neglectus* and *P. thornei* on the growth of faba bean. *Nematologia Mediterranea* 28, 261-265.

DI VITO, M., CATALANO, F. & ZACCHEO, G. (2002a). Reproduction of six populations of *Pratylenchus* spp. from the Mediterranean region on selected plant species. *Nematologia Mediterranea* 30, 103-105.

DI VITO, M., CATALANO, F., ZACCHEO, G., SADIKI, M. & KHARRAT, M. (2002b). Response of lines of faba bean to six populations of root-lesion nematodes (*Pratylenchus* spp.) from the Mediterranean region. *Nematologia Mediterranea* 30, 107-109.

DICKERSON, O.J. (1979). The effects of temperature on *Pratylenchus scribneri* and *P. alleni* populations on soybeans and tomatoes. *Journal of Nematology* 11, 23-26.

DICKERSON, O.J., DARLING, H.M. & GRIFFIN, G.D. (1964). Pathogenicity and population trends of *Pratylenchus penetrans* on potato and corn. *Phytopathology* 54, 317-322.

DICKLOW, M.B., ACOSTA, N. & ZUCKERMAN, B.M. (1993). A novel *Streptomyces* species for controlling plant-parasitic nematodes. *Journal of Chemical Ecology* 19, 159-173.

DICKSON, D.W. (1998). Peanut. In: Barker, K.R., Pederson, G.A. & Windham, G.L. (Eds). *Plant and nematode interactions.* Madison, WI, USA, American Society of Agronomy, Inc., Crop Science Society of America, Inc., Soil Science Society of America, Inc., pp. 523-566.

DICKSON, D.W. & DE WAELE, D. (2005). Nematode parasites of peanut. In: Luc, M., Sikora, R.A. & Bridge, J. (Eds). *Plant parasitic nematodes in subtropical and tropical agriculture*. Wallingford, UK, CABI Publishing, pp. 393-436.

DICKSON, D.W. & MCSORLEY, R. (1990). Interaction of three plant parasitic nematodes in corn and soybean. *Journal of Nematology* 22, 783-791.

DINARDO-MIRANDA, L.L. & FERRAZ, L.C.C.B. (1991). Patogenicidade de *Pratylenchus brachyurus* e *P. zeae* a duas variedades de cana de acucar (*Saccharum* sp.). *Nematologia Brasileira* 15, 9-16.

DINARDO-MIRANDA, L.L., MORELLI, J.L., DE LANDELL, M.G.A. & DE SILVA, M.A. (1996). Comportamento de genotipos de cana de acucar em relacao a *Pratylenchus zeae*. *Nematologia Brasileira* 20, 52-58.

DINARDO-MIRANDA, L.L., GIL, M.A. & MENEGATTI, C.C. (2003). Danos causados por nematoides a variedades de cana de acucar em cana planta. *Nematologia Brasileira* 27, 69-73.

DIRECTIVA COMUNITARIA 67 (1999). Condiciones que deben cumplir los materiales de reproducción de plantas ornamentales y las plantas ornamentales. 164, 78.

DMOWSKA, E. & ILIEVA, K. (1995). The effect of prolonged diverse mineral fertilization on nematodes inhabiting the rhizosphere of spring barley. *European Journal of Soil Biology* 31, 189-198.

DONALD, P.A. & HOSFORD JR, R.M. (1980). Plant parasitic nematodes of North Dakota. *Plant Disease* 64, 45-47.

DOUCET, M.E. (1988). Descripción de cuatro poblaciones de *Pratylenchus* (Nematoda: Tylenchida) provenientes de la provincia de Córdoba, Argentina. *Revista de Ciencias Agropecuarias* 6, 7-21.

DOUCET, M.E. & CAGNOLO, S. (1998). Variabilidad intra e inter-específica de caracteres morfométricos en poblaciones del orden Tylenchida (Nematoda) provenientes de Argentina. *Nematologia Mediterranea* 26, 231-236.

DOUCET, M.E. & LAX, P. (1997). Caracterización de una población y un aislado de *Pratylenchus vulnus* Allen et Jensen, 1951 (Nematoda: Tylenchida) provenientes de la Provincia de Córdoba, Argentina. *Nematologia Mediterranea* 25, 287-298.

DOUCET, M.E., PINOCHET, J. & DI RIENZO, J.A. (1996). Comparative analysis of morphological and morphometrical characters in six isolates of *Pratylenchus vulnus* Allen & Jensen, 1951 (Nemata: Tylenchida). *Fundamental and Applied Nematology* 19, 79-84.

DOUCET, M.E., LAX, P. & PINOCHET, J. (1998). Variability of some external characters in *Pratylenchus vulnus* Allen & Jensen, 1951 (Nematoda: Tylenchida). *Fundamental and Applied Nematology* 21, 205-206.

DOUCET, M.E., LAX, P., DI RIENZO, J.A., PINOCHET, J. & BAUJARD, P. (2001). Temperature-induced morphometrical variability in an isolate of *Pratylenchus vulnus* Allen & Jensen, 1951 (Nematoda: Tylenchida). *Nematology* 3, 1-8.

DOUCET, M.E., LORENZO, E., AZPILICUETA, C. & MAERO, E. (2005). Histopathology of *Rubus idaeus* roots infected by *Pratylenchus alleni*. *Nematologia Brasileira* 29, 311-313.

DOWE, A., DECKER, H., WALTER, A.M. & LUCKE, W. (1990). Damage caused by wandering root nematodes in winter barley. *Nachrichtenblatt des Deutschen Pflanzenschutzdienstes* 44, 95-96.

DOYLE, A.D., MCLEOD, R.W., WONG, P.T.W., HETHERINGTON, S.E. & SOUTHWELL, R.J. (1987). Evidence for the involvement of the root lesion nematode *Pratylenchus thornei* in wheat yield decline in northern New South Wales. *Australian Journal of Experimental Agriculture* 27, 563-570.

DUNCAN, L.W. (2005). Nematode parasites of citrus. In: Luc, M., Sikora, R.A. & Bridge, J. (Eds). *Plant parasitic nematodes in subtropical and tropical agriculture*. Wallingford, UK, CABI Publishing, pp. 437-466.

DUNCAN, L.W. & COHN, E. (1990). Nematode parasites of citrus. In: Luc, M., Sikora, R.A. & Bridge, J. (Eds). *Plant parasitic nematodes in subtropical and tropical agriculture*. Wallingford, UK, CABI Publishing, pp. 387-430.

DUNCAN, L.W. & MOENS, M. (2006). Migratory endoparasitic nematodes. In: Perry, R.N. & Moens, M. (Eds). *Plant nematology*. Wallingford, UK, CABI Publishing, pp. 123-152.

DUNCAN, L.W., INSERRA, R.N. & DUNN, D. (1998). Seasonal changes in citrus fibrous root starch concentration and body length of female *Pratylenchus coffeae*. *Nematropica* 28, 263-266.

DUNCAN, L.W., INSERRA, R.N., THOMAS, W.K., DUNN, D., MUSTIKA, L., FRISSE, MENDES, M.L., MORRIS, K. & KAPLAN, D.T. (1999). Molecular and morphological analysis of isolates of *Pratylenchus coffeae* and closely related species. *Nematropica* 29, 61-80.

DUNCAN, R.A., STAPLETON, J.J. & MCKENRY, M.V. (1992). Establishment of orchards with black polyethylene film mulching: effect on nematode and fungal pathogens, water conservation, and tree growth. *Journal of Nematology* 24, 681-687.

DUNN, R.A. (1972). Importance of depth in soil, presence of host roots, and the role of eggs as compared to vermiform stages in overwintering of *Pratylenchus penetrans* at Ithaca, New York. *Journal of Nematology* 4, 221-222. [Abstr.]

DUNN, R.A. (1973). Resistance in potato (*Solanum tuberosum*) to *Pratylenchus penetrans*. *Second International Congress of Plant Pathology, St Paul, MN, USA, September-5-12 1973*, p. 860.

EASTWOOD, D.R., SMITH, A. & WILSON, J. (1994). *Pratylenchus thornei* is causing yield losses in Victorian wheat crops. *Australasian Nematology Newsletter* 5, 10-13.

EDMUNDS, J.E. & MAI, W.F. (1966). Effect of *Trichoderma viride*, *Fusarium oxysporum*, and fungal enzymes upon the penetration of alfalfa roots by *Pratylenchus penetrans*. *Phytopathology* 56, 1132-1135.

EDMUNDS, J.E. & MAI, W.F. (1967). Effect of *Fusarium oxysporum* on movement of *Pratylenchus penetrans* towards alfalfa roots. *Phytopathology* 57, 468-471.

EDONGALI, E.A. (1996). Diseases of date palms (*Phoenix dactylifera* L.) of Libya. *Arab Journal of Plant Protection* 14, 41-43.

EDWARD, J.C., MISRA, S.L., RAI, B.B. & PETER, E. (1969). Association of *Pratylenchus chrysanthus* n. sp. with chrysanthemum root rot. *Allahabad Farmer* 43, 175-179.

EDWARDS, L.M. & KIMPINSKI, J. (1997). Relationships between soil penetration resistance and soil nematode burden in barley on Prince Edward Island. *Biology and Fertility of Soils* 24, 13-17.

EGUNJOBI, O.A. (1974). Nematodes and corn growth in Nigeria. I. Population dynamics of *Pratylenchus brachyurus* in and about the roots of corn and its effects on corn production at Ibadan. *Nematologica* 20, 181-186.

EGUNJOBI, O.A. (1977). Nematodes and corn growth in Nigeria. III. Effects of cocoa pod husk soil amendments on populations of *Pratylenchus brachyurus* and on the growth and yield of corn (*Zea mays* L.). *Nematologia Mediterranea* 5, 151-157.

EGUNJOBI, O.A. (1978). Effects of *Meloidogyne incognita* and *Pratylenchus brachyurus* singly and combined on growth of Nigerian tobacco (NTC 5). *Nematropica* 8, 62-66.

EGUNJOBI, O.A. & BOLAJI, E.T. (1979). Dry season survival of *Pratylenchus* spp. in corn fields in Western Nigeria. *Nematologia Mediterranea* 7, 129-135.

EGUNJOBI, O.A. & LARINDE, M.A. (1975). Nematodes and corn growth in Nigeria. II. Effects of some amendments on populations of *Pratylenchus brachyurus* and on the growth and production of corn (*Zea mays*) in Nigeria. *Nematologia Mediterranea* 3, 65-74.

EGUNJOBI, O.A., AKONDE, P.T. & CAVENESS, F.E. (1986). Interaction between *Pratylenchus sefaensis*, *Meloidogyne javanica* and *Rotylenchus reniformis* in sole and mixed crops of corn and cowpea. *Revue de Nématologie* 9, 61-70.

ELBADRI, G.A.A., BERT, W. & GERAERT, E. (2001). Tylenchida associated with different crops in Sennar State (Sudan). *Mededelingen Faculteit Landbouwkundige en Toegepaste Biologische Wetenschappen, Universiteit Gent* 66, 633-643.

ELEKCIOGLU, I.H. (1995). Occurrence of *Pasteuria* bacteria as parasites of plant-parasitic nematodes in the east Mediterranean region of Turkey. *Nematologia Mediterranea* 23, 213-215.

ELEKCIOGLU, I.H. & GOZEL, U. (1997). Effect of mixed population of *Paratrophurus acristylus*, *Pratylenchus thornei* and *Paratylenchus* sp. (Nematoda: Tylenchida) on yield parameters of wheat in Turkey. *International Journal of Nematology* 7, 217-220.

ELEKCIOGLU, I.H., OHNESORGE, B., LUNG, G. & UYGUN, N. (1994). Plant parasitic nematodes in the east Mediterranean region of Turkey. *Nematologia Mediterranea* 22, 59-63.

ELLIOTT, A.P. & BIRD, G.W. (1985). Pathogenicity of *Pratylenchus penetrans* to Navy Bean (*Phaseolus vulgaris* L.). *Journal of Nematology* 17, 81-85.

ELLIOTT, A.P., BIRD, G.W. & SAFIR, G.R. (1984). Joint influence of *Pratylenchus penetrans* (Nematoda) and *Glomus fasciculatum* (Phycomyceta) on the ontogeny of *Phaseolus vulgaris*. *Nematropica* 14, 111-119.

EL-MALEH, A. & EDONGALI, Z. (1994). Plant parasitic nematodes associated with almond in Libya. In: *Proceedings of the Second Afro-Asian Nematology Symposium, Menoufiya, Egypt, 18-22 December 1994*, pp. 30-32.

ELSEN, A., BAIMEY, H., SWENNEN, R. & DE WAELE, D. (2003a). Relative mycorrhizal dependency and mycorrhiza-nematode interaction in banana cultivars (*Musa* spp.) differing in nematode susceptibility. *Plant and Soil* 256, 303-313.

ELSEN, A., DECLERCK, S. & DE WAELE, D. (2003b). Use of root organ cultures to investigate the interaction between *Glomus intraradices* and *Pratylenchus coffeae*. *Applied and Environmental Microbiology* 69, 4308-4311.

ELSEN, A., GOOSSENS, B., BELPAIRE, B., NEYENS, A., SPEIJER, P. & DE WAELE, D. (2004). Recolonisation by nematodes of hot water treated cooking banana planting material in Uganda. *Nematology* 6, 215-221.

ENDO, B.Y. (1959). Responses of root-lesion nematodes, *Pratylenchus brachyurus* and *P. zeae*, to various plants and soil types. *Phytopathology* 49, 417-421.

ENDO, B.Y. (1967). Comparative population increase of *Pratylenchus brachyurus* and *P. zeae* in corn and in soybean varieties Lee and Perking. *Phytopathology* 57, 118-120.

ENDO, B.Y., ZUNKE, U. & WERGIN, W.P. (1997). Ultrastructure of the lesion nematode, *Pratylenchus penetrans* (Nemata: Pratylenchidae). *Journal of the Helminthological Society of Washington* 64, 59-95.

ENDO, B.Y., ZUNKE, U. & WERGIN, W.P. (1998). Ultrastructure of the male gonad and spermatogenesis in the lesion nematode, *Pratylenchus penetrans* (Nemata: Pratylenchidae). *Journal of the Helminthological Society of Washington* 65, 227-242.

ENDO, B.Y., ZUNKE, U. & WERGIN, W.P. (1999). Ultrastructure of the female reproductive system of the lesion nematode, *Pratylenchus penetrans* (Nemata: Pratylenchidae). *Journal of the Helminthological Society of Washington* 66, 155-174.

ERDAL, F., DURMUS, F., KEPENEKCI, I. & OKTEN, M.E. (2001). Preliminary list of Tylenchida (Nematoda) with cereals, pulses, industrial crops, vegetables, orchards, vineyards and citrus fields in Turkey. *Turkiye Entomoloji Dergisi* 25, 49-64.

EROSHENKO, A.S. (1978). [Pathogenic nematodes of pine plantations in the south of Sakhalin Island.] *Fitogel'mintologicheskie issledovaniya* 32, 39.

ESCUER, M. & BELLO, A. (2003). Nematodos fitoparásitos en los cultivos de Villena. In: Bello, A., López-Pérez, J.A. & García Álvarez, A. (Eds). *Biofumigación en agricultura extensiva de regadío. Producción integrada de hortícolas.* Agrícola Villena Sociedad Cooperativa, Horfres, Ayuntamiento de Villena y Caja Rural de Alicante, Alicante, Spain, pp. 153-190.

ESMENJAUD, D., MINOT, J.C., VOISIN, R. & DE GUIRAN, G. (1989). Effet de l'aldicarbe sur le rendement d'une culture de menthe poivrée (*Mentha piperita* Huds.) et résidus dans l'huile essentielle. *Mededelingen van de Faculteit Landbouwwetenschappen, Rijksuniversiteit Gent* 54, 193-198.

ESMENJAUD, D., RIVOAL, R. & MARZIN, H. (1990). Numbers of *Pratylenchus* spp. (Nematoda) in the field on winter wheat in different cereal rotations. *Nematologica* 36, 217-226.

ESPÁRRAGO, G. & NAVAS, A. (1995). Nematofauna fitoparásita asociada a cultivos hortícolas y tabaco en regadíos de Extremadura. *Boletín Sanidad Vegetal Plagas* 21, 303-317.

ESTORES, R.A. (1971). Interaction of *Pratylenchus penetrans* and *Meloidogyne incognita* as coinhabitants in tomato. *Dissertation Abstracts International* 32B, 3415.

ESTORES, R.A. & CHEN, T.A. (1972). Interactions of *Pratylenchus penetrans* and *Meloidogyne incognita* as coinhabitants in tomato. *Journal of Nematology* 4, 170-174.

EVANS, K. & WEBB, R.M. (1989). Nematode problems in oilseed rape. *Aspects of Applied Biology* 23, 303-310.

EVENHUIS, A., KORTHALS, G.W. & MOLENDIJK, L.P.G. (2004). *Tagetes patula* as an effective catch crop for long-term control of *Pratylenchus penetrans*. *Nematology* 6, 877-881.

FARKAS, K., HANGYA, L. & NEMETH, L. (1985). Nematological studies on chrysanthemums. *Novenyvedelem* 21, 529-537.

FARSI, M., VANSTONE, V.A., FISHER, J.M. & RATHJEN, A.J. (1995). Genetic variation in resistance to *Pratylenchus neglectus* in wheat and triticales. *Australasian Journal of Experimental Agriculture* 35, 597-602.

FELDMESSER, J., FEDER, W.A., REBOIS, R.V. & HUTCHINS, P.C. (1960). Longevity of *Radopholus similis* and *Pratylenchus brachyurus* in the greenhouse. *Anatomical Record* 137, 355.

FELIPE, A.J. (1989). *Patrones para frutales de pepita y hueso*. Barcelona, Spain, Ediciones Técnicas Europeas, 181 pp.

FERNANDEZ DIAZ-SILVEIRA, M. & ORTEGA HERRERA, J. (1998). An overview of nematological problems in Cuba. *Nematropica* 28, 151-164.

FERRAZ, L.C.C.B. (1995). Patogenicidade de *Pratylenchus brachyurus* a tres cultivares de soja. *Nematologia Brasileira* 19, 1-8.

FERRAZ, L.C.C.B. & MONTEIRO, A.R. (1983). Nova ocorrencia de *Pratylenchus penetrans* no Brasil. *Revista de Agricultura* 58, 301-303.

FERRAZ, L.C.C.B., MONTEIRO, A.R. & INOMOTO, M.M. (1989). Reaction of Barbados cherry to seven phytonematode species in Brazil. *Nematologia Brasileira* 13, 39-49.

FERRAZ, S. & DE FREITAS, L.G. (2004). Use of antagonistic plants and natural products. In: Cheng, Z.X., Chen, S.Y. & Dickson, D.W. (Eds). *Nematology advances and perspectives, volume 2. Nematode management and utilization.* Wallingford, UK, CABI Publishing, pp. 931-977.

FERRIS, H. (1981). Dynamic action thresholds for diseases induced by nematodes. *Annual Review of Phytopathology* 19, 427-436.

FERRIS, H., CARLSON, H.L. & WESTERDAHL, B.B. (1994). Nematode population changes under crop rotation sequences: Consequences for potato production. *Agronomy Journal* 86, 340-348.

FERRIS, J.M. (1970). Soil temperature effects on onion seedling injury by *Pratylenchus penetrans. Journal of Nematology* 2, 248-251.

FERRIS, V.R. (1961). A new species of *Pratylenchus* (Nemata-Tylenchida) from roots of soybeans. *Proceedings of the Helminthological Society of Washington* 28, 109-111.

FILIPJEV, I.N. (1934). The classification of free-living nematodes and their relations to parasitic nematodes. *Smithsonian Miscellaneous Collection* 89, 1-63.

FILIPJEV, I.N. (1936). On the classification of the Tylenchinae. *Proceedings of the Helminthological Society of Washington* 3, 80-82.

FILIPJEV, I.N. & SCHUURMANS STEKHOVEN JR, J.H. (1941). *A manual of agricultural helminthology*. Leiden, The Netherlands, E.J. Brill, 878 pp.

FILONOW, A.B. & RUSSELL, C.C. (1991). Nematodes and fungi associated with pod rot of peanuts in Oklahoma. *Nematologia Mediterranea* 19, 207-210.

FLORINI, D.A., LORIA, R. & KOTCON, J.B. (1987). Influence of edaphic factors and previous crop on *Pratylenchus* spp. population densities in potato. *Journal of Nematology* 19, 85-92.

FOGAIN, R. (2001). Nematodes and weevil of bananas and plantains in Cameroon: occurrence, importance and host susceptibility. *International Journal of Pest Management* 47, 201-205.

FOGAIN, R. & GOWEN, S.R. (1998). "Yangambi km5" (*Musa* AAA, Ibota subgroup): a possible source of resistance to *Radopholus similis* and *Pratylenchus goodeyi. Fundamental and Applied Nematology* 21, 75-80.

FOREY, P.L., HUMPHRIES, C.J., KITCHING, I.L., SCOTLAND, R.W., SIEBERT, D.J. & WILLIAMS, D.M. (1992). *Cladistics, a practical course in systematics*. New York, NY, USA, Oxford University Press, 191 pp.

FORGE, T., MUEHLCHEN, A., HACKENBERG, C., NEILSEN, G. & VRAIN, T. (2001). Effects of preplant inoculation of apple (*Malus domestica* Borkh.) with arbuscular mycorrhizal fungi on population growth of the root-lesion nematode, *Pratylenchus penetrans. Plant and Soil* 236, 185-196.

FORGE, T.A., DEYOUNG, R. & VRAIN, T.C. (1998). Temporal changes in the vertical distribution of *Pratylenchus penetrans* under raspberry. *Journal of Nematology* 30, 179-183.

FORGE, T.A., BITTMAN, S. & KOWALENKO, C.G. (2005). Impacts of sustained use of dairy manure slurry and fertilizers on populations of *Pratylenchus penetrans* under tall fescue. *Journal of Nematology* 37, 207-213.

FORTUNER, R. (1973). Description of *Pratylenchus sefaensis* n. sp. and of *Hoplolaimus clarissimus* n. sp. (Nematoda: Tylenchida). *Cahier ORSTOM, Série Biologie* 21, 25-34.

FORTUNER, R. (1976). *Pratylenchus zeae. CIH Descriptions of plant-parasitic nematodes*, Set 6, No. 77. St Albans, UK, Commonwealth Agricultural Bureaux, 3 pp.

FORTUNER, R. (1977). *Pratylenchus thornei. CIH Descriptions of plant-parasitic nematodes*, Set 7, No. 93. St Albans, UK, Commonwealth Agricultural Bureaux, 3 pp.

FORTUNER, R. (1985). Notes on nomenclature of plant nematodes. *Revue de Nématologie* 8, 77-83.

FORTUNER, R. & MAGGENTI, A.R. (1991). A statistical approach to the objective differentiation of *Hirschmanniella oryzae* from *H. belli* (Nemata: Pratylenchidae). *Revue de Nématologie* 14, 165-180.

FOURIE, H., MC DONALD, A.H. & LOOTS, G.C. (2001). Plant-parasitic nematodes in field crops in South Africa. 6. Soybean. *Nematology* 3, 447-454.

FRANCE, R.A. & BRODIE, B.B. (1995). Differentiation of two New York isolates of *Pratylenchus penetrans* based on their reaction on potato. *Journal of Nematology* 27, 339-345.

FRANCE, R.A. & BRODIE, B.B. (1996). Characterization of *Pratylenchus penetrans* from ten geographically isolated populations based on their reaction on potato. *Journal of Nematology* 28, 520-526.

FRECKMAN, D.W. & CASWELL, E.P. (1985). The ecology of nematodes in agroecosystems. *Annual Review of Phytopathology* 23, 275-296.

FREDERICK, J.J. & TARJAN, A.C. (1978). Variability in measurements made of same nematode specimen by various observers or by one observer on different days. *Nematologica* 24, 476-479.

FREDERICK, J.J. & TARJAN, A.C. (1989). A compendium of the genus *Pratylenchus* Filipjev, 1936 (Nemata: Pratylenchidae). *Revue de Nématologie* 12, 243-256.

GAIR, R., MATHIAS, P.L. & HARVEY, P.N. (1969). Studies of cereal nematode populations and cereal yields under continuous or intensive culture (*Heterodera avenae, Pratylenchus neglectus, Trichodorus primitivus*). *Annals of Applied Biology* 63, 503-512.

GALEANO, M., VERDEJO-LUCAS, S. & CIANCIO, A. (2003). Morphology and ultrastructure of a *Pasteuria* form parasitic in *Tylenchorhynchus cylindricus* (Nematoda). *Journal of Invertebrate Pathology* 83, 83-85.

GAO, F., GONG, Y.F. & ZHANG, P.B. (2000). Production and deployment of virus-free sweetpotato in China. *Crop Protection* 19, 105-111.

GAO, X., CHENG, H. & FANG, C. (1997). A new type of morphometrical variation of *Pratylenchus vulnus* on peach from China. *International Journal of Nematology* 7, 147-151.

GAO, X., CHENG, H. & FANG, C. (1999). New diagnostic characters for *Pratylenchus vulnus*. *Nematologia Mediterranea* 27, 9-13.

GAPASIN, R.M. (1995). Evaluacion de *Paecilomyces lilacinus* (Thom.) Samson para el control de *Pratylenchus* sp. en maiz. *Biocontrol* 1, 35-39.

GAUR, H.S., SINGH, J., SHARMA, S.N. & CHANDEL, S.T. (1996). Distribution and community analysis of plant parasitic nematodes in rice growing areas of Haryana, India. *Annals of Plant Protection Sciences* 4, 115-121.

GERAERT, E. (2006). *Functional and detailed morphology of the Tylenchida (Nematoda). Nematology Monographs and Perspectives, volume 4* (series editors Hunt, D.J. & Perry, R.N.). Leiden, The Netherlands, Brill Academic Publishers, 215 pp.

GERAERT, E. & DE GRISSE, A. (1981). The male copulatory system in tylenchid taxonomy (Nematoda). *Nematologica* 27, 432-442.

GERAERT, E., ZEPP, A. & BORAZANCI, N. (1975). Some plant nematodes from Turkey. *Mededelingen van de Faculteit Landbouwwetenschappen, Rijksuniversiteit Gent* 40, 511-515.

GERMANI, G., MUGNIER, J. & DOMMERGUES, Y. (1984). Influence of pathogenic nematodes on nodulation and seed yield of soybeans in Senegal. *Revue de Nématologie* 7, 335-340.

GIBLIN-DAVIS, R.M., WILLAMS, D.S., BEKAL, S., DICKSON, D.W., BRITO, J.A., BECKER, J.O. & PRESTON, J.F. (2003). 'Candidatus *Pasteuria usgae*' sp. nov., an obligate endoparasite of the phytoparasitic nematode *Belonolaimus longicaudatus*. *International Journal of Systematic and Evolutionary Microbiology* 53, 197-200.

GIRIBABU, P. & SAHA, M. (2003). Studies on nematode communities at different depths, altitudes and periods of year in coffee plantation. *Annals of Plant Protection Sciences* 11, 360-363.

GLAZER, J. & ORION, D. (1983). Studies on anhydrobiosis of *Pratylenchus thornei*. *Journal of Nematology* 15, 333-338.

GNANAPRAGASAM, N.C. (1997). Plant derivatives in managing nematodes attacking tea in Sri Lanka. *International Journal of Nematology* 7, 111-112.

GNANAPRAGASAM, N.C. (2002). Slow decline observed in nematode tolerant tea clone. *International Journal of Nematology* 12, 232-233.

GNANAPRAGASAM, N.C. & MANUELPILLAI, M.E.K. (1984). Influence of inoculum level and temperature on the population build-up and pathogenicity of the root-lesion nematode (*Pratylenchus loosi* Loof) of tea. *Tea Quarterly* 53, 19-22.

GNANAPRAGASAM, N.C. & MOHOTTI, K.M. (2005). Nematode parasites of tea. In: Luc, M., Sikora, R.A. & Bridge, J. (Eds). *Plant parasitic nematodes in subtropical and tropical agriculture*. Wallingford, UK, CABI Publishing, pp. 581-609.

GODFREY, G.H. (1929). A destructive root disease of pineapple and other plants due to *Tylenchus brachyurus*. *Phytopathology* 19, 611-629.

GOELLNER, M., WANG, X. & DAVIS, E.L. (2001). Endo-β-1,4-glucanase expression in compatible plant-nematode interactions. *Plant Cell* 13, 2241-2255.

GOFFART, H. (1929). Beobachtungen über Anguillulina pratensis de Man. *Zeitschrift für Parasitenkunde, Berlin* 2, 97-120.

GOFFART, H. (1963). Nematoden in maiblumen-kulturen. *Gartenwelt* 63, 380-381.

GOLDEN, A.M., LOPEZ, C.R. & VILCHEZ, R.H. (1992). Description of *Pratylenchus gutierrezi* n. sp. (Nematoda: Pratylenchidae) from coffee in Costa Rica. *Journal of Nematology* 24, 298-304.

GÓMEZ-BARCINA, A., GONZÁLEZ-PAÍS, M.A. & CASTILLO, P. (1989). Nematodos del orden Tylenchida Thorne, 1949 en España. *Revista Ibérica de Parasitología* 49, 85-109.

GÓMEZ BARCINA, A., FERNÁNDEZ VALDIVIA, J., GONZÁLEZ PAÍS, M.A. & CASTILLO, P. (1991). Análisis de la distribución de los nematodos fitoparásitos del orden Tylenchida Thorne, 1949 en diez comunidades vegetales. In: Del Moral de la Vega, J.M. (Ed.). *Estudios de fitopatología.* Badajoz, Spain, Sociedad Española de Fitopatología, pp. 256-265.

GOMMERS, F.J. & BAKKER, J. (1988). Physiological changes induced by plant responses or products. In: Poinar Jr, G.O. & Jansson, H.B. (Eds). *Diseases of nematodes, volume 1.* Boca Raton, FL, USA, CRC Press, Inc., pp. 3-22.

GOOD, J.M. & STANSELL, J.R. (1965). Effect of irrigation, soil fumigation and date of harvest on *Pratylenchus brachyurus* (Godfrey) infection of pods, peg rot and yield of peanuts. *Nematologica* 11, 38-39.

GOOD, J.M., BOYLE, L.W. & HAMMONS, R.O. (1958). Studies of *Pratylenchus brachyurus* on peanuts. *Phytopathology* 48, 530-535.

GOODEY, J.B. (1963). *Soil and freshwater nematodes.* London, UK, Methuen & Co, 544 pp.

GOODEY, J.B., FRANKLIN, M.T. & HOOPER, D.J. (1965). *The nematode parasites of plants catalogued under their hosts,* 3rd edition. Farnham Royal, UK, Commonwealth Agricultural Bureaux, 214 pp.

GOODEY, T. (1928). Observations on *Tylenchus musicola* Cobb, 1919, from diseased banana roots. *Journal of Helminthology* 6, 193-198.

GOODEY, T. (1932). The genus *Anguillulina* Gerv. & v. Ben., 1859 vel *Tylenchus* Bastian, 1865. *Journal of Helminthology* 10, 75-180.

GOODEY, T. (1937). On *Anguillulina mahogani* (Cobb, 1920). *Journal of Helminthology* 15, 133-136.

GOTOH, A. (1964). Studies on the control of root rot nematode disease of sweet potato caused by *Pratylenchus coffeae*. *Bulletin of the Miyazaki Agricultural Experiment Station* 5, 1-121.

GOTOH, A. (1970). The root-lesion nematodes found in uncultivated soils, mainly in natural grasslands in Japan. *Proceedings of the Association for Plant Protection of Kyushu* 16, 34-37.

GOTOH, A. (1974). Geographic distribution of *Pratylenchus* spp. (Nematoda: Tylenchida) in Japan. *Bulletin of the Kyushu Agricultural Experiment Station* 17, 139-224.

GOTOH, A. & OSHIMA, Y. (1963). *Pratylenchus* species and their geographical distribution in Japan (Nematoda: Tylenchida). *Japanese Journal of Applied Entomology and Zoology* 7, 187-199.

GOVAERTS, B., MEZZALAMA, M., SAYRE, K.D., CROSSA, J., NICOL, J.M. & DECKERS, J. (2006). Long-term consequences of tillage, residue management, and crop rotation on maize/wheat root rot and nematode populations in subtropical highlands. *Applied Soil Ecology* 32, 305-315.

GOWEN, S. & QUÉNÉHERVÉ, P. (1990). Nematode parasites of bananas, plantains and abaca. In: Luc, M., Sikora, R.A. & Bridge, J. (Eds). *Plant parasitic nematodes in subtropical and tropical agriculture.* Wallingford, UK, CABI Publishing, pp. 431-460.

GOWEN, S., QUÉNÉHERVÉ, P. & FOGAIN, R. (2005). Nematode parasites of bananas and plantains. In: Luc, M., Sikora, R.A. & Bridge, J. (Eds). *Plant parasitic nematodes in subtropical and tropical agriculture*, 2nd edition. Wallingford, UK, CABI Publishing, pp. 611-643.

GRAHAM, T.W. (1951). Nematode root rot of tobacco and other plants. *Bulletin 390. South Carolina Agricultural Experiment Station, Clemson Agricultural College*, 25 pp.

GRANDISON, G.S. & WALLACE, H.R. (1974). The distribution and abundance of *Pratylenchus thornei* in fields of strawberry clover (*Trifolium fragiferum*). *Nematologica* 20, 283-290.

GRECO, N., DI VITO, M., REDDY, M.V. & SAXENA, M.C. (1984). A preliminary report of survey of plant parasitic nematodes of leguminous crops in Syria. *Nematologia Mediterranea* 12, 87-93.

GRECO, N., DI VITO, M., SAXENA, M.C. & REDDY, M.V. (1988). Investigation on the root lesion nematode *Pratylenchus thornei*, in Syria. *Nematologia Mediterranea* 16, 101-105.

GRECO, N., DI VITO, M. & SAXENA, M.C. (1992). Plant parasitic nematodes of cool season food legumes in Syria. *Nematologia Mediterranea* 20, 37-46.

GREEN, C.D. (1979). Relationships between small numbers of *Pratylenchus penetrans* and field damage in a celery crop. *Plant Pathology* 28, 81-85.

GRIFFIN, G.D. (1991). Differential pathogenicity of four *Pratylenchus neglectus* populations on alfalfa. *Journal of Nematology* 25, 380-385.

GRIFFIN, G.D. (1993). Comparative response of alfalfa to *Pratylenchus penetrans* populations. *Journal of Nematology* 25, 461-465.

GRIFFIN, G.D. (1994). Effect of single and interplantings on pathogenicity of *Pratylenchus penetrans* and *P. neglectus* to alfalfa and crested wheatgrass. *Journal of Nematology* 26, 460-466.

GRIFFIN, G.D. (1996). Importance of soil texture to the pathogenicity of plant-parasitic nematodes on rangeland grasses. *Nematropica* 26, 27-37.

GRIFFIN, G.D. & GRAY, F.A. (1990). Biology and pathogenicity of *Pratylenchus neglectus* on alfalfa. *Journal of Nematology* 22, 546-551.

GRIFFIN, G.D. & JENSEN, K.B. (1997). Differential effects of *Pratylenchus neglectus* populations on single and interplantings of alfalfa and crested wheatgrass. *Journal of Nematology* 29, 82-89.

GRISLAES-LOPEZ, F. & LESCOT, T. (1999). Encuesta diagnostico multifactorial sobre platano en la zona cafetera central de Colombia. Una vision analitica de la realidad agronomica del cultivo del platano en la zona central. *Boletin Tecnico Cenicafe* 18, 66.

GRUJICIC, G. (1969). Prilog proucavanju parazitnih nematoda na psenici u Jugoslaviji. *Savremena Poljoprivreda* 17, 531-539.

GRUJICIC, G. (1974). [Studies on plant parasitic nematodes of corn plants.] *Biljna Zastita* 5, 193.

GRUJICIC, G. & JOVICIC, D. (1988). [Harmful nematofauna on tobacco fields in Serbia.] *Zastita Bilja* 39, 149-157.

GUBINA, V.G. (1973). [Parasitic nematodes of conifer seedlings.] *Trudy Gel'mintologicheskoi Laboratorii Ekologiya I taksonomiya gel'mintov* 23, 48-52.

GUEROUT, R. (1975). Nematodes of pineapple: a review. *Pest Articles and News Summaries* 21, 123-140.

HACKENBERG, C., MUEHLCHEN, A., FORGE, T. & VRAIN, T. (2000). *Pseudomonas chlororaphis* strain sm3, bacterial antagonist of *Pratylenchus penetrans*. *Journal of Nematology* 32, 183-189.

HACKNEY, R.W. & DICKERSON, O.J. (1975). Marigold, castor bean, and chrysanthemum as controls of *Meloidogyne incognita* and *Pratylenchus alleni*. *Journal of Nematology* 7, 84-90.

HAFEZ, S.L., MILLER, D. & SUNDARARAJ, P. (2000). Screening of alfalfa cultivars to the lesion nematode *Pratylenchus penetrans* for commercial release. *Nematologia Mediterranea* 28, 157-161.

HAIDAR, M.G. & PATHAK, K.N. (2001). Influence of cropping sequences on the population of phytonematodes. *Annals of Plant Protection Sciences* 9, 354-356.

HAN, Y.L., BA, Y.R. & GA, R.D. (1997). Taxonomic study on pasture parasitic nematodes. *Grassland of China* 5, 59-60.

HANDOO, Z.A. & GOLDEN, A.M. (1989). A key and diagnostic compendium to the species of the genus *Pratylenchus* Filipjev, 1936 (Lesion Nematodes). *Journal of Nematology* 21, 202-218.

HANDOO, Z.A., CARTA, L.K. & SKANTAR, A.M. (2001). Morphological and molecular characterisation of *Pratylenchus arlingtoni* n. sp., *P. convallariae* and *P. fallax* (Nematoda: Pratylenchidae). *Nematology* 3, 607-618.

HAQUE, M.M. (1966). Two new species of plant nematodes. *Doklady Timiryazevskoi Selskokhozyaistvennoi Akademii* 113, 209-219.

HARLEY, J.L. & SMITH, S.E. (1983). *Mycorrhizal symbiosis*. London, UK, Academic Press, 483 pp.

HARR, J., BASEL, S.A.G. & KLINGLER, J. (1976). Single and combined effect of *Pratylenchus penetrans* and *Thielaviopsis basicola* on the growth of cherry tree cutting. *Journal of Plant Diseases and Protection* 83, 615-619.

HARRISON, B.D. & HOOPER, D.J. (1963). Longevity of *Longidorus elongatus* (de Man) and other nematodes in soil kept in polythene bags. *Nematologica* 9, 159-160.

HARTMAN, K.M. & EISENBACK, J.D. (1991). Amended description of *Pratylenchus macrostylus* Wu, 1971 with SEM observations. *Journal of Nematology* 23, 104-109.

HASAN, A. (1988). Interaction between *Pratylenchus coffeae* and *Pythium aphanidermatum* and/or *Rhizoctonia solani* on chrysanthemum. *Journal of Phytopathology* 123, 227-232.

HASAN, N., BHASKAR, R.B., PANDEY, K.C. & MELKANIA, N.P. (2003). Ecofriendly integrated pest management in lucerne (*Medicago sativa* L.). *Range Management and Agroforestry* 24, 45-48.

HASHIM, Z. (1983). Plant parasitic nematodes associated with olive in Jordan. *Nematologia Mediterranea* 11, 27-32.

HASHIM, Z. (1984). Description of *Pratylenchus jordanensis* n. sp. (Nematoda: Tylenchida) and notes on other Tylenchida from Jordan. *Revue de Nématologie* 6, 187-192.

HASHMI, S. & HASHMI, G. (1989). Host status of several economic plants to *Pratylenchus zeae* from Pakistan. *Nematropica* 19, 195-197.

HAY, F.S. & PETHYBRIDGE, S.J. (2005). Nematodes associated with carrot production in Tasmania, Australia, and the effect of *Pratylenchus crenatus* on yield and quality of Kuroda-type carrot. *Plant Disease* 89, 1175-1180.

HEIL, M. (2001). The ecological concept of costs of induced systemic resistance (ISR). *European Journal of Plant Pathology* 107, 137-146.

HERMAN, M., HUSSEY, R.S. & BOERMA, H.R. (1988). Interactions between *Meloidogyne incognita* and *Pratylenchus brachyurus* on soybean. *Journal of Nematology* 20, 79-84.

HERNÁNDEZ, M.A., JORDANA, R., GOLDARACENA, A. & PINOCHET, J. (2000). SEM observations on nine species of the genus *Pratylenchus* Filipjev, 1936 (Nematoda: Pratylenchidae). *Journal of Nematode Morphology and Systematics* 3, 165-174.

HERNÁNDEZ-DORREGO, A., PINOCHET, J. & CALVET, C. (1999). Growth response of peach and plum rootstocks infected with *Pratylenchus vulnus* in microplots. *Journal of Nematology* 31, 656-661.

HERVE, G., BERTRAND, B., VILLAIN, L., LICARDIE, D. & CILAS, C. (2005). Distribution analyses of *Meloidogyne* spp. and *Pratylenchus coffeae sensu lato* in coffee plots in Costa Rica and Guatemala. *Plant Pathology* 54, 471-475.

HEYNS, J. (1962). Taxonomic studies on nematodes from South Africa. *Dissertation Abstracts International* 22, 2933.

HEYNS, J. (1971). *A guide to the plant and soil nematodes of South Africa.* Cape Town, South Africa, A.A. Balkema, 233 pp.

HILLOCKS, R.J. & SONGA, W. (1993). Root-knot and other nematodes associated with pigeonpea plants infected with *Fusarium udum* in Kenya. *Afro-Asian Journal of Nematology* 3, 143-147.

HILLOCKS, R.J., SIDDIQI, M.R. & KHONGA, E.B. (1995). Nematodes associated with subsistence crops in southern Malawi. *Afro-Asian Journal of Nematology* 5, 14-19.

HINTON, D.M. & BACON, C.W. (1985). The distribution and ultrastructure of the endophyte of toxic tall fescue. *Canadian Journal of Botany* 63, 36-42.

HIRATA, K., FERNANDO, K.G.I. & HEWAGE, L.C. (2000). Plant parasitic nematodes associated with crops in Sri Lanka. *Research Bulletin of the Plant Protection Service, Japan* 36, 61-66.

HOCKLAND, S., INSERRA, R.N., MILLAR, L. & LEHMAN, P.S. (2006). International plant health – putting legislation into practice. In: Perry, R.N. & Moens, M. (Eds). *Plant nematology.* Wallingford, UK, CABI Publishing, pp. 327-345.

HOLLAWAY, G.J. (2002). Effect of oat (*Avena sativa*) on the population density of *Pratylenchus thornei* in the field. *Australasian Plant Pathology* 31, 147-149.

HOLLAWAY, G.J., TAYLOR, S.P., EASTWOOD, R.F. & HUNT, C.H. (2000). Effect of field crops on density of *Pratylenchus* in southeastern Australia; Part 2. *P. thornei*. *Journal of Nematology* 32, 600-608.

HOLTERMAN, M., VAN DER WURFF, A., VAN DEN ELSEN, S., VAN MEGEN, H., BONGERS, T., HOLOVACHOV, O., BAKKER, J. & HELDER, J. (2006). Phylum-wide analysis of SSU rDNA reveals deep phylogenetic relationships among nematodes and accelerated evolution toward crown clades. *Molecular Biology and Evolution* 23, 1792-1800.

HOOPER, D.J., WEBB, R.M. & FLYNN, C. (1990). Plant parasitic nematodes associated with the roots of wind toppled trees at the Royal Botanic Gardens, Kew. *Annals of Applied Biology* 117, 407-414.

HOY, J.W. (1999). Yield decline of sugarcane in Louisiana. In: Magarey, R.C. (Ed.). *Proceedings of the first Australasian soilborne diseases symposium.* Brisbane, Australia, BSES Limited, pp. 17-18.

HUAN, J. & XU, M.Q. (1985). A report on the identification of plant parasitic nematodes of staple crops in Zhejiang Province. II. The five species of parasitic nematodes on sweet potatoes. *Acta Agriculture Universitatis Zhejiangensis* 11, 231-235.

HUANG, C.S. & CHIANG, Y.C. (1976). *Pratylenchus coffeae* found in Taiwan citrus orchard. *Plant Protection Bulletin Taiwan* 18, 75-78.

HUETTEL, R.N., FRANCL, L.J., REISE, R.W., MEYER, S.L.F. & HENN, R.A. (1991). Plant-parasitic nematodes in the potato growing areas of Maine. *American Potato Journal* 68, 345-354.

HUSAIN, S.I. (1976). Phylogeny and interrelationship of the superfamily Heteroderoidea (Skarbilovich, 1947) Golden, 1971. *Geobios* 3, 9-12.

HUSSEY, R.S. & BARKER, K.R. (1976). Influence of nematodes and light sources on growth and nodulation of soybean. *Journal of Nematology* 8, 48-52.

HUSSEY, R.S. & DAVIS, E.L. (2004). Nematode esophageal glands and plant parasitism. In: Chen, Z.X., Chen, S.Y. & Dickson, D.W. (Eds). *Nematology advances and perspectives, volume I. Nematode morphology, physiology and ecology.* Wallingford, UK, CABI Publishing, pp. 258-294.

HUSSEY, R.S. & RONCADORI, R.W. (1978). Interaction of *Pratylenchus brachyurus* and *Gigaspora margarita* on cotton. *Journal of Nematology* 10, 16-20.

HUTTON, D.G. (1979). Comparison of the performance of white yam plants growing from *Pratylenchus coffeae*-infested and disinfested planting pieces. *Bulletin Ministry of Agriculture Jamaica* 47-48.

HUTTON, D.G., WILKINSON, R.E. & MAI, W.F. (1973). Effect of two plant parasitic nematodes on Fusarium dry root rot of beans. *Phytopathology* 63, 749-751.

HYMAN, B.C. & POWERS, T.O. (1991). Integration of molecular data with systematics of plant parasitic nematodes. *Annual Review of Phytopathology* 29, 89-107.

IBRAHIM, I.K.A., REZK, M.A. & IBRAHIM, A.A.M. (1988). Plant parasitic nematodes associated with gramineous plants in northern Egypt. *Pakistan Journal of Nematology* 6, 31-37.

IBRAHIM, S.K., PERRY, R.N. & WEBB, R.M. (1995). Use of isoenzyme and protein phenotypes to discriminate between six *Pratylenchus* species from Great Britain. *Annals of Applied Biology* 126, 317-327.

IGLESIAS, J., GARCIA-MINA, J.M., HERNANDEZ, M., RODRIGUEZ-RODRIGUEZ, R. & JORDANA, R. (1999). Evaluation of Coactyl on yield, plant growth and nematode communities in banana plant in the Canary Islands. *International Journal of Nematology* 9, 34-42.

INAGAKI, H. & POWELL, N.T. (1969). Influence of root-lesion nematode on black shank symptom development in flue-cured tobacco. *Phytopathology* 59, 1350-1355.

INGHAM, R.E., HAMM, P.B., RIGA, E. & MERRIFIELD, K.J. (2005). First report of stunting and root rot of potato associated with *Pratylenchus penetrans* in the Columbia basin of Washington. *Plant Disease* 89, 207.

INGRAM, E.G. & RODRÍGUEZ-KÁBANA, R. (1980). Nematodes parasitic on peanuts in Alabama and evaluation of methods for detection and study of population dynamics. *Nematropica* 10, 21-30.

INOMOTO, M.M., OLIVEIRA, C.M.G., MAZZAFERA, P. & GONÇALVES, W. (1998). Effects of *Pratylenchus brachyurus* and *P. coffeae* on seedlings of *Coffea arabica*. *Journal of Nematology* 30, 362-367.

INOMOTO, M.M., GOULART, A.M.C., MACHADO, A.C.Z. & MONTEIRO, A.R. (2001). Effect of population densities of *Pratylenchus brachyurus* on the growth of cotton plants. *Fitopatologia Brasileira* 26, 192-196.

INSERRA, R., VOVLAS, N., LAMBERTI, F. & BLEVE-ZACHEO, T. (1976). Plant parasitic nematodes associated with declining olive trees in southern Italy. *Poljoprivredna Znanstvena Smotra, Agriculturae Conspectus Scientificus* 39, 419-424.

INSERRA, R.N. & VOVLAS, N. (1974). Danni da *Pratylenchus vulnus* su arancio amaro in Puglia. *Nematologia Mediterranea* 2, 183-185.

INSERRA, R.N. & VOVLAS, N. (1977). Effects of *Pratylenchus vulnus* on the growth of sour orange. *Journal of Nematology* 9, 154-157.

INSERRA, R.N. & VOVLAS, N. (1981). Indagine sulla distribuzione geografica dei nematodi parassiti dell'olivo in Italia. *Informatore Fitopatologico* 31, 117-119.

INSERRA, R.N., VOVLAS, N. & BRANDONISIO, A. (1978). Nematodi endoparassiti associati a colture di cereali in deperimento nell'Italia meridionale. *Nematologia Mediterranea* 6, 163-174.

INSERRA, R.N., ZEPP, A. & VOVLAS, N. (1979). *Pratylenchus* dell'italia meridionale. *Nematologia Mediterranea* 7, 137-162.

INSERRA, R.N., O'BANNON, J.H. & DUNCAN, L.W. (1990). Native hosts of *Pratylenchus coffeae* in Florida. *Nematology Circular, Florida Department of Agriculture & Consumer Services* No. 182, 3 pp.

INSERRA, R.N., OOSTENDORP, M. & DICKSON, D.W. (1992). *Pasteuria* sp. parasitizing *Trophonema okamotoi* in Florida. *Journal of Nematology* 24, 36-39.

INSERRA, R.N., DUNCAN, L.W., VOVLAS, N. & LOOF, P.A.A. (1996). *Pratylenchus loosi* from pasture grasses in Central Florida. *Nematologica* 42, 159-172.

INSERRA, R.N., DUNCAN, L.W., DUNN, D., KAPLAN, D.T. & POROZINSKA, D. (1998). *Pratylenchus pseudocoffeae* from Florida and its relationship with *P. gutierrezi* and *P. coffeae*. *Nematologica* 42, 683-712.

INSERRA, R.N., DUNCAN, L.W., TROCCOLI, A., DUNN, D., DOS SANTOS, J.M., KAPLAN, D. & VOVLAS, N. (2001). *Pratylenchus jaehni* sp. n. from citrus in Brazil and its relationship with *P. coffeae* and *P. loosi* (Nematoda: Pratylenchidae). *Nematology* 3, 653-665.

INSERRA, R.N., DUNCAN, L.W., DUNN, D., HANDOO, Z.A., TROCCOLI, A. & ROWE, J. (2005). *Pratylenchus jordanensis* a junior synonym of *P. zeae*. *Nematropica* 35, 161-170.

INSERRA, R.N., TROCCOLI, A., GOZEL, U., BERNARD, E.C., DUNN, D. & DUNCAN, L.W. (2007). *Pratylenchus hippeastri* n. sp. (Nematoda: Pratylenchidae) from amaryllis in Florida with notes on *P. scribneri* and *P. hexincisus*. *Nematology* 9, 25-42.

IRWIN, J.A.G. & JONES, R.M. (1977). The role of fungi and nematodes as factors associated with death of white clover (*Trifolium repens*) stolons over summer in south-eastern Queensland. *Australian Journal of Experimental Agriculture and Animal Husbandry* 17, 789-794.

ISHIGURO, K., YAMAKAWA, O., KUMAGAI, T., YOSHINAGA, M., HIDAKA, M., KAI, Y., KOMAKI, K. & KUKIMURA, H. (2004a). Tamaotome: new sweet potato cultivar for dried products after steaming. *Bulletin of the National Agricultural Research Center for Kyushu Okinawa Region* 43, 27-58.

ISHIGURO, K., YAMAKAWA, O., KUMAGAI, T., YOSHINAGA, M., KAI, Y. & HIDAKA, M. (2004b). Benimasari: new sweet potato cultivar for table use. *Bulletin of the National Agricultural Research Center for Kyushu Okinawa Region* 43, 59-85.

ISLAM, S., KHAN, A. & BILQEES, F.M. (1994). Nematodes associated with apple trees in Swat, N.W.F.P., Pakistan. *Proceedings of Pakistan Congress of Zoology* 14, 317-320.

ISLAM, S., KHAN, A. & BILQEES, F.M. (1996). Survey of stylet-bearing nematodes associated with apple (*Malus pumila* Mill.) in swat district, N.W.F.P., Pakistan. *Proceedings of Parasitology* 21, 1-9.

ISUTSA, D.K. & MERWIN, I.A. (2000). *Malus* germplasm varies in resistance or tolerance to apple replant disease in a mixture of New York orchard soils. *HortScience* 35, 262-268.

IVANOVA, L. & CHOLEVA, B. (1999). Preliminary observations on apricot decline. *Acta Horticulturae* 488, 683-685.

IVANOVA, T.S. (1968). [*Nematodes of cereals from the Zeravshan Valley of Tadzhikistan.*] Dushanbe, Tadzhikistan, Izdatelstvo Donish, 84 pp.

IWAHORI, H. & SANO, Z. (2003). Distribution of main plant-parasitic nematodes in sweet potato and taro fields in Kyushu and Okinawa, Japan. 3.

Survey in the northern part in Kyushu Island (Fukuoka, Saga, Nagasaki and Ohita Prefs.). *Kyushu Plant Protection Research* 49, 83-87.

JAFFEE, B.A., ABAWI, G.S. & MAI, W.F. (1982). Role of soil microflora and *Pratylenchus penetrans* in an apple replant disease. *Phytopathology* 72, 247-251.

JAIZME-VEGA, M.C. & PINOCHET, J. (1997). Growth response of banana to three mycorrhizal fungi in *Pratylenchus goodeyi* infested soil. *Nematropica* 27, 69-76.

JAKOBSEN, J. (1974). Migratory nematodes as the causes of damage to rose crops under glass. *Nordisk Jordbrugsforskning* 56, 417-418.

JAKOBSEN, J. (1975). Plant parasitic nematodes on roses in Danish glasshouses. *Tidsskrift for Planteavl* 79, 489-494.

JANSSON, H.B., DACKMAN, C. & ZUCKERMAN, B.M. (1987). Adhesion and infection of plant parasitic nematodes by the fungus *Drechmeria coniospora*. *Nematologica* 33, 480-487.

JAUHARI, R.K. & LAL, M. (2002). Effect of soil organisms on the population of migratory nematode, *Pratylenchus penetrans* (Cobb, 1917) (Nematoda: Hoplolaimidae) on tea plantations in Doon Valley. *Journal of Experimental Zoology, India* 5, 45-48.

JAUMOT, M., PINOCHET, J. & FERNÁNDEZ, C. (1997). Protein analysis of root-lesion nematodes using SDS-PAGE. *Nematropica* 27, 33-39.

JELIC, A. (1992). A contribution to the study of phytoparasitic nematodes in wheat. *Znanost i Praksa u Poljoprivredi i Prehrambenoj Tehnologiji* 22, 229-236.

JENSEN, H.J., MARTIN, J.P., WISMER, C.A. & KOIKE, K. (1959). Nematodes associated with varietal yield decline of sugar cane in Hawaii. *Plant Disease Reporter* 43, 253-260.

JIMENEZ, N., CROZZOLI, R., PETIT, P. & GRECO, N. (2001). Nematodos fitoparasiticos asociados con el cultivo de la pina, *Ananas comosus*, en los estados Lara y Trujillo, Venezuela. *Nematologia Mediterranea* 29, 13-17.

JOHNSON, A.W. & CHALFANT, R.B. (1973). Influence of organic pesticides on nematode and corn earworm damage and on yield of sweet corn. *Journal of Nematology* 5, 177-180.

JOHNSON, A.W., DOWLER, C.C. & HAUSER, E.W. (1974). Seasonal population dynamics of selected plant-parasitic nematodes on four monocultured crops. *Journal of Nematology* 6, 187-190.

JOHNSON, A.W., DOWLER, C.C. & HAUSER, E.W. (1975). Crop rotation and herbicide effects on population densities of plant-parasitic nematodes. *Journal of Nematology* 7, 158-168.

JOHNSON, A.W., DOWLER, C.C., BAKER, S.H. & HANDOO, Z.A. (1998). Crop yields and nematode population densities in triticale-cotton and triticale-soybean rotations. *Journal of Nematology* 30, 353-361.

JOHNSON, C.S. (1998). Tobacco. In: Barker, K.R., Pederson, G.A. & Windham, G.L. (Eds). *Plant and nematode interactions.* Madison, WI, USA, American Society of Agronomy, Inc., Crop Science Society of America, Inc., Soil Science Society of America, Inc., pp. 485-522.

JOHNSON, D.A. & SANTO, G.S. (2001). Development of wilt in mint in response to infection by two pathotypes of *Verticillium dahliae* and co-infection by *Pratylenchus penetrans*. *Plant Disease* 85, 1189-1192.

JOHNSON, P.W. & MCCLANAHAM, R.J. (1974). Nematode control and other effects of soil treatment of greenhouse rosebeds with aldicarb and oxamyl. *Plant Disease Reporter* 58, 730-732.

JONES, J.T., PHILLIPS, M.S. & ARMSTRONG, M.R. (1997). Molecular approaches in plant nematology. *Fundamental and Applied Nematology* 20, 1-14.

JONES, J.T., FURLANETTO, C. & KIKUCHI, T. (2005). Horizontal gene transfer from bacteria and fungi as a driving force in the evolution of plant parasitism in nematodes. *Nematology* 7, 641-646.

JONES, M.L. & HILLOCKS, R.J. (1995). Host status for *Pratylenchus zeae* of food crops and associated weed species in Malawi. *Afro-Asian Journal of Nematology* 5, 120-126.

JORDAAN, E.M. & DE WAELE, D. (1988). Host status of five weed species and their effects on *Pratylenchus zeae* infestation of corn. *Journal of Nematology* 20, 620-624.

JORDAAN, E.M., DE WAELE, D. & VAN ROOYEN, P.J. (1989). Endoparasitic nematodes in corn roots in the western Transvaal as related to soil texture and rainfall. *Journal of Nematology* 21, 356-360.

JORDAAN, E.M., VAN DEN BERG, E. & DE WAELE, D. (1992). Plant-parasitic nematodes on field crops in South Africa. 5. Wheat. *Fundamental and Applied Nematology* 15, 531-537.

JOSHI, L.M., RENFRO, B.L., SAARI, E.E., WILCOXSORI, R.D. & RAYCHAUDHURI, S.P. (1970). Diseases of wheat in India other than rusts and smuts. *Plant Disease Reporter* 54, 594-597.

JOTHI, G. & SUNDARABABU, R. (1997). Studies on the management of root lesion nematode, *Pratylenchus zeae* with the endomycorrhizal fungus, *Glomus fasciculatum* on corn. *Indian Journal of Nematology* 27, 264-266.

JOTHI, G., SUNDARABABU, R., RAMAKRISHNAN, S. & RAJENDRAN, G. (2004). Management of root lesion nematode, *Pratylenchus delattrei* in crossandra using oil cakes. *Bioresource Technology* 93, 257-259.

KABLE, P.F. & MAI, W.F. (1968). Influence of soil moisture on *Pratylenchus penetrans*. *Nematologica* 14, 101-122.

KALYVIOTIS-GAZELAS, C. (1981). Nematodes and host-plants reported for the first time in Greece. *Annales de l'Institut Phytopathologique Benaki, Nouveau Série* 13, 30-35.

KANWAR, R.S., RANA, B.P., WALIA, K.K. & WALIA, R.K. (1996). Studies on the pathogenic potential of *Heterodera zeae* and *Pratylenchus zeae* on corn in Haryana. *Indian Journal of Nematology* 26, 7-14.

KAPLAN, D.T. (1985). Plant-parasitic nematodes affecting citrus. *Fruit Varieties Journal* 39, 26-30.

KAPLAN, D.T. (1994). Partial characterization of a *Pasteuria* sp. attacking the citrus nematode, *Tylenchulus semipenetrans*, in Florida. *Fundamental and Applied Nematology* 17, 509-512.

KAPLAN, D.T. & DAVIS, E.L. (1987). Mechanisms of plant incompatibility with nematodes. In: Veech, A. & Dickson, D.W. (Eds). *Vistas on nematology.* Hyattsville, MD, USA, Society of Nematologists, pp. 267-276.

KAPLAN, D.T. & TIMMER, L.W. (1982). Effects of *Pratylenchus coffeae-Tylenchulus semipenetrans* interactions on nematode population dynamics in citrus. *Journal of Nematology* 14, 368-373.

KAPOOR, M. (1983). *Taxonomic studies on nematodes of some medicinal and aromatic plants of North India.* Ph.D. Thesis, Panjab University, Chandigarh, India, 418 pp.

KARIMOVA, M.M. (1986). [Nematodes of vegetable and cucurbit crops in the Izbaskan district of the Andizhan region.] *Uzbekskii Biologicheskii Zhurnal* 5, 64-65.

KARSSEN, G. & BOLK, R.J. (2000). An additional character useful for the identification of *Pratylenchus crenatus* Loof, 1960 (Nematoda: Pratylenchidae). *Nematology* 2, 695-697.

KARSSEN, G. & BRINKMAN, H. (2003). On the male of *Pratylenchus crenatus* Loof, 1960 (Nematoda: Pratylenchidae). *Nematology* 5, 479.

KARSSEN, G., WAEYENBERGE, L. & MOENS, M. (2000). *Pratylenchus brzeskii* sp. nov. (Nematoda: Pratylenchidae), root-lesion nematode from European coastal dunes. *Annales Zoologici (Warszawa)* 50, 255-261.

KARSSEN, G., VAN AELST, A.C., WAEYENBERGE, L. & MOENS, M. (2001). Observations on *Pratylenchus penetrans* Cobb, 1917 parasitizing the coastal dune grass *Ammophila arenaria* (L.) Link in The Netherlands. *Journal of Nematode Morphology and Systematics* 4, 1-9.

KASHAIJA, I.N., MCINTYRE, B.D., SSALI, H. & KIZITO, F. (2004). Spatial distribution of roots, nematode populations and root necrosis in highland banana in Uganda. *Nematology* 6, 7-12.

KASIMOVA, G.A. & ATAKISHIEVA, Y.Y. (1980). [The nematode fauna of *Crocus satiyus* in Apsheron.] *Izvestiya Akademii Nauk Azerbaidzhanskoi SSR Azarbajcan SSR Elmlar Akademijasynyn Habarlari, Biologicheskie Nauki* 1, 94-98.

KASIMOVA, G.A. & ATAKISHIEVA, Y.Y. (1981). [Nematode fauna and nematode dynamics on winter wheat in Apsheron.] *Parazitologicheskie issledodovaniya v Azerbaidzhane* 1, 119-134.

KATALAN-GATEVA, SH. (1980). [Structural analysis of the nematode coenosis of fruit trees in the district of Blagoevgrad. III. Nematode parasites of the roots of peaches (*Persica vulgaris*).] *Khelmintologiya* 9, 11-17.

KATALAN-GATEVA, SH. & BAICHEVA, O. (1978). [The nematode fauna of tobacco Oriental-261 in the Blagoevgrad region. II.] *Khelmintologiya, Sofia* 5, 47-59.

KATALAN-GATEVA, SH. & BUDUROVA, L. (1979). [Species of plant nematodes new for Bulgaria.] *Rastitelna Zashchita* 27, 27-28.

KATALAN-GATEVA, SH. & MILKOVA, M.TS. (1982). The nematode fauna of *Dianthus caryophyllus*. *Acta Zoologica Bulgarica* 20, 84-88.

KATALAN-GATEVA, SH.D. & NEDELCHEV, S.L. (1983). [New species of plant nematodes for Bulgaria.] *Acta Zoologica Bulgarica* 22, 76-81.

KATAN, J. (1981). Solar heating (solarization) of soil for control of pests. *Annual Review of Phytopathology* 19, 211-236.

KATAYAMA, K., KUMAGAI, T., KAI, Y., ISHIGURO, K., YOSHINAGA, M. & NAKAZAWA, Y. (2004). Daichinoyume: new sweetpotato cultivar for starch production. *Sweetpotato Research Front* 18, 4.

KATHIRESAN, T. & MEHTA, U.K. (2002a). Induction of peroxidase and acid phosphatase activities in resistant and susceptible sugarcane clones infected with *Pratylenchus zeae*. *International Journal of Nematology* 12, 98-105.

KATHIRESAN, T. & MEHTA, U.K. (2002b). Penetration, multiplication and histopathological response of *Pratylenchus zeae* in resistant and susceptible sugarcane clones. *International Journal of Nematology* 12, 189-196.

KATHIRESAN, T. & MEHTA, U.K. (2003). Changes of catalase activity in sugarcane due to early and post infection of *Pratylenchus zeae*. *Indian Journal of Nematology* 33, 116-119.

KATHIRESAN, T. & MEHTA, U.K. (2005). Effect of *Pratylenchus zeae* infection on the expression of isozyme activities in resistant and susceptible sugarcane clones. *Nematology* 7, 677-688.

KELIC, A. (1991). [Root-lesion nematodes in soybean fields.] *Znanost u Proizvodnji* 18/19, 100-105.

KEPENEKCI, I. (2001). Taxonomic investigations on the species of Tylenchida (Nematoda) in sunflower (*Helianthus annuus* L.) fields in Marmara Region. *Bitki Koruma Bulteni* 41, 101-134.

KEPENEKCI, I. (2002). A survey of Tylenchida (Nematoda) found in hazelnut (*Corylus* sp.) orchards in the west Black Sea region of Turkey. *Nematropica* 32, 83-85.

KEPENEKCI, I. (2003). Preliminary list of Tylenchida (Nematoda) associated with anise (*Pimpinella anisum* L.) in Turkey. *Pakistan Journal of Nematology* 21, 37-40.

KEPENEKCI, I. & AKGUL, H.C. (1999). Plant parasitic nematodes associated with tea (*Camellia sinensis* L.) in Rize region, Turkey. *Pakistan Journal of Nematology* 17, 181-184.

KEPENEKCI, I. & OZTURK, G. (2002). Plant parasitic nematodes of Tylenchidae (nematoda) associated with groundnut (*Arachis hypogaea*) fields in the Mediterranean region of Turkey. *Phytoparasitica* 30, 288-289.

KHAKIMOVA, L. (1978). [Parasitic nematodes of flowers.] *Uzbekskii Biologicheskii Zhurnal* 2, 64-67.

KHAN, A., KHANZADA, A.K. & ASLAM, M. (1987). Plant-parasitic nematodes associated with datepalm in Baluchistan. *Pakistan Journal of Nematology* 5, 15-17.

KHAN, A., SHAUKAT, S.S. & IFTIKHAR, A. (2001). Management of plant parasitic nematodes associated with wheat var. Kiran with oil-cakes and their effect on growth parameters. *International Journal of Nematology* 11, 229-232.

KHAN, A., WILLIAMS, K.L. & NEVALAINEN, H.K.M. (2006). Infection of plant-parasitic nematodes by *Paecilomyces lilacinus* and *Monacrosporium lysipagum*. *BioControl* 51, 659-678.

KHAN, E. & SINGH, D.B. (1975). Five new species of *Pratylenchus* (Nematoda: Pratylenchidae) from India. *Indian Journal of Nemato*logy 4, 199-211.

KHAN, F.A. & MISARI, S.M. (1992). Plant-parasitic nematodes associated with groundnut crop in four ecological zones of Nigeria. *Journal of African Zoology* 106, 263-271.

KHAN, H.A., KHAN, S.A., QAMAR, F. & SHAH, A.H. (2001). Comparative efficacy of chemicals in the control of nematodes in banana. *Pakistan Journal of Scientific and Industrial Research* 44, 231-233.

KHAN, M.L. (2000). Occurrence of root-knot nematode (*Meloidogyne incognita*) and other plant parasitic species on Kiwi fruit (*Actinidia delicious* Chev.) in Himachal Pradesh. *Indian Journal of Nematology* 30, 245.

KHAN, M.L. (2003). Plant parasitic nematodes associated with strawberry in Himachal Pradesh and varietal screening. *Indian Journal of Nematology* 33, 83-84.

KHAN, M.L. & SHARMA, G.C. (1990). Effect of temperature and moisture on population fluctuation of nematodes in an apple orchard. *Indian Journal of Nematology* 20, 1-13.

KHAN, M.L. & SHARMA, G.C. (1992a). Seasonal fluctuations of plant parasitic nematodes in almond (*Prunus amygdalus*) rhizosphere. *Indian Journal of Hill Farming* 5, 145-146.

KHAN, M.L. & SHARMA, N.K. (1992b). A description of *Pratylenchus manaliensis* sp. n. associated with apple in India (Nematoda: Pratylenchinae). *Nematologia Mediterranea* 19, 275-277.

KHAN, M.Q. & HUSSAIN, A. (2004). Plant parasitic nematodes in seed potato crop of northern areas of Pakistan. *International Journal of Biology and Biotechnology* 1, 365-367.

KHAN, M.W. (ED.) (1993). *Nematode interactions*. London, UK, Chapman & Hall, 377 pp.

KHAN, S.A., KHAN, H.A., SAEED, M. & SHAKIR, M.A. (1989). Nematodes associated with nurseries in Karachi. Part II Croton (*Codiaeum variegatum* L.) A.H.L. Juss. *Pakistan Journal of Scientific and Industrial Research* 32, 603-607.

KHAN, S.A., KHAN, H.A., QAMAR, F. & SEEMA, N. (1992). Nematodes associated with coconut nurseries in Karachi and adjoining areas. Part I. *Pakistan Journal of Scientific and Industrial Research* 35, 342-344.

KHAN, S.A., QAMAR, F., KHAN, H.A., SHAKIR, M.A. & ABID, A. (1997). Plant-parasitic nematodes associated with floricultural crops in nurseries of Karachi and their control. *Bangladesh Journal of Scientific and Industrial Research* 32, 422-426.

KHANNA, A.S. & KHAN, M.L. (1990). Nematodes associated with some ornamental plants in India. *International Nematology Network Newsletter* 7, 28-29.

KHANNA, A.S. & JEEVAN, J. (1999). Nematode fauna associated with carnation in Himachal Pradesh. *Pest Management and Economic Zoology* 7, 79-83.

KHEIRI, A. (1972). Plant parasitic nematodes (Tylenchida) from Iran. *Biologischeskie Jaarboek Dodonaea* 40, 224-239.

KHEIRI, A., BAROOTI, S. & KARIMIPOUR, H. (2002a). Tylenchida associated with field crops in Tehran and Central Provinces of Iran. *Mededelingen Faculteit Landbouwkundige en Toegepaste Biologische Wetenschappen, Universiteit Gent* 67, 707-713.

KHEIRI, A., BORHANI, A., OKHOVAT, M. & POURJAM, E. (2002b). Interaction between root lesion nematode *Pratylenchus vulnus* and two species of *Fusarium* on growth and development of maple seedlings. *Mededelingen Faculteit Landbouwkundige en Toegepaste Biologische Wetenschappen, Universiteit Gent* 67, 703-706.

KHONGA, E.B. & HILLOCKS, R.J. (1996). Soilborne diseases in corn-based cropping systems in southern Malawi: incidence and farmers' perceptions. *African Plant Protection* 2, 131-138.

KIM, J.H., JEON, Y.H., PARK, H., LEE, B.D., CHO, D.H., PARK, B.Y., KHAN, Z. & KIM, Y.H. (2006). The root-lesion nematode, *Praty-*

lenchus subpenetrans, on ginseng (*Panax ginseng*) in Korea. *Nematology* 8, 637-639.

KIM, J.I. & MINAGAWA, N. (1996). Classification of root-lesion nematodes occurring in vinyl-houses planted horticultural crops. *RDA Journal of Agricultural Science, Crop Protection* 38, 530-538.

KIMENJU, J.W., WAUDO, S.W., MWANG'-OMBE, A.W., SIKORA, R.A. & SCHUSTER, R.P. (1998). Distribution of lesion nematodes associated with corn in Kenya and susceptibility of corn cultivars to *Pratylenchus zeae*. *African Crop Science Journal* 6, 367-375.

KIMMONS, C.A., GWINN, K.D. & BERNARD, E.C. (1990). Nematode reproduction on endophyte-infected and endophyte-free tall fescue. *Plant Disease* 74, 757-761.

KIMPINSKI, J. (1982). The effect of nematicides on *Pratylenchus penetrans* and potato yields. *American Potato Journal* 59, 327-335.

KIMPINSKI, J. & DUNN, R.A. (1985). Effect of low temperatures in the field and laboratory on survival of *Pratylenchus penetrans*. *Plant Disease* 69, 526-527.

KIMPINSKI, J. & MARTIN, R.A. (1994). Nematode parasites of alfalfa on Prince Edward Island. *Forage Notes* 37, 11-12.

KIMPINSKI, J. & MCRAE, K.B. (1988). Relationship of yield and *Pratylenchus* spp. population densities in Superior and Russet Burbank potato. *Journal of Nematology* 2, 34-37.

KIMPINSKI, J. & SANDERSON, K. (2004). Effects of crop rotations on carrot yield and on the nematodes *Pratylenchus penetrans* and *Meloidogyne hapla*. *Phytoprotection* 85, 13-17.

KIMPINSKI, J. & SMITH, E.M. (1988). Nematodes in potato soils in New Brunswick. *Canadian Plant Disease Survey* 68, 147-148.

KIMPINSKI, J. & WILLIS, C.B. (1981). Influence of soil temperature and pH on *Pratylenchus penetrans* and *P. crenatus* in alfalfa and timothy. *Journal of Nematology* 13, 333-358.

KIMPINSKI, J., WALLACE, H.R. & CUNNINGHAM, R.B. (1976). Influence of some environmental factors on populations of *Pratylenchus minyus* in wheat. *Journal of Nematology* 8, 310-314.

KIMPINSKI, J., KUNELIUS, H.T. & WILLIS, C.B. (1984). Plant parasitic nematodes in temperate forage grass and legume species in Prince Edward Island. *Canadian Journal of Plant Pathology* 6, 160-164.

KIMPINSKI, J., ARSENAULT, W.J. & STURZ, A.V. (2001). Differential effect of nematicide treatments on tuber yields in early- and late-maturing potato cultivars. *Plant Pathology* 50, 509-514.

KIRKEGAARD, J., GARDNER, P., ANGUS, J. & KOETZ, E. (1994). Effect of *Brassica* break crops on the growth and yield of wheat. *Australian Journal of Agricultural Research* 45, 529-545.

KISIEL, M., HIMMELHOCH, S. & ZUCKERMAN, B.M. (1972). Fine structure of the body wall and vulva area of *Pratylenchus penetrans*. *Nematologica* 18, 234-238.

KISIEL, M., HIMMELHOCH, S. & ZUCKERMAN, B.M. (1976). Fine structure of the esophagus of *Pratylenchus penetrans*. *Journal of Nematology* 8, 218-228.

KISIEL, M.J., HIMMELHOCH, S., CASTILLO, J.M. & ZUCKERMAN, B.M. (1974). Fine structure of the intestine of *Pratylenchus penetrans*. *Nematologica* 20, 262-264.

KLINDIC, O. & PETROVIC, D. (1975). A contribution to the data on plant-parasitic nematodes associated with the roots of corn in Bosnia and Hercegovina. *Agronomski Glasnik* 37, 149-152.

KNIGHT, K.W.L. (2001). Plant parasitic nematodes associated with six subtropical crops in New Zealand. *New Zealand Journal of Crop and Horticultural Science* 29, 267-275.

KNOBLOCH, N.A. & LAUGHLIN, C.W. (1973). A collection of plant parasitic nematodes (Nematoda) from Mexico with descriptions of three new species. *Nematologica* 19, 205-217.

KNUTH, P. (2002). Nematoden im mehrjahrigen Maisanbau: Schadlinge haufig Nutzniesser schlechter Wachstumsbedingungen. *Mais* 30, 107-109.

KO, M.P., BERNARD, E.C., SCHMITT, D.P. & SIPES, B.S. (1995). Occurrence of *Pasteuria*-like organisms on selected plant-parasitic nematodes of pineapple in the Hawaiian Islands. *Journal of Nematology* 27, 395-408.

KOCH, B.L., COVEY JR, R.P. & HAGLUND, W.A. (1980). Growth response of apple and pear seedlings to pear soil fumigation with chloropicrin. *HortScience* 80, 598-600.

KOEN, H. (1967). Notes on the host range, ecology and population dynamics of *Pratylenchus brachyurus*. *Nematologica* 13, 118-124.

KOEN, H. (1969). Thermotherapeutic control of *Pratylenchus brachyurus* in potato tubers. *Phytophylactica* 1, 67-70.

KOEN, H. & HOGEWIND, W.L. (1967). Symptoms and characteristics of *Pratylenchus brachyurus* infestation on stored potatoes. *South African Journal of Agricultural Science* 10, 543-550.

KOENNING, S.R. & BARKER, K.R. (2004). Influence of poultry litter applications on nematode communities in cotton agroecosystems. *Journal of Nematology* 36, 524-533.

KOENNING, S.R., SCHMITT, D.P. & BARKER, K.R. (1985). Influence of selected cultural practices on winter survival of *Pratylenchus brachyurus* and subsequent effects on soybean yield. *Journal of Nematology* 17, 464-469.

KOLIOPANOS, C.N. & KALYVIOTIS-GAZELAS, C. (1969). List of nematodes and their host-plants as identified in Greece during 1965-1968. *Annales de l'Institut Phytopathologique Benaki* 9, 30-32.

KOLIOPANOS, C.N. & VOVLAS, N. (1977). Records of some plant parasitic nematodes in Greece with morphometrical descriptions. *Nematologia Mediterranea* 5, 207-215.

KORNOBIS, S. (1983). [Nematodes on corn crops in Wilkopolska.] *Zeszyty Problemowe Postepow Nauk Rolniczych* 278, 139-148.

KORNOBIS, S. (1987). [Plant parasitic nematodes associated with poor growth of red clover in the region Wielkopolska.] *Prace Naukowe Instytutu Ochrony Roslin* 29, 109-117.

KORNOBIS, S. (1990). [Plant-parasitic nematodes associated with retarded growth of lucerne in Wielkopolska (Poland).] *Zeszyty Problemowe Postepow Nauk Rolniczych* 391, 89-96.

KORNOBIS, S. (2000). Effects of six year herbicides use on population dynamics of migratory plant parasitic nematodes in the field. *Journal of Plant Protection Research* 40, 68-72.

KORNOBIS, S. & DOBOSZ, R. (1997). Some species of Tylenchida (Nemata) from North Korea. *Journal of Plant Protection Research* 37, 113-116.

KORNOBIS, S. & WOLNY, S. (1997). Occurrence of plant parasitic nematodes on weeds in agrobiocenosis in the Wielkopolska region in Poland. *Fundamental and Applied Nematology* 20, 627-632.

KOTCON, J.B., LORIA, R. & WIXTED, D.J. (1987). *Pratylenchus penetrans* populations dynamics on three potato cultivars. *Journal of Nematology* 19, 361-368.

KOZLOWSKA, J. & WASILEWSKA, L. (1972). [Nematodes parasites of fruit-trees in nurseries in central and northern Poland.] *Acta Agraria et Silvestria, Series Agraria* 12, 17-35.

KRAUS, W. (1995). Biologically active ingredients: azadirachtin and other triterpenoids. In: Schmutterer, H. (Ed.). *The neem tree,* Azadirachtica indica *A. Juss., and other meliaceous plants: sources of unique natural products for integrated pest management, medicine, industry and other purposes.* Weinheim, Germany, VCH, pp. 35-74.

KREIS, H.A. (1930). Freilebende terrestrische Nematoden aus der Umgebung von Peking (China). II. *Zoologischer Anzeiger* 87, 67-87.

KRISHNAMOORTHY, V. & KUMAR, N. (2004). Screening of banana hybrids (4X) against *Pratylenchus coffeae* under field conditions. *Indian Journal of Nematology* 34, 5-8.

KRUCHINA, S.N. & CHIZHOV, V.N. (1984). [Pathogenicity of root parasitic nematodes to some graminaceous fodder crops.] *Byulleten' Vsesoyuznogo Instituta Gel'mintologii im KI Skryabina* 36, 21-26.

KRUSBERG, L.R. (1960). Hydrolytic and respiratory enzymes of species of *Ditylenchus* and *Pratylenchus*. *Phytopathology* 50, 9-22.

KRUSBERG, L.R. (1963). Host response to nematode infection. *Annual Review of Phytopathology* 1, 219-240.

KUBO, R.K., INOMOTO, M.M., SILVA, R.A. & ANTEDOMENICO, S.R. (2001). Symptoms caused by *Meloidogyne incognita, Pratylenchus brachyurus* and *P. coffeae* on tubers of potato cv. Monalisa and cv. Bintje. *Summa Phytopathologica* 27, 331-333.

KUBO, R.K., SILVA, R.A., TOMAZINI, M.D., OLIVEIRA, C.M.G., MAZZAFERA, P. & INOMOTO, M.M. (2002). Patogenicidade de *Pratylenchus coffeae* em plântulas de cafeeiro cv. Mundo Novo. *Fitopatologia Brasileira* 28, 41-48.

KÜHN, J. (1890). Neure Erfahrungen auf dem Gebiete der Zuckerrübenkultur. *Jahrbuch Deutchland Landbouwkundige Gesellschaft für 1889* 4, 93-94.

KUMAR, A.C. & VISWANATHAN, P.R.K. (1972). Studies on physiological races of *Pratylenchus coffeae*. *Journal of Coffee Research* 2, 10-15.

KURPPA, S. (1988). Distribution of migratory plant parasitic nematodes in cultivated Finnish soils. *Annales Agriculturae Fenniae* 27, 315-322.

KURPPA, S. & VRAIN, T.C. (1985). Penetration and feeding behaviour of *Pratylenchus penetrans* in strawberry roots. *Revue de Nématologie* 8, 273-276.

KUSANO, M., KOSHINO, H., UZAWA, J., FUJIOKA, S., KAWANO, T. & KIMURA, Y. (2000). Nematicidal alkaloids and related compounds produced by the fungus *Penicillium* cf. *simplicissimum*. *Bioscience, Biotechnology and Biochemistry* 64, 2559-2568.

KWON, T.Y., KIM, D.G., CHOI, C.D. & YOON, J.T. (2004). Survey on plant-parasitic nematodes in peach orchards in Gyeongbuk Province. *Korean Journal of Applied Entomology* 43, 81-84.

LAL, A. & KHAN, E. (1990). Species of *Pratylenchus* Filipjev, 1936 and *Helicotylenchus* Steiner, 1945 (Nematoda: Tylenchida) found associated with forest trees in northern India. *Indian Journal of Nematology* 19, 44-50.

LAMBERTI, F. (1969). Nematodi parassiti dell'olivo. *Informatore Fitopatologico* 15-16, 1-4.

LAMBERTI, F., TACCONI, R., MARINARI, A., D'ERRICO, F.P. & BASILE, M. (1987). I principali nematodi parassiti delle colture floricole in Italia relativa lotta. *Difesa delle Piante* 10, 77-84.

LAMBERTI, F., CICCARESE, F., SASANELLI, N., AMBRICO, A., D'ADDABBO, T. & SCHIAVONE, D. (2001). Relationships between plant parasitic nematodes and *Verticillium dahliae* on olive. *Nematologia Mediterranea* 29, 3-9.

LAMBERTI, F., CICCARESE, F., SASANELLI, N., AMBRICO, A., D'ADDABBO, T. & SCHIAVONE, D. (2002). Relationships between plant parasitic nematodes and *Verticillium dahliae* on olive. *Acta Horticulturae* 586, 749-752.

LAMONDIA, J.A. (1999a). The effects of *Pratylenchus penetrans* on strawberry vigor and yield. *Journal of Nematology* 31, 418-423.

LAMONDIA, J.A. (1999b). Influence of rotation crops on the strawberry pathogens *Pratylenchus penetrans*, *Meloidogyne hapla*, and *Rhizoctonia fragariae*. *Journal of Nematology* 31, 650-655.

LAMONDIA, J.A. (2002). Seasonal populations of *Pratylenchus penetrans* and *Meloidogyne hapla* in strawberry roots. *Journal of Nematology* 34, 409-413.

LAMONDIA, J.A. (2003). The interaction of *Pratylenchus penetrans* and *Rhizoctonia fragariae* in strawberry black root rot. *Journal of Nematology* 35, 17-22.

LAMONDIA, J.A. & COWLES, R.S. (2002). Effect of entomopathogenic nematodes and *Trichoderma harzianum* on the strawberry black root rot pathogens *Pratylenchus penetrans* and *Rhizoctonia fragariae*. *Journal of Nematology* 34, 351-357.

LAMONDIA, J.A. & COWLES, R.S. (2005). Comparison of *Pratylenchus penetrans* infection and *Maladera castanea* feeding on strawberry root rot. *Journal of Nematology* 37, 131-135.

LANGE, B., MEYER, J. & BURMEISTER, P. (1964). VorEiufige ermittlungen über schäden an tabak durch freilebende nematoden und versuche zu ihrer Bekämpfung. *Mitteilungen aus der Biologischen Bundesanstalt* 111, 85-93.

LAWN, D.A. & SAYRE, K.D. (1992). Soilborne pathogens on cereals in a highland location of Mexico. *Plant Disease* 76, 149-154.

LAWN, D.A., NOEL, G.R. & SINCLAIR, J.B. (1988a). Plant-parasitic nematodes and *Neocosmospora vasinfecta* var. *africana* associated with soybeans in the Republic of Zambia. *Nematropica* 18, 33-34.

LAWN, D.A., NOEL, G.R. & SINCLAIR, J.B. (1988b). Plant-parasitic nematodes associated with sunflower and corn in the Republic of Zambia. *Nematropica* 18, 143-154.

LAX, P., DOUCET, M., DI RIENZO, J.A., PINOCHET, J. & BAUJARD, P. (2004). Inter-population variability in *Pratylenchus vulnus* Allen & Jensen, 1951 (Nematoda: Tylenchida). *Nematology* 6, 257-260.

LAYNE, R.E. (1987). Peach rootstocks. In: Rom, R.C. & Carlson, R.F. (Eds). *Rootstocks for fruit crops*. New York, NY, USA, Wiley & Sons, pp. 385-416.

LEE, J.K., CHOI, D.R., PARK, B.Y. & KIM, D.G. (2006). Occurrence of root-lesion nematodes *Pratylenchus* spp., from chrysanthemum in Korea. *Journal of Nematology* 38, 279. [Abstr.]

LEHMAN, P.S. (1996). Nematology section. *Triology Technical Report* 35, 10-11.

LEVINE, A., TEHAKEN, R., DIXON, R. & LAMB, C. (1994). H_2O_2 from the oxidative burst orchestrates the plant hypersensitive disease resistance response. *Cell* 79, 583-593.

LEWIS, S.A. (1987). Nematode-plant compatibility. In: Veech, A. & Dickson, D.W. (Eds). *Vistas on nematology*. Hyattsville, MD, USA, Society of Nematologists, pp. 246-252.

LEWIS, S.A., SKIPPER, H.D. & MUSEN, H.L. (1977). Nematode and nodulation studies in coastal plain soybean fields of South Carolina. *Journal of Nematology* 9, 275.

LEYNS, L., BORGONIE, G., ARNAUT, G. & DE WAELE, D. (1995). Nematicidal activity of *Bacillus thuringiensis* isolates. *Fundamental and Applied Nematology* 18, 211-218.

LI, D.F., ZHANG, X.Y. & WANG, M.X. (1996). Test on hot water treatment to eradicate parasitic nematodes in imported corm of *Cycas revoluta*. *Plant Protection* 22, 44-45.

LI, D.Z. (1985). Description of some species of parasitic nematodes of the genus *Pratylenchus* on plant roots in Sichuan Province. *Journal of Southwest Agricultural College* 2, 49-51.

LINDSEY, D.W. & CAIRNS, E.J. (1971). Pathogenicity of the lesion nematode, *Pratylenchus brachyurus*, on six soybean cultivars. *Journal of Nematology* 3, 220-226.

LINFORD, M.B. & OLIVEIRA, J.M. (1937). The feeding of hollow-spear nematodes on other nematodes. *Science* 85, 295-297.

LISKOVA, M. (1980). [Influence of the soil types on the composition of nematode communities in the rhizosphere of *Vitis vinifera* L. in eastern Slovakia.] *Biologia, Bratislava* 35, 561-566.

LISKOVA, M., SABOVA, M. & VALOCKA, B. (1988). [Distribution and abundance of parasitic nematodes in cereal areas of the Slovak Socialist Republic.] *Pol'nohospodarstvo* 34, 842-848.

LIU, G. & ZHANG, S. (1999). [Identification of parasitic nematodes on longan in Fujian, China.] *Journal of Fujian Agricultural University* 28, 59-65.

LIU, Z.M. & FENG, Z.X. (1995). [Six new records of plant nematodes in China.] *Journal of Guanxi Agricultural University* 14, 121-124.

LOBANOV, A.L., STEPANJANTS, S.D. & DIANOV, M.B. (1996). Dialogue computer system BIKEY as applied to diagnostics of Cnidaria (illustrated on example of hydroids of the genus *Symplectoscyphus*). *Scientia Marina, Special volume. Advances in Hydrozoan Biology* 60, 211-220.

LOCASCIO, S.J., GILREATH, J.P., DICKSON, D.W., KUCHAREK, T.A., JONES, J.P. & NOLING, J.W. (1997). Fumigant alternatives to methyl bromide for polyethylene-mulched tomato. *HortScience* 32, 1208-1211.

LOLAS, M.A., MERRIFIELD, K.J., PINKERTON, J.K. & INGHAM, R.E. (1992). Effect of fenamiphos on population dynamics of *Pratylenchus penetrans* and *Xiphinema americanum* in Oregon red raspberry fields. *Journal of Nematology* 24, 604. [Abstr.]

LOOF, P.A.A. (1960). Taxonomic studies on the genus *Pratylenchus* (Nematoda). *Tijdschrift ober Plantenziekten* 66, 29-90.

LOOF, P.A.A. (1964). Free-living and plant parasitic nematodes from Venezuela. *Nematologica* 10, 201-300.

LOOF, P.A.A. (1974). *Pratylenchus pratensis. CIH Descriptions of plant-parasitic nematodes*, Set 4, No. 52. St Albans, UK, Commonwealth Agricultural Bureaux, 2 pp.

LOOF, P.A.A. (1978). The genus *Pratylenchus* Filipjev, 1936 (Nematoda: Pratylenchidae): a review of its anatomy, morphology, distribution, systematics and identification. *Sveriges Landbruksuniversiteit Växtskyddsrapporter* 5, 1-12.

LOOF, P.A.A. (1985). *Pratylenchus scribneri. CIH Descriptions of plant-parasitic nematodes*, Set 8, No. 110. St Albans, UK, Commonwealth Agricultural Bureaux, 4 pp.

LOOF, P.A.A. (1991). The family Pratylenchidae Thorne, 1949. In: Nickle, W.R. (Ed.). *Manual of agricultural nematology*. New York, NY, USA, Marcel Dekker, pp. 363-421.

LOOF, P.A.A. & YASSIN, A.M. (1971). Three new plant-parasitic nematodes from the Sudan, with notes on *Xiphinema basiri* Siddiqi, 1959. *Nematologica* 16, 537-546.

LOOS, C.A. (1953). Eelworms. *Tea Quarterly* 24, 34-38.

LÓPEZ, A., PINOCHET, J., FERNÁNDEZ, C., CALVET, C. & CAMPRUBÍ, A. (1997). Growth response of OHF-333 pear rootstock to arbuscular mycorrhizal fungi, phosphorus nutrition and *Pratylenchus vulnus* infection. *Fundamental and Applied Nematology* 20, 87-93.

LOPEZ, R. & SALAZAR, L. (1990). Morfología de algunos *Pratylenchus* spp. (Nemata: Pratylenchidae) encontrados en Costa Rica, vista con el microscopio electrónico de rastreo. *Agronomia Costarricense* 14, 189-195.

LORDELLO, L.G.E. (1956). Sobre un nematódeo do género *Pratylenchus*, parasito das raízes de *Allium cepa*. *Revista de Agricultura* 31, 181-188.

LORDELLO, L.G.E. (1967). A root-lesion nematode attacking Eucalyptus trees. *Revista de Agricultura* 42, 11-13.

LORDELLO, L.G.E. (1968). Nematoides associados a uma doenca do cacaueiro. *Revista de Agricultura* 43, 1.

LORDELLO, L.G.E. (1973). Nematoide nocivo ao pessegueiro. *Revista de Agricultura* 48, 128.

LORDELLO, L.G.E. & MEHLO FILHO, A.T. (1969a). A capim pangola difunde nematoides. *Revista de Agricultura* 44, 1.

LORDELLO, L.G.E. & MEHLO FILHO, A.T. (1969b). Capins gordura e jaragua, hospedeiros novos de um nematoide migrador. *Revista o Solo* 61, 27.

LORDELLO, L.G.E. & MEHLO FILHO, A.T. (1970). Mais um nematoide ataca a cafeeiro. *Revista de Agricultura* 45, 1.

LORDELLO, L.G.E. & MENDONCA, M.M. (1970). Nematodes associated with roots of sugar cane in Piaui, Brazil. *Revista Brasileira de Biolgia* 30, 617-618.

LORDELLO, L.G.E., ZAMITH, A.P.L. & BOOCK, O.J. (1954). Novo nematódeo parasito da batatinha. *Bragantia* 13, 141-149.

LORDELLO, L.G.E., ZAMITH, A.P.L. & ARRUDA, H.V. (1958). Nematode parasites of soybean and cotton roots and its implication in crop rotation. *Revista de Agricultura* 33, 161-166.

LORDELLO, L.G.E., ZAMITH, A.P.L. & BOOCK, O.J. (1961). Two nematodes found attacking potato in Cochabamba, Bolivia. *Anais de Academia Brasileira de Ciencias* 33, 209-215.

LORDELLO, L.G.E., MONTEIRO, A.R. & D'ARCE, R.D. (1968). Distribuicao geografica dos nematoides nocivos ao cafeeiro. *Revista de Agricultura* 43, 79-82.

LORDELLO, R.R.A., LORDELLO, A.I.L., SAWAZAKI, E. & ALOISI SOBRINHO, J. (1985). Resistance of corn in the field to *Pratylenchus* infection. *Nematologia Brasileira* 9, 163-173.

LORDELLO, R.R.A., LORDELLO, A.I.L. & SAWAZAKI, E. (1992). Flutacao e controle de *Pratylenchus* spp. em milho. *Summa Phytopathologica* 18, 146-152.

LOWNSBERY, B.F. & SHER, E.F. (1963). Fruit and nut tree rootstocks as hosts for a root-lesion nematode, *Pratylenchus vulnus*. *Proceedings of the American Society for Horticultural Science* 82, 250-254.

LOWNSBERY, B.F., MOODY, E.H. & BRAUN, A.L. (1974). Plant-parasitic nematodes in California prune orchards. *Plant Disease Reporter* 58, 633-636.

LUC, M. (1958). Les nématodes et le flétrissement des cotonniers dans le Sud-Ouest de Madagascar. *Coton et Fibres Tropicales* 13, 1-18.

LUC, M. (1987). A reappraisal of Tylenchina (Nemata). 7. The family Pratylenchidae Thorne, 1949. *Revue de Nématologie* 10, 203-218.

LUC, M., BALDWIN, J.G. & BELL, A.H. (1986). *Pratylenchus morettoi* n. sp. (Nemata: Pratylenchidae). *Revue de Nématologie* 9, 119-123.

MACDONALD, D.H. (1966). The relationships between the mineral nutrition of the host and the buildup of *Pratylenchus penetrans*. *Disease Abstracts* 26, 6286-6287.

MACGOWAN, J.B. (1978). The lesion nematode, *Pratylenchus brachyurus*, infecting citrus. *Nematology Circular. Florida Department of Agriculture & Consumer Services No. 42*, 2 pp.

MACGUIDWIN, A. E. (1989). Distribution of *Pratylenchus scribneri* between root and soil habitats. *Journal of Nematology* 21, 409-415.

MACGUIDWIN, A.E. & FORGE, T.A. (1991). Winter Survival of *Pratylenchus scribneri*. *Journal of Nematology* 23, 198-204.

MACGUIDWIN, A.E. & ROUSE, D.I. (1990). Role of *Pratylenchus penetrans* in the potato early dying disease of Russet Burbank potato. *Phytopathology* 80, 1077-1082.

MACGUIDWIN, A.E. & STANGER, B.A. (1991). Changes in vertical distribution of *Pratylenchus scribneri* under potato and corn. *Journal of Nematology* 23, 73-81.

MACHADO, A.C.Z. & INOMOTO, M.M. (2001). Reacción de dieciocho hortalizas a *Pratylenchus brachyurus*. *Nematropica* 31, 259-265.

MACHON, J.E. & HUNT, D.J. (1985). *Pratylenchus goodeyi*. *CIH Descriptions of plant-parasitic nematodes*, Set 8, No. 120. St Albans, UK, Commonwealth Agricultural Bureaux, 3 pp.

MAGAREY, R.C. & BULL, J.I. (1998). Replant disease of sugarcane. *Acta Horticulturae* 477, 135-142.

MAGGENTI, A.R. & VIGLIERCHIO, D.R. (1975). *Sequoia sempervirens* and *Sequoiadendron giganteum*: hosts of common plant-parasitic nematodes of California. *Plant Disease Reporter* 59, 116-121.

MAHARAJU, D. & DAS, V.M. (1981). *Pratylenchus nizamabadensis* n. sp. (Nematoda: Tylenchidae) from Andhra Pradesh. *Proceedings of the Indian Academy of Parasitology* 2, 24-25.

MAI, W.F. & ABAWI, G.S. (1981). Controlling replant diseases of pome and stone fruits in northeastern United States by preplant fumigation. *Plant Disease* 65, 859-864.

MAI, W.F. & ABAWI, G.S. (1987). Interactions among root-knot nematodes and fusarium wilt fungi on host plants. *Annual Review of Phytopathology* 25, 317-338.

MAI, W.F. & PARKER, K.G. (1972). Root diseases of fruit trees in New York State. IV. Influence of preplant treatment with a nematicides, charcoal from burned brush, and complete mineral nutrition on growth and yields of sour cherry trees and numbers of *Pratylenchus penetrans*. *Plant Disease Reporter* 56, 141-145.

MAI, W.F., CRITTENDEN, H.W. & JENKINS, W.R. (1960). *Distribution of stylet-bearing nematodes in the north-eastern United States*. Bulletin New Jersey Agricultural Experiment Station 795, 62 pp.

MAI, W.F., BLOOM, J.R. & CHERT, T.A. (1977). *Biology and ecology of the plant-parasitic nematode* Pratylenchus penetrans. Pennsylvania State University Bulletin No. 815, University Park, 64 pp.

MAI, W.F., MERWIN, I.A. & ABAWI, G.S. (1994). Diagnosis, etiology and management of replant disorders in New York cherry and apple orchards. *Acta Horticulturae* 363, 33-41.

MALO, S.E. (1964). Pathogenic nematodes of peach trees in Florida and others of potential importance. *Proceedings of the Florida Station of Horticulture* 76, 377-379.

MAMIYA, Y. (1969). Plant parasitic nematodes associated with coniferous seedlings in forest nurseries in eastern Japan. *Bulletin of the Government Forest Experiment Station* 219, 95-119.

DE MAN, J.G. (1880) Die einheimischen, frei in der reinen Erde und im süssen Wasser lebenden Nematoden. Vorläufiger Bericht und descriptiv-systematischer Theil. *Tijdschrift der Nederlandsche Dierkundige Vereeniging* 5, 1-104.

DE MAN, J.G. (1884). *Die frei in der reinen Erde und im süssen Wasser lebenden Nematoden der niederländischen Fauna*. Leiden, The Netherlands, Brill, 206 pp.

MANCINI, G. & MORETTI, F. (1974). The infestation of *Lactuca sativa* and *Galinsoga parviflora* by *Pratylenchus penetrans*. *Informatore Fitopatologico* 24, 23-26.

MANI, A. (1998). Plant parasitic nematodes intercepted from imported plants and organic materials. *Nematologia Mediterranea* 26, 93-95.

MANI, A. (1999). Survival of the root-lesion nematode *Pratylenchus jordanensis* Hashim in a fallow field after harvest of alfalfa. *Nematology* 1, 79-84.

MANI, A. & AL-HINAI, M.S. (1997). Plant-parasitic nematodes associated with alfalfa and fluctuations of *Pratylenchus jordanensis* populations in the Sultanate of Oman. *Fundamental and Applied Nematology* 20, 443-447.

MANI, A. & AL-HINAI, M.S. (1998). Toxicity of harmal, *Rhazya stricta*, to *Meloidogyne incognita* and *Pratylenchus jordanensis*. *Nematologia Mediterranea* 26, 27-30.

MANI, A. & AL-HINAI, M.S. (2003). Observations on sampling tool, vertical distribution and weed hosts of *Pratylenchus jordanensis* in an alfalfa field. *Nematologia Mediterranea* 31, 155-156.

MANI, A., AL-HINAI, M.S. & HANDOO, Z.A. (1997). Occurrence, population density, and distribution of root-lesion nematodes, *Pratylenchus* spp., in the Sultanate of Oman. *Nematropica* 27, 209-219.

MANI, A., AL-HINAI, M.S. & HANDOO, Z.A. (1998). Plant-parasitic nematodes of crops in Dhofar Governorate, Sultanate of Oman. *Nematropica* 28, 61-69.

MANI, A., HANDOO, Z.A. & LIVINGSTON, S. (2005). Plant-parasitic nematodes associated with date palm trees (*Phoenix dactylifera*) in the Sultanate of Oman. *Nematropica* 35, 135-143.

MANZANILLA-LÓPEZ, R., EVANS, K. & BRIDGE, J. (2004). Plant diseases caused by nematodes. In: Cheng, Z.X., Chen, S.Y. & Dickson, D.W. (Eds). *Nematology advances and perspectives, volume 2. Nematode management and utilization.* Wallingford, UK, CABI Publishing, pp. 637-716.

MARBÁN-MENDOZA, N. & FLORES, L. (1993). Prospeccion fitonematologica en ornamentales de follaje de Costa Rica. *Manejo Integrado de Plagas* 32, 1-4.

MARCINOWSKI, K. (1909). Parasitisch und semiparasitisch an Pflanzen lebende Nematoden. *Arbeiten aus der Kaiserlichen Biologischen Anstalt für Land-und Forstwirtschaft* 7, 1-192.

MARTIN, G.C. (1972). Potatoes and plant parasitic nematodes. *Stocks and Crops* 3, 38-39.

MARTIN, M.J., RIEDEL, R.M. & ROWE, R.C. (1982). *Verticillium dahliae* and *Pratylenchus penetrans*: interactions in the early dying complex of potato in Ohio. *Phytopathology* 72, 640-644.

MARTIN, W.J., NEWSON, L.D. & SCHWEGMANN, J. (1951). Nematode root rot of cotton in Louisiana. *Plant Disease Reporter* 35, 388.

MARTINEZ, A.A., LORDELLO, L.G.E. & LORDELLO, R.R.A. (1972). Nota sobre os nematoides que atacam a seringueira no estado de Sao Paulo. *Revista de Agricultura* 47, 159-160.

MARULL, J. & PINOCHET, J. (1991). Host suitability of *Prunus* rootstocks to four *Meloidogyne* species and *Pratylenchus vulnus* in Spain. *Nematropica* 21, 185-195.

MATHUR, V.K. & MCLEOD, R.W. (1977). Plant parasitic nematodes in rice soils in New South Wales. *Search* 8, 130.

MAUZA, B.E. & WEBSTER, J.M. (1982). Suppression of alfalfa growth by concomitant populations of *Pratylenchus penetrans* and two *Fusarium* species. *Journal of Nematology* 14, 364-367.

MAYER, A., KILIAN, M., HOSTER, B., STERNER, O. & ANKE, H. (1999). *In-vitro* and *in-vivo* nematicidal activities of the cyclic dodecapeptide omphalotin A. *Pesticide Science* 55, 27-30.

MAYER, E. (1969). *Principles of systematic zoology*. New York, NY, USA, McGraw-Hill, 428 pp.

MAZZAFERA, P., KUBO, R.K. & INOMOTO, M.M. (2004). Carbon fixation and partitioning in coffee seedlings infested with *Pratylenchus coffeae*. *European Journal of Plant Pathology* 110, 861-865.

MAZZOLA, M. (1998). Elucidation of the microbial complex having a causal role in the development of apple replant disease in Washington. *Phytopathology* 88, 930-938.

MAZZOLA, M. & GU, Y.H. (2000). Impact of wheat cultivation on microbial communities from replant soils and apple growth in greenhouse trials. *Phytopathology* 90, 114-119.

MAZZOLA, M., GRANATSTEIN, D.M., ELFVING, D.C. & MULLINIX, K. (2001). Suppression of specific apple root pathogens by *Brassica napus* seed meal amendment regardless of glucosinolate content. *Phytopathology* 91, 673-679.

MAZZOLA, M., GRANATSTEIN, D.M., ELFVING, D.C., MULLINIX, K. & GU, Y.H. (2002). Cultural management of microbial community structure to enhance growth of apple in replant soils. *Phytopathology* 92, 1363-1366.

MBWANA, A.S.S., WAUDO, S.W. & SESHU-REDDY, K.V. (1995). Host-range of the lesion nematode, *Pratylenchus goodeyi*, commonly found in highland bananas of East Africa. *International Journal of Pest Management* 41, 46-49.

MC DONALD, A.H. & NICOL, J.M. (2005). Nematode parasites of cereals. In: Luc, M., Sikora, R.A. & Bridge, J. (Eds). *Plant parasitic nematodes in subtropical and tropical agriculture*, 2nd edition. Wallingford, UK, CABI Publishing, pp. 131-191.

MC DONALD, A.H. & VAN DEN BERG, E.H. (1993). Effect of watering regimen on injury to corn and grain sorghum by *Pratylenchus* species. *Journal of Nematology* 25, 654-658.

MCINTYRE, J.L. & MILLER, P.M. (1978). Protection of tobacco against *Phytophthora parasitica* var. *nicotianae* by cultivar-nonpathogenic races, cell-free sonicates, and *Pratylenchus penetrans*. *Phytopathology* 68, 235-239.

MCKENRY, M.V. & KRETSCH, J. (1987). Survey of nematodes associated with almond production in California. *Plant Disease* 71, 71-73.

MCKENRY, M.V. & KRETSCH, J.O. (1984). Nematodes in pistachio orchards. *California Agriculture* 38, 21.

MCKENRY, M.V. (1989). Nematodes. In: LaRue, J.H. & Scout Johnson, R. (Eds). *Peaches, plums and nectarines. Growing and handling for fresh market*. Oakland, CA, USA, Publication 3331, Cooperative Extension Service, University of California, Division of Agriculture and Natural Resources, pp. 139-147.

MCLEOD, R.W. (1978). A survey of plant parasitic nematodes in vineyards in the Murrumbidgee Irrigation Areas. *Plant disease survey 1976-77, Biology Branch Articles and notes on the occurrence of plant diseases in New South Wales for the twelve months ending 30th June*, pp. 26-30.

MCSORLEY, R. (2001). Multiple cropping systems for nematode management: A review. *Soil and Crop Science Society of Florida Proceedings* 60, 132-142.

MCSORLEY, R. & CAMPBELL, C.W. (1980). Relationship between nematode density and weed density in avocado groves. *Nematropica* 10, 96-102.

MCSORLEY, R. & DICKSON, D.W. (1989). Effects and dynamics of a nematode community on corn. *Journal of Nematology* 21, 462-471.

MCSORLEY, R. & DICKSON, D.W. (1990a). Vertical distribution of plant-parasitic nematodes in sandy soil under soybean. *Journal of Nematology* 22, 90-96.

MCSORLEY, R. & DICKSON, D.W. (1990b). Vertical distribution of plant-parasitic nematodes in sandy soil under corn. *Plant and Soil* 123, 95-100.

MEAGHER, J.W. (1969). Nematodes and their control in vineyards in Victoria, Australia. *International Pest Control (Sept-Oct. 1969)*, 5 pp.

MEHTA, U.K. & SUNDARARAJ, P. (1990). Distribution of seven species of *Pratylenchus* in sugarcane fields of Tamil Nadu. *Indian Journal of Nematology* 20, 211-213.

MEHTA, U.K. & SUNDARARAJ, P. (1995). Efficacy of some new soil amendments for the control of the lesion nematode in sugarcane. *Nematologia Mediterranea* 23, 321-323.

MEHTA, U.K. & SUNDARARAJ, P. (1999). Field evaluation of organic amendments for control of *Pratylenchus zeae* in sugarcane. *International Journal of Nematology* 9, 87-89.

MEHTA, U.K., SUNDARARAJ, P. & NATESAN, N. (1992). Horizontal and vertical distribution of *Pratylenchus zeae* for five consecutive years in a south Indian sugarcane field. *Afro-Asian Journal of Nematology* 2, 67-73.

MEHTA, U.K., NATESAN, N. & SUNDARARAJ, P. (1994). Screening of sugarcane cultivars to *Pratylenchus zeae* fro commercial release. *Afro-Asian Journal of Nematology* 4, 109-111.

MEJIA-ANAYA, R. & CANTO SAENZ, M. (1985). Interaction between *Pratylenchus flakkensis* and *Pseudomonas solanacearum* race 3 on potato. *Nematropica* 15, 120. [Abstr.]

MELAKEBERHAN, H. (2004). Physiological interactions between nematodes and their host plants. In: Cheng, Z.X., Chen, S.Y. & Dickson, D.W. (Eds). *Nematology advances and perspectives, volume 2. Nematode management and utilization.* Wallingford, UK, CABI Publishing, pp. 771-794.

MELAKEBERHAN, H. (2006). Fertiliser use efficiency of soybean cultivars infected with *Meloidogyne incognita* and *Pratylenchus penetrans*. *Nematology* 8, 129-137.

MELAKEBERHAN, H. & DEY, J. (2003). Competition between *Heterodera glycines* and *Meloidogyne incognita* or *Pratylenchus penetrans*: independent infection rate measurements. *Journal of Nematology* 35, 1-6.

MELAKEBERHAN, H. & WEBSTER, J.M. (1993). The phenology of plant-nematode interaction and yield loss. In: Khan, M.W. (Ed.). *Nematode interactions*. London, UK, Chapman & Hall, pp. 26-41.

MELAKEBERHAN, H., BIRD, G.W. & GORE, R. (1997). Impact of plant nutrition on *Pratylenchus penetrans* infection of *Prunus avium* rootstocks. *Journal of Nematology* 29, 381-388.

MELAKEBERHAN, H., BIRD, G.W. & PERRY, R. (1994). Plant-parasitic nematodes associated with cherry rootstocks in Michigan. *Journal of Nematology* 26, 767-772.

MELAKEBERHAN, H., JONES, A.L. & BIRD, G.W. (2000). Effects of soil pH and *Pratylenchus penetrans* on the mortality of 'Mazzard' cherry seedlings and their susceptibility to *Pseudomonas syringae* pv. *syringae*. *Canadian Journal of Plant Pathology* 22, 131-137.

MELLO FILHO, A.T. & LORDELLO, L.G.E. (1970). Causas do declino da cultura do feijao no norte do Parana. *O Solo (Piracicaba)* 62, 15.

MENDES, M.L., PRIA, M.D., GEUS-BOUWMAN, D.M., VICENTE, F.R. & TOMASELLI, F. (2001). Ocorrencia de *Pratylenchus penetrans* (Cobb, 1917)

Chitwood & Oteifa, 1952, em mandioquinha-salsa (*Arracacia xanthorrhiza* Bancr.) no municipio de Castro, PR. *Nematologia Brasileira* 25, 85-87.

MENNAN, S. & HANDOO, Z.A. (2006). Plant-parasitic nematodes associated with cabbages (*Brassica* spp.) in Samsun (Middle Black Sea Region), Turkey. *Nematropica* 36, 99-104.

MERNY, G. (1970). Les nématodes phytoparasites des rizières inondées en Côte d'Ivoire 1. – les espèces observées. *Cahier ORSTOM, Série Biologie* 11, 3-43.

MERRIFIELD, K.J. & INGHAM, R.E. (1996). Population dynamics of *Pratylenchus penetrans*, *Pratylenchus* sp., and *Criconemella xenoplax* on Western Oregon peppermint. *Journal of Nematology* 28, 557-564.

MERZHEEVSKAYA, O.I. (1951). [New species of nematodes.] *Sbornik Nauchnykh Trudov Akademii Nauk Belorusskoi SSR, Institut Bioloogia* 2, 112-120.

MESKINE, M. & ABBAD ANDALOUSI, F. (1993). Importance et repartition des nématodes phythoparasites associés aux cultures de blé et orge au Maroc. *Al-Awamia* 80, 123-134.

MEYER, A.J. (1985). Mass culture of *Pratylenchus zeae* (Nematoda; Pratylenchinae) on excised corn roots growing on sterile nutrient agar. *Phytophylactica* 16, 259-261.

MEYL, A.H. (1954). Die bisher in Italien defundenen freilebenden Erd- und Süsswasser-Nematoden. *Archivii di Zoologia Italiana* 39, 161-264.

MEYL, A.H. (1961). Die freilebenden Erd- und Süsswasser-Nematoden (Fadenwürmer). In: *Die Tierwelt Mitteleuropas*. Leipzig, Germany, Verlag Quelle & Meyer, 164 pp.

MIAN, J.H., GODOY, G., SHELBY, R.A., RODRÍGUEZ-KÁBANA, R. & MORGAN-JONES, G. (1982). Chitin amendments for control of *Meloidogyne arenaria* in infested soil. *Nematropica* 12, 71-84.

MILINKOVITCH, M.C. & LYONS-WEILER, J. (1998). Finding optimal ingroup topologies and convexities when the choice of outgroups is not obvious. *Molecular Phylogenetics and Evolution* 9, 348-357.

MILLER, P.M. (1975). Effect of *Pratylenchus penetrans* on subsequent growth of tomato plants. *Plant Disease Reporter* 59, 866-867.

MILLER, P.M. (1977). Interaction of plastic, hay and grass mulches, and metam-sodium on control of *Pratylenchus penetrans* in tomatoes. *Journal of Nematology* 9, 350-351.

MILLER, P.M. (1978). Reproduction, penetration, and pathogenicity of *Pratylenchus penetrans* on tobacco, vegetables, and cover crops. *Phytopathology* 68, 1502-1504.

MILLER, P.M. & RICH, S. (1980). Injury to sweet corn by a *Longidorus* sp. in Connecticut. *HortScience* 15, 29-30.

MINAGAWA, N. (1982). Descriptions of *Pratylenchus gibbicaudatus* n. sp. and *P. macrostylus* Wu, 1971 (Tylenchida: Pratylenchidae) from Kyushu. *Japanese Journal of Applied Entomology and Zoology* 17, 418-423.

MINAGAWA, N. (1991). Descriptions of two new species of *Pratylenchus* (Tylenchida: Pratylenchidae) from Japan. *Japanese Journal of Nematology* 21, 18-25.

MINAGAWA, N. & MAESO-TOZZI, D. (1990). Plant-parasitic nematodes of Uruguay: a preliminary report. *Japanese Journal of Nematology* 20, 44-50.

MINAGAWA, N., AIBA, S., KATAYAMA, K. & MIURA, K. (2004). Effect of field solarization on control of soil nematodes. *Bulletin of the National Agricultural Research Center* 4, 25-34.

MINTON, N.A. (1965). Reaction of white clover and five other crops to *Pratylenchus scribneri*. *Plant Disease Reporter* 49, 856-859.

MINTON, N.A., CAIRNS, E.J., MINTON, E.B. & HOPPER, B.E. (1963). Occurrence of plant-parasitic nematodes in Alabama. *Plant Disease Reporter* 47, 743-745.

MINTON, N.A., CSINOS, A.S. & MORGAN, L.W. (1990). Relationship between tillage and nematicide, fungicide, and insecticide treatments on pests and yield of peanuts double-cropped with wheat. *Plant Disease* 74, 1025-1029.

MINZ, G. (1957). Free-living plant-parasitic and possible plant-parasitic nematodes in Israel. *Plant Disease Reporter* 41, 92-94.

MISHRA, C. & SASMAL, B.C. (1988). Pathogenicity of *Pratylenchus coffeae* on jute. *Nematologia Mediterranea* 16, 133.

MITREVA, M., ELLIN, A.A., DANTE, M., KLOEK, A.P., KALYANARAMAN, A., ALURU, S., CLIFTON, S.W., BIRD, D.M., BAUM, T.J. & MCCARTER, J.P. (2004). A survey of SL1-spliced transcripts from the root-lesion nematode *Pratylenchus penetrans*. *Molecular Genetics and Genomics* 272, 138-148.

MIZUKUBO, T. (1992). Morphological and statistical differentiation of *Pratylenchus coffeae* complex in Japan (Nematoda: Pratylenchidae). *Applied Entomology and Zoology* 27, 213-224.

MIZUKUBO, T. (1995). Evidence for *Pratylenchus coffeae* races in differential reproduction on fifteen cultivars (Nematoda: Pratylenchidae). *Japanese Journal of Nematology* 25, 85-93.

MIZUKUBO, T. & ADACHI, H. (1997). Effect of temperature on *Pratylenchus penetrans* development. *Journal of Nematology* 29, 306-314.

MIZUKUBO, T. & MINAGAWA, N. (1991). Morphometric differentiation of *Pratylenchus neglectus* (Rensch, 1924) and *P. gotohi* n. sp. (Nematoda: Pratylenchidae). *Japanese Journal of Nematology* 21, 26-42.

MIZUKUBO, T., TOIDA, Y., KEEREEWAN, S. & YOSHIDA, M. (1990). *Pratylenchus subranjani* n. sp. (Nematoda: Pratylenchidae) from corn in Thailand. *Applied Entomology and Zoology* 25, 311-318.

MIZUKUBO, T., ORUI, Y. & MINAGAWA, N. (1997). Morphology and molecular characteristics of *Pratylenchus japonicus* (Ryss, 1988) n. stat. (Nematoda, Pratylenchidae). *Esakia* 37, 203-214.

MIZUKUBO, T., ORUI, Y., HANADA, K. & SANO, Z. (2003). Microevolutionary trend in *Pratylenchus coffeae* sensu stricto (Nematoda: Pratylenchidae): the diversity in PCR-RFLP phenotype, compatibility on host plants and reproductive segregation. *Japanese Journal of Nematology* 33, 57-76.

MOHOTTI, K.M., NAVARATNE, N. & NUGALIYADDA, M. (1997). New record on the occurrence of walnut root lesion nematode, *Pratylenchus vulnus* on strawberry in Sri Lanka. *International Journal of Nematology* 7, 225-226.

MOJTAHEDI, H., SANTO, G.S. & KRAFT, J.M. (1988). First report of *Pratylenchus thornei* on dry land wheat in Washington State. *Plant Disease* 72, 175.

MOJTAHEDI, H., SANTO, G.S. & KRAFT, J.M. (1992). *Pratylenchus neglectus* on dryland wheat in Washington. *Plant Disease* 76, 323.

MONTASSER, S.A., MOUSSA, F.F., YOUSSEF, M.M.A., ABOUL-SOOUD, A.B. & MOHAMED, M.M.M. (2002). Response of certain cultivars of sugarcane to infection by *Pratylenchus zeae*, the root-lesion nematode. *Pakistan Journal of Nematology* 20, 47-55.

MONTEIRO, A.R. & LORDELLO, L.G.E. (1972). Nematoides parasitos do abacaxizeiro. *Revista de Agricultura, Piracicaba* 47, 163.

MONTEIRO, A.R. & LORDELLO, L.G.E. (1976). Nematoides parasitos associados a framboesa no Brasil. *Revista de Agricultura, Piracicaba* 51, 122.

MONTIEL, A., SOSA, L., MEDRANO, C. & ROMERO, D. (1997). Nematodes fitoparasitos en plantaciones de platano (*Musa* AAB) de la margen izquierda del rio Chama, Estado Zulia, Venezuela. *Revista de la Facultad de Agronomia, Universidad del Zulia* 14, 245-251.

MOODY, E.H., LOWNSBERY, B.F. & AHMED, J.M. (1973). Culture of the root-lesion nematode *Pratylenchus vulnus* on carrot disks. *Journal of Nematology* 5, 225-226.

MORGAN, G.D., MACGUIDWIN, A.E., ZHU, J. & BINNING, L.K. (2002a). Population dynamics and distribution of root lesion nematode (*Pratylenchus penetrans*) over a three-year potato crop rotation. *Agronomy Journal* 94, 1146-1155.

MORGAN, G.D., STEVENSON, W.R., MACGUIDWIN, A.E., KELLING, K.A., BINNING, L.K. & ZHU, J. (2002b). Plant pathogen population dynamics in potato fields. *Journal of Nematology* 34, 189-193.

MORGAN, G.T. & MACLEAN, A.A. (1968). Influence of soil pH on an introduced population of *Pratylenchus penetrans*. *Nematologica* 14, 311-312.

MORGAN, G.T. & MCALLAN, J.W. (1962). Hydrolytic enzymes in plant-parasitic nematodes. *Nematologica* 8, 209-215.

MORI, M., HYEON, S.B., KIMURA, Y. & SUZUKI, A. (1982). The nematicidal activity of acetylene compounds. *Agricultural and Biological Chemistry* 46, 309-311.

MOTALAOTE, B., STARR, J.L., FREDERIKSEN, R.A. & MILLER, F.R. (1987). Host status and susceptibility of sorghum to *Pratylenchus* species. *Revue de Nématologie* 10, 81-86.

MOTSINGER, R.E. & MINTON, N.A. (1986). First report of *Pratylenchus sefaensis* in the United States. *Plant Disease* 70, 259.

MOUNTAIN, W.B. (1954). Studies of nematodes in relation to brown root rot of tobacco in Ontario. *Canadian Journal of Botany* 32, 737-759.

MOUNTAIN, W.B. (1960). Acceptable standards of proof and approaches for evaluating plant-nematode relationships. In: Sasser, J.N. & Jenkins, W.R. (Eds). *Nematology*. Chapel Hill, NC, USA, University of North Carolina Press, pp. 422-425.

MOUNTAIN, W.B. & PATRICK, Z.A. (1959). The peach replant problem in Ontario. VII. The pathogenicity of *Pratylenchus penetrans* (Cobb, 1917) Filip. & Stek. 1941. *Canadian Journal of Botany* 37, 459-470.

MOURA, R.M., PEDROSA, E.M.R., LIRA, R.V., MENEZES, M., FREIRE, F.C.O. & CARDOSO, J.E. (1998). A etiologia da morte subita da gravioleira (*Annona muricata*). *Fitopatologia Brasileira* 23, 173-175.

MOURA, R.M., PEDROSA, E.M.R., MARANHAO, S.R.V.L., MOURA, A.M., MACEDO, M.E.A. & SILVA, E.G. (1999). Nematoides associados a cana-de-acucar no estado de Pernambuco, Brasil. *Nematologia Brasileira* 23, 92-99.

MOURA, R.M., PEDROSA, E.M.R. & GUIMARAES, L.M.P. (2001). Novos dados sobre a etiologia da easca preta do inhame no nordeste do Brasil. *Nematologia Brasileira* 25, 235-237.

MUDIOPE, J. (1999). *Response of yam landraces and hybrid lines to nematode attack in Uganda.* M.Sc. Thesis, Makerere University, Kampala, 92 pp.

MUDIOPE, J., COYNE, D.L., ADIPALA, E. & SIKORA, R.A. (2004). Monoxenic culture of *Pratylenchus sudanensis* on carrot disks, with evidence of differences in reproductive rates between geographical isolates. *Nematology* 6, 617-619.

MUKHERJEE, B., NATH, R.C. & DASGUPTA, M.K. (2000). Plant parasitic nematode communities in rubber nurseries and plantations in Tripura. *Indian Journal of Nematology* 30, 170-174.

MUSABYIMANA, T. & SAXENA, R.C. (1999). Efficacy of neem seed derivatives against nematodes affecting banana. *Phytoparasitica* 27, 43-49.

MYERS, R.F. (1965). Amylase, cellulase, invertase and pectinase in several free-living and plant-parasitic nematodes. *Nematologica* 11, 441-448.

MYERS, R.F. (1979). Interaction of yield and nutritional status of tomatoes with *Pratylenchus penetrans*. *Journal of Nematology* 11, 308-309.

NAKAHARA, S., KUSANO, M., FUJIOKA, S., SHIMADA, A. & KIMURA, Y. (2004). Penipratynolene, a novel nematicide from *Penicillium bilaiae* Chalabuda. *Bioscience, Biotechnology and Biochemistry* 68, 257-259.

NAKASONO, K., TSURUMACHI, M. & TERUYA, R. (1984). Plant parasitic nematodes associated with vegetables and sugarcane in the subtropical region of Japan, Okinawa Miyako, and Ishigaki islands. *Proceedings of the Association for Plant Protection of Kyushu* 30, 126-129.

NAMAGANDA, J.M., KARAMURA, E., NAMANYA, P., BRIDGE, J. & GOWEN, S.R. (2000). Studies on host range of the banana lesion nematode, *Pratylenchus goodeyi*. *Acta Horticulturae* 540, 419-425.

NANDAKUMAR, C. & KHERA, S. (1969). Feedings habits of *Pratylenchus muli* n. sp. on pearl millet roots. *Indian Science Congress* 3, 508.

NANDAKUMAR, C. & KHERA, S. (1970). A new nematode species, *Pratylenchus mulchandi*, from millets of Rajasthan. *Indian Phytopathology* 22, 359-363.

NANDAKUMAR, C. & KHERA, S. (1974). Life cycle and reproduction in female *Pratylenchus mulchandi* with emphasis on the development of gonad. *Zoologischer Anzeiger* 193, 287-296.

NAVEDA, I., CROZZOLI, R., GRECO, N. & ZAMBRANO, B. (1999). Nematodos fitoparasiticos asociados con cucurbitaceas en la Peninsula de Paraguana, estado Falcon, Venezuela. *Fitopatologia Venezolana* 12, 14-17.

NESTEROV, P.I. & LIZOGUBOVA, L.P. (1972). [Nematode fauna of the biocoenosis of corn in the Moldavian SSR.] *Parazity Zhivotnykh i Rastenii* 8, 122-132.

NIBLACK, T.L. (1992). *Pratylenchus*, *Paratylenchus*, *Helicotylenchus*, and other nematodes on soybean in Missouri. *Journal of Nematology* 24, 738-744.

NICO, A.I., VOVLAS, N., TROCCOLI, A. & CASTILLO, P. (1999). Reproduction of the banana root-lesion nematode, *Pratylenchus goodeyi*, in monoxenic cultures. *Nematologia Mediterranea* 27, 187-192.

NICO, A.I., RAPOPORT, H.F., JIMÉNEZ-DÍAZ, R.M. & CASTILLO, P. (2002). Incidence and population density of plant-parasitic nematodes associated with olive planting stocks at nurseries in Southern Spain. *Plant Disease* 86, 1075-1079.

NICO, A.I., JIMÉNEZ-DÍAZ, R.M. & CASTILLO, P. (2003). Host suitability of the olive cultivars Arbequina and Picual for plant-parasitic nematodes. *Journal of Nematology* 35, 29-34.

NICOL, J.M. & ORTIZ-MONASTERIO, I. (2004). Effects of the root-lesion nematode, *Pratylenchus thornei*, on wheat yields in Mexico. *Nematology* 6, 485-493.

NICOL, J.M., DAVIES, K.A., HANCOCK, T.W. & FISHER, J.M. (1999a). Yield loss caused by *Pratylenchus thornei* on wheat in South Australia. *Journal of Nematology* 31, 367-376.

NICOL, J.M., STIRLING, G.R., ROSE, B.J., MAY, B.J. & HEESWIJCK, R.V. (1999b). Impact of nematodes on grapevine growth and productivity: current knowledge and future directions, with special reference to Australian viticulture. *Australian Journal of Grape and Wine Research* 5, 109-127.

NICOTRA, A., SIMEONE, A.M. & DI VITO, M. (2003). Research on kiwifruit source of genetic resistance to root-knot and lesion nematodes. *Acta Horticulturae* 610, 449-451.

NIMROD, I., MICHAEL, B., YAAKOV, G. & EVGENI, K. (1999). Nematodes in Israeli apple orchards. *Alon Hanotea* 53, 193-197.

NJUGUNA, L.K. & BRIDGE, J. (1998). Plant parasitic nematodes of Irish potatoes (*Solanum tuberosum*) in Central Province and sweet potatoes (*Ipomoea batatas*) in Central, Nyanza and Coastal Provinces of Kenya. *International Journal of Nematology* 8, 21-26.

NOMBELA, G. & ROMERO, D. (1999). Host response to *Pratylenchus thornei* of a wheat line carrying the *Cre2* gene for resistance to *Heterodera avenae*. *Nematology* 1, 381-388.

NOMBELA, G. & ROMERO, M.D. (2001). Field response to *Pratylenchus thornei* of a wheat line with the *CreAet* gene for resistance to *Heterodera avenae*. *European Journal of Plant Pathology* 107, 749-755.

NOMBELA, G., NAVAS, A. & BELLO, A. (1993). Spatial and temporal variation of the nematofauna in representative soils of the central region of the Iberian peninsula. *Nematologica* 39, 81-91.

NOMBELA, G., NAVAS, A. & BELLO, A. (1998). Effects of crop rotations of cereals with vetch and fallow on soil nematofauna in central Spain. *Nematologica* 44, 63-80.

NORTON, D.C. (1984). *Distribution of plant-parasitic nematode species in North America*. Hyattsville, MD, USA, Society of Nematologists, 199 pp.

NORTON, D.C. (1989). Host efficiencies of *Zea diploperennis* and *Z. perennis* for *Pratylenchus* spp. *Journal of Nematology* 21, 547-548.

NORTON, D.C. & BURNS, N. (1971). Colonization and sex ratios of *Pratylenchus alleni* in soybean roots under two soil moisture regimes. *Journal of Nematology* 3, 374-377.

NORTON, D.C. & HINZ, P.N. (1976). Relationship of *Hoplolaimus galeatus* and *Pratylenchus hexincisus* to reduction of corn yields in sandy soils in Iowa. *Plant Disease Reporter* 60, 197-200.

NORTON, D.C., IRVINE, W.A. & CASTANER, D. (1964). Nematodes associated with plants in Iowa. *Proceedings of the Iowa Academy of Science* 71, 109-116.

NORTON, D.C., EDWARDS, J. & HINZ, P.N. (1985). Nematode populations in corn and related species. *Maydica* 30, 67-74.

NOUPADJA, P. (1997). Plantain/corn intercropping. *InfoMusa* 6, 24-26.

NOVARETTI, W.R.T. (1992). Nematodes in sugarcane and their control. *Informe Agropecuario Belo Horizonte* 16, 37-42.

NOVARETTI, W.R.T., MONTEIRO, A.R. & BARBOSA-FERRAZ, L.C.C. (1998). Chemical control of *Meloidogyne incognita* and *Pratylenchus zeae* on sugarcane through carbofuran or terbufos application. *Nematologia Brasileira* 22, 60-74.

NYCZEPIR, A.P. (1991). Nematode management strategies in stone fruits in the United States. *Journal of Nematology* 23, 334-341.

NYCZEPIR, A.P. & BECKER, J.O. (1998). Fruit and citrus trees. In: Barker, K.R., Pederson, G.A. & Windham, G.L. (Eds). *Plant and nematode interactions.* Madison, WI, USA, American Society of Agronomy, Inc., Crop Science Society of America, Inc., Soil Science Society of America, Inc., pp. 637-684.

NYCZEPIR, A.P. & HALBRENDT, J.M. (1993). Nematode pests of deciduous fruit and nut trees. In: Evans, K., Trudgill, D.L. & Webster, J.M. (Eds). *Plant parasitic nematodes in temperate agriculture.* Wallingford, UK, CABI Publishing, pp. 381-425.

NYCZEPIR, A.P. & PINOCHET, J. (2001). Assessment of Guardian peach rootstock for resistance to two isolates of *Pratylenchus vulnus*. *Journal of Nematology* 33, 302-305.

NYCZEPIR, A.P., RITCHIE, D. & MILLER, R.W. (2000). Nematode control on peaches. In: Horton, D., Gorsuch, C. & Ritchie, D. (Eds). *Southern peach, nectarine, and plum pest management and culture guide*. Athens, Georgia, Bulletin 1171, University of Georgia Cooperative Extension Service, pp. 27.

O'BANNON, J.H. & TOMERLIN, A.T. (1973). Citrus tree decline caused by *Pratylenchus coffeae*. *Journal of Nematology* 5, 311-316.

O'BANNON, J.H., RADEWALD, J.D. & TOMERLIN, A.T. (1972). Population fluctuation of three parasitic nematodes in Florida citrus. *Journal of Nematology* 4, 194-199.

O'BANNON, J.H., RADEWALD, J.D., TOMERLIN, A.T. & INSERRA, R.N. (1976). Comparative influence of *Radopholus similis* and *Pratylenchus coffeae* on citrus. *Journal of Nematology* 8, 58-63.

O'BRIEN, P.C. (1983). A further study on the host range of *Pratylenchus thornei*. *Australian Plant Pathology* 12, 1-3.

OEVER, H.A.M. & MANGANE, S.E. (1992). A survey of nematodes on various crops in Mozambique. *Afro-Asian Journal of Nematology* 2, 74-79.

OEVER, R., VAN DEN BERG, E. & CHIRRUCO, J.A. (1998). Plant parasitic nematodes associated with crops grown by smallholders in Mozambique. *Fundamental and Applied Nematology* 21, 645-654.

OGIGA, I.R. & ESTEY, R.H. (1973). Seasonal changes in population density, sex ratios, and vertical distribution of nematodes in orchard soil. *Phytoprotection* 54, 57-68.

OHKAWA, K. & SAIGUSA, T. (1981). Resistance of rose rootstocks to *Meloidogyne hapla*, *Pratylenchus penetrans* and *P. vulnus*. *HortScience* 16, 590-560.

OKIE, W.R. (1987). Plum rootstocks. In: Rom, R.C. & Carlson, R.F. (Eds). *Rootstocks for fruit crops*. New York, NY, USA, Wiley & Sons, pp. 321-360.

OLIVEIRA, C.M.G., INOMOTO, M.M., VIEIRA, A.M.C. & MONTEIRO, A.R. (1999a). Effect of initial population densities of *Pratylenchus brachyurus* on the growth of *Coffea arabica* cv. Mundo Novo and *C. canephora* cv. Apota. *Nematropica* 29, 215-221.

OLIVEIRA, C.M.G., MONTEIRO A.R., ANTEDOMÊNICO, S.R. & INOMOTO, M.M. (1999b). Host reaction of *Coffea* spp. to *Pratylenchus brachyurus*. *Nematropica* 29, 241-244.

OLOWE, T. & CORBETT, D.C.M. (1976). Aspects of the biology of *Pratylenchus brachyurus* and *P. zeae*. *Nematologica* 22, 202-211.

OLOWE, T. & CORBETT, D.C.M. (1983). Morphology and morphometrics of *Pratylenchus brachyurus* and *P. zeae*. I. Effect of fixative and processing. *Indian Journal of Nematology* 13, 141-154.

OLOWE, T. & CORBETT, D.C.M. (1984a). Morphology and morphometrics of *Pratylenchus brachyurus* and *P. zeae*. II. Influence of environmental factors. *Indian Journal of Nematology* 14, 6-17.

OLOWE, T. & CORBETT, D.C.M. (1984b). Morphology and morphometrics of *Pratylenchus brachyurus* and *P. zeae*. III. Influence of geographical location. *Indian Journal of Nematology* 14, 30-35.

OLTHOF, T.H.A. (1968). Races of *Pratylenchus penetrans*, and their effect on black root rot resistance of tobacco. *Nematologica* 14, 482-488.

OLTHOF, T.H.A. (1971). Seasonal fluctuations in population densities of *Pratylenchus penetrans* under a rye-tobacco rotation in Ontario. *Nematologica* 17, 453-459.

OLTHOF, T.H.A. (1986). Reaction of six *Solanum tuberosum* cultivars to *Pratylenchus penetrans*. *Journal of Nematology* 18, 54-58.

OLTHOF, T.H.A. (1987). Effects of fumigants and systemic pesticides on *Pratylenchus penetrans* and potato yield. *Journal of Nematology* 19, 424-430.

OLTHOF, T.H.A. (1989). Effects of fumigant and non-fumigant nematicides on *Pratylenchus penetrans* and yield of potato. *Supplement to the Journal of Nematology* 21, 645-649.

OLTHOF, T.H.A. & HOPPER, B.E. (1973). Distribution of *Pratylenchus* spp. and other stylet-bearing nematode genera in soils in the flue-cured tobacco area of southern Ontario. *Canadian Plant Disease Survey* 53, 31-33.

OLTHOF, T.H.A. & POTTER, J.W. (1973). The relationship between population densities of *Pratylenchus penetrans* and crop losses in summer-maturing vegetables in Ontario. *Phytopathology* 63, 577-582.

OLTHOF, T.H.A. & TOWNSHEND, J.L. (1991). Effect of oxamyl treatment of potato seed pieces on *Pratylenchus penetrans* and yield. *Journal of Nematology* 23, 699-705.

OLTHOF, T.H.A. & WOLYNETZ, M.S. (1991). *Pratylenchus penetrans* and *P. neglectus* in tubers of potato (*Solanum tuberosum*) in Ontario. *Canadian Journal of Plant Science* 71, 1251-1256.

OLTHOF, T.H.A. & YU, Q. (1999). Reduction of root-lesion nematodes (*Pratylenchus penetrans*) in tubers of potato (*Solanum tuberosum*) during cold storage. *Canadian Journal of Plant Pathology* 21, 154-158.

OLTHOF, T.H.A., MARKS, C.F. & ELLIOT, J.M. (1973). Relationship between population densities of *Pratylenchus penetrans* and crop losses in flue-cured tobacco in Ontario. *Journal of Nematology* 5, 158-162.

OLTHOF, T.H.A., ANDERSON, R.V. & SQUIRE, S. (1982). Plant-parasitic nematodes associated with potatoes (*Solanum tuberosum* L.) in Simcoe County, Ontario. *Canadian Journal of Plant Pathology* 4, 389-391.

OLTHOF, T.H.A., MCGARVEY, B.D. & CHIBA, M. (1985). Oxamyl in the control of *Pratylenchus penetrans* on potatoes. *Canadian Journal of Plant Pathology* 7, 155-160.

ONIFADE, A.K. & EGUNJOBI, O.A. (1996). Response of *Sorghum bicolor* to different population densities of *Pratylenchus zeae*. *Global Journal of Pure and Applied Sciences* 2, 23-27.

ORAMAS, D. & ROMÁN, J. (1982). Plant parasitic nematodes associated with plantain (*Musa acuminata* × *M. balbisiana*, AAB) in Puerto Rico. *Journal of Agriculture of the University of Puerto Rico* 66, 52-59.

ORAMAS, N.D. (2002). Seed treatment of yam (*Dioscorea rotundata*) with oxamil L during storage for the control of *Pratylenchus coffeae*. *Journal of Agriculture of the University of Puerto Rico* 86, 169-172.

ORION, D. & SHLEVIN, E. (1989). Nematicide seed dressing for cyst and lesion nematode control in wheat. *Journal of Nematology* 21, 629-631.

ORION, D., AMIR, J. & KRIKUN, J. (1984). Field observations on *Pratylenchus thornei* and its effects on wheat under arid conditions. *Revue de Nématologie* 7, 341-345.

ORION, D., SHELVIN, E. & YANIV, A. (1988). Controlling the migratory nematode *Pratylenchus mediterraneus* improves carrot yield and quality. *Hassadh* 69, 72-74.

ORION, D., NACHMIAS, A., LAPID, D. & ORENSTEIN, J. (1995). Observations on the parasitic behavior of *Pratylenchus mediterraneus* on excised potato roots. *Nematropica* 25, 71-74.

ORNAT, C., VERDEJO-LUCAS, S., SORRIBAS, F.J. & TZORTZAKIS, E.A. (1999). Effect of fallow and root destruction on survival of root-knot and root-lesion nematodes in intensive vegetable cropping systems. *Nematropica* 29, 5-16.

ORUI, Y. (1996). Discrimination of the main *Pratylenchus* species (Nematoda: Pratylenchidae) in Japan by PCR-RFLP Analysis. *Applied Entomology and Zoology* 31, 505-514.

ORUI, Y. & MIZUKUBO, T. (1999a). Geographical distribution of *Pratylenchus* species in tobacco fields in Eastern Japan. *Japanese Journal of Applied Entomology and Zoology* 43, 75-79.

ORUI, Y. & MIZUKUBO, T. (1999b). Species composition of *Pratylenchus* (Nematoda: Tylenchida) in crop fields of central Japan, along with an observation of the geographic border lying between *Pratylenchus coffeae* and its closely allied species. *Japanese Journal of Nematology* 29, 1-6.

OSSENI, B., SARAH, J.L. & HUGON, R. (1997). Effect of soil pH on the population development of *Pratylenchus brachyurus* (Godfrey) in pineapple roots and on the growth and yield of the plant. *Acta Horticulturae* 425, 423-433.

OSTEEN, C.D., MOFFITT, L.J. & JOHNSON, A.D. (1988). Risk efficient action thresholds for nematode management. *Journal of Production Agriculture* 1, 332-338.

OSTREC, L.J. & GRUBISIC, D. (2003). Effects of soil solarization on nematodes in Croatia. *Journal of Pesticide Science* 76, 139-144.

OTEIFA, B.A. (1962). Species of root-lesion nematodes commonly associated with economic crops in the Delta of the U.A.R. *Plant Disease Reporter* 46, 572-575.

OTEIFA, B.A. & EL-SHARKAWI, S.E. (1965). Species identity of some Egyptian parasitic nematodes of onion. *Bulletin of the Zoological Society of Egypt* 20, 55-62.

OTEIFA, B.A., RUSHDI, M. & SALEM, A. (1963). Nematodes associated with sugar-cane. *Bulletin of Science and Technology Assiut University* 6, 251-261.

PAETZOLD, D. (1955). Untersuchungen an freilebenden nematoden der salzwiese bei aseleben. *Wissenschaftliche Zeitschrift der Universität Halle* 4, 1057-1090.

PAETZOLD, D. (1958). Beiträge zur nematodenfauna mitteldeutscher Salzstellen im Raum von Halle. *Wissenschaftliche Zeitschrift der Universität Halle* 8, 17-48.

PALMISANO, A.M. (1992). Nematodi fitoparassiti associati a colture di grano duro nell'Italia centale e meridionale. *Redia* 75, 501-515.

PANACCIONE, D.G., KOTCON, J.B., SCHARDL, C.L., JOHNSON, R.D. & MORTON, J.B. (2006). Ergot alkaloids are not essential for endophytic fungus-associated population suppression of the lesion nematode, *Pratylenchus scribneri*, on perennial ryegrass. *Nematology* 8, 583-590.

PANDEY, R. (2003). Root-knot disease of safed musli. *Indian Journal of Nematology* 33, 64.

PANKHURST, C.E., MAGAREY, R.C., STIRLING, G.R., BLAIR, B.L., BELL, M.J. & GARSIDE, A.L. (2003). Management practices to improve soil health and reduce the effects of detrimental soil biota associated with yield decline of sugarcane in Queensland, Australia. *Soil and Tillage Research* 72, 125-137.

PANKHURST, C.E., BLAIR, B.L., MAGAREY, R.C., STIRLING, G.R. & GARSIDE, A.L. (2005a). Effects of biocides and rotation breaks on soil organisms associated with the poor early growth of sugarcane in continuous monoculture. *Plant and Soil* 268, 255-269.

PANKHURST, C.E., STIRLING, G.R., MAGAREY, R.C., BLAIR, B.L., HOLT, J.A., BELL, M.J. & GARSIDE, A.L. (2005b). Quantification of the effects of rotation breaks on soil biological properties and their impact on yield decline in sugarcane. *Soil Biology and Biochemistry* 37, 1121-1130.

PARK, S., ZAKAULLAH, K., KIM, J.C., KIM, S., KIM, K.J. & CHOI, B. (1999). Association and density of plant parasitic nematodes in fruit orchards in Gyeongbuk province, Korea. *International Journal of Nematology* 9, 185-190.

PARK, S.D., KHAN, Z., YEON, I.K. & SHIN, Y.S. (2002). Plant parasitic nematodes associated with oriental melon (*Cucumis melo* L.) in Gyeongbuk province of Korea. *International Journal of Nematology* 12, 155-158.

PARK-BYEONG, Y., CHOI, D., LEE, J., CHOI, Y.E. & SHIN, G.H. (2002). An unrecorded species of *Pratylenchus loosi* Loof (Tylenchida: Pratylenchidae) from tea in Korea. *Korean Journal of Applied Entomology* 41, 299-303.

PAVLYUK, L.V. (1972). [Analysis of the nematode fauna of *Valeriana officinalis* L., cultivated in the Moscow region.] *Vestnik Zoologii* 6, 30-34.

PAYAN, L.A. & DICKSON, D.W. (1990). Comparison of populations of *Pratylenchus brachyurus* based on isozyme phenotypes. *Journal of Nematology* 22, 538-545.

PELSMAEKER, M. & COOMANS, A. (1987). Nematodes in potato fields and the relation to some biotic and abiotic factors. *Mededelingen van de Faculteit Landbouwwetenschappen, Rijksuniversiteit Gent* 52, 561-569.

DE LA PEÑA, E., MOENS, M., VAN AELST, A. & KARSSEN, G. (2006a). Description of *Pratylenchus dunensis* sp. n. (Nematoda: Pratylenchidae), a root-lesion nematode associated with the dune grass *Ammophila arenaria* (L.) Link. *Nematology* 8, 79-88.

DE LA PEÑA, E., RODRIGUEZ-ECHEVERRIA, S., VAN DER PUTTEN, W.H., FREITAS, H. & MOENS, M. (2006b). Mechanism of control of root-feeding nematodes by mycorrhizal fungi in the dune grass *Ammophila arenaria. New Phytologist* 169, 829-840.

PENG, Y. & MOENS, M. (1999). Effects of surface sterilisation and cold storage on *in vitro* behaviour of *Pratylenchus penetrans. Nematology* 1, 647-653.

PENG, Y. & MOENS, M. (2002a). Host suitability of rose rootstocks to *Pratylenchus penetrans. Nematology* 4, 387-394.

PENG, Y. & MOENS, M. (2002b). Tolerance of *Rosa* rootstocks and species to *Pratylenchus penetrans. Nematology* 4, 395-401.

PENG, Y. & MOENS, M. (2003). Host resistance and tolerance to migratory plant-parasitic nematodes. *Nematology* 5, 145-177.

PENG, Y., CHEN, W. & MOENS, M. (2003). Resistance of *Rosa* species and cultivars to *Pratylenchus penetrans. HortScience* 38, 560-564.

PEÑA-SANTIAGO, R. (1990). Plant-parasitic nematodes associated with olive (*Olea europea* L.) in the province of Jaén, Spain. *Revue de Nématologie* 13, 113-115.

PEÑA-SANTIAGO, R., CASTILLO, P., ESCUER, M., GUERRERO, P., TALAVERA, M. & VIEIRA, P. (2003). *Tylenchid species (Nematoda, Tylenchida) recorded in the Iberian Peninsula and the Balearic Islands: A compendium.* Collection 'Monographic Papers on Nematology' n°2. Servicio de Publicaciones, Universidad de Jaén, España (Spain), 127 pp.

PEREGRINE, W.T.H. & BRIDGE, J. (1992). The lesion nematode *Pratylenchus goodeyi*, an important pest of Ensete in Ethiopia. *Tropical Pest Management* 38, 325-326.

PEREZ, B.J.G., VAN GUNDY, S.D., STOLZY, L.H., THOMASON, I.J. & LAIRD, R.J. (1970). *Pratylenchus thornei*, a nematode pest of wheat in Sonora, Mexico. *Phytopathology* 60, 1307.

PERICHI, G., CROZZOLI, R., GRECO, N. & RIVAS, J.T. (2002). Nematodos fitoparasiticos asociados a la cana de azucar en Venezuela. Estado Carabobo. *Fitopatologia Venezolana* 15, 26-29.

PERRY, R.N. (1996). Chemoreception in plant parasitic nematodes. *Annual Review of Phytopathology* 34, 181-189.

PERRY, R.N. (2005). An evaluation of types of attractants enabling plant-parasitic nematodes to locate plant roots. *Russian Journal of Nematology* 13, 83-88.

PERRY, R.N., PLOWRIGHT, R.A. & WEBB, R.M. (1980). Mating between *Pratylenchus penetrans* and *P. fallax* in sterile culture. *Nematologica* 26, 125-129.

PETIT, P. & CROZZOLI, R. (1995). Nematodos fitoparasiticos asociados a cultivos de ornamentales en Venezuela. *Fitopatologia Venezolana* 8, 41-44.

PHILIS, J. (1976). Occurrence and control of nematodes affecting carrot crops in Cyprus. *Nematologia Mediterranea* 4, 7-12.

PHILIS, J. (1997). *Heterodera latipons* and *Pratylenchus thornei* attacking barley in Cyprus. *Nematologia Mediterranea* 25, 305-309.

PINKERTON, J. & FINN, C.E. (2005). Responses of strawberry species and cultivars to the root-lesion and northern root-knot nematodes. *HortScience* 40, 33-38.

PINKERTON, J.N., IVORS, K.L., MILLER, M.L. & MOORE, L.W. (2000). Effect of soil solarization and cover crops on populations of selected soilborne plant pathogens in Western Oregon. *Plant Disease* 84, 952-960.

PINOCHET, J. (1998). A review of banana attacking nematodes in the subtropics with emphasis on *Pratylenchus goodeyi* in the Canary Islands. *Acta Horticulturae* 490, 353-359.

PINOCHET, J. & ROWE, P.R. (1978). Reaction of two banana cultivars to three different nematodes. *Plant Disease Reporter* 62, 727-729.

PINOCHET, J., RASKI, D.J. & GOHEEN, A.C. (1976). Effects of *Pratylenchus vulnus* and *Xiphinema index* singly and combined on vine growth of *Vitis vinifera*. *Journal of Nematology* 8, 330-335.

PINOCHET, J., SANCHEZ, L. & LAFFITE, R. (1978). Plant parasitic nematodes associated with citrus in Honduras. *FAO Plant Protection Bulletin* 26, 58-62.

PINOCHET, J., VERDEJO-LUCAS, S. & MARULL, J. (1991). Host suitability of eight *Prunus* spp. and one *Pyrus communis* rootstocks to *Pratylenchus vulnus*, *P. neglectus*, and *P. thornei*. *Journal of Nematology* 23, 570-575.

PINOCHET, J., VERDEJO, S., SOLER, A. & CANALS, J. (1992). Host range of the lesion nematode *Pratylenchus vulnus* in commercial fruit, nut tree, citrus, and grape rootstocks in Spain. *Journal of Nematology* 24, 693-698.

PINOCHET, J., CAMPRUBÍ, A. & CALVET, C. (1993a). Effects of the root lesion nematode *Pratylenchus vulnus* and the mycorrhizal fungus *Glomus mosseae* on the growth of EMLA-26 apple rootstock. *Mycorrhiza* 4, 79-83.

PINOCHET, J., FERNÁNDEZ, C., ESMENJAUD, D. & DOUCET, M. (1993b). Effects of six *Pratylenchus vulnus* isolates on the growth of peach-almond hybrid and apple rootstocks. *Journal of Nematology* 25, 843-848.

PINOCHET, J., MARULL, J., RODRÍGUEZ-KABANA, R., FELIPE, A. & FERNÁNDEZ, C. (1993c). Pathogenicity of *Pratylenchus vulnus* on plum rootstocks. *Fundamental and Applied Nematology* 16, 375-380.

PINOCHET, J., CENIS, J.L., FERNANDEZ, C., DOUCET, M. & MARULL, J. (1994). Reproductive fitness and random amplified polymorphic DNA variation among isolates of *Pratylenchus vulnus*. *Journal of Nematology* 26, 271-277.

PINOCHET, J., CALVET, C., CAMPRUBÍ, A. & FERNÁNDEZ, C. (1995a). Growth and nutritional response of Nemared peach rootstock infected with *Pratylenchus vulnus* and the mycorrhizal fungus *Glomus mosseae*. *Fundamental and Applied Nematology* 18, 205-210.

PINOCHET, J., CALVET, C., CAMPRUBÍ, A. & FERNÁNDEZ, C. (1995b). Interaction between the root-lesion nematode *Pratylenchus vulnus* and the mycorrhizal association of *Glomus intraradices* and Santa Lucia 64 cherry rootstock. *Plant and Soil* 170, 323-329.

PINOCHET, J., ANGLÉS, M., DALMAU, E., FERNÁNDEZ, C. & FELIPE, A. (1996). *Prunus* rootstock evaluation to root-knot and lesion nematodes in Spain. *Journal of Nematology* 28, 616-623.

PINOCHET, J., JAIZME, M.C., FERNÁNDEZ, C., JAUMOT, M. & DE WAELE, D. (1998). Screening bananas for root-knot (*Meloidogyne* spp.) and lesion nematode (*Pratylenchus goodeyi*) resistance for the Canary Islands. *Fundamental and Applied Nematology* 21, 17-23.

PINOCHET, J., FERNÁNDEZ, C., CALVET, C., HERNÁNDEZ DORREGO, A. & FELIPE, A. (2000). Selection against *Pratylenchus vulnus* populations attacking *Prunus* rootstocks. *HortScience* 35, 1333-1337.

PINOCHET, J., FERNÁNDEZ, C., CUNILL, M., TORRENTS, J., FELIPE, A., LÓPEZ, M.M., LASTRA, B. & PENYALVER, R. (2002). Response of new interspecific hybrids for peach to root-knot and lesion nematodes, and crown gall. *Acta Horticulturae* 592, 707-716.

PITCHER, R.S. (1963). The role of plant parasitic nematodes in bacterial diseases. *Phytopathology* 53, 35-39.

PITCHER, R.S., PATRICK, Z.A. & MOUNTAIN, W.B. (1960). Studies on the host-parasite relations of *Pratylenchus penetrans* (Cobb) to apple seedlings. I. Pathogenicity under sterile conditions. *Nematologica* 5, 309-314.

PLOWRIGHT, R.A. & HUNT, D.J. (1994). Plant parasitic nematodes of upland, hydromorphic and inland swamp rice ecosystems in Côte d'Ivoire. *Afro-Asian Journal of Nematology* 4, 61-67.

POINAR JR, G.O. (1983). *The natural history of nematodes*. New Jersey, USA, Prentice Hall, 323 pp.

POORNIMA, K. & VADIVELU, S. (1999). Occurrence and seasonal population behaviour of phytonematodes in turmeric (*Curcuma longa* L.). *Pest Management in Horticultural Ecosystems* 5, 42-45.

PORTER, I.J. & MERRIMAN, P.R. (1985). Evaluation of soil solarization for control of root diseases of row crops in Victoria. *Plant Pathology* 34, 108-118.

POTTER, J.W. & DALE, A. (1991). Root-lesion nematode tolerance in wild and cultivated strawberry. In: Dale, A. & Luby, J.J. (Eds). *The strawberry into the 21st century*. Portland, Oregon, USA, Timber Press, pp. 202-208.

POTTER, J.W. & DALE, A. (1994). Wild and cultivated strawberries can tolerate or resist root-lesion nematode. *HortScience* 29, 1074-1077.

POTTER, J.W. & MCKEOWN, A.W. (2002). Inhibition of *Pratylenchus penetrans* by intercropping of *Rudbeckia hirta* and *Lycopersicon esculentum* in pot cultivation. *Phytoprotection* 83, 115-120.

POTTER, J.W. & OLTHOF, T.H.A. (1974). Yield losses in fall-maturing vegetables relative to population densities of *Pratylenchus penetrans* and *Meloidogyne hapla*. *Phytopathology* 64, 1072-1075.

POTTER, J.W. & OLTHOF, T.H.A. (1977). Analysis of crop losses in tomato due to *Pratylenchus penetrans*. *Journal of Nematology* 9, 290-295.

POTTER, J.W., DIRKS, V.A., JOHNSON, P.W., OLTHOF, T.H.A., LAYNE, R.E.C. & MCDONNELL, M.M. (1984). Response of peach seedlings to infection by the root lesion nematode *Pratylenchus penetrans* under controlled conditions. *Journal of Nematology* 16, 317-322.

POTTER, M.J., DAVIES, K. & RATHJEN, A.J. (1998). Suppressive impact of glucosinolates in *Brassica* vegetative tissues on root lesion nematode *Pratylenchus penetrans*. *Journal of Chemical Ecology* 24, 67-80.

POTTER, M.J., VANSTONE, V.A., DAVIES, K.A., KIRKEGAARD, J.A. & RATHJEN, A.J. (1999). Reduced susceptibility of *Brassica napus* to *Pratylenchus neglectus* in plants with elevated root levels of 2-phenylethyl glucosinolate. *Journal of Nematology* 31, 291-298.

POURJAM, E., KHEIRI, A. & GERAERT, E. (1997). The genus *Pratylenchus* Filip'jev, 1936 (Tylenchida: Pratylenchidae) from North of Iran. *Mededelingen Faculteit Landbouwkundige en Toegepaste Biologische Wetenschappen, Universiteit Gent* 62, 741-756.

POURJAM, E., KHEIRI, A., GERAERT, E. & ALIZADEH, A. (1999a). Variations in Iranian population of *Pratylenchus neglectus* and *P. thornei* (Nematoda: Pratylenchidae). *Iranian Journal of Plant Pathology* 35, 67.

POURJAM, E., WAEYENBERGE, L., MOENS, M. & GERAERT, E. (1999b). Morphological, morphometrical and molecular study of *Pratylenchus coffeae* and *P. loosi* (Nematoda: Pratylenchidae). *Mededelingen Faculteit Landbouwkundige en Toegepaste Biologische Wetenschappen, Universiteit Gent* 64, 391-401.

POWELL, N.T. (1971). Interactions between nematodes and fungi in disease complexes. *Annual Review of Phytopathology* 9, 253-274.

POWERS, T.O., TOOD, T.C., BURNELL, A.M., MURRAY, P.C.B., FLEMING, C.C., SZALANSKI, A.L., ADAMS, B.A. & HARRIS, T.S. (1997). The rDNA internal transcribed spacer region as a taxonomic marker for nematodes. *Journal of Nematology* 29, 441-450.

PRASAD, D. & RANGAPPA, T. (1994). Distribution and estimation of avoidable loss due to nematodes associated with Kalahasti malady of groundnut, *Arachis hypogaea. Annals of Plant Protection Sciences* 2, 37-42.

PRASAD, J.S. & RAO, S.Y. (1978). Influence of crop rotations on the population densities of the root-lesion nematode *Pratylenchus indicus* in rice and rice soils. *Annales de Zoologie et Écologie Animale* 10, 627-634.

PRASAD, J.S., SESHU-REDDY, K.V. & SIKORA, R.A. (1995). Hosts of the banana root-lesion nematode, *Pratylenchus goodeyi* in east Africa. *Nematologia Mediterranea* 23, 253-254.

PRASAD, J.S., SESHU-REDDY, K.V. & SIKORA, R.A. (1999). Life cycle of *Pratylenchus goodeyi* in banana roots. *Nematologia Mediterranea* 27, 111-114.

PRICE, N.S. & BRIDGE, J. (1995). *Pratylenchus goodeyi* (Nematoda: Pratylenchidae): a plant parasitic nematode of the montane highlands of Africa. *African Journal of Zoology* 109, 435-442.

PUDASAINI, M.P., SCHOMAKER, C.H., BEEN, T.H. & MOENS, M. (2006a). Vertical distribution of the plant-parasitic nematode, *Pratylenchus penetrans*, under four field crops. *Phytopathology* 96, 226-233.

PUDASAINI, M.P., VIAENE, N. & MOENS, M. (2006b). Effect of marigold (*Tagetes patula*) on population dynamics of *Pratylenchus penetrans* in a field. *Nematology* 8, 477-484.

QASIM, M. & HASHMI, S. (1988). Seasonal population fluctuation of nematodes on pistachio in Baluchistan. *International Nematology Network Newsletter* 5, 50-53.

QIU, J., WILLIAMSON, V.M. & WESTERDAHL, B.B. (2005). Identification of *Pratylenchus* species using PCR with species-specific primers. In: *Proceedings 44th Annual Meeting of the Society of Nematologists, Fort Lauderdale, Florida, USA*, p. 62. [Abstr.]

QUÉNEHÉRVÉ, P., VAN DEN BERG, E., TOPART, P. & HOSTACHY, B. (1997). Analyse écologique de la specificité parasitaire des nématodes phytoparasites associés à quelques plantes ornementales cultivées à la Martinique. *Nematologica* 43, 214-227.

QURAISHI, M.A. (1982). One new species of the genus *Pratylenchus* from the grape vineyards of Hyderabad City (Andhra Pradesh). *Indian Journal of Nematology* 12, 208-210.

RAHM, G.F. (1928). Alguns nematodes parasitos e semiendoparasitos das plantas culturaes do Brasil. *Archives do Instituto de Biologia* 2, 67-136.

RAJAB, K.A., SALIM, S.S. & SPEIJER, P.R. (1999). Plant-parasitic nematodes associated with Musa in Zanzibar. *African Plant Protection* 5, 105-110.

RAMA, K. & DASGUPTA, M.K. (1998). Population ecology and community structure of plant parasitic nematodes associated with ginger in West Bengal. *Indian Journal of Nematology* 28, 10-14.

RAMA, K. & DASGUPTA, M.K. (2000a). Population ecology and community structure of plant parasitic nematodes associated with ginger in northern West Bengal. *Indian Journal of Nematology* 30, 42-45.

RAMA, K. & DASGUPTA, M.K. (2000b). Population ecology and community structure of plant parasitic nematodes associated with coconut and arecanut in northern West Bengal. *Indian Journal of Nematology* 30, 175-182.

RAMAKRISHNAN, S. & VADIVELU, S. (1995). Nematodes associated with chrysanthemum and their management. *South Indian Horticulture* 43, 174-175.

RAMMING, D.W. & COCIU, V. (1990). Peaches (*Prunus*). *Acta Horticulturae* 290, 175-182.

RAMSDELL, D.C., BIRD, G.W., WARNER, F.W., DAVENPORT, J.F., DIAMOND, C.J. & GILLET, J.M. (1996). Field pathogenicity studies of four species of plant-pathogenic nematodes on French-American hybrid grapevine cultivars in Michigan. *Plant Disease* 80, 334-338.

RAO, G.V. (1973). Effects of nematode-trapping fungi on the biology of the lesion nematode, *Pratylenchus penetrans* (Cobb) Filipjev & Shuurmans Stekhoven. *Dissertation Abstracts International* 34B, 486-487.

RASHID, A. (1974). A new species of the genus *Pratylenchus* Filipjev, 1934 (Nematoda: Pratylenchidae) from India. *Proceedings of the 61st Indian Science Congress* 61, 65.

RASHID, A. (1997). Plant parasitic nematodes associated with sugarcane in western Uttar Pradesh. *Indian Phytopathology* 50, 133-136.

RASHID, A. & KHAN, A.M. (1975). Studies on the lesion nematode, *Pratylenchus coffeae*, associated with chrysanthemum. *Indian Phytopathology* 28, 22-28.

RASHID, A. & KHAN, A.M. (1976). Morphometric studies on *Pratylenchus coffeae* with description of *Pratylenchus typicus* Rashid, 1974. *Indian Journal of Nematology* 6, 63-72.

RASKI, D.J. & KRUSBERG, L.R. (1984). Nematode parasites of grapes and other small fruits. In: Nickle, W.R. (Ed.). *Plant and insect nematodes*. New York, NY, USA, Marcel Dekker, pp. 457-506.

RAUT, S.P. & SETHI, C.L. (1986). Plant parasitic nematodes associated with soybean. *Indian Journal of Nematology* 16, 126-128.

RAZHIVIN, A.A. & O'RELLY, KH.P. (1976). *Pratylenchus cubensis* sp. n. (Nematoda, Pratylenchidae) from the rhizosphere of sugar cane in Cuba. *Zoologichesky Zhurnal* 55, 135-136.

REBOIS, R.V. & GOLDEN, A.M. (1978). Nematode occurrences in soybean fields in Mississippi and Louisiana. *Plant Disease Reporter* 62, 433-437.

REBOIS, R.V. & GOLDEN, A.M. (1985). Pathogenicity and reproduction of *Pratylenchus agilis* in field microplots of soybeans, corn, tomato, or corn-soybean cropping systems. *Plant Disease* 69, 927-929.

REBOIS, R.V. & HUETTEL, R.N. (1986). Populations dynamics, root penetration, and feeding behavior of *Pratylenchus agilis* in monoxenic root cultures of corn, tomato, and soybean. *Journal of Nematology* 18, 392-397.

REID, A.P., MANZANILLA-LÓPEZ, R.H. & HUNT, D.J. (2003). *Nacobbus aberrans* (Thorne, 1935) Thorne & Allen, 1944 (Nematoda: Pratylenchidae); a nascent species complex revealed by RFLP analysis and sequencing of the ITS-rDNA region. *Nematology* 5, 441-451.

RENSCH, D. (1924). *Aphelenchus neglectus* sp. n. eine neue parasitäre Nematodenart. *Sonderabdruck aus dem Zoologischen Anzeiger* 59, 277-280.

REVELO-MORAN, J.A., CID DEL PRADO VERA, I., ZAVALETA-MEJIA, E., RENFRO, B. & ZARATE DE LARA, G. (1993). Influencia de *Pratylenchus pratensis* en el desarrollo de la pudricion de la raiz del maiz por *Fusarium moniliforme* var. *subglutinans*. *Revista Mexicana de Fitopatologia* 11, 13-18.

REYNOLDS, L.B., OLTHOF, T.H.A. & POTTER, J.W. (1992). Effects of fumigant nematicides on yield and quality on paste tomatoes grown in southwestern Ontario. *Journal of Nematology (Supplement)* 24, 656-661.

REYNOLDS, L.B., POTTER, J.W. & BALL-COELHO, B.R. (2000). Crop rotation with *Tagetes* sp. is an alternative to chemical fumigation for control of root-lesion nematodes. *Agronomy Journal* 92, 957-966.

RICH, J.R. & DUNN, R.A. (1982). Pathogenicity and control of nematodes affecting sunflower in North Central Florida. *Plant Disease* 66, 297-298.

RICH, J.R., KEEN, N.T. & THOMASON, I.J. (1977a). Association of coumestans with the hypersensitivity of Lima bean roots to *Pratylenchus scribneri*. *Physiological Plant Pathology* 10, 105-116.

RICH, J.R., THOMASON, I.J. & O'MELIA, F.C. (1977b). Host-parasite interactions of *Pratylenchus scribneri* on selected crop plants. *Journal of Nematology* 9, 131-135.

RICH, J.R., JOHNSON, J.T. & HODGE, C.H. (1985). Corn response to subsoiling and nematicide application. *Journal of Nematology* 17, 404-407.

RICH, J.R., ARNETT, J.D., SHEPHERD, J.A. & WATSON, M.C. (1989). Chemical control of nematodes on flue-cured tobacco in Brazil, Canada, United States, and Zimbabwe. *Journal of Nematology* 21, 609-611.

RICH, J.R., DUNN, R.A. & NOLING, J.W. (2004). Nematicides: Past and present uses. In: Chen, Z.X., Chen, S.Y. & Dickson, D.W. (Eds). *Nematology advances and perspectives, volume 2*. Wallingford, UK, CABI Publishing, pp. 1179-1220.

RIEDEL, R.M., ROWE, R.C. & MARTIN, M.J. (1985). Differential interactions of *Pratylenchus crenatus*, *P. penetrans*, and *P. scribneri* with *Verticillium dahliae* in potato early dying disease. *Phytopathology* 75, 419-422.

RIGA, E., HOOPER, C. & POTTER, J. (2005). *In vitro* effect of marigold seed exudates on plant parasitic nematodes. *Phytoprotection* 86, 31-35.

RILEY, I.T. & KELLY, S.J. (2001). *Radopholus sativus* (Nematoda: Pratylenchidae), a potential economic pest of wheat in Western Australia. *Nematology* 3, 25-30.

RILEY, I.T. & KELLY, S.J. (2002). Endoparasitic nematodes in cropping soils of Western Australia. *Australian Journal of Experimental Agriculture* 42, 49-56.

RILEY, I.T. & WOUTS, W.M. (2001). *Pratylenchus* and *Radopholus* species in agricultural soils and native vegetation in Southern Australia. *Transactions of the Royal Society of South Australia* 125, 147-153.

RIVOAL, R. & COOK, R. (1993). Nematode pests of cereals. In: Evans, K., Trudgill, D.L. & Webster, J.M. (Eds). *Plant parasitic nematodes in temperate agriculture.* Wallingford, UK, CABI Publishing, pp. 259-303.

RIVOAL, R., LASERRE, F. & COOK, R. (1995). Consequences of long-term cropping with resistant cultivars on the population dynamics of the endoparasitic nematodes *Heterodera avenae* and *Pratylenchus neglectus* in a cereal production ecosystem. *Nematologica* 41, 516-529.

RIZAEVA, S.M. (1980). [The nematodes of the principal vegetable crops and of potato in the Andizhan region (Uzbek SSR).] *Gel'mintozy cheloveka, zhivotnykh, rastenii I mery bor'by s nimi Tezisy dokladov konferentsii vsesoyuznogo obshchestva gel'mintologov AN SSSR Moskva, 27-29 Yanvarya 1981 g. 1980*, 150-151.

ROBBINS, R.T., RIGGS, R.D. & VON STEEN, D. (1987). Results of annual phytoparasitic nematode surveys of Arkansas soybean fields, 1978-1986. *Annals of Applied Nematology* 1, 50-55.

ROBBINS, R.T., RIGGS, R.D. & VON STEEN, D. (1989a). Phytoparasitic nematode surveys of Arkansas cotton fields, 1986-88. *Journal of Nematology* 21, 619-623.

ROBBINS, R.T., RIGGS, R.D. & VON STEEN, D. (1989b). Phytoparasitic nematode surveys of Arkansas wheat fields, 1986-88. *Journal of Nematology* 21, 624-628.

ROBERTS, P.A. (2002). Concept and consequence of resistance. In: Starr, J.L., Bridge, J. & Cook, R. (Eds). *Plant resistance to parasitic nematodes.* Wallingford, UK, CABI Publishing, pp. 23-41.

ROBERTS, P.A., MAGYAROSY, A.C., MATTHEWS, W.C. & MAY, D.M. (1988). Effects of metam-sodium applied by drip irrigation on root-knot nematodes, *Pythium ultimum* and *Fusarium* sp. in soil and on carrot and tomato roots. *Plant Disease* 72, 213-217.

RODRÍGUEZ-KÁBANA, R. (1986). Organic and inorganic nitrogen amendments to soil as nematode suppressants. *Journal of Nematology* 18, 129-135.

RODRÍGUEZ-KÁBANA, R. & INGRAM, E.G. (1978). Susceptibilidad del gandul a especies de fitonematodos en Alabama. *Nematropica* 8, 32-34.

RODRÍGUEZ-KÁBANA, R., MORGAN JONES, G. & CHET, I. (1987). Biological control of nematodes: Soil amendments and microbial antagonists. *Plant and Soil* 100, 237-247.

ROMABATI, N., MOHILAL, N. & DHANACHAND, C. (1999). Effect of aqueous rhizome extract of *Acorus calamus* L. on larvae of *Pratylenchus thornei* Sher and Allen, 1953. *Indian Journal of Nematology* 29, 97.

ROMÁN, J. & GRULLON, L. (1975). Nematodes associated with sugarcane in the Dominican Republic. *Journal of Agriculture of the University of Puerto Rico* 59, 138-140.

ROMÁN, J. & HIRSCHMANN, H. (1969a). Morphology and morphometrics of six species of *Pratylenchus. Journal of Nematology* 1, 363-386.

ROMÁN, J. & HIRSCHMANN, H. (1969b). Embryogenesis and postembryogenesis in species of *Pratylenchus* (Nematoda: Tylenchidae). *Proceedings of the Helminthological Society of Washington* 36, 164-174.

ROMÁN, J. & TRIANTAPHYLLOU, A.C. (1969). Gametogenesis and reproduction of seven species of *Pratylenchus. Journal of Nematology* 1, 357-362.

ROMÁNIKO, V.I. (1960). [A new species of nematode of leguminous crops in southern Ural.] *Zoologicheskyi Zhurnal* 39, 1256-1257.

ROMÁNIKO, V.I. (1977). [New species of parasitic nematodes on wheat in the Southern Urals and Trans-Ural region.] *Fauna, sistematika, biologiya I ekologiya gel'mintov I ikh promezhutochnykh khozyaev Respublikanskii Sbornik* 60-66.

ROSSI, C.E. & MONTEIRO, A.R. (2001). Alcachofra: nova planta hospedeira de *Pratylenchus penetrans* no Estado de São Paulo, Brasil. *Ciencia Rural* 31, 703-704.

ROSSI, C.E., CALDARI-JUNIOR, P. & MONTEIRO, A.R. (2000a). Ocorrencia de *Pratylenchus vulnus* em rosa no Estado de Minas Gerais. *Arquivos do Instituto Biologico São Paulo* 67, 147-148.

ROSSI, C.E., FERRAZ, L.C.C.B., BERTI, A.J., ARAUJO, A.C.M. & MARCON, L. (2000b). Pratilencoses do feijoeiro: levantamento de ocorrencia no Sudoeste Paulista e hospedabilidade de cultivares a *Pratylenchus zeae. Revista de Agricultura Piracicaba* 75, 395-404.

RÖSSNER, J. (1971). Einfluss der Austrocknung des Bodens auf wandernde Wurzelnematoden. *Nematologica* 17, 127-144.

RÖSSNER, J. (1972). Vertical distribution of migratory root nematodes in the soil in relation to water content and extent of root system. *Nematologica* 18, 360-372.

ROTENBERG, D., MACGUIDWIN, A.E., SAEED, I.A.M. & ROUSE, D.I. (2004). Interaction of spatially separated *Pratylenchus penetrans* and *Verticillium dahliae* on potato measured by impaired photosynthesis. *Plant Pathology* 53, 294-302.

ROWE, R.C. & POWELSON, M.L. (2002). Potato early dying: management challenges in a changing production environment. *Plant Disease* 86, 1184-1193.

ROWE, R.C., RIEDEL, R.M. & MARTIN, M.J. (1985). Synergistic interactions between *Verticillium dahliae* and *Pratylenchus penetrans* in potato early dying disease. *Phytopathology* 75, 412-418.

RUEHLE, J.L. (1969). Nematodes parasitic on forest trees. *Nematologica* 15, 76-80.

RUEHLE, J.L. (1971). Nematodes parasitic on forest trees: III. Reproduction on selected hardwoods. *Journal of Nematology* 3, 170-173.

RUEHLE, J.L. (1973). Influence of plant-parasitic nematodes on longleaf pine seedlings. *Journal of Nematology* 5, 7-9.

RYSS, A. (1992). [Some records of plant parasitic nematodes (Nematoda, Tylenchida) from Estonia.] *Eesti Teaduste Akadeemia Toimetised, Bioloogia* 41, 72-76.

RYSS, A., BAICHEVA, O. & STOYANOV, D. (1991). A new phytonematode for Bulgaria *Pratylenchus pinguicaudatus* Corbett, 1969 and morphological description of *Pratylenchus thornei* Sher et Allen, 1953 and *Zygotylenchus guevarai* Tobar Jimenez, 1963. *Khelmintologiya* 30, 3-8.

RYSS, A., BOSTRÖM, S. & SOHLENIUS, B. (2005). Tylenchid nematodes found on the Nunatak Base, East Antarctica. *Annales Zoologici* 55, 315-324.

RYSS, A.Y. (1982). [New species of plant nematodes from the genus *Pratylenchus* in Estonia.] *Eesti NSV Teaduste Akadeemia Toimetised Izvestiya Akademii Nauk Estonskoi SSR, Bioloogia* 31, 22-29.

RYSS, A.Y. (1983). [Anisometric growth of seven species from the family Pratylenchidae (Nematoda: Tylenchida).] *Proceedings of the Zoological Institute USSR Academy* 121, 114-128.

RYSS, A.Y. (1984). [Identification of six species of nematodes of the genus *Pratylenchus* (Tylenchida) by larvae.] *Proceedings of the Zoological Institute USSR Academy* 126, 57-66.

RYSS, A.Y. (1986). [The influence of some soil properties on the host-parasite system '*Pratylenchus flakkensis-Poa pratensis*'.] *Trudy Zoologicheskogo Instituta Morfologiya, sistematika i faunistika paraziticheskikh zhivotnykh* 155, 25-40.

RYSS, A.Y. (1988). [*World fauna of the root parasitic nematodes of the family Pratylenchidae (Tylenchida).*] Leningrad, USSR, 367 pp.

RYSS, A.Y. (2002a). Genus *Pratylenchus* Filipjev: multientry and monoentry keys and diagnostic relationships (Nematoda: Tylenchida: Pratylenchidae). *Zoosystematica Rossica* 10, 241-255.

RYSS, A.Y. (2002b). Phylogeny and evolution of the genus *Pratylenchus* according to morphological data (Nematoda: Tylenchida). *Zoosystematica Rossica* 10, 257-273.

RYSS, A.Y. & FAM-TKHAN, B. (1989). [Plant-parasitic nematodes of the genus *Pratylenchus* from Vietnam.] *Trudy Zoologicheskogo Instituta, Akademiya Nauk SSSR* 194, 60-64.

SAADABI, A.M. (1988). Plant-parasitic nematodes associated with sugarcane at Kenana Sugar Estate of Sudan. *International Nematology Network Newsletter* 5, 28-30.

SAADABI, A.M. (1993). Plant-parasitic nematodes associated with some ornamental plants in Libya. *Pakistan Journal of Nematology* 11, 49-51.

SAADABI, A.M., YASSIN, A.M. & TAYEB, Y.M.EL. (1987). Comparative morphology of *Pratylenchus sudanensis*, *P. zeae* and *P. neglectus* with emphasis on variation within the *P. sudanensis* population. *International Nematology Network Newsletter* 4, 39-50.

SABIR, N. (2000). Population fluctuation of important nematodes in the rhizosphere of papaya in Lucknow. *Indian Journal of Nematology* 30, 261-263.

SAEED, I.A.M., MACGUIDWIN, A.E. & ROUSE, D.I. (1997). Synergism of *Pratylenchus penetrans* and *Verticillium dahliae* manifested by reduced gas exchange in potato. *Phytopathology* 87, 435-439.

SAEED, I.A.M., MACGUIDWIN, A.E. & ROUSE, D.I. (1998). Effect of initial nematode population density on the interaction of *Pratylenchus penetrans* and *Verticillium dahliae* on 'Russet Burbank' potato. *Journal of Nematology* 30, 100-107.

SAEED, M., KHAN, H.A., KHAN, S.A. & QAMAR, F. (1986). A survey of the nematode problem of tobacco in Pakistan: (A) Nurseries. *Pakistan Journal of Scientific and Industrial Research* 29, 279-283.

SAEED, M., KHAN, H.A., KHAN, S.A. & QAMAR, F. (1988). Nematodes associated with nurseries in Karachi, part I. Rose. *Pakistan Journal of Scientific and Industrial Research* 31, 729-730.

SAEKI, Y., KAWANO, E., YAMASHITA, C., AKAO, S. & NAGATOMO, Y. (2003). Detection of plant parasitic nematodes, *Meloidogyne incognita* and *Pratylenchus coffeae* by multiplex PCR using specific primers. *Soil Science and Plant Nutrition* 49, 291-295.

SAHOO, C.R. & SAHU, S.C. (1993). Pathogenicity of *Pratylenchus zeae* on rice. *Nematologia Mediterranea* 21, 177-178.

SAIGUSA, T. & ANDO, Y. (1973). Notes on the bionomics and control of *Pratylenchus convallariae* on German lily-of-the-valley. In: *17th Annual*

Meeting of the Japanese Society of Applied Entomology and Zoology, Nagano, Japan, 3-5 April 1973, p. 245.

SAKWE, P.N. & GERAERT, E. (1994). Species of the genus *Pratylenchus* Filip'jev, 1936 (Nematode: Tylenchida) from Cameroon. *Fundamental and Applied Nematology* 17, 161-173.

SALEM, A.A., EL-MORSHEDY, M.F. & EL-ZAWAHRY, A.M. (1994). Nematodes associated with soybean (*Glycine max*) in upper Egypt. *Fundamental and Applied Nematology* 17, 401-404.

SAMAC, D.A. & KINKEL, L.L. (2001). Suppression of the root-lesion nematode (*Pratylenchus penetrans*) in alfalfa (*Medicago sativa*) by *Streptomyces* spp. *Plant and Soil* 235, 35-44.

SAMIBAEVA, K.KH. (1966). [New species of nematodes in tobacco and adjacent soils in Urgutskom Province of the Samarkand Territory.] *Samarkand Governamental University of Sciences and Work* 156, 121-128.

SAMSOEN, L. & GERAERT, E. (1975). La faune nématologique des rizières du Cameroun. I. Ordre des Tylenchides. *Revue de Zoologie Africaine* 89, 536-554.

SAN MARTÍN, R. & MAGUNACELAYA, J.C. (2005). Control of plant-parasitic nematodes with extracts of *Quillaja saponaria*. *Nematology* 7, 577-585.

SANCHEZ DE VIALA, S., BRODIE, B., RODRIGUEZ, E. & GIBSON, D.M. (1998). The potential of thiarubrine C as a nematicidal agent against plant-parasitic nematodes. *Journal of Nematology* 30, 192-200.

SANDOVAL-HERNANDEZ, J. & TÉLIZ-ORTÍZ, D. (1990). Nematodos asociados al cultivo de la fresa en Zamora, Mich. e Irapuato, Gto. *Revista Chapingo* 15, 98-101.

SANTO, G.S. & LEAR, B. (1976). Influence of *Pratylenchus vulnus* and *Meloidogyne hapla* on the growth of rootstocks of rose. *Journal of Nematology* 8, 18-23.

SANTO, G.S. & WILSON, J.H. (1990). Effects of fenamiphos on *Pratylenchus penetrans* and growth of apple. *Journal of Nematology* 22, 779-782.

SARAH, J.-L. (1987). Influences of nematicide treatment on pineapple sucker production. *Fruits* 42, 71-75.

SARAH, J.-L. (1989). Banana nematodes and their control in Africa. *Nematropica* 19, 199-216.

SARAH, J.-L., LASSOUDIÈRE, A. & GUÉROUT, R. (1983). La jachère nue et l'immersion du sol: deux méthodes interessantes de lutte intégrée contre *Radopholus similis* (Cobb) dans les bananeraies de sols tourbeux de Côte d'Ivoire. *Fruits* 38, 35-42.

SARATHCHANDRA, S., WATSON, R.N., COX, N.R., DIMENNA, M.E., BROWN, J.A., BURCH, G. & NEVILLE, F.J. (1996). Effects of chitin amendment of soil on microorganisms, nematodes, and growth of white clover (*Trifolium repens* L.) and perennial ryegrass (*Lolium perenne* L.). *Biology and Fertility of Soils* 22, 221-226.

SARATHCHANDRA, S.U., GHANI, A., YEATES, G.W., BURCH, G. & COX, N.R. (2001). Effect of nitrogen and phosphate fertilisers on microbial and nematode diversity in pasture soils. *Soil Biology and Biochemistry* 33, 953-964.

SASSER, J.N. & FRECKMAN, D.W. (1987). A world perspective on Nematology: The role of the Society. In: Veech, J.A. & Dickson, D.W. (Eds). *Vistas on nematology*. Hyattsville, MD, USA, Society of Nematologists, pp. 7-14.

SAUERBORN, J., SAXENA, M.C. & MASRI, H. (1990). Control of faba bean nematodes by soil solarization in Syria. *Arab Journal of Plant Protection* 8, 38-40.

SAVKINA, E.V. (1989). [Photosynthetic activity of pine seedlings infected with the fungus *Lophodermium seditiosum* and the nematode *Pratylenchus teres*.] *Trudy Gel'mintologicheskoi Laboratorii* 37, 137-140.

SAWAZAKI, E., LORDELLO, A.I.L. & LORDELLO, R.R.A. (1987). Heranca da resistencia de milho a *Pratylenchus* spp. *Bragantia* 46, 27-33.

SAYRE, R.M. (1964). Cold-hardiness of nematodes. I. Effects of rapid freezing on the eggs and larvae of *Meloidogyne incognita* and *M. hapla*. *Nematologica* 10, 168-179.

SAYRE, R.M. & STARR, M.P. (1989). Genus *Pasteuria* Metchnikoff, 1888. In: Williams, S.T., Sharpe, M.E. & Holt, J.G. (Eds). *Bergey's manual of systematic bacteriology*. Baltimore, MD, USA, Williams & Wilkins, pp. 2601-2615.

SCHENCK, S. & SCHMITT, D.P. (1992). Survey of nematodes on coffee in Hawaii. *Journal of Nematology* 24, 771-775.

SCHEPMAN, M.A. (1994). Effects of crop rotation of nursery stock on the population of the nematode *Pratylenchus penetrans* Cobb. *Mededelingen Faculteit Landbouwkundige en Toegepaste Biologische Wetenschappen, Universiteit Gent* 59, 763-769.

SCHMITT, D.P. (1976). Relative suitability of soybean cultivars to *Pratylenchus brachyurus*. *Journal of Nematology* 8, 302. [Abstr.]

SCHMITT, D.P. & BARKER, K.R. (1981). Damage and reproductive potentials of *Pratylenchus brachyurus* and *P. penetrans* on soybean. *Journal of Nematology* 13, 327-332.

SCHNEIDER, J.H.M., S'JACOB, J.J. & VAN DE POL, P.A. (1995). *Rosa multiflora* 'Ludiek', a rootstock with resistant features to the root lesion nematode *Pratylenchus vulnus*. *HortScience* 63, 37-45.

SCHNEIDER, W. (1939). Würmer oder vermes. II. Fadenwürmer oder nematoden. I. Freilebende und planzenparasitische nematoden. *Tierwelt Deutschlands (Dahl)* 36, 1-260.

SCHOLTE, K. (2000). Effect of potato used as a trap crop on potato cyst nematodes and other soil pathogens and on the growth of a subsequent main potato crop. *Annals of Applied Biology* 136, 229-238.

SCOTTO LA MASSESE, C. (1971). Pathogenic nematodes of fruit and forest nurseries. *Pépinièristes-Horticulteurs-Maraîchers, Special Pépinières* 115, 11 pp.

SCOTTO LA MASSESE, C., MINOT, J.C. & VOISIN, R. (1981). Nematodes phytophages associes a la culture du colza en France. *Comptes rendus des 6'èmes Journées de Phytiatrie et de Phytopharmacie Circum Méditerranéennes, Perpignan, France, 25-28 mai 1981*, pp. 109-115.

SCOTTO LA MASSESE, C., DEPORTES, L., MERCIER, S. & ROGER, J.P. (1984). Les principaux ennemis du figuier: les nématodes et les maladies. *Phytoma Défense des Cultures* 354, 39-41.

SCURRAH, M.I., NIERE, B. & BRIDGE, J. (2005). Nematode parasites of Solanum and sweet potatoes. In: Luc, M., Sikora, R.A. & Bridge, J. (Eds). *Plant parasitic nematodes in subtropical and tropical agriculture*, 2nd edition. Wallingford, UK, CABI Publishing, pp. 193-219.

SEBASTIAN, S. & GUPTA, P. (1996). Evaluation of three plant latices as seed treatment against *Pratylenchus thornei* on chickpea. *International Chickpea and Pigeonpea Newsletter* 3, 39-40.

SEINHORST, J.W. (1959). Two new species of *Pratylenchus*. *Nematologica* 4, 83-86.

SEINHORST, J.W. (1965). The relation between nematode density and damage to plants. *Nematologica* 11, 137-154.

SEINHORST, J.W. (1968). Three new *Pratylenchus* species with a discussion of the structure of the cephalic framework and of the spermatheca in this genus. *Nematologica* 14, 497-510.

SEINHORST, J.W. (1971). The structure of the glandular part of the oesophagus of Tylenchidae. *Nematologica* 17, 431-443.

SEINHORST, J.W. (1977a). *Pratylenchus loosi*. *CIH Descriptions of plant-parasitic nematodes*, Set 7, No. 98. St Albans, UK, Commonwealth Agricultural Bureaux, 2 pp.

SEINHORST, J.W. (1977b). *Pratylenchus fallax*. *CIH Descriptions of plant-parasitic nematodes*, Set 7, No. 100. St Albans, UK, Commonwealth Agricultural Bureaux, 2 pp.

SEINHORST, J.W. (1979). Nematodes and growth of plants: formulation of the nematode-plant system. In: Lamberti, F. & Taylor, C.E. (Eds). *Root-knot nematodes (*Meloidogyne *species) systematics, biology and control*. London, UK, Academic Press, pp. 231-256.

SEINHORST, J.W. (1998). The common relation between population density and plant weight in pot and microplot experiments with various nematode plant combinations. *Fundamental and Applied Nematology* 21, 459-468.

SETHI, C.L. & SWARUP, G. (1971). Plant parasitic nematodes of North-Western India. III. The genus *Pratylenchus*. *Indian Phytopathology* 24, 410-412.

SETTERQUIST, R.A., SMITH, G.K., JONES, R. & FOX, G.E. (1996). Diagnostic probes targeting the major sperm protein gene that may be useful in the molecular identification of nematodes. *Journal of Nematology* 28, 414-421.

SHAGALINA, L.M., KRALL, E.L. & BARYSHEVA, I.N. (1988). [*Pratylenchus neglectus* parasitic on annual cereals in the Karakum desert.] *Izvestiya Akademii Nauk Turkmenskoi SSR, Seriya Biologicheskikh Nauk* 4, 37-39.

SHAHINA, F. & MAQBOOL, M.A. (1996). Two new species of the genus *Radopholus* Thorne, 1949 (Nematoda: Pratylenchidae) from Pakistan. *Fundamental and Applied Nematology* 19, 289-292.

SHANER, G., STROMBERG, E.L. & LACY, G.H. (1992). Nomenclature and concepts of pathogenicity and virulence. *Annual Review of Phytopathology* 30, 47-66.

SHARMA, N.N., EDWARD, J.C., MISRA, S.L. & CHANDRASHEKAR, M. (1992). Studies on phytonematodes associated with important rice growing tracts of India. *Current Nematology* 3, 107-109.

SHARMA, R.D. & AMABILE, R.F. (1998a). Fitonematoides associados a genotipos de girassol em condicoes de Cerrado. *Nematologia Brasileira* 23, 88-92.

SHARMA, R.D. & AMABILE, R.F. (1998b). Fitonematoides associados a genotipos de gergelim em condicoes de Cerrado. *Nematologia Brasileira* 23, 84-87.

SHARMA, R.D., DA SILVA, J.E., RESCK, D.V.S. & GOMES, A.C. (2000). Dinamica de Populacao de fitonematoides em solo tratado com lodo de esgoto em cultivos de milho. *Nematologia Brasileira* 24, 37-40.

SHARMA, R.D., CAVACALCANTE, M. DE J.B. & VALENTIM, J.F. (2001). Nematoides associados ao capim *Brachiaria brizantha* cv. Marandu no Estado do Acre, Brasil. *Nematologia Brasileira* 25, 217-222.

SHARMA, R.D., CAVALCANTE, M. DE J.B., MOURA, G. DE M. & VALENTIM, J.F. (2002). Nematoides associados a genotipos de soja cultivados no Acre, Brasil. *Nematologia Brasileira* 26, 109-111.

SHARMA, S.B. & DAVIES, K.G. (1996). Comparison of two sympatric *Pasteuria* populations isolated from a tropical vertisol soil. *World Journal of Microbiology and Biotechnology* 12, 361-366.

SHARMA, S.B., REDDY, B.M.R. & KRISHNAPPA, K. (1992). Incidence and importance of plant-parasitic nematodes on pigeon pea and groundnut in Karnataka State in southern India. *Afro-Asian Journal of Nematology* 2, 21-26.

SHARMA, S.B., SIDDIQI, M.R., VAN, N.V. & HONG, N.X. (1994). Plant-parasitic nematodes associated with groundnut in North Vietnam. *Afro-Asian Journal of Nematology* 4, 185-189.

SHARMA, V.K. & NOWAK, J. (1998). Enhancement of verticillium wilt resistance in tomato transplants by *in vitro* co-culture of seedlings with a plant growth promoting rhizobacterium (*Pseudomonas* sp. strain PsJN). *Canadian Journal Microbiology* 44, 528-536.

SHERBAKOFF, C.D. & STANLEY, W.W. (1943). The more important diseases and insect pests of crops in Tennessee. *Tennessee Agriculture Experiment Station Bulletin* 186, 1-142.

SHER, S.A. (1948). *Revision of the genus* Pratylenchus *(Nematoda: Tylenchidae).* Ph.D. Thesis, University of California, Riverside, CA, USA, 64 pp.

SHER, S.A. (1957). A disease of roses caused by the root-lesion nematode, *Pratylenchus vulnus*. *Phytopathology* 47, 703-706.

SHER, S.A. (1959). A root-lesion nematode disease on *Cymbidium* orchids. *Phytopathology* 49, 458-460.

SHER, S.A. & ALLEN, M.W. (1953). Revision of the genus *Pratylenchus* (Nematoda: Tylenchidae). *University of California Publications in Zoology* 57, 441-447.

SHER, S.A. & BELL, A.H. (1965). The effect of soil type and soil temperature on root lesion nematode disease of roses. *Plant Disease Reporter* 49, 849-851.

SHUKLA, B.N., VADHERA, I. & TIWARI, S.P. (2001). Impact of deep summer ploughing, seed dressing and soaking with nematicide on *Pratylenchus thornei*. *JNKVV Research Journal* 35, 82-83.

SHUKLA, P.K., HASEEB, A. & SRIVASTAVA, N.K. (1998). Influence of pH on reproduction and damage potential of *Pratylenchus thornei* on *Mentha piperita*. *Fundamental and Applied Nematology* 21, 103-105.

SHURTLEFF, M.C. & AVERRE, C.W. (2000). *Diagnosing plant diseases caused by nematodes*. St Paul, MN, USA, The American Phytopathological Society, 187 pp.

SIDAM, A.K. (1989). The distribution, damage and control of the root-knot nematode, *Meloidogyne incognita* (Kofoid and White) Chitwood on tobacco in Malaysia. *Berita Penyelidikan* 3, 7.

SIDDIQI, M.R. (1963). On the classification of the Pratylenchidae (Thorne, 1949) nov. grad. (Nematoda: Tylenchida) with a description of *Zygotylenchus browni* nov. gen. et nov. sp. *Zeitschrift fur Parasitenkunde* 23, 390-396.

SIDDIQI, M.R. (1964). Studies on nematode root-rot of citrus in Uttar Pradesh, India. *Proceedings of the Zoological Society of Calcutta* 17, 67-75.

SIDDIQI, M.R. (1971). Structure of the oesophagus in the classification of the superfamily Tylenchoidea (Nematoda). *Indian Journal of Nematology* 1, 25-43.

SIDDIQI, M.R. (1972). *Pratylenchus coffeae. CIH Descriptions of plant-parasitic nematodes*, Set 1, No. 6. St. Albans, UK, Commonwealth Institute of Helminthology, 3 pp.

SIDDIQI, M.R. (1974). Plant-parasitic nematodes of sugarcane in northwestern Venezuela. *Nematropica* 4, 6.

SIDDIQI, M.R. (1986). *Tylenchida parasites of plants and insects*. Slough, UK, Commonwealth Agricultural Bureaux, 645 pp.

SIDDIQI, M.R. (1994). Nematodes of tropical rainforests. 4. Two new species of *Pratylenchus*. *Afro-Asian Journal of Nematology* 4, 190-193.

SIDDIQI, M.R. (2000). *Tylenchida parasites of plants and insects*, 2nd edition. Wallingford, UK, CABI Publishing, 833 pp.

SIDDIQI, M.R., DABUR, K.R. & BAJAJ, H.K. (1991). Descriptions of three new species of *Pratylenchus* Filipjev, 1936 (Nematoda: Pratylenchidae). *Nematologia Mediterranea* 19, 1-7.

SIDDIQUI, I.A., SHER, S.A. & FRENCH, A.M. (1973). *Distribution of plant parasitic nematodes in California*. Sacramento, CA, USA, Department of Food and Agriculture, Division of Plant Industry, 324 pp.

SIJMONS, P.C., ATKINSON, H.J. & WYSS, U. (1994). Parasitic strategies of root nematodes and associated host cell responses. *Annual Review of Phytopathology* 32, 235-259.

SIKORA, R.A., GRECO, N. & VELOSO SILVA, J.F. (2005). Nematode parasites of food legumes. In: Luc, M., Sikora, R.A. & Bridge, J. (Eds). *Plant parasitic nematodes in subtropical and tropical agriculture*, 2nd edition. Wallingford, UK, CABI Publishing, pp. 259-318.

SIKORA, R.A., TAYLOR, D.P., MALEK, R.B. & EDWARDS, D.I. (1972). Interaction of *Meloidogyne naasi*, *Pratylenchus penetrans*, and *Tylenchorhynchus agri* on creeping bentgrass. *Journal of Nematology* 4, 162-165.

SILVA, R.A. & INOMOTO, M.M. (2002). Host-range characterization of two *Pratylenchus coffeae* isolates from Brazil. *Journal of Nematology* 34, 135-139.

SILVA, R.A., INOMOTO, M.M., KUBO, R.K., MAZZAFERA, P. & TOMAZINI, M.D. (2001). Efeito de dois aislados de *Pratylenchus coffeae* na fotosintesis e no crescimento de cafeeiros cv. Mundo novo. In: *Proceedings Congresso Brasileiro de Nematologia, Marilia, São Paulo*, 87 pp.

SIMEONE, A.M., NICOTRA, A. & DI VITO, M. (1995). Susceptibility of progenies of *Actinidia deliciosa*, *A. chinensis*, and *A. arguta* to four species of *Meloidogyne* and to *Pratylenchus vulnus*. *Nematologia Mediterranea* 23, 85-89.

SINGH, B.P. (1999). Plant parasitic nematodes associated with *Leucaena leucocephala* (Lam.) and *Albizia procera* L. Benth. *Annals of Forestry* 7, 153-154.

SINGH, N.D. (1973). Preliminary report of plant parasitic nematodes associated with important crops in Trinidad. *Nematropica* 3, 56-61.

SINGH, N.D. (1974). Effects of chemicals on nematode populations and yield of sweet pepper (*Capsicum annuum* L. var. *grossum* Sendt.). *Tropical Agriculture* 51, 81-84.

SINGH, N.D. & GILL, J.S. (1986). *Pratylenchus hyderabadensis* nomen novum for *Pratylenchus capitatus* Das & Sultana, 1979. *Indian Journal of Nematology* 16, 139.

SINGH, S.P. & JAIN, V.K. (1984). Two new species of the genus *Pratylenchus* Filipjev, 1936 from Lucknow. *Bulletin of Entomology* 25, 179-185.

SIPES, B.S., CASWELL-CHEN, E.P., SARAH, J.-L. & APT, W.J. (2005). Nematode parasites of pineapple. In: Luc, M., Sikora, R.A. & Bridge, J. (Eds). *Plant parasitic nematodes in subtropical and tropical agriculture*, 2nd edition. Wallingford, UK, CABI Publishing, pp. 709-731.

SITARAMAIAH, K. & PATHAK, K.N. (1993). Nematode bacterial disease interactions. In: Khan, M.W. (Ed.). *Nematode interactions*. London, UK, Chapman & Hall, pp. 232-250.

SIVAKUMAR, M. & VADIVELU, S. (1997). Parasitic nematodes associated with medicinal and aromatic plants. *Indian Journal of Nematology* 27, 58-62.

SIYANAND, SESHADRI, A.R. & DASGUPTA, D.R. (1982). Investigation on the life-cycles of *Tylenchorhynchus vulgaris*, *Pratylenchus thornei* and *Hoplolaimus indicus* individually and in combined infestations in corn. *Indian Journal of Nematology* 12, 272-276.

SLABAUGH, W.B. (1974). The nature of resistance in tomatoes to three nematodes. *Dissertation Abstracts International* 34B, 5280.

SLOOTWEG, A.F.G. (1956). Root rot of bulbs caused by *Pratylenchus penetrans* and *Hoplolaimus* spp. *Nematologica* 1, 192-201.

SMILEY, R.W., WHITTAKER, R.G., GOURLIE, J.A. & EASLEY, S.A. (2003). Genetic tolerance to root-lesion nematode (*Pratylenchus neglectus* and *P. thornei*) in wheat. *Phytopathology* 93, 580. [Abstr.]

SMILEY, R.W., MERRIFIELD, K., PATTERSON, L.M., WHITTAKER, R.G., GOURLIE, J.A. & EASLEY, S.A. (2004). Nematodes in dryland field crops in the semiarid Pacific Northwest United States. *Journal of Nematology* 36, 54-68.

SMILEY, R.W., WHITTAKER, R.G., GOURLIE, J.A. & EASLEY, S.A. (2005a). *Pratylenchus thornei* associated with reduced wheat yield in Oregon. *Journal of Nematology* 37, 45-54.

SMILEY, R.W., WHITTAKER, R.G., GOURLIE, J.A. & EASLEY, S.A. (2005b). Suppression of wheat growth and yield by *Pratylenchus neglectus* in the Pacific Northwest. *Plant Disease* 89, 958-968.

SMITH, G.S. (1988). The role of phosphorus nutrition in interactions of vesicular-arbuscular mycorrhizal fungi with soilborne nematodes and fungi. *Phytopathology* 78, 371-374.

SMITH, O.D., BOSWELL, T.E. & THAMES, W.H. (1978). Lesion nematode resistance in peanuts. *Crop Science* 18, 1008-1011.

SMITH, S.E. & READ, D.J. (1997). *Mycorrhizal symbiosis*. San Diego, California, USA, Academic Press, 605 pp.

SMOLIK, J.D. (1978). Influence of previous insecticidal use on ability of carbofuran to control nematode populations in corn and effect on corn yield. *Plant Disease Reporter* 62, 95-99.

SMOLIK, J.D. & EVENSON, P.D. (1987). Relationship of yields and *Pratylenchus* spp. Population densities in dryland and irrigated corn. *Annals of Applied Nematology* 1, 71-73.

SMOLIK, J.D. & WICKS, Z.W. (1987). Reproduction of *Pratylenchus hexincisus* and *P. scribneri* in corn inbreds. *Annals of Applied Nematology* 1, 29-31.

SNEATH, P.H.A. & SOKAL, R.R. (1973). *Numerical taxonomy*. San Francisco, USA, W.H. Freeman & Co., 573 pp.

SOHLENIUS, B. & SANDOR, A. (1987). Vertical distribution of nematodes in arable soil under grass (*Festuca pratensis*) and barley (*Hordeum distichum*). *Biology and Fertility of Soils* 3, 19-25.

SOLTWEDEL, F. (1888). Mededeelingen van het Proefstation voor Midden-Java te Semarang, April 1887. *Tijdschrift voor Land-en Tuinbouw en Boschkultuur Nederlandsch Oost-Indië* 3, 51-57, 158-163, 168, plates 1-3.

SONNEMANN, I., FINKHAEUSER, K. & WOLTERS, V. (2002). Does induced resistance in plants affect the belowground community? *Applied Soil Ecology* 21, 179-185.

SORIANO, I.R., RILEY, I.T., POTTER, M.J. & BOWERS, W.S. (2004). Phytoecdysteroids: A novel defense against plant-parasitic nematodes. *Journal of Chemical Ecology* 30, 1885-1899.

SPEIJER, P.R. & DE WAELE, D. (2001). Nematodes associated with East African Highland cooking bananas and cv. Pisang Awak (*Musa* spp.) in Central Uganda. *Nematology* 3, 535-541.

SPEIJER, P.R., ROTIMI, M.O. & DE WAELE, D. (2001). Plant parasitic nematodes associated with plantain (*Musa* spp., AAB-group) in southern Nigeria and their relative importance compared to other biotic constraints. *Nematology* 3, 423-436.

SRINIVASAN, A. & MUTHUKRISHNAN, T.S. (1975). Pathogenicity of the lesion nematode, *Pratylenchus delattrei* Luc, 1958 on *Crossandra undulaefolia* Salisb. *Indian Journal of Nematology* 5, 222-225.

STALIN, N., SALESSES, G., MINOT, J.C., VOISIN, R. & ESMENJAUD, D. (1998). Comparative host suitability to *Meloidogyne* spp. and *Praty-*

lenchus vulnus in Myrobalan plum (*Prunus cerasifera*). *Plant Pathology* 47, 211-215.

STANTON, J.M., SIDDIQI, M.R. & LENNE, J.M. (1989). Plant-parasitic nematodes associated with tropical pastures in Colombia. *Nematropica* 19, 169-175.

STANTON, J.M., PATTISON, A.B. & KOPITTKE, R.A. (2001). A sampling strategy to assess banana crops for damage by *Radopholus similis* and *Pratylenchus goodeyi*. *Australian Journal of Experimental Agriculture* 41, 675-679.

STAPLETON, J.J. & DEVAY, J.E. (1984). Thermal components of soil solarization as related to changes in soil and root microflora and increased growth response. *Phytopathology* 74, 255-259.

STAPLETON, J.J., FERGUSON, L., MCKENRY, M.V., DOUGHERTY, D.S. & STAPLETON, S.C. (1999). Using solarization to disinfest soil for olive nursery production. *Acta Horticulturae* 474, 589-591.

STARR, J.L. (1984). Expression of resistance in peanuts to *Pratylenchus brachyurus*: impact on screening for resistance. *Journal of Nematology* 16, 404-406.

STARR, J.L., BRIDGE, J. & COOK, R. (2002). Resistance to plant parasitic nematodes: history, current use and future potential. In: Starr, J.L., Bridge, J. & Cook, R. (Eds). *Plant resistance to parasitic nematodes*. Wallingford, UK, CABI Publishing, pp. 1-22.

STARR, M.P. & SAYRE, R.M. (1988). *Pasteuria thornei* sp. nov. and *Pasteuria penetrans sensu stricto* emend., mycelial and endospore-forming bacteria parasitic, respectively, on plant-parasitic nematodes of the genera *Pratylenchus* and *Meloidogyne*. *Annals of the Institut Pasteur* 139, 11-31.

STEINER, G. (1928). *Tylenchus pratensis* and various other nemas attacking plants. *Journal of Agricultural Research* 35, 961-981.

STEINER, G. (1949). Plant nematodes the grower should know. *Proceedings of the Soil Science Society of Florida* 4B, 72-117.

STIRLING, A.M. & STIRLING, G.R. (2002). Root rot of *Hippeastrum* sp. caused by lesion nematode (*Pratylenchus jordanensis*). *Australasian Plant Pathology* 31, 427.

STIRLING, G.R. (1975). A survey of the plant-parasitic nematodes in Riverland peach orchards. *Agricultural Record* 2, 11-13.

STIRLING, G.R. (1991). *Biological control of plant parasitic nematodes*. Brisbane, Australia, CAB International, 282 pp.

STIRLING, G.R., DULLAHIDE, S.R. & NIKULIN, A. (1995). Management of lesion nematode (*Pratylenchus jordanensis*) on replanted apple trees. *Australian Journal of Experimental Agriculture* 35, 247-258.

STOEEN, M. (1974). Nematodes on roses in Norway. *Nordisk Jordbrugsforskning* 56, 418-419.

STOFFELEN, R., VU, T.T.T., SWENNEN, R.L. & DE WAELE, D. (1999). Host plant response of banana (*Musa* spp.) cultivars from Southeast Asia to nematodes. *International Journal of Nematology* 9, 130-136.

STOFFELEN, R., VERLINDEN, R., XUYEN, N.T., SWENNEN, R. & DE WAELE, D. (2000). Host plant response of *Eumusa* and *Australimusa* bananas (*Musa* spp.) to migratory endoparasitic and root-knot nematodes. *Nematology* 2, 907-916.

STOYANOV, D. (1977). [Parasitic species of nematodes on corn.] *Rastitelna Zashchita* 25, 23-24.

STOYANOV, D. (1979). [Endoparasites of plant nematodes found in Bulgaria.] *Rastitelna Zashchita* 26, 29-31.

STURHAN, D. (1985). Untersuchungen uber verbreitung und wirte des nematodenparasiten *Bacillus penetrans. Mitteilungen aus der Biologischen Bundesanstalt fur Land und Forstwirtschaft Berlin Dahlem* 226, 75-93.

STURZ, A.V. & KIMPINSKI, J. (1999). Effects of fosthiazate and aldicarb on populations of plant-growth-promoting bacteria, root-lesion nematodes and bacteria-feeding nematodes in the root zone of potatoes. *Plant Pathology* 48, 26-32.

STURZ, A.V. & KIMPINSKI, J. (2004). Endoroot bacteria derived from marigolds (*Tagetes* spp.) can decrease soil population densities of root-lesion nematodes in the potato root zone. *Plant and Soil* 262, 241-249.

SUATMADJI, R.W. (1988). *Pratylenchus penetrans* and *Rotylenchus robustus* on thirty herbaceous ornamental species. *Australian Plant Pathology* 17, 97-98.

SUATMADJI, R.W. & MARKS, G.C. (1983). *Pratylenchus penetrans* in *Pinus radiata* in Victoria. *Australasian Plant Pathology* 12, 29-31.

SUBBOTIN, S.A., STURHAN, D., CHIZHOV, V.N., VOVLAS, N. & BALDWIN, J.G. (2006). Phylogenetic analysis of Tylenchida Thorne, 1949 as inferred from D2 and D3 expansion fragments of the 28S rRNA gene sequences. *Nematology* 8, 455-474.

SUBRAMANIAN, S. & SIVAKUMAR, C.V. (1988). Morphometric and allometric variations in *Pratylenchus delattrei* Luc, 1958 from Tamil Nadu, India. *Indian Journal of Nematology* 18, 392-394.

SUBRAMANIYAN, S. & SIVAKUMAR, C.V. (1991). New species of *Pratylenchus* Filipjev 1936 (Pratylenchidae: Nematoda). *Current Research University of Agricultural Sciences Bangalore* 20, 89-90.

SUMNER, D.R. & MINTON, N.A. (1987). Interaction of Fusarium wilt and nematodes in Cobb soybean. *Plant Disease* 71, 20-23.

SUNDARABABU, R. & VADIVELU, S. (1988). Studies on the nematodes associated with rose. *Indian Journal of Nematology* 18, 139.

SUNDARABABU, R., SANKARANARAYANAN, C. & JOTHI, G. (1998). Management of *Pratylenchus zeae* on corn by biofertilizers and VAM. *Indian Journal of Nematology* 28, 77-80.

SUNDARABABU, R., RAMAKRISHNAN, S., JOTHI, G. & AJENDRAN, G. (2004). Optimization of commercial formulation of VAM (*Glomus mosseae*) against *Pratylenchus delattrei* in crossandra. *Indian Journal of Nematology* 34, 89-90.

SUNDARAM, R., VELAYUTHAM, B. & VADIVELU, S. (1987). Seasonal fluctuation of nematodes affecting garden bean. *International Nematology Network Newsletter* 4, 22-24.

SUNDARARAJU, P. & CANNAYANE, I. (2002). Antinemic activity of plant extracts against *Pratylenchus coffeae* infecting banana. *Indian Journal of Nematology* 32, 121-124.

SUNDARARAJU, P. & JEYABASKARAN, K.J. (2003). Evaluation of different soil types on multiplication of *Pratylenchus coffeae* and growth of banana seedlings var, Nendran. *Nematologia Mediterranea* 31, 151-153.

SUNDARARAJU, P. & RATNAKARAN, K. (2002). Factors influencing the prevalence of the root-lesion nematode, *Pratylenchus coffeae* on oil palm (*Elaeis guineensis* Jacq.). *Indian Journal of Nematology* 32, 16-19.

SUNDARARAJU, P., MUSTAFFA, M.M., KUMAR, V., CANNAYANE, I. & PRIYA, B.T. (2002). Effect of organic farming on plant parasitic nematodes infesting banana cv. Karpuravalli. *Current Nematology* 13, 39-43.

SUNDARARAJU, P., PADMANABAN, B. & SATHIAMOORTHY, S. (2003). Efficacy of certain botanicals against root-lesion nematode, *Pratylenchus coffeae* in banana. *Nematologia Mediterranea* 31, 201-205.

SWANEPOEL, V.S. & LOOTS, G.C. (1988). The influence of temperature on anhydrobiosis of *Pratylenchus zeae*. *Phytophylactica* 20, 31-33.

SZCZYGIEL, A. (1974). [Plant parasitic nematodes associated with strawberry plantations in Poland.] *Zeszyty Problemowe Postepow Nauk Rolniczych* 154, 9-132.

SZCZYGIEL, A. (1982). Tolerance limit of *Pratylenchus penetrans* for strawberry plants as affected by experimental conditions. *Nematologica* 28, 176.

SZCZYGIEL, A. & HASIOR, H. (1972). Vertical distribution of plant-parasitic nematodes in the soil of strawberry plantations. *Ekologia Polska* 20, 493-506.

SZCZYGIEL, A. & ZEPP, A.L. (1998). An occurrence and importance of apple replant disease in Polish orchards. *Acta Horticulturae* 477, 99-102.

TACCONI, R. & TALAME, M. (1995). Nematodi parassiti di drupacee, pomacee e piante ortive. *Rivista di Frutticoltura e di Ortofloricoltura* 57, 63-71.

TACCONI, R., SANTI, R., MENICHETTI, P. & POLA, R. (1985). Presenza di *Pratylenchus scribneri* Steiner su radici di granturco (*Zea mays*) e di sorghetta (*Sorghum halepense*). *Informatore Fitopatologico* 35, 35-37.

TACCONI, R., GIORDANI, G. & TODERI, G. (1988a). Effeto di concimazione e di avvicendamenti su una populazione di *Pratylenchus thornei*. *Informatore Fitopatologico* 38, 47-50.

TACCONI, R., SANTI, R. & POLA, R. (1988b). *Pratylenchus neglectus* su granturco (*Zea mays*) in Emilia Romagna. *Informatore Fitopatologico* 38, 59-61.

TAHERI, A., HOLLAMBY, G.J., VANSTONE, V.A. & NEATE, S.M. (1994). Interaction between root lesion nematode, *Pratylenchus neglectus* (Rensch 1924) Chitwood and Oteifa 1952, and root rotting fungi of wheat. *New Zealand Journal of Crop and Horticultural Science* 22, 181-185.

TALAVERA, M. & NAVAS, A. (2002). Incidence of plant-parasitic nematodes in natural and semi-natural mountain grassland and the host status of some common grass species. *Nematology* 4, 541-552.

TALAVERA, M. & TOBAR JIMÉNEZ, A. (1997). Plant parasitic nematodes from unirrigated fields in Alhama, southeastern Spain. *Nematologia Mediterranea* 25, 73-81.

TALAVERA, M. & VALOR, H. (2000). Influence of the previous crop on the anhydrobiotic ability of *Pratylenchus thornei* and *Merlinius brevidens*. *Nematologia Mediterranea* 28, 77-81.

TALAVERA, M. & VANSTONE, V.A. (2001). Monitoring *Pratylenchus thornei* densities in soil and roots under resistant (*Triticum turgidum durum*) and susceptible (*Triticum aestivum*) wheat cultivars. *Phytoparasitica* 29, 29-35.

TALAVERA, M., MAGUNACELAYA, J.C. & TOBAR, A. (1999). Plant parasitic nematodes from a forest tree nursery in southern Spain with some notes about the influence of soil storage on the quantitative recovery of *Meloidogyne arenaria*. *Nematology* 1, 261-266.

TALAVERA, M., ITOU, K. & MIZUKUNO, T. (2001). Reduction of nematode damage by root colonization with arbuscular mycorrhiza (*Glomus* spp.) in tomato-*Meloidogyne incognita* (Tylenchida: Meloidogynidae) and carrot-*Pratylenchus penetrans* (Tylenchida: Pratylenchidae) pathosystems. *Applied Entomology and Zoology* 36, 387-392.

TALWANA, H.A.L., SPEIJER, P.R. & DE WAELE, D. (2000). Spatial distribution of nematode population densities and nematode damage in roots of three banana cultivars in Uganda. *Nematropica* 30, 19-31.

TALWANA, H.A.L., SPEIJER, P.R., GOLD, C.S., SWENNEN, R.L. & DE WAELE, D. (2003). A comparison of the effects of the nematodes *Radopholus similis* and *Pratylenchus goodeyi* on growth, root health and yield of an East African highland cooking banana (*Musa* AAA-group). *International Journal of Pest Management* 49, 199-204.

TARTÉ, R. & MAI, W.F. (1976). Morphological variation in *Pratylenchus penetrans*. *Journal of Nematology* 8, 185-195.

TARTÉ, R., CERRUD, D., RODRIGUEZ, I. & OSORIO, J.M. (1977). Presencia y parasitismo de *Pratylenchus zeae* en cana de azucar en Panama con indicaciones sobre la susceptibilidad relativa de algunos cultivares. *Turrialba* 27, 259-266.

TAYLOR, D.P. & JENKINS, W.R. (1957). Variation within the nematode genus *Pratylenchus,* with the descriptions of *P. hexincisus* n. sp. and *P. subpenetrans* n. sp. *Nematologica* 2, 159-174.

TAYLOR, S.P. & EVANS, M.L. (1998). Vertical and horizontal distribution of and soil sampling for root lesion nematodes (*Pratylenchus neglectus* and *P. thornei*) in South Australia. *Australasian Plant Pathology* 27, 90-96.

TAYLOR, S.P., VANSTONE, V.A., WARE, A.H., MCKAY, A.C., SZOT, D. & RUSS, M.H. (1999). Measuring yield loss in cereals caused by root lesion nematodes (*Pratylenchus neglectus* and *P. thornei*) with and without nematicide. *Australian Journal of Agricultural Research* 50, 617-622.

TAYLOR, S.P., HALLAWAY, G.J. & HUNT, C.H. (2000). Effect of field crops on population densities of *Pratylenchus neglectus* and *P. thornei* in southeastern Australia; part 1: *P. neglectus*. *Journal of Nematology* 32, 591-599.

TEBEN'KOVA, T.M. & IVANOVA, T.S. (1989). [Plant-parasitic nematodes of fruits in Tadzhikistan. 1. Nematodes of cultivated fruits in the lowland zone.] *Izvestiya Akademii Nauk Tadzhiskoi SSR, Biologicheskikh Nauk* 3, 7-11.

TELIZ, D. & GOHEEN, A.C. (1968). Diseases of grapevines in Mexico. *Plant Disease Reporter* 52, 372-373.

THAMES, W.H. (1982). The genus *Pratylenchus.* In: Riggs, R.D. (Ed.). *Nematology in the southern region of the United States.* Arkansas, Fayetteville, USA, Southern Cooperative Series Bulletin 276, pp. 108-126.

THIES, J.A., BARNES, D.K., RABAS, D.L., SHEAFFER, C.C. & WICOXSON, R.D. (1992). Seeding date, carbofuran, and resistance to root lesion nematode affect alfalfa stand establishment. *Crop Science* 32, 786-792.

THIES, J.A., BASIGALUP, D. & BARNES, D.K. (1994). Inheritance of resistance to *Pratylenchus penetrans* in alfalfa. *Journal of Nematology* 26, 452-459.

THIES, J.A., PETERSEN, A.D. & BARNES, D.K. (1995). Host suitability of forage grasses and legumes for the root-lesion nematode *Pratylenchus penetrans*. *Crop Science* 35, 1647-1651.

THOMAS, S.H. (1978). Population densities of nematodes, under seven tillage regimes. *Journal of Nematology* 10, 24-27.

THOMASON, I.J., RADEWALD, J.D., WELCH, N.C. & MCKINNEY, H.E. (1964). Onion yield increases following application of nematocides to control stubby-root, root-knot and lesion nematodes. *Plant Disease Reporter* 48, 409-413.

THOMASON, I.J., RICH, J.R. & O'MELIA, F.C. (1976). Pathology and histopathology of *Pratylenchus scribneri* infecting snap bean and lima bean. *Journal of Nematology* 8, 347-352.

THOMASON, J. & O'MELIA, F.C. (1962). Pathogenicity of *Pratylenchus scribneri* to crop plants. *Phytopathology* 52, 755. [Abstr.]

THOMPSON, J.P. (1993). Assessing damage to wheat by *Pratylenchus thornei*. In: *Proceedings of the* Pratylenchus *workshop, 9th biennial Australasian plant pathology conference, Hobart, Tasmania, July 1993*.

THOMPSON, J.P. & CLEWETT, T.G. (1986). Research on root-lesion nematode. In: *Queensland Wheat Research Institute Biennial Report 1982-1984*. Toowoomba, Queensland, Australia, Queensland Wheat Research Institute, pp. 32-35.

THOMPSON, J.P. & HAAK, M.I. (1997). Resistance to root-lesion nematode (*Pratylenchus thornei*) in *Aegilops tauschii* Coss., the D-genome donor to wheat. *Australian Journal of Agricultural Research* 48, 553-559.

THOMPSON, L.S. & WILLIS, C.B. (1975). Influence of fensulfothion and fenamiphos on root lesion nematode numbers and yield of forage legumes. *Canadian Journal of Plant Science* 55, 727-735.

THORNE, G. (1935). Some plant parasitic nemas with descriptions of three new species. *Journal of Agricultural Research* 49, 755-763.

THORNE, G. (1940). *Duboscqia penetrans*, n. sp. (Sporozoa, Microsporidia, Nosemmatidae), a parasite of the nematode *Pratylenchus pratensis* (de Man) Filipjev. *Proceedings of the Helminthological Society of Washington* 7, 50-53.

THORNE, G. (1949). On the classification of the Tylenchida, new order (Nematoda, Phasmidia). *Proceedings of the Helminthological Society of Washington* 16, 37-73.

THORNE, G. (1961). *Principles of nematology*. New York, USA, McGraw-Hill Book Co. Inc., 533 pp.

THORNE, G. & ALLEN, M.W. (1944). *Nacobbus dorsalis*, nov. gen., nov. sp. (Nematoda: Tylenchidae) producing galls on the roots of alfileria, *Erodium cicutarium* (L.) L'Hér. *Proceedings of the Helminthological Society of Washington* 11, 27-31.

THORNE, G. & MALEK, R.B. (1968). Nematodes of the Northern Great Plains. I. Tylenchida. *South Dakota Agricultural Experimental Station, Technical Bulletin* 31, 111 pp.

TIHOHOD, D., FERRAZ, L.C.C.B. & BARBOSA, J.C. (1991). Manejo de fitonematoides do algodoeiro 1. Comparacao entre metodos de amostragem. *Nematologia Brasileira* 15, 112-120.

TIMPER, P. & BRODIE, B.B. (1993). Infection of *Pratylenchus penetrans* by nematode-pathogenic fungi. *Journal of Nematology* 25, 297-302.

TIMPER, P. & BRODIE, B.B. (1994). Effect of *Hirsutella rhossiliensis* on infection of potato by *Pratylenchus penetrans*. *Journal of Nematology* 26, 304-307.

TIMPER, P. & BRODIE, B.B. (1997). First report of *Pratylenchus neglectus* in New York. *Plant Disease* 81, 228.

TIMPER, P. & HANNA, W.W. (2004). Host status of pearl millet for sting, stubby-root, and lesion nematodes. *Journal of Nematology* 36, 349.

TIMPER, P. & HANNA, W.W. (2005). Reproduction of *Belonolaimus longicaudatus*, *Meloidogyne javanica*, *Paratrichodorus minor*, and *Pratylenchus brachyurus* on pearl millet (*Pennisetum glaucum*). *Journal of Nematology* 37, 214-219.

TIMPER, P., GATES, R.N. & BOUTON, J.H. (2005). Response of *Pratylenchus* spp. in tall fescue infected with different strains of the fungal endophyte *Neotyphodium coenophialum*. *Nematology* 7, 105-110.

TINER, J.D. (1960). Cultures of the plant parasitic nematode genus *Pratylenchus* on sterile excised roots. I. Their establishment and maintenance. *Experimental Parasitology* 9, 121-126.

TIWARI, S.P., VADHERA, I., SHUKLA, B.N. & BHATT, J. (1992). Studies on the pathogenicity and relative reactions on chickpea lines to *Pratylenchus thornei*. *Indian Journal of Mycology and Plant Pathology* 22, 255-259.

TIXIER, P., RISÈDE, J.M., DOREL, M. & MALÉZIEUX, E. (2006). Modelling population dynamics of banana plant-parasitic nematodes: A contribution to the design of sustainable cropping systems. *Ecological Modelling* 198, 321-331.

TIYAGI, S.A. & SHAMIM, A. (2004). Biological control of plant parasitic nematodes associated with chickpea using oil cakes and *Paecilomyces lilacinus*. *Indian Journal of Nematology* 34, 44-48.

TOBAR, A., VALOR, H. & TALAVERA, M. (1995). Effect of different cultivars, mainly of wheat, on the population densities of *Pratylenchus thornei* and *Merlinius brevidens* in dry soils in Spain. *Nematologica* 41, 642-644.

TOBAR, A., VALOR, H. & TALAVERA, M. (1996). On the field anhydrobiotic ability of *Pratylenchus thornei* and *Merlinius brevidens*. *Fundamental and Applied Nematology* 19, 43-46.

TODD, T.C. & OAKLEY, T.R. (1995). Evaluation of nematicides for lesion nematode control in corn, 1994. *Fungicide and Nematicide Tests* 50, 188.

TOIDA, Y., TANGCHITSOMKID, N., KEEREEWAN, S. & MIZUKUBO, T. (1996). Nematode species attacking crops in Thailand with measurements of second-stage juveniles of *Meloidogyne* spp. *JIRCAS Journal* 3, 59-68.

TOKOBAEV, M.M. & MATYASHOV, V.D. (1986). [Study of the nematode fauna of *Juglans regia* in the Sary-Chelekskii nature reserve.] *Svobodnozhivushchie i paraziticheskie skoletsidy fauny Kirgizii*, 28-37.

TORRES, M.S. & CHAVES, E.J. (1999). Description of some Pratylenchidae (Nemata) from Argentina. *Nematologia Mediterranea* 27, 281-289.

TOWNSHEND, J.L. (1962). The root-lesion nematode, *Pratylenchus penetrans* (Cobb, 1917) Filip. & Stek. 1941, in celery. *Canadian Journal of Plant Science* 42, 314-322.

TOWNSHEND, J.L. (1963a). The pathogenicity of *Pratylenchus penetrans* to celery. *Canadian Journal of Plant Science* 43, 70-74.

TOWNSHEND, J.L. (1963b). The pathogenicity of *Pratylenchus penetrans* to strawberry. *Canadian Journal of Plant Science* 43, 75-78.

TOWNSHEND, J.L. (1973). Survival of *Pratylenchus penetrans* and *P. minyus* in two Ontario soils. *Nematologica* 19, 35-42.

TOWNSHEND, J.L. (1984a). Inoculum densities of five plant parasitic nematodes in relation to alfalfa seedling growth. *Canadian Journal of Plant Pathology* 6, 309-312.

TOWNSHEND, J.L. (1984b). Anhydrobiosis in *Pratylenchus penetrans. Journal of Nematology* 16, 282-289.

TOWNSHEND, J.L. (1987). Anhydrobiosis in *Pratylenchus penetrans* and *Tylenchorhynchus* n. sp. in cultivated soils cropped to winter rye. *Journal of Nematology* 19, 164-171.

TOWNSHEND, J.L. (1989a). Population densities of four species of root-lesion nematodes (*Pratylenchus*) in the oat cultivars, Saia and OAC Woodstock. *Canadian Journal of Plant Science* 69, 903-905.

TOWNSHEND, J.L. (1989b). Efficacy of oxamyl coated on alfalfa seed with a polymer sticker in *Pratylenchus* and *Meloidogyne* infested soils. *Journal of Nematology* 21, 242-246.

TOWNSHEND, J.L. (1990). Growth of 'Bartlett' pear seedlings in response to number of root-lesion nematodes and temperature. *HortScience* 25, 318-320.

TOWNSHEND, J.L. (1991). Morphological observations of *Pratylenchus penetrans* from celery and strawberry in Southern Ontario. *Journal of Nematology* 23, 205-209.

TOWNSHEND, J.L. & ANDERSON, R.V. (1976). *Pratylenchus neglectus. CIH Descriptions of plant-parasitic nematodes*, Set 6, No. 82. Farnham Royal, UK: Commonwealth Agricultural Bureaux, 4 pp.

TOWNSHEND, J.L. & OLTHOF, T.H.A. (1988). Growth of potato and control of *Pratylenchus penetrans* with oxamyl-treated seed-pieces in greenhouse studies. *Journal of Nematology* 20, 405-409.

TOWNSHEND, J.L. & STOBBS, L. (1981). Histopathology and histochemistry of lesions caused by *Pratylenchus penetrans* in roots of forage legumes. *Canadian Journal of Plant Pathology* 3, 123-128.

TOWNSHEND, J.L. & WEBBER, L.R. (1971). Movement of *Pratylenchus penetrans* and the moisture characteristics of three Ontario soils. *Nematologica* 17, 47-57.

TOWNSHEND, J.L. & WOLYNETZ, M.S. (1991). Penetration of celery and alfalfa roots by *Pratylenchus penetrans* as affected by temperature. *Journal of Nematology* 23, 194-197.

TOWNSHEND, J.L., MARKS, C.F. & NEFF, A.E. (1975). Effects of fumigation on growth of grape rootstocks on soil infested with root-lesion nematodes. *Plant Disease Reporter* 59, 777-779.

TOWNSHEND, J.L., POTTER, J.W. & WILLIS, C.B. (1978a). Ranges of distribution of species of *Pratylenchus* in northeastern North America. *Canadian Plant Disease Survey* 58, 80-82.

TOWNSHEND, J.L., TARTE, R. & MAI, W.F. (1978b). Growth response of three vegetables to smooth- and crenate-tailed females of three species of *Pratylenchus*. *Journal of Nematology* 10, 259-263.

TOWNSHEND, J.L., STOBBS, L. & CARTER, R. (1989). Ultrastructural pathology of cells affected by *Pratylenchus penetrans* in alfalfa roots. *Journal of Nematology* 21, 530-539.

TOWNSON, A.J. & LEAR, B. (1982a). Effect of temperature on reproduction and motility of *Pratylenchus vulnus*. *Journal of Nematology* 14, 602.

TOWNSON, A.J. & LEAR, B. (1982b). Control of nematodes in rose plants by hot-water treatment preceded by heat-hardening. *Nematologica* 28, 339-353.

TRETT, M.W. & PERRY, R.N. (1985a). Effects of the carbamoyloxime, aldicarb, on the ultrastructure of the root-lesion nematode, *Pratylenchus penetrans* (Nematoda: Pratylenchidae). *Nematologica* 31, 321-334.

TRETT, M.W. & PERRY, R.N. (1985b). Functional and evolutionary implications of the anterior sensory anatomy of species of root-lesion nematode (genus *Pratylenchus*). *Revue de Nématologie* 8, 341-355.

TREVATHAN, L.E., CUAREZMA-TERÁN, J.A. & GOURLEY, L.M. (1985). Relationship of plant-nematodes and edaphic factors in colombian grain sorghum production. *Nematropica* 15, 145-153.

TRIANTAPHYLLOU, A.C. (1987). Genetics of nematode parasitism of plants. In: Veech, A. & Dickson, D.W. (Eds). *Vistas on nematology*. Hyattsville, MD, USA, Society of Nematologists, pp. 354-371.

TRIANTAPHYLLOU, A.C. & HIRSCHMANN, H. (1980). Cytogenetics and morphology in relation to evolution and speciation of plant-parasitic nematodes. *Annual Review of Phytopathology* 18, 333-359.

TRIFONOVA, Z. & PENEVA, A. (2003). Influence of imazethapyr on the population densities of plant parasitic nematodes on bean (*Phaseolus vulgaris* L.). *Bulgarian Journal of Agricultural Science* 9, 515-520.

TROCCOLI, A. (1995). Particolarità diagnostiche del genere *Pratylenchus* (Nemata: Pratylenchidae). *Nematologia Mediterranea* 23, 25-33.

TROCCOLI, A. & DI VITO, M. (2002). Root lesion and stem nematodes associated with faba bean in North Africa. *Nematologia Mediterranea* 30, 79-81.

TROCCOLI, A., LAMBERTI, F. & GRECO, N. (1992). *Pratylenchus* species occurring in Algeria (Nematoda, Pratylenchidae). *Nematologia Mediterranea* 20, 97-103.

TROCCOLI, A., COSTILLA, M.A. & LAMBERTI, F. (1996a). Note morfobiometriche su *Pratylenchus zeae* Graham, 1951. *Nematologia Mediterranea* 24, 43-47.

TROCCOLI, A., PESTANA, M., VOVLAS, N., ABRANTES, I.M.O. & DE SANTOS, M.S.N.A. (1996b). Endoparasitic nematode species, *Pratylenchus goodeyi*, from banana root in Madeira island orchards. In: *Proceedings of the II Symposium 'Fauna and Flora of The Atlantic Islands', February, 1996, Las Palmas de Gran Canaria*, p. 125.

TRUDGILL, D.L. (1991). Resistance to and tolerance of plant parasitic nematodes in plants. *Annual Review of Phytopathology* 29, 167-192.

TSAI, B.Y. & YEH, H.L. (1995). Effect of *Steinernema carpocapsae* Weiser on the infectivity of *Pratylenchus coffeae* (Zimmermann) Filipjev & Schuurmans Stekhoven and *Meloidogyne javanica* (Treub) Chitwood. *Plant Pathology Bulletin* 4, 106-110.

TSAY, T.T., CHENG, Y.H., LIN, Y.Y. & CHENG, C.F. (1994). Nematode diseases of root and tuber crops in Taiwan. *Plant Protection Bulletin Taipei* 36, 225-238.

TSKITISHVILI, T.D. (1983). [Nematodes of citrus crops in Abkhaziya (Georgian SSR).] *Parazitologicheskii Sbornik, Tbilisi, USSR* 5, 114-119.

TU, C.M., MARKS, C.F., JOHNSON, P.W. & ELLIOT, J.M. (1995). Effects of nematicides on populations of soil microflora and *Pratylenchus penetrans*, soil enzymes and flue cured tobacco. *Journal of Environmental Science and Health Part B, Pesticides, Food Contaminants, and Agricultural Wastes* 30, 95-111.

TU, C.M., MARKS, C.F. & ELLIOT, J.M. (1996). Effects of nematicides on *Pratylenchus penetrans*, soil nitrification, and growth of flue-cured tobacco. *Bulletin of Environmental Contamination and Toxicology* 57, 924-931.

TURNER, D.R. & CHAPMAN, R.A. (1972). Infection of seedlings of alfalfa and red clover by concomitant populations of *Meloidogyne incognita* and *Pratylenchus penetrans*. *Journal of Nematology* 4, 280-286.

UEHARA, T. (2003). Molecular phylogenetic analysis and characterization of secretory enzyme genes of plant parasitic nematodes. *Journal of Genetic Plant Pathology* 69, 422.

UEHARA, T., MIZUKUBO, T., KUSHIDA, A. & MOMOTA, Y. (1998a). Identification of *Pratylenchus penetrans* (Cobb) by PCR using ITS-based species-specific primers. *Japanese Journal of Nematology* 28, 1-7.

UEHARA, T., MIZUKUBO, T., KUSHIDA, A. & MOMOTA, Y. (1998b). Identification of *Pratylenchus coffeae* and *P. loosi* using specific primers for amplification of ribosomal DNA. *Nematologica* 44, 357-368.

UEHARA, T., KUSHIDA, A. & MOMOTA, Y. (2001). PCR-based cloning of two 4-1,4-endoglucanases from the root-lesion nematode *Pratylenchus penetrans*. *Nematology* 3, 335-341.

UMESH, K.C. & FERRIS, H. (1992). Effects of temperature on *Pratylenchus neglectus* and on its pathogenicity to barley. *Journal of Nematology* 24, 504-511.

UMESH, K.C. & FERRIS, H. (1994). Influence of temperature and host plant on the interaction between *Pratylenchus neglectus* and *Meloidogyne chitwoodi*. *Journal of Nematology* 26, 65-71.

UNNY, K.L. & JERATH, M.L. (1965). Parasitic nematodes on *Dioscorea* spp. in eastern Nigeria. *Plant Disease Reporter* 49, 875-876.

UREK, G., ŠIRCA, S. & KARSSEN, G. (2003). A review of plant-parasitic and soil nematodes in Slovenia. *Nematology* 5, 391-403.

USMANOVA, A.Z. & ADYLOVA, N.A. (1976). [The nematode fauna of potato in the Ferganskaya region.] *Sbornik Nauchnykh Trudov, Tashkentskogo Universiteta Biologiya i pochvovedenie* 514, 73-74.

VAAST, P., CASWELL-CHEN, E.P. & ZASOSKI, R.J. (1998). Influences of a root-lesion nematode, *Pratylenchus coffeae*, and two arbuscular mycorrhizal fungi, *Acaulospora mellea* and *Glomus clarum* on coffeae (*Coffeae arabica* L.). *Biology and Fertility of Soils* 26, 130-135.

VADIVELU, S. & MUTHUKRISHNAN, T.S. (1979). Nematodes associated with crossandra crop in Tamil Nadu. *Indian Journal of Nematology* 9, 68-69.

VALENZUELA, A. & RASKI, D.J. (1985). *Pratylenchus australis* n. sp. and *Eutylenchus fueguensis* n. sp. (Nematoda: Tylenchina) from southern Chile. *Journal of Nematology* 17, 330-336.

VALLOTTON, R. (1989). Nématodes des petits fruits et principales méthodes de lutte contre ces ravageurs. *Revue Suisse de Viticulture, d'Arboriculture et d'Horticulture* 21, 121-122.

VALOCKA, B. & SABOVA, M. (1978). [Nematodes of cereal crops: species new for the fauna of Czechoslovakia.] *Biologia, Bratislava, B, Zoologia* 33, 686-688.

VAN BILJON, E.R. & MEYER, A.J. (2000). Reproduction of *Pratylenchus zeae* and *P. delattrei* on crops resistant to *Meloidogyne javanica* and *M. incognita* races 2 and 4. *African Plant Protection* 6, 43-45.

VAN DAM, N.M., RAAIJMAKERS, C.E. & VAN DER PUTTEN, W.H. (2005). Root herbivory reduces growth and survival of the shoot feeding specialist *Pieris rapae* on *Brassica nigra*. *Entomologia Experimentalis et Applicata* 115, 161-170.

VAN DEN BERG, E. (1976). Some species of Tylenchoidea (Nematoda) from South Africa, with descriptions of two new species. *Phytophylactica* 8, 55-64.

VAN DEN BERG, E. & DE WAELE, D. (1989). Further observations on nematodes associated with rice in South Africa. *Phytophylactica* 21, 125-130.

VAN DEN BERG, E. & QUÉNÉHERVÉ, P. (2000). *Hirschmanniella caribbeana* sp. n. and new records of *Pratylenchus* spp. (Pratylenchidae. Nematoda) from Guadeloupe, French West Indies. *Nematology* 2, 179-190.

VAN DEN BERG, E., MARAIS, M., KANDJI, S., MOUNPORT, D., N'DIAYE, B. & CADET, P. (2001). Plant-parasitic nematodes associated with corn-bean intercropping systems in Kenya. *African Plant Protection* 7, 81-84.

VAN DEN BERG, E., NGUYET, D.T.M., TUYET, N.T., NHI, H.H. & DE WAELE, D. (2002). Screening of Vietnamese *Musa* germplasm for resistance to root knot and root lesion nematodes in the greenhouse. *Australasian Plant Pathology* 31, 363-371.

VAN DEN BERG, W. & ROSSING, W.A.H. (2005). Generalized linear dynamics of a plant-parasitic nematode population and the economic evaluation of crop rotations. *Journal of Nematology* 37, 55-65.

VAN DEN BERGH, I., NGUYET, D.T.M., TUYET, N.T., NHI, H.H. & DE WAELE, D. (2006a). Influence of *Pratylenchus coffeae* and *Meloidogyne* spp. on plant growth and yield of banana (*Musa* spp.) in Vietnam. *Nematology* 8, 265-271.

VAN DEN BERGH, I., TUYET, N.T., NGUYET, D.T.M., NHI, H.H. & DE WAELE, D. (2006b). Population dynamics of *Pratylenchus coffeae* on banana in North Vietnam. *Nematology* 7, 891-900.

VAN GUNDY, S.D. (1965). Factors in survival of nematodes. *Annual Review of Phytopathology* 3, 43-68.

VAN GUNDY, S.D., PEREZ, J.G., STOLZY, L.H. & THOMASON, I.J. (1974). A pest management approach to the control of *Pratylenchus thornei* on wheat in Mexico. *Journal of Nematology* 6, 107-116.

VANSTONE, V.A. & RUSS, M.H. (2001a). Ability of weeds to host the root lesion nematodes *Pratylenchus neglectus* and *P. thornei* I. Grass weeds. *Australasian Plant Pathology* 30, 245-250.

VANSTONE, V.A. & RUSS, M.H. (2001b). Ability of weeds to host the root lesion nematodes *Pratylenchus neglectus* and *P. thornei* II. Broad-leaf weeds. *Australasian Plant Pathology* 30, 251-258.

VEGH, I. & BOURGEOIS, M. (1975). Observations préliminaires sûr l'étiologie du dépérissement des conifères d'ornement dans les pépinières françaises. Rôle du *Phytophthora cinnamomi* Rands. *Pépiniéristes Horticulteurs Maraîchers* 153, 39-49.

VENTER, C., DE WAELE, D. & VAN EEDEN, C.F. (1992). Plant-parasitic nematodes on field crops in South Africa. 4. Groundnut. *Fundamental and Applied Nematology* 15, 7-14.

VERSCHOOR, B.C., DE GOEDE, R.G.M., DE HOOP, J.W. & DE VRIES, F.W. (2001). Seasonal dynamics and vertical distribution of plant-feeding nematode communities in grasslands. *Pedobiologia* 45, 213-233.

VIAENE, N.M., SIMOENS, P. & ABAWI, G.S. (1997). SeinFit, a computer program for the estimation of the Seinhorst equation. *Journal of Nematology* 29, 474-477.

VIGLIERCHIO, D.R. (1979). Response of *Pinus ponderosa* seedlings to stylet-bearing nematodes. *Journal of Nematology* 11, 377-387.

VIGLIERCHIO, D.R., MAGGENTI, A.R., SCHMITT, R.V. & PAXMAN, G.A. (1977). Nematicidal injection: targeted control of plant-parasitic nematodes of trees and vines. *Journal of Nematology* 9, 307-311.

VILLAIN, L., MOLINA, A., SIERRA, S., DECAZY, B. & SARAH, J.-L. (2000). Effect of grafting and nematicide treatments on damage by the root-lesion nematode *Pratylenchus* spp. on *Coffea arabica* L. in Guatemala. *Nematropica* 30, 87-100.

VILLAIN, L., ANZUETO, F. & SARAH, J.-L. (2002). Resistance to root-lesion nematodes in *Coffea canephora.* In: Cook, R. & Hunt, D.J. (Eds). *Nematology Monographs and Perspectives, volume 2. Proceedings of the Fourth International Congress of Nematology, 8-13 June 2002, Tenerife, Spain.* Leiden, The Netherlands, Brill, pp. 157-158.

VILLENAVE, C. & CADET, P. (1998). Interactions of *Helicotylenchus dihystera, Pratylenchus pseudopratensis*, and *Tylenchorhynchus gladiolatus* on two plants from the Soudano-Sahelian zone of West Africa. *Nematropica* 28, 31-39.

VLADIMIROVA, L.A. (1972). [Nematodes of strawberries in the Alma-Ata region.] *Kul'tura zemlyaniki v SSSR Doklady simpoziuma, 28 iyunya 1 iyulya-1971*, 437-441.

VOLOSSIOUK, T., ROBB, E.J. & NAZAR, R.N. (2003). Avoiding false positives in PCR-based identification methods for nonsterile plant pathogens. *Canadian Journal of Plant Pathology* 25, 192-197.

VOVLAS, N. & ROCA, F. (1981). Histopathology of artichoke roots infected by *Pratylenchus penetrans* in Greece. *Nematologia Mediterranea* 9, 213-216.

VOVLAS, N. & TROCCOLI, A. (1990). Histopathology of broad bean roots infected by the lesion nematode *Pratylenchus penetrans. Nematologia Mediterranea* 18, 239-242.

VOVLAS, N., AVGELIS, A., GOUMAS, D. & FRISULLO, S. (1994). A survey of banana diseases in sucker propagated plantations in Crete. *Nematologia Mediterranea* 22, 101-107.

VRAIN, T., UONG, R.D., HALL, J. & FREYMAN, S. (1996). Cover crops resistant to root-lesion nematodes in raspberry. *HortScience* 31, 1195-1198.

VRAIN, T.C. (1987). Effect of *Ditylenchus dipsaci* and *Pratylenchus penetrans* on Verticillium wilt of alfalfa. *Journal of Nematology* 19, 379-383.

VRAIN, T.C. & BELAIR, G. (1981). Symptoms induced by the lesion nematode, *Pratylenchus penetrans* on carrot taproots in organic soil. *Phytoprotection* 62, 79-81.

VRAIN, T.C. & COPEMAN, R.J. (1987). Interactions between *Agrobacterium tumefaciens* and *Pratylenchus penetrans* in the roots of two red raspberry cultivars. *Canadian journal of Plant Pathology* 9, 236-240.

VRAIN, T.C. & DAUBENY, H.A. (1986). Relative resistance of red raspberry and related genotypes to the root lesion nematode. *HortScience* 21, 1435-1437.

VRAIN, T.C. & KENG, J.C.W. (1986). Applications of non-volatile nematicides through trickle irrigation system to control *Pratylenchus penetrans* in raspberries. *Canadian Journal of Plant Pathology* 8, 97-101.

VRAIN, T.C. & ROUSSELLE, G.L. (1980). Distribution of plant-parasitic nematodes in Quebec apple orchards. *Plant Disease* 64, 582-583.

VRAIN, T.C., DAUBENY, H.A., HALL, J.W., DE YOUNG, R.M. & ANDERSON, A.K. (1994). Inheritance of resistance to root lesion nematode in red raspberry. *HortScience* 29, 1340-1341.

VRAIN, T.C., FORCE, T.A. & YOUNG, R.D. (1997). Population dynamics of *Pratylenchus penetrans* parasitizing raspberry. *Fundamental and Applied Nematology* 20, 29-36.

WAEYENBERGE, L., RYSS, A., MOENS, M., PINOCHET, J. & VRAIN, T.C. (2000). Molecular characterisation of 18 *Pratylenchus* species using rDNA restriction fragment length polymorphism. *Nematology* 2, 135-142.

WALIA, R.K. & SESHADRI, A.R. (1985). Chemical control of *Pratylenchus thornei* on chickpea through seed treatment. *International Chickpea Newsletter* 13, 32-34.

WALIULLAH, M.I.S. & BHAT, M.A. (1990). Occurrence of mixed population of *Pratylenchus* Filipjev, 1936 on potato (*Solanum tuberosum* L.) in Srinagar, Kashmir. *Indian Journal of Nematology* 20, 239-241.

WALKER, J.T., SPECHT, C.H. & BEKKER, J.F. (1966). Nematocidal activity to *Pratylenchus penetrans* by culture fluids from actinomycetes and bacteria. *Canadian Journal of Microbiology* 12, 347-351.

WALLACE, H.R. (1973). *Nematode ecology and plant disease*. London, UK, Edward Arnold, 228 pp.

WALLACE, H.R. (1983). Interactions between nematodes and other factors on plants. *Journal of Nematology* 15, 221-227.

WALLACE, H.R. (1987a). Effects of nematode parasite on photosynthesis. In: Veech, J.A. & Dickson, D.W. (Eds). *Vistas on nematology*. Hyattsville, MD, USA, Society of Nematologists, Inc., pp. 253-259.

WALLACE, H.R. (1987b). A perception of tolerance. *Nematologica* 33, 419-432.

WALLACE, M.K. & MACDONALD, D.H. (1979). Plant-parasitic nematodes in Minnesota apple orchards. *Plant Disease Reporter* 63, 1063-1067.

WALLER, J.M. & BRIDGE, J. (1978). Plant diseases and nematodes in the Sultanate of Oman. *PANS (Pest Articles and News Summaries)* 24, 313-326.

WALTERS, S.A., BOND, J.P., BABADOOST, M., EDWARDS, D.I. & HANDOO, Z.A. (2004). Plant-parasitic nematodes associated with horseradish in Illinois. *Nematropica* 34, 191-197.

WANG, K.H., SIPES, B.S. & SCHMITT, D.P. (2002). *Crotalaria* as a cover crop for nematode management: a review. *Nematropica* 32, 35-57.

WANG, R.X. (1993). The identification of nematodes on fruit trees in Shaanxi Province. *Acta Agriculturae Boreali Occidentalia Sinica* 2, 81-86.

WARDOJO, S., HIJINK, J.J. & OOSTENBRINK, M. (1963). Schade big witte Klaver door inokulatie met *Heterodera trifolii*, *Meloidogyne hapla*, en *Pratylenchus penetrans*. *Mededelingen van de Landbonwhogeschool en de Opzoekingsstations van de Staat te Gent* 28, 672-678.

WAUDO, S.W. & NORTON, D.C. (1986). Pathogenic effects of *Pratylenchus scribneri* in maize inbreds and related cultivars. *Plant Disease* 70, 636-638.

WAUDO, S.W., SESHU-REDDY, S. & LUBEGA, M.C. (1990). Occurrence and prevalence of banana nematodes in Kenya. *Nematologica* 36, 400. [Abstr.]

WAUDO, S.W., SESHU-REDDY, K.V. & LUBEGA, M.C. (1998). Incidence and distribution of banana nematodes and weevils in Kenya. *Discovery and Innovation* 10, 164-169.

WAY, J.I. (1973). Phytothermotherapy for nematode control. *Nematological Society of Southern Africa Newsletter* 4, 4.

WEBB, R.M. (1990). Effects of the nematode *Pratylenchus fallax* on roots of oilseed rape (*Brassica napus* var. *oleifera*). *Revue de Nématologie* 13, 115-117.

WEBB, R.M. (1996). *In vitro* studies of six species of *Pratylenchus* (Nematoda: Pratylenchidae) on four cultivars of oilseed rape (*Brassica napus* var. *oleifera*). *Nematologica* 42, 89-95.

WEHUNT, E.J., GOLDEN, A.M. & ROBBINS, R.T. (1989). Plant nematodes occurring in Arkansas. *Journal of Nematology* 21, 677-681.

WEHUNT, E.J., GOLDEN, A.M., CLARK, J.R., KIRKPATRICK, T.L., BAKER, E.C. & BROWN, M.A. (1991). Nematodes associated with blackberry in Arkansas. *Journal of Nematology* 23, 620-623.

WEN, G.Y. & CHEN, T.A. (1976). Ultrastructure of the spicules of *Pratylenchus penetrans*. *Journal of Nematology* 8, 69-74.

WERGIN, W.P., YAKLICH, R.W., CARTA, L.K., ERBE, E.F. & MURPHY, C.A. (2000). Effect of an ice-nucleating activity agent on subzero survival of nematode juveniles. *Journal of Nematology* 32, 198-204.

WESTCOTT III, S.W. & KLUEPFEL, D.A. (1993). Inhibition of *Criconemella xenoplax* egg hatch by *Pseudomonas aureofaciens*. *Phytopathology* 83, 1245-1249.

WESTERDAHL, B.B., GIRAUD, D., ETTER, S., RIDDLE, L.J., RADEWALD, J.D., ANDERSON, C.A. & DARSO, J. (2003). Management options for *Pratylenchus penetrans* in Easter lily. *Journal of Nematology* 35, 443-449.

WHEELER, T.A., MADDEN, L.V., RIEDEL, R.M. & ROWE, R.C. (1994). Distribution and yield-loss relations of *Verticillium dahliae*, *Pratylenchus penetrans*, *P. scribneri*, *P. crenatus* and *Meloidogyne hapla* in commercial potato fields. *Phytopathology* 84, 843-852.

WHITEHEAD, A.G. (1998). *Plant nematode control*. Wallingford, UK, CABI Publishing, 384 pp.

WILLIAMS, D.D. (1982). The known Pratylenchidae (Nematoda) of Iowa. *Iowa State Journal of Research* 56, 419-424.

WILLIS, C.B. (1972). Effects of soil pH on reproduction of *Pratylenchus penetrans* and forage yield of alfalfa. *Journal of Nematology* 4, 291-295.

WILLIS, C.B. & THOMPSON, L.S. (1969). Effect of the root-lesion nematode on yield of four forage legumes under greenhouse conditions. *Canadian Journal of Plant Science* 59, 505-509.

WILLUT, J.M. & MALEK, R.B. (1979). Spatial distribution of plant-parasitic nematodes associated with corn. *Phytopathology* 69, 544.

WINDHAM, G.L. (1998). Corn. In: Barker, K.R., Pederson, G.A. & Windham, G.L. (Eds). *Plant and nematode interactions*. Madison, WI, USA, American Society of Agronomy, Inc., Crop Science Society of America, Inc., Soil Science Society of America, Inc., pp. 335-357.

WINFIELD, A.L. (1974). Observations on the occurrence, pathogenicity and control of *Pratylenchus vulnus*, *P. thornei* and *Xiphinema diversicaudatum* associated with glasshouse roses. *Annals of Applied Biology* 77, 297-307.

WINKELHEIDE, R. & STURHAN, D. (1996). Untersuchungen zur wirtsspezifitat eines *Pasteuria* isolates von *Heterodera goettingiana*. *Mitteilungen aus der Biologischen Bundesanstalt fur Land und Forstwirtschaft Berlin Dahlem* 317, 46-53.

WINOTO SUATMADJI, R. (1988). *Pratylenchus penetrans* and *Rotylenchus robustus* on thirty herbaceous ornamental species. *Australasian Plant Pathology* 17, 97-98.

WIRYADIPUTRA, S. (1996). Resistance of Robusta coffee to coffee root lesion nematode, *Pratylenchus coffeae*. *Pelita Perkebunan* 12, 137-148.

WOJTOWICZ, M. & SOBIO, J. (1994). [Parasitic nematofauna in rose nurseries in Poland.] *Zeszyty Naukowe Instytutu Sadownictwa Kwiaciarstwa w Skierniewicach* 1, 147-153.

WORTMANN, C.S., MCINTYRE, B.D. & KAIZZI, C.K. (2000). Annual soil improving legumes: agronomic effectiveness, nutrient uptake, nitrogen fixation and water use. *Field Crop Research* 68, 75-83.

WRATHER, J.A., NIBLACK, T.L. & MILAM, M.R. (1992). Survey of plant-parasitic nematodes in Missouri cotton fields. *Journal of Nematology* 24, 779-782.

WU, H.Y., TSAY, T.T. & LIN, Y.Y. (2002). Identification and biological study of *Pratylenchus* spp. isolated from the crops in Taiwan. *Plant Pathology Bulletin* 11, 123-136.

WU, L.Y. (1971). *Pratylenchus macrostylus* n. sp. (Pratylenchidae: Nematoda). *Canadian Journal of Zoology* 49, 487-489.

WU, S.F. (1993). Investigation and identification of root parasitic nematodes of peach (*Amygdalus persica* L.) in Zhejiang Province. *Acta Agriculturae Zhejiangensis* 5, 245-247.

YAMADA, E., HASHIZUME, K. & TAKAHASHI, M. (2003). Antagonistic effect of leguminous green manure crops on *Heterodera glycines* and the effect of these crops on *Pratylenchus penetrans*. *Japanese Journal of Nematology* 33, 1-13.

YAMAKAWA, O., KUMAGAI, T., YOSHINAGA, M., ISHIGURO, K., HIDAKA, M., KOMAKI, K. & KUKIMURA, H. (1999). 'Sunny Red': new sweetpotato cultivar for powder. *Bulletin of the Kyushu National Agricultural Experiment Station* 35, 19-40.

YANG, Y.Z. & ZHAO, J.H. (1992). A new species of *Pratylenchus* (Tylenchida: Pratylenchidae). *Acta Zootaxonomica Sinica* 17, 161-162.

YANG, Y.Z., DENG, X.M. & LIU, G.Z. (1992). Studies on species and genera of plant parasitic nematodes in cotton fields in Sichuan. *Journal of Southwest Agricultural University* 14, 292-295.

YEATES, G.W. (1976). Effect of fertilizer treatment and stocking rate on pasture nematode populations in a yellow-grey earth. *New Zealand Journal of Agricultural Research* 19, 405-408.

YIN, K.C. (1991). The identification of *Pratylenchus* species in Guangdong Province. *Plant Quarantine* 5, 253-254.

YIN, Y.Q. (1994). The investigation and identification of nematodes on mango in Guangdong. *Guangdong Agricultural Sciences* 2, 39-41.

YIN, Y.Q. (1995). Surveys of parasitic nematodes on mango in Guangdong, China. *Acta Phytopathologica Sinica* 25, 42.

YIN, Y.Q., GAO, X.B. & FENG, Z.X. (1994). Investigations of parasitic nematodes on lychee in Guangdong province. *Journal of South China Agricultural University* 15, 22-27.

YOKOO, T. & IKEGEM, Y. (1966). Some observations on growth of the new host plant, Snapdragon *(Antirrhinum majus* L.) attacked by the root lesion nematode, *Pratylenchus coffeae*, and control-effect of some nematicides. *Agricultural Bulletin of Saga University* 22, 83-92.

YOUNG, L.D. (1998). Breeding for nematode resistance and tolerance. In: Barker, K.R., Pederson, G.A. & Windham, G.L. (Eds). *Plant and nematode interactions.* Madison, WI, USA, American Society of Agronomy, Inc., Crop Science Society of America, Inc., Soil Science Society of America, Inc., pp. 187-207.

YOUNG, T.W. & RUEHLE, G.D. (1955). The role of the burrowing and meadow nematodes in avocado decline. *Plant Disease Reporter* 39, 815-817.

YU, Q. (1997). First report of *Pratylenchus thornei* from spring wheat in southern Ontario. *Canadian Journal of Plant Pathology* 19, 289-292.

YU, Q., POTTER, J.W. & GILBY, G.A. (1997). First report of *Pratylenchus fallax* on turfgrass in Ontario. *Plant Disease* 81, 1331.

YU, Q., POTTER, J.W. & GILBY, G.A. (1998). Plant parasitic nematodes associated with turfgrass in golf courses in Southern Ontario. *Canadian Journal of Plant Pathology* 20, 304-307.

ZACHEO, G., BLEVE-ZACHEO, T., MELILLO, M.T. & PERRINO, P. (1990). Some biochemical properties of *Pisum sativum* susceptible and resistant to *Heterodera goettingiana. Nematologia Mediterranea* 18, 253-259.

ZAKI, F.A., KAUL, V. & WALIULLAH, M.I.S. (1991). Plant parasitic nematodes associated with potato-turnip rotation and other vegetable crops in Kashmir. *Indian Journal of Nematology* 21, 174-175.

ZARINA, B. & MAQBOOL, M.A. (1998). Descriptions and observations on two new and two known species of the genus *Pratylenchus* Filipjev, 1936 (Nematoda: Pratylenchidae) from Pakistan. *Pakistan Journal of Nematology* 16, 13-24.

ZEIDAN, A.B. & GERAERT, E. (1991). *Pratylenchus* from Sudan, with the description of two new species (Nemata: Tylenchida). *Revue de Nématologie* 14, 261-275.

ZEPP, A.L. & SZCZYGIEL, A. (1986). [Pathogenicity of five populations of *Pratylenchus penetrans* to three fruit tree seedling rootstocks.] *Zeszyty Problemowe Postepow Nauk Rolniczych* 323, 73-90.

ZHANG, S., XIAO, R.F., LIN, N.Q. & AI-HONG, M. (2002). Identification of parasitic nematode species from Chinese olive in Fujian, China. *Journal of Fujian Agricultural and Forestry University* 31, 445-451.

ZHENG, J.W. & CHEN, P.S. (1994). Primary report on a new species in the genus *Pratylenchus. Journal Shanxi Agricultural University* 14, 387-390.

ZIMMERMANN, A.W.P. (1898). De nematoden der koffiewortels. Deel I. *Mededelingen uit's Lands Plantentuin* 27, 1-64.

ZINNO, Y. & IGARASHI, Y. (1972). Distribution of plant parasitic nematodes associated with coniferous seedlings in Shikoku, Japan. *Bulletin of the Government Forestry Experimental Station, Tokyo* 246, 11-20.

ZIRAKPARVAR, M.E. (1980). Host range of *Pratylenchus hexincisus* and its pathogenicity on corn, soybean, and tomato. *Phytopathology* 70, 749-753.

ZIRAKPARVAR, M.E. (1982). Susceptibility of soybean cultivars and lines to *Pratylenchus hexincisus*. *Journal of Nematology* 14, 217-220.

ZUCKERMAN, B.M., MILLER, C.W. & DEUBERT, K. (1966). Phenylalanine deaminase in plant parasitic nematodes. *Nematologica* 12, 428-430.

ZUNKE, U. (1990a). Observation on the invasion and endoparasitic behavior of the root lesion nematode *Pratylenchus penetrans*. *Journal of Nematology* 22, 309-320.

ZUNKE, U. (1990b). Ectoparasitic feeding behavior of the root lesion nematode, *Pratylenchus penetrans*, on root hairs of different host plants. *Revue de Nématologie* 13, 331-337.

ZWART, R.S., THOMPSON, J.P. & GODWIN, I.D. (2004). Genetic analysis of resistance to root-lesion nematode *Pratylenchus thornei* in wheat. *Plant Breeding* 123, 209-212.

ZWART, R.S., THOMPSON, J.P. & GODWIN, D.I. (2005). Identification of quantitative trait loci for resistance to two species of root-lesion nematode (*Pratylenchus thornei* and *P. neglectus*) in wheat. *Australian Journal of Agricultural Research* 56, 345-352.

ZYUBIN, B.N. (1966). [A new species *Pratylenchus montanus* n. sp. (Nematoda: Pratylenchidae) in a culture of the opium poppy in Kirgizia.] In: Tokobaev, M.M. (Ed.). *Helminths of animals in Kirgizia and adjacents territories*. Izdanie Ilim, Frunze, Kirgizia, pp. 147-151.

Subject index

www.ingramcontent.com/pod-product-compliance
Lightning Source LLC
LaVergne TN
LVHW011100300926
849160LV00008B/239
9789004155640